COLOSS
BEEBOOK

Volume II:
Standard Methods
for *Apis mellifera*
Pest and Pathogen Research

Edited by
Vincent Dietemann
James D Ellis
Peter Neumann

INTERNATIONAL BEE
RESEARCH ASSOCIATION

Jointly published by: The International Bee Research Association, a
Company Limited by Guarantee, 91 Brinsea Road, Congresbury,
Bristol BS49 5JJ (UK) & Northern Bee Books, Scout Bottom Farm,
Mytholmroyd, Hebden Bridge HX7 5JS (UK).

Obtainable from:
www.ibrabee.org.uk & www.northernbeebooks.co.uk

© 2013 The International Bee Research Association.
Reprinted 2017

ISBN 978-0-86098-282-1

IBRA

INTERNATIONAL BEE
RESEARCH ASSOCIATION

From its origins as a mere idea borne out of frustrations caused by a lack of standardisation of laboratory techniques expressed at an early meeting of the COLOSS (Prevention of honey bee COlony LOSSes) Network in Bern, Switzerland in 2009, this project has grown to what you see today. This is the result of work by more than 234 authors from 34 countries, together with a large number of others who have reviewed the papers. Without the voluntary efforts of these, a large proportion of the world's bee scientists, the BEEBOOK would not have been possible. In particular our thanks are due to Vincent Dietemann, Jamie Ellis and Peter Neumann, who selected, approached and then cajoled the authors into writing the chapters. They then solicited reviews and edited the papers. The unique way that BEEBOOK has evolved also allowed constructive criticisms from many more people. Thanks to Pilar de la Rua for translating the summaries into Spanish, and to Shi Wei and Huo Qing Zheng for the Mandarin Chinese summary translations. The mammoth task of production of the papers for publication was carried out by Tony Gruba, Sarah Jones, Julian Rees and especially Diane Griffiths at the IBRA office in Cardiff. I must thank the scientific members of IBRA Council, most of whom have acted as authors or reviewers, and Ivor Davis, IBRA Vice-Chairman for realising the potential importance of the BEEBOOK from the outset, and for providing unwavering support for the project. Ralf Bünemann, COLOSS webmaster has produced the html version.

The COLOSS network was funded through the COST Action FA0803. COST (European Cooperation in Science and Technology) is a unique means for European researchers to jointly develop their own ideas and new initiatives across all scientific disciplines through trans-European networking of nationally funded research activities. Based on a pan-European intergovernmental framework for cooperation in science and technology, COST has contributed since its creation more than 40 years ago to closing the gap between science, policy makers and society throughout Europe and beyond. COST is supported by the EU Seventh Framework Programme for research, technological development and demonstration activities (Official Journal L 412, 30 December 2006). The European Science Foundation as implementing agent of COST provides the COST Office through an EC Grant Agreement. The Council of the European Union provides the COST Secretariat. The COLOSS network is now supported by the Ricola Foundation - Nature & Culture.

Norman L Carreck.
IBRA Science Director and Senior Editor, *Journal of Apicultural Research*.
July 2013.

COST is supported by the
EU Framework Programme

ESF provides the COST Office through
a European Commission contract

The COLOSS *BEEBOOK*, Volume II: Standard methods for *Apis mellifera* pest and pathogen research.	
Foreword	Marie-Pierre Chauzat, Alberto Laddomada
Standard methods for *Apis mellifera* pest and pathogen research	Vincent Dietemann, James D Ellis, Peter Neumann
Standard epidemiological methods to understand and improve *Apis mellifera* health	Dennis vanEngelsdorp, Eugene Lengerich, Angela Spleen, Benjamin Dainat, James Cresswell, Kathy Baylis, Kim Bach Nguyen, Victoria Soroker, Robyn Underwood, Hannelie Human, Yves Le Conte, Claude Saegerman
Standard survey methods for estimating colony losses and explanatory risk factors in *Apis mellifera*	Romée van der Zee, Alison Gray, Céline Holzmann, Lennard Pisa, Robert Brodschneider, Róbert Chlebo, Mary F Coffey, Aykut Kence, Preben Kristiansen, Franco Mutinelli, Bach Kim Nguyen, Adjlane Noureddine, Magnus Peterson, Victoria Soroker, Grażyna Topolska, Flemming Vejsnæs, Selwyn Wilkins
Standard methods for small hive beetle research	Peter Neumann, Christian W W Pirk, M O Schäfer, Jay D Evans, Jeffrey S Pettis, Gina Tanner, Geoffrey Williams, James D Ellis
Standard methods for tracheal mites research	Diana Sammataro, Lilia De Guzman, Sherly George, Ron Ochoa, Gard Otis
Standard methods for *Tropilaelaps* mites research	Denis L Anderson, John M K Roberts
Standard methods for varroa research	Vincent Dietemann, Francesco Nazzi, Stephen J Martin, Denis L Anderson, Barbara Locke, Keith S Delaplane, Quentin Wauquiez, Cindy Tannahill, Bettina Ziegelmann, Peter Rosenkranz, James D Ellis
Standard methods for wax moth research	James D Ellis, Jason R Graham, Ashley Mortensen
Standard methods for American foulbrood research	Dirk C de Graaf, Adriana M Alippi, Karina Antúnez, Katherine A Aronstein, Giles E Budge, Dieter De Koker, Lina De Smet, Douglas W Dingman, Jay D Evans, Leonard J Foster, Anne Fünfhaus, Eva Garcia Gonzalez, Aleš Gregorc, Hannelie Human, K Daniel Murray, Bach Kim Nguyen, Lena Poppinga, Marla Spivak, Dennis vanEngelsdorp, Selwyn Wilkins, Elke Genersch

The COLOSS *BEEBOOK*, Volume II: Standard methods for *Apis mellifera* pest and pathogen research.	
Standard methods for European foulbrood research	Eva Forsgren, Giles E Budge, Jean-Daniel Charrière, Michael A Z Hornitzky
Standard methods for fungal brood disease research	Annette Bruun Jensen, Kathrine Aronstein, José Manuel Flores, Svjetlana Vojvodic, María Alejandra Palacio, Marla Spivak
Standard methods for *Nosema* research	Ingemar Fries, Marie-Pierre Chauzat, Yan-Ping Chen, Vincent Doublet, Elke Genersch, Sebastian Gisder, Mariano Higes, Dino P McMahon, Raquel Martín-Hernández, Myrsini Natsopoulou, Robert J Paxton, Gina Tanner, Thomas C Webster, Geoffrey R Williams
Standard methods for virus research in *Apis mellifera*	Joachim R de Miranda, Lesley Bailey, Brenda V Ball, Philippe Blanchard, Giles E Budge, Nor Chejanovsky, Yan-Ping Chen, Laurent Gauthier, Elke Genersch, Dirk C de Graaf, Magali Ribière, Eugene Ryabov, Lina De Smet, Jozef J M van der Steen

Journal of Apicultural Research 52(4): (2013)
DOI 10.3896/IBRA.1.52.4.17

Notes and Comments

The COLOSS *BEEBOOK* Volume 2 - Foreword

Marie-Pierre Chauzat [1*] and Alberto Laddomada [2]

[1]ANSES, Sophia-Antipolis Laboratory, Les Templiers, 105, route des Chappes, BP 111 06902 Sophia-Antipolis, France.
[2]Unit G2 Animal Health, Directorate General for Health and Consumers (DG SANCO), European Commission, Brussels, Belgium.

*Corresponding author: Email: marie-pierre.CHAUZAT@anses.fr

Honey bees have fascinated mankind since the Stone Age. The oldest evidence of honey collection by humans dates from *c.* 6000 BC in a cave near Valencia in Spain. Subsequent to the interest in their honey, the only sweetener available in the wild, honey bees have also captured the attention of people because of their social organization. In 1973, the Nobel Prize in physiology and medicine was awarded to Karl von Frisch, Nikolaas Tinbergen and Konrad Lorenz "for their discoveries concerning organisation and elicitation of individual and social behaviour patterns" especially the description of the honey bee dance language by Karl von Frisch. More recently, honey bees have drawn the attention of the public and stakeholders because of higher than normal mortalities of colonies reported by beekeepers in many countries for which the exact causes remain largely unknown. There is, however, a scientific consensus that there is no single cause for these losses, which are undoubtedly caused by a combination of a number of factors.

Honey bees are affected by various pathogens, and some have long been known. In *c.* 300 BC, the Greek philosopher Aristotle reported the presence of "rust" in honey bee colonies, which was probably what we now know as American foulbrood, a serious bacterial disease. It is also worth noting that the gut parasite *Nosema apis* has been the subject of studies for more than a century.

However in recent decades, honey bees have been subjected to new threats. The discovery and exploration of "new worlds" by Europeans made possible the introduction and establishment of the western honey bee (*Apis mellifera*) outside its natural distribution area. This movement suppressed the natural geographical segregation established for millions for years between the different species of honey bees. The Asian honey bee (*Apis cerana*) has evolved together with its pathogens for thousands of years reaching a sustainable equilibrium. The introduction of the western honey bee in Asia made possible the jump of pathogens from *A. cerana* to *A. mellifera*. The current major threat of honey bees worldwide is the ectoparasite *Varroa destructor* which originated in Asia as a parasite of *A. cerana*. The mite feeds on the honey bee haemolymph by piercing the host cuticle. The wounds created facilitate the exchange of viruses between the host and the parasite. Although honey bee viruses had been identified in Europe before the introduction of *V. destructor*, they were of limited economic importance. The deleterious effect of *V. destructor* on honey bee colonies has now been studied for more than 30 years, yet the pathogenicity mechanisms still remain largely unknown, because they involve complex interactions between the host, pathogens and parasite.

The worldwide trade in honey bees and hive products has accelerated the diffusion of 'new' pathogens, predators and pests to other parts of world. A new species of *Nosema* (*N. ceranae*), also initially described from the Asian honey bee, has lately been widely detected on *A. mellifera*. This parasite has apparently silently spread throughout the world for decades, yet only recently has it supplanted *N. apis* in honey bee colonies in warm climates. American foulbrood (AFB) and European foulbrood are two bacterial diseases affecting honey bees. Caused respectively by *Paenibacillus larvae* and *Melissococcus plutonius*, the diseases affect the brood of *A. mellifera*. Because *P. larvae* is a spore forming bacteria, AFB is extremely contagious.

Beekeeping activity is highly dependent of the environment, more so than any other animal keeping or food production industry. It is therefore crucial to assess the influence all environmental factors on honey bee health. These can include, *inter alia*, the insufficient availability and diversity of pollen and nectar as food, the presence of natural bee predators or competitors and the effects of pesticides on honey bee health, not to mention the role of honey bee genetic diversity and the bees' natural resistance mechanisms against pathogens.

The need for more efficient standardisation in honey bee experimentation is widely recognised, but until now has not been widely adopted. Standardisation is vital for harmonious veterinary standards, and for daily routine diagnostics that are essential to ensure confidence in the comparability of results from different laboratories. Researchers, policy makers, veterinary services, beekeepers and other stakeholders all have the same goal: to contribute to a healthy and productive apicultural sector in its quest for long-term sustainability both in the EU and worldwide.

In terms of the regulatory aspects of honey bee pathogens, under the WTO (World Trade Organisation) agreement on the application of Sanitary and Phytosanitary Measures (SPS Agreement), the OIE (World Organisation for Animal Health) publishes sanitary standards for international trade in animals and animal products. As part of this activity, OIE develops international standards and guidelines for diagnostic tests and vaccines as well as for veterinary laboratories. Six honey bee diseases are listed by the OIE in the Manual of Diagnostic Tests and Vaccines for Terrestrial Animals to provide internationally agreed diagnostic laboratory methods.

In the framework of European honey bee health rules, the European Commission designated the ANSES Sophia-Antipolis laboratory as the European Union Reference Laboratory (EU RL) for bee health on 29 October 2010. The EU RL for honey bee health, among its duties and functions, has to coordinate, in consultation with the Commission (more specifically with the Directorate General for Health and Consumers), the methods employed in the Member States for diagnosing the relevant bee diseases, specifically by typing, storing and, where appropriate, supplying strains of the pathogenic agents to facilitate the diagnostic service in the EU. The EU RL will organise periodic comparative tests of diagnostic procedures at EU level with the NRLs (National Reference Laboratories) designated by the Member States, in order to provide information on the methods of diagnosis used and the result of the test carried out in the EU.

The COLOSS *BEEBOOK* provides many additional protocols which can improve pest and pathogen research and its potential impact in these times of pollinator declines. It is therefore a pleasure to acknowledge the work of the *BEEBOOK* editors and the authors. The relevant specialists in each field have been gathered to constructively list and to criticise existing research protocols. The *BEEBOOK* is thus an important step forward to the establishment of standardised protocols for the study of honey bee diseases.

Journal of Apicultural Research 52(4): (2013)
DOI 10.3896/IBRA.1.52.4.16

GUEST EDITORIAL

The COLOSS *BEEBOOK* Volume II, Standard methods for *Apis mellifera* pest and pathogen research: Introduction

IBRA
INTERNATIONAL BEE
RESEARCH ASSOCIATION

Vincent Dietemann[1,2*], James D Ellis[3] and Peter Neumann[4,2]

[1]Swiss Bee Research Centre, Agroscope Liebefeld-Posieux Research Station ALP-Haras, Bern, Switzerland.
[2]Social Insect Research Group, Department of Zoology & Entomology, University of Pretoria, Pretoria, South Africa.
[3]Honey bee Research and Extension Laboratory, Department of Entomology and Nematology, University of Florida, Gainesville, Florida, USA.
[4]Institute of Bee Health, Vetsuisse Faculty, University of Bern, Bern, Switzerland.

Received 29 July 2013, accepted for publication 30 July 2013.

*Corresponding author: Email: vincent.dietemann@agroscope.admin.ch

Keywords: COLOSS, *BEEBOOK,* honey bee, *Apis mellifera,* research, standard methods, laboratory, field, pests, disease, diagnosis

The COLOSS *BEEBOOK* is a practical manual compiling standard methods in all fields of research on the western honey bee, *Apis mellifera*. The COLOSS network was founded in 2008 as a consequence of the heavy and frequent losses of managed honey bee colonies experienced in many regions of the world (Neumann and Carreck, 2010). As many of the world's honey bee research teams began to address the problem, it soon became obvious that a lack of standardized research methods was seriously hindering scientists' ability to harmonize and compare the data on colony losses obtained internationally. In its second year of activity, during a COLOSS meeting held in Bern, Switzerland, the idea of a manual of standardized honey bee research methods emerged. The manual, to be called the COLOSS *BEEBOOK,* was inspired by publications with similar purposes for fruit fly research (Lindsley and Grell, 1968; Ashburner, 1989; Roberts, 1998; Greenspan, 2004).

Production of the *BEEBOOK* began after recruiting international experts to lead the compilation of each research domain. These senior authors (first in the author list) were tasked with recruiting a suitable team of contributors to select the methods to be used as standards and then to report them in a user-friendly manner (Williams *et al.,* 2012).

The initial *BEEBOOK* project is divided into three volumes: The COLOSS *BEEBOOK,* Volume I: Standard methods for *Apis mellifera* research; The COLOSS *BEEBOOK,* Volume II: Standard methods for *Apis mellifera* pest and pathogen research; and The COLOSS *BEEBOOK,* Volume III: Standard methods for *Apis mellifera* product research.

Papers in the *BEEBOOK* are organized according to research topics. The authors have compiled those methods selected as the 'best' in each domain of research. These methods are for both laboratory and field research. We recognize that it is often necessary to use methods from several domains of research to complete a given experiment with honey bees. Whenever there is a need for multi-disciplinary approach, the manual describes the specific instructions

necessary for a given method, and cross references all general methods from other papers as necessary. For example, identifying a subspecies of honey bee can be done using genetic tools. The general instructions to use microsatellites are given in the molecular methods paper (Evans *et al.,* 2013), whereas the specific method appropriate for subspecies identification is described in the paper on ecotypes and subspecies identification (Meixner *et al.,* 2013). Consequently, one would visit the ecotypes paper to determine how to identify a given subspecies. That paper will then refer to the molecular methods paper when discussing microsatellites specifically.

The reader may wonder about the difference between the *BEEBOOK* and existing standards provided by the Office International des Epizooties (OIE), and the European Organisation for Economic Co-operation and Development (OECD). In the *BEEBOOK,* we often refer to OIE, OECD, and other standards, since they describe methods to diagnose pests and diseases (OIE) or to perform, for example, routine analyses for toxicity tests (OECD). The *BEEBOOK,* however, goes well beyond diagnosis and routine analyses by describing the methods to perform research on the honey bee and associated organisms. Where necessary, the *BEEBOOK* recognizes existing standards such as those provided by the OIE and OECD, and presents a harmonized compendium of research methods, written and reviewed by an international team of scientists.

In addition to producing a bench-friendly manual, and in an effort to make the methods broadly available, every paper forming the *BEEBOOK* is also available as open access articles in several special issues of the *Journal of Apicultural Research.*

To further build on the availability of digital media, a novel concept was developed around the manual. An online version of the manual was created, where each method can be discussed and improvements suggested. Development work on the online *BEEBOOK* platform started in 2009, and the current iteration can be found at www.coloss.org/beebook. On the platform, each webpage describing a method has a comment field, which can be used to suggest changes or additions. Users can thus assist with the improvement and further development of the *BEEBOOK*. Once sufficient updates have accumulated online, a new print version of the manual can then be edited and published. Such a Wiki-like tool is especially useful for following fast evolving fields, such as for molecular protocols.

The *BEEBOOK* is a tool for all who want to do research on honey bees. It was written in such a way that those new to honey bee research can use it to start research in a field with which they may not be familiar. Of course, such an endeavour is often limited by the availability of complex and expensive machinery and other equipment. However, provided access to and training on the necessary equipment are secured, the instructions provided in the *BEEBOOK* can be followed by everyone, from undergraduate student to experienced researcher. All details on how to implement instructions are given.

The editors and author team hope that the *BEEBOOK* will serve as a reference tool for honey bee and other researchers globally. As with the original *Drosophila* book that evolved into a journal where updates and new methods are published, we hope that the honey bee research community will embrace this tool and work to improve it. The online platform is open for everyone to use and further contribute to the development of our research field.

The study of honey bee pests and pathogens is globally relevant and remarkably varied. The major impact of bee pests and pathogens on the well-being of honey bees has led to considerable research in this field. Consequently, the editorial and author team felt it necessary to develop a volume focused solely on protocols related to studying the pests and pathogens of western honey bees.

This effort resulted in the production of Volume II of the COLOSS *BEEBOOK*: Standard methods for *Apis mellifera* pest and pathogen research. Here, 98 international scientists from 22 countries have collaborated to produce 12 chapters including over 500 protocols related to studying honey bee pests and pathogens. The chapters are grouped into three sections. The first section focuses on protocols related to honey bee epidemiology. They include: epidemiology itself (vanEngelsdorp *et al.*, 2013) and survey methods for estimating colony losses (van der Zee *et al.*, 2013). The second section includes chapters focused on honey bee pests and they include: small hive beetles (Neumann *et al.*, 2013), tracheal mites (Sammataro *et al.*, 2013), *Tropilaelaps* mites (Anderson *et al.*, 2013), varroa (Dietemann *et al.*, 2013), and wax moths (Ellis *et al.*, 2013). The third and final section includes chapters listing protocols for studying honey bee pathogens. Included in this section are chapters on: American foulbrood (de Graaf *et al.*, 2013), European foulbrood (Forsgren *et al.*, 2013), fungi (Jensen *et al.*, 2013), nosema (Fries *et al.*, 2013), and viruses (de Miranda *et al.*, 2013). It was our intention to be exhaustive when working with senior authors to develop the chapters included in this volume. We

hope that we have included all of the relevant research pests and pathogens (including emerging ones such as *Tropilaelaps* spp.), but recognize that, as with any undertaking of such size, we may have overlooked important topics, and other pests and pathogens may emerge in the future. If so, this can be addressed via the online *BEEBOOK* platform (www.coloss.org/beebook), leading to an improved version in the future. We hope that the information provided herein will assist everyone interested in investigating the pests and pathogens of the honey bee.

The western honey bee remains a fascinating research topic, and one of timeless significance considering the bee's importance for food production and ecosystem sustainability. We hope that we and our team of international colleagues have produced a resource that will be useful into perpetuity. We also hope that you will find research on honey bees, their pests and pathogens to be professionally rewarding and intellectually stimulating.

Acknowledgements

The COLOSS (Prevention of honey bee COlony LOSSes) network aims to explain and prevent massive honey bee colony losses. It was funded through the COST Action FA0803. COST (European Cooperation in Science and Technology) is a unique means for European researchers to jointly develop their own ideas and new initiatives across all scientific disciplines through trans-European networking of nationally funded research activities. Based on a pan-European intergovernmental framework for cooperation in science and technology, COST has contributed since its creation more than 40 years ago to closing the gap between science, policy makers and society throughout Europe and beyond. COST is supported by the EU Seventh Framework Programme for research, technological development and demonstration activities (Official Journal L 412, 30 December 2006). The European Science Foundation as implementing agent of COST provides the COST Office through an EC Grant Agreement. The Council of the European Union provides the COST Secretariat. The COLOSS network is now supported by the Ricola Foundation - Nature & Culture.

References

ASHBURNER, M (1989) *Drosophila: a laboratory manual.* Cold Spring Harbor Laboratory Press; Cold Spring Harbor, USA. 434 pp.

ANDERSON, D L; ROBERTS, J M K (2013) Standard methods for *Tropilaelaps* mites research. In *V Dietemann; J D Ellis; P Neumann (Eds) The COLOSS BEEBOOK, Volume II: standard methods for* Apis mellifera *pest and pathogen research. Journal of Apicultural Research* 52(4): http://dx.doi.org/10.3896/IBRA.1.52.4.21

DE GRAAF, D C; ALIPPI, A M; ANTÚNEZ, K; ARONSTEIN, K A; BUDGE, G; DE KOKER, D; DE SMET, L; DINGMAN, D W; EVANS, J D; FOSTER, L J; FÜNFHAUS, A; GARCIA-GONZALEZ, E; GREGORC, A; HUMAN, H; MURRAY, K D; NGUYEN, B K; POPPINGA, L; SPIVAK, M; VANENGELSDORP, D; WILKINS, S; GENERSCH, E (2013) Standard methods for American foulbrood research. In *V Dietemann; J D Ellis; P Neumann (Eds) The COLOSS* BEEBOOK, *Volume II: standard methods for* Apis mellifera *pest and pathogen research. Journal of Apicultural Research* 52(1): http://dx.doi.org/10.3896/IBRA.1.52.1.11

DE MIRANDA, J R; BAILEY, L; BALL, B V; BLANCHARD, P; BUDGE, G; CHEJANOVSKY, N; CHEN, Y-P; GAUTHIER, L; GENERSCH, E; DE GRAAF, D; RIBIÈRE, M; RYABOV, E; DE SMET, L; VAN DER STEEN, J J M (2013) Standard methods for virus research in *Apis mellifera*. In *V Dietemann; J D Ellis; P Neumann (Eds) The COLOSS* BEEBOOK, *Volume II: standard methods for* Apis mellifera *pest and pathogen research. Journal of Apicultural Research* 52(4): http://dx.doi.org/10.3896/IBRA.1.52.4.22

DIETEMANN, V; NAZZI, F; MARTIN, S J; ANDERSON, D; LOCKE, B; DELAPLANE, K S; WAUQUIEZ, Q; TANNAHILL, C; FREY, E; ZIEGELMANN, B; ROSENKRANZ, P; ELLIS, J D (2013) Standard methods for varroa research. In *V Dietemann; J D Ellis; P Neumann (Eds) The COLOSS* BEEBOOK, *Volume II: standard methods for* Apis mellifera *pest and pathogen research. Journal of Apicultural Research* 52(1): http://dx.doi.org/10.3896/IBRA.1.52.1.09

ELLIS, J D; GRAHAM, J R; MORTENSEN, A (2013) Standard methods for wax moth research. In *V Dietemann; J D Ellis; P Neumann (Eds) The COLOSS* BEEBOOK, *Volume II: standard methods for* Apis mellifera *pest and pathogen research. Journal of Apicultural Research* 52(1): http://dx.doi.org/10.3896/IBRA.1.52.1.10

EVANS, J D; SCHWARZ, R S; CHEN, Y P; BUDGE, G; CORNMAN, R S; DE LA RUA, P; DE MIRANDA, J R; FORET, S; FOSTER, L; GAUTHIER, L; GENERSCH, E; GISDER, S; JAROSCH, A; KUCHARSKI, R; LOPEZ, D; LUN, C M; MORITZ, R F A; MALESZKA, R; MUÑOZ, I; PINTO, M A (2013) Standard methodologies for molecular research in *Apis mellifera*. In *V Dietemann; J D Ellis; P Neumann (Eds) The COLOSS* BEEBOOK, *Volume I: standard methods for* Apis mellifera *research. Journal of Apicultural Research* 52(4): http://dx.doi.org/10.3896/IBRA.1.52.4.11

FORSGREN, E; BUDGE, G E; CHARRIÈRE, J-D; HORNITZKY, M A Z (2013) Standard methods for European foulbrood research. In *V Dietemann; J D Ellis; P Neumann (Eds) The COLOSS* BEEBOOK, *Volume II: Standard methods for* Apis mellifera *pest and pathogen research. Journal of Apicultural Research* 52(1): http://dx.doi.org/10.3896/IBRA.1.52.1.12

FRIES, I; CHAUZAT, M-P; CHEN, Y-P; DOUBLET, V; GENERSCH, E; GISDER, S; HIGES, M; MCMAHON, D P; MARTÍN-HERNÁNDEZ, R; NATSOPOULOU, M; PAXTON, R J; TANNER, G; WEBSTER, T C; WILLIAMS, G R (2013) Standard methods for *Nosema* research. In *V Dietemann; J D Ellis; P Neumann (Eds) The COLOSS* BEEBOOK, *Volume II: Standard methods for* Apis mellifera *pest and pathogen research. Journal of Apicultural Research* 52(1): http://dx.doi.org/10.3896/IBRA.1.52.1.14

GREENSPAN, R J (2004) *Fly pushing: the theory and practice of* Drosophila *genetics (Second Ed.)*. Cole Spring Harbor Laboratory Press; Cold Spring Harbor, USA. 191 pp.

JENSEN, A B; ARONSTEIN, K; FLORES, J M; VOJVODIC, S; PALACIO, M A; SPIVAK, M (2013) Standard methods for fungal brood disease research. In *V Dietemann; J D Ellis; P Neumann (Eds) The COLOSS* BEEBOOK, *Volume II: Standard methods for* Apis mellifera *pest and pathogen research. Journal of Apicultural Research* 52(1): http://dx.doi.org/10.3896/IBRA.1.52.1.13

LINDSLEY, D L; GRELL, E H (1968) *Genetic variations of* Drosophila melanogaster. Carnegie Institute of Washington; Washington, USA. 472 pp.

MEIXNER, M D; PINTO, M A; BOUGA, M; KRYGER, P; IVANOVA, E; FUCHS, S (2013) Standard methods for characterising subspecies and ecotypes of *Apis mellifera*. In *V Dietemann; J D Ellis; P Neumann (Eds) The COLOSS* BEEBOOK, *Volume I: standard methods for* Apis mellifera *research. Journal of Apicultural Research* 52(4): http://dx.doi.org/10.3896/IBRA.1.52.4.05

NEUMANN, P; CARRECK, N L (2010) Honey bee colony losses. *Journal of Apicultural Research* 49(1): 1-6. http://dx.doi.org/ 10.3896/IBRA.1.49.1.01

NEUMANN, P; EVANS, J D; PETTIS, J S; PIRK, C W W; SCHÄFER, M O; TANNER, G; ELLIS, J D (2013) Standard methods for small hive beetle research. In *V Dietemann; J D Ellis; P Neumann (Eds) The COLOSS* BEEBOOK, *Volume II: Standard methods for* Apis mellifera *pest and pathogen research. Journal of Apicultural Research* 52(4): http://dx.doi.org/10.3896/IBRA.1.52.4.19

ROBERTS, D B (Ed.) (1998) Drosophila: *a practical approach*. Oxford University Press; Oxford, UK. 389 pp.

SAMMATARO, D; DE GUZMAN, L; GEORGE, S; OCHOA, R; OTIS, G (2013) Standard methods for tracheal mites research. In *V Dietemann; J D Ellis; P Neumann (Eds) The COLOSS* BEEBOOK, *Volume II: Standard methods for* Apis mellifera *pest and pathogen research. Journal of Apicultural Research* 52(4): http://dx.doi.org/10.3896/IBRA.1.52.4.20

VAN DER ZEE, R; GRAY, A; HOLZMANN, C; PISA, L; BRODSCHNEIDER, R; CHLEBO, R; COFFEY, M F; KENCE, A; KRISTIANSEN, P; MUTINELLI, F; NGUYEN, B K; ADJLANE, N; PETERSON, M; SOROKER, V; TOPOLSKA, G; VEJSNÆS, F; WILKINS, S (2012) Standard survey methods for estimating colony losses and explanatory risk factors in *Apis mellifera*. In *V Dietemann; J D Ellis; P Neumann (Eds) The COLOSS BEEBOOK, Volume II: Standard methods for* Apis mellifera *research. Journal of Apicultural Research* 52(4): http://dx.doi.org/10.3896/IBRA.1.52.4.18

VANENGELSDORP, D; LENGERICH, E; SPLEEN, A; DAINAT, B; CRESSWELL, J; BAYLISS, K; NGUYEN, K B; SOROKER; V; UNDERWOOD, R; HUMAN, H; LE CONTE, Y; SAEGERMAN, C (2013) Standard epidemiological methods to understand and improve *Apis mellifera* health. In *V Dietemann; J D Ellis; P Neumann (Eds) The COLOSS* BEEBOOK, *Volume II: Standard methods for* Apis mellifera *pest and pathogen research. Journal of Apicultural Research* 52(4): http://dx.doi.org/10.3896/IBRA.1.52.4.15

WILLIAMS, G R; DIETEMANN, V; ELLIS, J D; NEUMANN P (2012) An update on the COLOSS network and the "*BEEBOOK*: standard methodologies for *Apis mellifera* research". *Journal of Apicultural Research* 51(2): 151-153. http://dx.doi.org/10.3896/IBRA.1.51.2.01

Journal of Apicultural Research 52(4): (2013)
DOI 10.3896/IBRA.1.52.4.15

REVIEW ARTICLE

Standard epidemiological methods to understand and improve *Apis mellifera* health

Dennis vanEngelsdorp[1*], **Eugene Lengerich**[2], **Angela Spleen**[2], **Benjamin Dainat**[3], **James Cresswell**[4], **Kathy Baylis**[5], **Bach Kim Nguyen**[6], **Victoria Soroker**[7], **Robyn Underwood**[1], **Hannelie Human**[8], **Yves Le Conte**[9] and **Claude Saegerman**[10]

[1]Department of Entomology, 3136 Plant Sciences, University of Maryland, College Park, MD 20742, USA.
[2]Department of Public Health Sciences, College of Medicine, The Pennsylvania State University, Hershey, PA, 17033, USA.
[3]Swiss Bee Research Centre, Agroscope Liebefeld-Posieux ALP-Haras, Schwarzenburgstrasse 161, 3003 Bern, Switzerland.
[4]Biosciences, College of Life & Environmental Sciences, University of Exeter, Hatherly Laboratories, Prince of Wales Road, Exeter, EX4 4PS, UK.
[5]Department of Agriculture and Consumer Economics, University of Illinois, Urbana, IL 61801, USA.
[6]Department of Functional and Evolutionary Entomology, University of Liege, Gembloux Agro-Bio Tech, B-5030 Gembloux, Belgium.
[7]Department of Entomology, Agricultural Research Organization, The Volcani Center P.O.B. 6, Bet Dagan 50250, Israel.
[8]Department of Zoology and Entomology, University of Pretoria, Pretoria, South Africa.
[9]INRA, UR 406 Abeilles et Environnement, Site Agroparc, Domaine St Paul, 84914 Avignon Cedex 9, France.
[10]Research Unit in Epidemiology and Risk Analysis Applied to Veterinary Sciences (UREAR-ULg), Department of Infectious and Parasitic Diseases, Faculty of Veterinary Medicine, University of Liège, Boulevard de Colonster 20, B42, 4000 Liège, Belgium.

Received 1 June 2012, accepted subject to revision 26 October 2012, accepted for publication 16 July 2013.

*Corresponding author: Email: dennis.vanengelsdorp@gmail.com

Summary

In this paper, we describe the use of epidemiological methods to understand and reduce honey bee morbidity and mortality. Essential terms are presented and defined and we also give examples for their use. Defining such terms as disease, population, sensitivity, and specificity, provides a framework for epidemiological comparisons. The term population, in particular, is quite complex for an organism like the honey bee because one can view "epidemiological unit" as individual bees, colonies, apiaries, or operations. The population of interest must, therefore, be clearly defined. Equations and explanations of how to calculate measures of disease rates in a population are provided. There are two types of study design; observational and experimental. The advantages and limitations of both are discussed. Approaches to calculate and interpret results are detailed. Methods for calculating epidemiological measures such as detection of rare events, associating exposure and disease (Odds Ratio and Relative Risk), and comparing prevalence and incidence are discussed. Naturally, for beekeepers, the adoption of any management system must have economic advantage. We present a means to determine the cost and benefit of the treatment in order determine its net benefit. Lastly, this paper presents a discussion of the use of Hill's criteria for inferring causal relationships. This framework for judging cause-effect relationships supports a repeatable and quantitative evaluation process at the population or landscape level. Hill's criteria disaggregate the different kinds of evidence, allowing the scientist to consider each type of evidence individually and objectively, using a quantitative scoring method for drawing conclusions. It is hoped that the epidemiological approach will be more broadly used to study and negate honey bee disease.

Métodos estándar epidemiológicos para entender y mejorar la salud de *Apis mellifera*

Resumen

En este trabajo se detalla el uso de métodos epidemiológicos para entender y reducir la morbilidad y la mortalidad de las abejas. Se presentan y definen algunos términos esenciales y también se ponen ejemplos de su uso. La definición de términos tales como enfermedad, población,

Footnote: Please cite this paper as: VANENGELSDORP, D; LENGERICH, E; SPLEEN, A; DAINAT, B; CRESSWELL, J; BAYLISS, K, NGUYEN, K B; SOROKER; V; UNDERWOOD, R; HUMAN, H; LE CONTE, Y; SAEGERMAN, C (2013) Standard epidemiological methods to understand and improve *Apis mellifera* health. In *V Dietemann; J D Ellis, P Neumann (Eds) The COLOSS BEEBOOK: Volume II: Standard methods for* Apis mellifera *pest and pathogen research. Journal of Apicultural Research* 52(4): http://dx.doi.org/10.3896/IBRA.1.52.4.15

sensibilidad y especificidad, proporciona un marco de referencia para las comparaciones epidemiológicas. El término población, en particular, es muy complejo en un organismo como la abeja de la miel, porque uno puede ver la "unidad epidemiológica" como las abejas individuales, las colonias, los colmenares o incluso, determinadas operaciones. La población de interés debe, por lo tanto, estar claramente definida. Se proporcionan además ecuaciones y explicaciones sobre cómo calcular las medidas de la tasas de enfermedad en una población.

研究和改善西方蜜蜂健康的标准流行病学研究方法

本文详述了如何应用流行病学的研究方法，探明和降低蜜蜂发病率及死亡率。同时还对一些关键术语进行了定义，并举例和说明了它们的用途。定义了诸如：疾病、群体、敏感性和特异性等术语，为流行病学比较研究提供了框架。"群体"的含义在蜜蜂学研究中是比较复杂的，研究者可将一个蜜蜂的个体、一个蜂群、一个蜂场或某项实验定义为一个"流行病学研究单位"。所以，在开展流行病学研究时必须对所研究的群体加以明确定义。本文还阐述了如何评价群体的发病程度，并给出了相关的计算公式和注解。

Key words: COLOSS, *BEEBOOK*, honey bee *Apis mellifera*, epidemiology, disease, case definition confidence interval, odds ratio, relative risk, Hills Criteria

1. Basic epidemiological terms and calculations

Epidemiology is traditionally defined as the study of the distribution and determinants of disease within a human population (Woodward, 2005). To accomplish this, epidemiological studies attempt to identify factors which may explain or contribute to disease outbreak. Once identified, these factors not only inform future clinical etiological studies, but also, and perhaps more importantly, they inform disease prevention and control programmes (Mausner and Kramer, 1985). The success of epidemiologists in reducing the occurrence of human disease over the last century is undeniable. The identification of factors that contribute to the occurrence of diseases such as lung cancer (smoking), sexually transmitted diseases (unprotected sex), and cardiovascular disease (high blood pressure) have permitted targeted community health initiatives aimed at preventing or controlling risk factor exposure. These initiatives, in turn, have helped reduce the rate of disease in targeted populations (Mausner and Kramer, 1985; Koepsell and Weiss, 2003; Woodward, 2005).

Considering the success of human epidemiology, it is not surprising that epidemiological methods have been adopted by those wishing to understand and reduce disease outbreak in non-human animals (epizootiology) (Nutter, 1999). The term epidemiology is now widely adopted by those studying disease and disease determinants in non-human organisms, including honey bees, and will be the term used in this paper. Nutter (1999) argued that the application of epidemiological methods for understanding disease occurrence in plant, human, and animal populations involves the implementation of six common steps which include defining disease in quantitative terms and quantifying state and rate variables of the disease system. An alternative way to look at this process is to consider the "virtuous epidemiological cycle" (Fig. 1) which outlines the various steps involved in quantifying disease in a population, determining risk factors contributing to disease occurrence, determining methods to

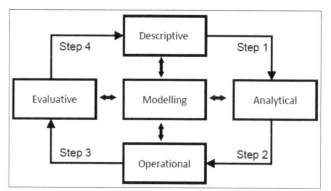

Fig. 1. The virtuous circle of epidemiology: Step 1. describe health characteristics of the population in space and time (descriptive epidemiology); Step 2. analyse data and mechanisms of development of the disease to understand behaviour (analytical epidemiology); Step 3. produce, select and apply control or preventive measures (operational epidemiology); Step 4. give necessary information that permits the follow up of measures (evaluative epidemiology); In addition, changing epidemiological methods should be supported by theoretical epidemiology (modelling).

reduce disease occurrence and then evaluating the effectiveness of these methods (Toma *et al.*, 1991).

A comprehensive review of all of these steps is well beyond the scope of this paper. Similarly, much of the data used by epidemiologists are derived from surveillance efforts, a discussion of which is also beyond the scope of this paper, but has received attention in other recent work (Hendrikx *et al.*, 2009, vanEngelsdorp *et al.*, 2013). Instead, we focus on presenting and defining the vocabulary needed to implement epidemiological studies, and then outline study design, analysis and interpretation. It is also the intent of this paper to present a framework for understanding and initiating ongoing and future studies of honey bee health. Unless otherwise noted, the following terms and concepts have been adapted from Koepsell and Weiss (2003).

1.1. Disease

To successfully develop tools which either quantify the rate of disease development in a population or quantify the factors which may contribute to disease occurrence, the "disease" of interest must be clearly defined. Broadly speaking, disease is any departure from perfect health. When applied to specific studies, a precise definition - the case definition - must be developed which unambiguously allows subjects to be classified as a case or not.

1.1.1. Case definition

The case definition is the operating definition of a disease for study purposes. Aristotle identified two crucial components that made for a good case definition: 1. it specifies characteristics common to all diseased individuals; and 2. it specifies how diseased individuals differ from non-diseased individuals (Koepsell and Weiss, 2003). Ideally, the characteristics used to identify the disease should be simple and recognizable by independent observers in different geographies. Because characteristics cannot always be recognized in the field, case confirmation by laboratory analyses is sometimes necessary. Case definitions, especially for emerging or newly identified diseases, often suffer from having limited specificity. Further, case definitions for a disease can evolve as understanding of a disease changes and / or the diagnostic tests performed to determine a diagnosis are refined. An outline of different classifications of case definitions has been provided by the World Health Organization (WHO, 1999). When applied to apiculture, it is important to define the "epidemiological unit" for which the case definition is being applied (discussed in greater detail in section 1.2). Epidemiological units are the groups which make up the population of interest, and can range from individual bees, colonies, apiaries, and operations.

1.1.2. Test sensitivity and specificity

Many case definitions are based on laboratory or clinical tests, but tests in themselves are prone to errors either by misidentifying disease positive cases, i.e. individuals that have the disease, incorrectly as negative cases, or disease-free, or disease negative cases as positive cases. The accuracy of a test is primarily given as sensitivity and specificity.

Sensitivity

Sensitivity is the probability that a human or animal will have a positive test result if indeed the human or animal does have a disease. This is expressed as: $P(T+|D+)$, where P is the probability, T+ is a positive test result and D+ is a disease being present. In applied epidemiology, sensitivity is often expressed as a proportion, and thus expressed as equation 1.1.2.a.

Equation 1.1.2.a

$$Sensitivity = \frac{Number\ of\ disease\ positives\ testing\ positive}{Number\ of\ disease\ positives\ testing\ positive\ +\ Number\ of\ disease\ positives\ testing\ negative}$$

Specificity

Similarly, specificity is the probability that a human/animal will have a negative test result if indeed it is disease free. This is expressed as: $P(T-|D-)$, where P is the probability, T- is a negative test result and D- is the disease not being present. In applied epidemiology, specificity is often expressed as the proportion of non-diseased (healthy) animals that test negative, expressed as equation 1.1.2.b.

Equation 1.1.2.b

$$Specificity = \frac{Number\ of\ disease\ negatives\ testing\ negative}{Number\ of\ disease\ negatives\ testing\ negative\ +\ Number\ of\ disease\ negatives\ testing\ positive}$$

1.1.2.1. Calculating confidence intervals for a proportion

Sensitivity and specificity are based upon a sample of test results around which there is uncertainty. In epidemiology, uncertainty can be expressed as a confidence interval (CI). Typically, they are expressed as a 95% confidence interval (95% CI). Briefly, confidence intervals indicate the precision of the estimate where a wide confidence interval indicates that the estimate is not very precise. In statistical terms, if we were to repeat the procedure using 100 different samples of the same size from the same population, the true proportion would be expected to lie within 95 of the 100 resulting confidence intervals. Implicit in presenting 95% CI is the assumption that the sample from which the CI is derived is representative of the population from which the sample was drawn. Representativeness is best achieved when the sample is randomly drawn from the population of interest. As long as the sample size is greater than 30, the 95% CI can be calculated using equation 1.1.2.c.

Equation 1.1.2.c

$$95\%\ CI\ for\ p\ =\ \hat{p}\ \pm Z_\alpha(\ s.e.(\hat{p}))$$

Where Z_α is the $(1-\alpha/2)$ percentile of the standard normal distribution (Z_α = 1.96 for 95% CI) and s.e. is the standard error.

$$s.e.(\hat{p}) = \sqrt{\frac{\hat{p}(1-\hat{p})}{n}}$$

In cases where the sample size is smaller than 30, where np < 5, n(1-p) < 5 or the proportion estimate is close to 0 or 1.0, standard statistical software tools (e.g. SAS JMP) will use the binomial distribution to calculate the CI. Estimates can also be determined by replacing Z_α in equation 1.1.2.c above with the critical value from a published binomial statistical table.

1.1.3. Positive and negative predictive values

While sensitivity and specificity primarily measure a test's accuracy, epidemiologists use two other measures, positive and negative predictive values, to help describe the certainty of a specific test result. A Positive Predictive Value (PPV) is the probability that a person/animal with a positive test result truly has a disease $P(D+|T+)$. PPV is typically expressed as a proportion (Equation 1.1.3.a).

Equation 1.1.3.a

$$PPV = \frac{Number\ of\ disease\ positives\ testing\ positive}{Total\ number\ testing\ positive}$$

A Negative Predictive Value (NPV) is the probability that a person/ animal with a negative test result truly does not have disease P (D-/T-). NPV is typically expressed as a proportion (Equation 1.1.3.b).

Equation 1.1.3.b

$$NPV = \frac{Number\ of\ disease\ negatives\ testing\ negative}{Total\ number\ testing\ negative}$$

If sensitivity and specificity remain constant, as the prevalence of a disease increases so does the PPV while the NPV decreases.

1.2. Population

Defining the population under study is a critical component of all epidemiological studies. Like case definitions, the population under study must have characteristics which set its members apart from non-members. These members can then be categorized into smaller groups for the purposes of comparing disease levels between different sub-groups within the study population. Defining the population of interest in apiculture represents a unique challenge as there is a hierarchy of population units, each of which could be considered "individual members" (Table 1). In apicultural terms there are several levels of potential interest, thus there are several different definitions for what makes up the individual of interest.

- Individual bees within a colony
- A group of colonies located within one area make up an apiary
- One or more groups of apiaries owned or managed by one beekeeper make up an operation
- Apiaries contained within a defined geography make up a region

Characteristics that commonly define sub-groups within any of these given populations often differ according to hierarchal classification of the population, but broadly include individual attributes, such as: age (i.e. bee cohort at the colony level (Giray *et al.*, 2000); genetics (i.e. patriline at the colony level (Estoup *et al.*, 1994), queen type at the apiary level); size of operations; production objectives; and management style (at the regional level) (Table 1). Once the defining criteria for a population have been established, the membership (epidemiological unit) of that population can be quantified. However, size may change over time because new members are added or existing members are removed.

1.3. Measures of disease in a population

Comparing frequency of disease between sub-groups of a population underpins most epidemiological research (see study design in Section 2.0). As such, various ways to quantify disease frequency have been developed.

Box 1.

Over the inspection season of 2004 and 2005, Pennsylvania state bee inspectors preformed 107 Holst's milk tests on suspect cases of clinical American foulbrood disease (for more information about this test, see the BEEBOOK paper on American foulbrood (de Graaf *et al.*, 2013)). Ninety samples tested positive with the Holst's milk test (Holst, 1945), of which 89 were confirmed in the laboratory to be AFB infection. Confirmation of diagnosis was performed by culturing a smear of diseased larvae sampled from the same colony. The Holst's milk test resulted in 14 negative and three inconclusive results. The latter were discarded. Six of the negative samples were later diagnosed to have had AFB when companion samples were cultured (vanEngelsdorp, unpublished data). The sensitivity and specificity as well as the positive and negative predictive value of this test can be calculated as follows:

In summary:

		Condition (as determined by AFB Culture)		
		Positive	Negative	Total
Holst's Milk Test	Test Positive	89	1	90
	Test Negative	6	8	14
	Total	95	9	104

Therefore:

$$Sensitivity = \frac{89}{89+6} = 0.94 = 94\%$$

$$where\ 95\%\ CI = 0.94 \pm 1.96\ x\ \sqrt{\frac{0.94(1-0.94)}{95}}$$

$$= 0.94 \pm 0.048$$

$$= (0.89 - 0.99) = 89 - 99\%$$

$$Specificity = \frac{8}{8+1} = 0.89 = 89\%$$

Because the denominator is less than 30, the normal approximation of the binomial distribution cannot be assumed and for the calculation of the 95% CI we used the binomial tables. Thus, the

$$95\%\ CI = 0.89 \pm 0.37$$
$$= (0.52-1.00) = 52-100\%$$

And

$$PPV = \frac{89}{89+1} = 0.99 = 99\%$$

And

$$NPV = \frac{8}{8+6} = 0.53 = 53\%$$

Thus, when a Holst's milk test is performed and comes back positive we are 99% certain the sample does contain American Foulbrood spores, while if the Holst's milk test comes back negative we are 53% sure that the sample does not have American foulbrood spores.

Table 1. Hierarchy of possible populations of interest, types of members, and common groupings or sub-categories for comparing members within the same population in honey bee epidemiological studies.

Population	Members	Common groupings / subcategories for comparisons
Colony	Bees	Caste (worker *vs* drone) Cohort (foragers, nurses, pupae, etc.)
Apiary	Colonies	Queen stock Treatment groups Micro –environment (shade *vs* full sun)
Operation	Apiaries	Region/microclimate Management system Disease history
Region	Operations	Operation size Management practices Geographic region

1.3.1. Point prevalence

Point prevalence is the frequency of ongoing disease in a defined population at a certain point in time (Equation 1.3.1).

Equation 1.3.1

$$Prevalence = \frac{Number\ of\ cases}{Size\ of\ population} \times 100\%$$

The method for calculating the 95% confidence interval for point prevalence is outlined in section 1.1.2.1. Again, it is important to stress that calculating the CI assumes the sample pool is representative of the population as a whole, this is best achieved if the sample was randomly drawn from the population. The estimate of the point prevalence is affected by the likelihood that a disease will be detected during a given inspection. Diseases which occur for only short periods of time are less likely to be observed during an inspection than are diseases that are more chronic (Box 2).

1.3.1.1. True versus apparent prevalence.

As can be inferred from the discussion above, the reported point prevalence of disease is influenced by the case definition and the test employed to determine a case's outcome. It is conceivable that for some diseases, in-field examination for phenotypic expression of disease may be negative while laboratory tests determine disease presence (i.e. deformed wing virus). In such cases, two types of prevalence can be specified; true prevalence with all cases of disease existing at a specific point in time, and apparent prevalence that is determined by test results (i.e. in-field examination, molecular test, etc.). The apparent prevalence is subject to the accuracy of the test (sensitivity and specificity).

1.3.2. Incidence rate

Incidence is the occurrence of a new case if a disease and is best calculated if the exact period of time at risk for each participant is known. The incidence rate is the proportion of incident cases in a population at risk of becoming an incident case during a specified period of time (Equation 1.3.2.a).

Equation 1.3.2.a

$$Incidence\ Rate = \frac{Number\ of\ incident\ cases}{Number\ at\ risk\ of\ experience} \times 100\%$$

Box 2.

In the summer of 2006 apiary inspectors in Pennsylvania inspected a sub-set of beekeeping operations in the state. In total, 1,706 apiaries were inspected containing 11,285 colonies. Clinical signs of Chalkbrood (CB) disease were found in a total of 384 colonies located in 156 apiaries (vanEngelsdorp, unpublished data).

$$Prevalence\ of\ CB\ in\ inspected\ apiaries = \frac{156}{1706} = 0.09 = 9.0\%$$

Where 95% CI for estimation of the prevalence in the total population

$$= 0.09 \pm 1.96\ x\ \sqrt{\frac{0.09\ (1 - 0.09)}{1706}}$$

$$= 0.09 \pm 0.014$$

$$= (0.076 - 0.104) = 7.6 - 10.4\%$$

$$Prevalence\ of\ CB\ in\ colonies = \frac{384}{11285} = 0.034 = 3.4\%$$

where 95% CI for estimation of prevalence in the total population

$$= 0.034 \pm 1.96\ x\ \sqrt{\frac{0.034\ (1 - 0.034)}{11285}}$$

$$= 0.034 \pm 0.0033$$

$$= (0.0307 - 0.0373) = 3.1 - 3.7\%$$

Thus, assuming that the inspected apiaries were representative of the entire Pennsylvanian population, Chalkbrood was present in 9% of all apiaries (95% CI: 7.6 - 10.4%) while 3.4% (95% CI 3.1 – 3.7%) of all colonies had clinical signs of the disease.

The incidence rate (IR) accounts for the fact that the number of incident cases is dependent on the size of the population observed and the time period over which individuals were observed. Because IRs are measured over time, the population under observation may change. Where precise data on the population at risk of becoming an incident case over the period is not available, the average population of individuals at risk for the time period is commonly used as the denominator. This technique is particularly useful when attempting to calculate the incidence rate of a condition which is very likely to be self-reported in a large population. IRs are presented as a proportion per time, or per unit-time if the exact time at risk is known for each member of the population.

1.3.2.1. Calculating confidence intervals for incidence rates

The confidence interval for an IR can be calculated for a population with the same time at risk using the method described in section 1.3.3.2.1 below, where Z_α is based on the Poisson distribution and n is an individual-time constant. In reality the IR is often not homogenous within a population. For instance, a random sample of honey bee colonies would express hygienic behaviour differently. As highly hygienic colonies are more likely to resist brood diseases, these colonies would be less likely to be diagnosed with the condition. Conversely, it is conceivable that the diagnosis of a certain brood disease in a given colony is a marker for increased susceptibility for the disease. Therefore, in comparison to disease-free colonies, a second diagnosis is more likely to occur in colonies that were previously diseased. This phenomenon is referred to as extra-Poisson variation and if left uncorrected will result in a confidence interval that is too narrow. To address this, a multivariate logistic regression model with terms for previous disease should be employed.

Just as the IR is not the same for all individuals in a population, it is also not likely to be constant over time. The prevalence of many bee diseases changes over time, thus affecting 95% CI calculation. This problem can be overcome by restricting analysis to sub-periods or "time bands" so that differences in IR over time are not a factor. Alternatively, time itself can be used as a predictor of disease when performing a multivariate analysis (Koepsell and Weiss, 2003).

1.3.3. Special cases of incidence

Over the last few years considerable effort has been placed on documenting winter losses in different regions of the world. As a result, different methods to calculate and report winter losses have been developed including Total Loss and Average Loss (vanEngelsdorp *et al.*, 2011).

1.3.3.1. Total colony loss (TL) (the cumulative incidence of mortality)

This is the percentage of colonies lost in a specific group over a fixed period of time. This figure is the most accurate snap shot of loss in a defined group, such as in an operation or geographic region. If all colonies in a region were enumerated it would give a precise figure for the proportion of all colonies that died in that region. However, within the population of interest, operations with large numbers of colonies will have a greater influence on the total colony loss metric than will the operations with only few colonies. Total Colony Loss in an operation or in a defined group is calculated by dividing the total number of colonies that died over a given time period (T_{dead}) by the total number of colonies at risk of dying in a given time period ($T_{Colonies\ at\ risk}$) and multiplying the quotient by 100% (Equation 1.3.3.1).

Equation 1.3.3.1

$$Total\ Loss = \frac{T_{Dead}}{T_{Colonies\ at\ risk\ of\ dying}} \times 100\%$$

Where the total number of colonies at risk of dying ($T_{Colonies\ at\ risk\ of\ dying}$) over a period was calculated by adding the number of colonies at the start of the period (T_{Start}) with the number of splits made by the beekeepers over the period (T_{Splits}) and the number of colonies purchased over the period ($T_{Purchased}$) and then subtracting the number of colonies removed (sold or given away) over the period ($T_{Removed}$).

$$T_{Colonies\ at\ risk\ of\ dying} = T_{Start} + T_{Splits} + T_{Purchased} - T_{Removed}$$

And where the total number of colonies that died (T_{Dead}) was calculated by subtracting the total number of colonies at the end of a period (T_{End}) from the total number of colonies at risk of dying for the period ($T_{colonies\ at\ risk\ of\ dying}$).

$$T_{Dead} = T_{Colonies\ at\ risk\ of\ dying} - T_{End}$$

Where period was the defined period of time for which colony loss was analysed. The unit of time, is the period defined by the time between T_{Start} and T_{End}. This unit is often not reported and is often loosely defined by the season encompassed by that time period (e.g. winter).

And where, respondents in a specific group are the group of respondents for whom valid loss data was collected.

1.3.3.1.1. 95% CI for total loss

Because total loss is a proportion, theoretically its confidence interval can be calculated using equation 1.1.2.c. This approach is valid when calculating a 95% CI for losses within one operation. However, if all the colonies in an operation are measured, one's sample is the whole population, there is no need to calculate the CI. When total losses are calculated for a region, the losses of several operations are being combined, using the previously mentioned equation to calculate the 95% CI is inappropriate because the assumption of independence is not meet. Across operations the chances that a colony will die is not the same for every colony. In such cases the quasi-binomial family is introduced to take into account the increased standard error introduced by dependence within the data (vanEngelsdorp *et al.*, 2011). An R script example which calculates both the corrected CI from the quasi-binomial model and the uncorrected CI from the Wald model (equation 1.1.2.c.) are given in Box 3.

1.3.3.2. Average loss (AL)

Average loss is the mean % of the total colony loss experienced by respondents in a defined group over a defined period of time. This metric is most appropriately used to compare groups partitioned by different risk factor exposures (see study design in Section 2.1.1.3). Usually average loss calculations are heavily influenced by smaller beekeeper operations as they often compose a larger portion of the response population. Average Loss is calculated by dividing the summed total colony loss of respondents (TLi) within a specified group by the number of respondents in that group (N) and then multiplying the quotient by 100%. Equation 1.3.3.2

Equation 1.3.3.2

$$Average\ Loss = \frac{\sum TL_i}{N}$$

1.3.3.2.1. 95% CI for average loss

Like other proportions, average loss confidence intervals can be calculated using equation 1.1.2.c. As mentioned previously, average losses are often skewed by smaller operations resulting in a Poisson distribution of losses rather than a normal distribution. When the number of respondents exceeds 100, the Poisson distribution resembles a normal distribution so adjustment in the equation 1.1.2.c is not needed. However, when the number of respondents is less than 100, the rate multiplier for the 95% CI can be determined by looking up the lower and higher rate multiplier in an appropriate table (e.g. Paoli *et al.*, 2002) (Box 3.).

2. Study design

Epidemiologists endeavour to reduce disease occurrence in a population. To achieve this one must quantify disease at the population level and determine risk factors that contribute to disease occurrence. Two study designs can be used to determine the association of exposure with a health outcome: observational and experimental. In an experimental design, the exposure is determined by the investigator, whereas in an observational design, the exposure is not determined by the investigator or the study (i.e. exposure is under the control of the study participants or the participant's environment). For example, if an investigator determines which hives are treated for Nosema and which are not, then the study design would be considered an experimental design. In an observational study, the investigator would observe the Nosema responses for beekeepers who applied and who did not apply treatment for Nosema, wherein this case, the application of the treatment is determined by the beekeeper.

2.1. Observational study designs

2.1.1. Cross-sectional studies

Cross-sectional studies are a point-in-time study, such as a one-time disease surveillance survey, and are typically used to estimate disease prevalence or the simultaneous association between a risk factor and a disease. In this design, the exposure and outcome for each subject in the study are ascertained simultaneously. This simultaneity often leads to difficulty in conclusively establishing the temporal relationship between the exposure and the outcome. It is also important to note that chronic conditions are more likely to be identified in a survey because they are more likely to persist in a population and are more

---

Box 3.

In the dialogue below, text starting with # describes the R script which follows. Text in bold is R script and text in italics is output

R code to calculate losses and CI

import data (format csv)

ruchers = read.csv("ruchers.CSV")

summary(ruchers)

```
   Rucher        nCol          nDead         nAlive
Min.   : 1.00  Min.   : 1.00  Min.   : 0.00  Min.   : 0.00
1st Qu.: 44.75  1st Qu.: 2.75  1st Qu.: 1.00  1st Qu.: 0.00
Median : 88.50  Median : 5.00  Median : 3.00  Median : 2.00
Mean   : 88.50  Mean   : 96.51  Mean   : 33.60  Mean   : 62.91
3rd Qu.:132.25  3rd Qu.: 12.00  3rd Qu.: 8.25  3rd Qu.: 5.00
Max.   :176.00  Max.   :6000.00  Max.   :2000.00  Max.   :5000.00
```

attach(ruchers)

general linear model, family quasi-binomial

ruchers.glm1 <- glm(cbind(nDead, nAlive)~1, family= quasibinomial, data=ruchers)

generate confidence intervals via quasi-binomial model

require(boot)
CI_GLM = inv.logit(coef(ruchers.glm1)+c(-1,1)*1.96*sqrt (vcov(ruchers.glm1)))

Verification : 'raw' confidence intervals (Wald formula as in equation 1.1.2.c)
For Total Loss: based on the number of colonies (this approach underestimates the 95% CI)

nColonies = sum(ruchers$nCol)
prop = with(ruchers, sum(nDead)/sum(nCol))
CI_Wald=prop+(c(-1,1)*1.96*sqrt(prop*(1-prop)/ nColonies))

#The output generates 1. total loss, 2. the standard error (based on the quasi-binomial),
#3. the CI from the quasi-binomial model and 4. the CI from the Wald formula.

titles=c("Total Loss","SE")
titles2=c("   Model-based Conf. Int.","")
titles3=c("   Wald Conf. Int.","")
stats=c(prop,sqrt(vcov(ruchers.glm1)))
output=rbind(titles,stats,titles2,CI_GLM,titles3,CI_Wald)
print(output)

```
          [,1]                      [,2]
titles    "Total Loss"              "SE"
stats     "0.348110208406923"       "0.0794099959221504"
titles2   "    Model-based Conf. Int."  ""
CI_GLM    "0.31367368043281"        0.384210926105528
titles3   "    Wald Conf. Int."     ""
CI_Wald   "0.340946201742426"       0.355274215071421"
```

Thus this table states the total loss was 34.8% with a standard error of 7.9%, giving a 95% CI of 31.4-38.4%.

common. Therefore this study design is less useful for studies of rare exposures and rare outcomes. However, cross-sectional studies can be inexpensive, relatively quick to conduct, and are used to identify potential associations between exposures and outcomes that warrant further research with more rigorous population-based study designs. An example of a cross-sectional study is when a bee inspector examines hives in an apiary for characteristics, such as size, strength, activity, and disease and then uses these data to generate estimates of the prevalence of hives with a particular disease (e.g., Chalkbrood) in a region.

2.1.1.1 Detection of rare events

Epidemiological surveys are often designed to detect (or not detect) relatively rare events in a population. It is often impractical or impossible to prove that a disease or pest organism is not found in a region with 100% certainty. However, a properly designed disease surveillance system can give a set level of confidence that a disease or pest species is not present in a defined population at a predefined prevalence level. These results, by extension, can help to declare a region as free from a particular disease or parasite which may have important implications for policy makers.

In most cases, disease prevalence in individual members (i.e. colonies) will be categorical, that is the disease will either be present or absent (Fosgate, 2009). The number of individuals that would need to be examined (n) in an infinite population (where the number of individuals exceeds 1,000 members) given a minimum disease prevalence (P) is given by equation 2.1.1.1.a (Fosgate, 2009).

Equation 2.1.1.1.a
$$n = \frac{\ln(\alpha)}{\ln(1 - P)}$$

Where α is the 1-confidence with which one wants to be certain the disease is detected. In finite populations (< 1,000) with a population size of N, the number of individuals that need to be examined (n) to be certain to detect at least one positive case at a defined 1-confidence (α), where the minimum prevalence of disease in the population (P) is given by equation 2.1.1.1.b.

Equation 2.1.1.1.b
$$n = \left(1 - \alpha^{1/D}\right)[N - \frac{D - 1}{2}]$$

Where D = N × P

Both of these approaches assume tests which are 100% sensitive, which is often unrealistic. In cases where sensitivity is imperfect but known (S), the number of individuals that would need to be examined (n) in an infinite population to be 1-confident (α) of detecting at least one diseased case with a disease prevalence of P is given by equation 2.1.1.1.c (Fosgate, 2009).

Equation 2.1.1.1.c
$$n = \frac{\ln(\alpha)}{\ln(1 - P * S)}$$

Box 4.

The bump technique is a new method meant to detect the presence of *Tropilaelaps* mites (Anderson *et al.*, 2013). This test, when applied to colonies that have an average infestation of 4.6 ± 0.06 mites per 100 brood cells, has a sensitivity of 36% (Pettis, Rose, and vanEngelsdorp, unpublished data). How many colonies need to be tested in a region with more than 1,000 colonies in order to detect one infected colony with 95% Confidence, assuming that 5% of colonies are infested?

$$n = \frac{\ln(0.05)}{\ln(1 - 0.05 * 0.36)} = \frac{-2.996}{-0.01816} = 165$$

Thus, 165 randomly selected colonies would need to be tested to be 95% confident of detecting at least one positive colony given a 5% infestation rate.

2.1.1.2 Data analysis and interpretation: making associations between exposure and disease in cross-sectional studies

When cross-sectional studies collect information on disease prevalence and simultaneous exposure to factors that may contribute to disease, Odds Ratios (ORs) can be used to calculate the degree of association between concurrent exposure and disease state. We can calculate the odds of exposure among cases compared to the odds of exposure among non-cases (controls). The OR is the odds of exposure in an individual who was diseased divided by the odds of exposure in an individual who was disease free.

Equation 2.1.1.2.a
$$Odds\ Ratio\ (OR) = \frac{\frac{a}{c}}{\frac{b}{d}} = \frac{ad}{bc}$$

Where a,b,c,d are defined by the Table 2. The Confidence Intervals for Odds Ratio can be calculated using Equation 2.1.1.2 b.

Equation 2.1.1.2.b
$$95\%\ CI\ of\ OR = e^{\ln(OR) \pm 1.96\sqrt{\frac{1}{a}+\frac{1}{b}+\frac{1}{c}+\frac{1}{d}}}$$

2.1.1.3. Significance of odds ratio measures

Generally speaking OR (and Relative Risk see below) values greater than 1 indicates that a disease is more likely to occur in an exposed group as compared to an unexposed group. Conversely, an OR value less than 1 means that a disease event is less likely to occur in an exposed groups compared to unexposed group. An OR that has a 95% CI that overlaps with 1 is indicative of an OR that is not a significant (Box 5).

Table 2. Structure of data for calculation of odds ratio.

Exposure	Disease Present	Absent	All Individuals
Yes	a	b	a+b
No	c	d	c+d

Box 5.

Between 1996 and 2007 the apiary inspection programme in the Commonwealth of Pennsylvania inspected 19,933 apiaries for clinical signs of chalkbrood and sacbrood disease. Over all inspections, 1,831 apiaries were found to have at least one colony with chalkbrood, and 547 colonies were found to have sacbrood. 212 apiaries had colonies infected with chalkbrood and sacbrood at the same time (vanEngelsdorp, unpublished data). Was there an association between the presence of chalkbrood and sacbrood?

	apiaries	sacbrood positive	negative	total
chalkbrood	positive	212	1,619	1,831
	negative	335	17,767	18,102
	total	547	19,386	19,933

$$Odds\ Ratio = \frac{\frac{212}{335}}{\frac{1619}{17767}} = 6.94$$

Thus, apiaries infected with chalkbrood are 6.9 times more likely to be infected with sacbrood when compared to apiaries not infected with chalkbrood.

The 95 % confidence interval for the Odds Ratio = $e^{\ln\ (OR)\ \pm\ 1.96\ x\ s.e.}$

Where s.e. $= \sqrt{\frac{1}{a} + \frac{1}{b} + \frac{1}{c} + \frac{1}{d}}$

Thus, the 95% confidence interval in this example is = 5.8 - 8.3. The confidence interval does not include 1.0, therefore the relationship between Sacbrood and Chalkbrood is statistically significant, and is unlikely due to chance.

2.1.1.3 Comparing prevalence / incidence rates

Some cross sectional studies may collect information on presumptive risk factors as well as health outcomes. For instance, winter loss surveys may collect information on management practices utilized in addition to health outcome (mortality). When the study permits the population to be divided based on different "exposures", the measures of disease outcomes (prevalence or incidence rates) can be compared. When prevalence is the measure of comparison, differences in exposure between two groups separated by risk factor exposure can be compared using a Chi-Square test, or in cases where fewer than 5 cases were expected in a given cell, the Fisher's exact test. Resulting from this approach is a p value, which simply provides a goal post by which we can assert that the populations differ significantly (typically when a $p \leq 0.05$ is calculated, the prevalence rates in two populations are considered to be significantly different). However, this approach does not give any indication as to the size of the effect of exposure to the risk factor. The magnitude of this effect can be gleaned by comparing the 95% CI of the point prevalence estimates. Generally speaking, populations that have point estimates with overlapping 95% CI are not significantly different, while those who do not have overlapping populations are. More importantly, the 95% CI aid in the interpretation of any exposure effect in that it puts

the upper and lower bounds on possible magnitude of any effect (Gardner and Altman, 1986).

When cross sectional studies result in incidence rates (e.g. from winter loss surveys), rates between groups separated by exposure can be compared using ANOVA and other basic parametric tests. As is the case for the non-parametric tests mentioned in the above paragraph, these will result in a P value which indicates if the incidence rates in the populations differ. This result is of limited value because not only is it of interest that the populations are different; the magnitude of the difference is of note. Calculating and comparing 95% CI for the point estimate of Incidence rates has more meaning than stating that the two groups within a population are different or not based on a statistical test (Box 6).

2.1.1.4 Multiple regression models

While comparing exposure prevalence in sub-groups of a population may have benefits in elucidating exposures that have pronounced effects on disease, often, several factors may contribute to disease outcomes. In these cases, multivariate regression analysis can be conducted to highlight exposure factors that differ between groups. If the outcome is at the individual level, a multivariate logit or probit may be appropriate. If the outcome is at a group level, a multivariate logistic regression may be preferred, although if most ratios or percentages range between 0.3 and 0.7, a linear regression can often give a good fit. Standard statistical packages (SAS, R, etc.) permit fairly straightforward disease modelling for datasets that are complete, that is have all the needed exposure measures present for each "diseased" and "non-diseased" epidemiological unit. However, frequently, cross sectional studies have incomplete data.

2.1.1.5. Classification and regression tree (CART) analysis

This analysis is useful for modelling diseases that have multiple contributing factors and an incomplete data set for quantifying possible risk factors in both the disease and disease-free populations. The CART analysis is a non-linear and non-parametric model, fitted by binary recursive partitioning of multidimensional co-variate space (Breiman *et al.*, 1984, Saegerman *et al.*, 2004, Speybroeck *et al.*, 2004). Using CART 6.0 software (Salford Systems; San Diego, USA), the analysis successively splits the data set into increasingly homogeneous subsets until it is stratified and meets specified criteria. The Gini index is normally used as the splitting method, and a ten-fold cross-validation is used to test the predictive capacity of the trees obtained. The CART analysis performs cross-validation by growing maximal trees on subsets of data, then calculating error rates based on unused portions of the data set.

The consequence of this complex process is a set of fairly reliable estimates of the independent predictive accuracy of the tree, even when some data for independent variables are incomplete and/or comparatively scarce. Further details about CART are presented in previously published articles (Saegerman *et al.*, 2011).

2.1.2 Cohort studies

Cohort studies allow an investigator to estimate the disease incidence rate because the study measures the time that participants don't have the disease. As compared to cross-sectional studies, cohort studies are better able to assess causality because the temporal relationship of exposure preceding outcome is not subject to question. This design is implemented through three steps. First, exposed and unexposed individuals who are free of the outcome of interest are identified and become the cohort. Next, each cohort is observed for a minimum period of time to determine if the outcome of interest develops. The risk of developing the outcome is calculated separately for the exposed group and for the unexposed group. Finally, the risk for the exposed and unexposed study subjects is compared, often by estimating the relative risk. Essentially, the incidence of disease over time is measured in exposed and unexposed individuals to determine the risk of disease in relation to exposure to a factor of interest. These studies can be performed retrospectively, where a post-hoc study is executed on previously collected data, or prospectively, where study subjects who do not have the outcome of interest are followed forward through time. Examples of cohort studies in honey bees

---

Box 6.

A winter loss survey was conducted to determine the winter mortality (Oct 1 – April 1) of US beekeepers over the winter of 2010-2011 (vanEngelsdorp *et al.*, 2012). A subset of these respondents also answered various questions regarding their management practices. In all 1,074 beekeepers indicated they had used a known varroa mite control product in a majority of their hives over the previous year, while 1,675 responding beekeepers reported not using any known varroa mite control product in any of their hives. Beekeepers who used a known varroa mite control product suffered an average loss of 29.5% (95% CI 27.5 - 31.4%) of their colonies, while those who did not indicate they used a known varroa mite control product suffered an average loss of 36.7% (95% CI 34.9 - 38.55) (BeeInformed.org Report 30).

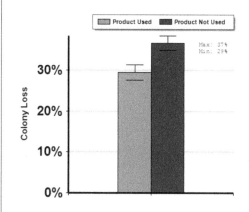

As the two confidence intervals do not overlap we know the two populations are different, we can say that beekeepers who treated with a known varroa control product lost 7 fewer overwintering colonies per 100 than those who did not; in other words beekeepers who treated with a known varroa control product lost 20% (difference in average loss rate / average loss rate in those not treating = 7/37*100 %) fewer colonies than those who did not.

---

include Genersch *et al.* (2010), Gisder *et al.* (2010) and vanEngelsdorp *et al.* (2013).

2.1.2.1 Data analysis and interpretation: making associations between exposure and disease in cohort studies

If the investigator knows the exact time that each participant was at risk, it is possible to calculate the incidence rate. Incidence rates can be compared between different groups within a population in the same way as prevalence rates can, that is using standard statistical tests, and/or, (perhaps more appropriately) comparing 95% CI between two groups in a population. Another valuable tool that can be used to highlight possible associations between disease outcome and risk factor exposure is the calculation of relative risk.

Box 7.

A longitudinal study was set up to monitor colonies for mortality and other factors as they moved up and down the east coast to pollinate crops. Forty nine colonies were examined in June of 2007, and 20 of them were found to have entombed pollen during the examination. In January 2008, 15 of the colonies that had entombed pollen were dead, as compared to the 6 colonies that died in the cohort without entombed pollen (vanEngelsdorp *et al.*, 2009a).

	Colonies	Outcome (January 2008)		
		Dead	Alive	Total
Entombed pollen (June 2007)	Positive	15	5	20
	Negative	6	23	29
	Total	21	28	49

$$Relative\ Risk\ (RR) = \frac{\frac{a}{a+b}}{\frac{c}{c+d}} = \frac{0.75}{0.21} = 3.6$$

95% CI = 1.68 - 7.61

As the RR is greater than 1 and the 95% CI do not overlap with 1, we can say that the increased risk of mortality associated with entombed pollen is significant. For every colony that died by January that did not have entombed pollen in June, 3.6 colonies died that did have entombed pollen.

2.1.2.2. Relative risk

The Relative risk is a measure of the chance of developing a disease after a particular exposure. It is calculated by dividing the incidence rate in an exposed population (I_e) by the incidence rate in an unexposed population (I_o).

Equation 2.1.2.2

$$Relative\ Risk\ (RR) = \frac{I_e}{I_o} = \frac{\frac{a}{(a+b)}}{\frac{c}{(c+d)}}$$

Where a, b, c, d are determined by Table 3.

2.1.2.3. The confidence intervals for relative risk

The Confidence Intervals for Relative Risk can be calculated using the equation given in Equation 2.1.2.3.

Equation 2.1.2.3.

$$95\%\ CI\ of\ RR = e^{\ln(RR)\pm1.96\sqrt{\frac{b}{a(a+b)}+\frac{d}{c(c+d)}}}$$

There are numerous online RR calculators (e.g. http://faculty.vassar.edu/lowry/VassarStats.html). Common statistical packages often give RR and associated CI when performing tests on 2x2 contingency tables. Caution should be used, however, to ensure that the data entered in such packages are in keeping with the layout presented in Table 3.

2.1.2.4. Significance of relative risk measures

Generally speaking RR (and Odds Ratio) values greater than 1 indicates that a disease is more likely to occur in an exposed group as compared to an unexposed group. Conversely, a RR value less than 1 means that a disease event is less likely to occur in an exposed group compared to unexposed group. The confidence that a RR value is a measure of a real increased measurable risk, and not a consequence of chance, is dependent on several factors: 1. the size of the population; 2. the variability in the responding population; and 3. the intensity of the effect. All of these attributes are accounted for in the calculation of the 95% CI. Thus, to gauge if a RR measure truly does indicate an increase or decrease in risk of disease after exposure, one should examine a RR 95% CI. If the interval overlaps with 1, the RR cannot be considered significant (Box 7).

2.1.3. Case-control studies

In contrast to cohort studies where participants are identified by exposure status, participants in case-control studies are identified by their disease or outcome status. Cases are participants who have developed the outcome of interest. Controls are subjects who do not have the outcome of interest and provide an estimate of the frequency of exposure in the population at risk. In this retrospective study design, cases and controls are first identified. Subsequently, the

Table 3. Structure of data for calculation of Relative Risk. Both disease outcome and risk factor exposure are dichotomous.

	Outcome		
Exposure	**Present**	**Absent**	**All Individuals**
Yes	a	b	a+b
No	c	d	c+d

exposure to the factor of interest is ascertained, for each case and control. Lastly, an odds ratio for the outcome of interest (in relation to exposure status) is calculated. Case-control studies are retrospective because they seek to determine previous exposure after the outcome has been established. Thus, they are subject to recall or information bias. Case-control studies are also subject to sampling bias because it is difficult to select controls which are (ideally) perfectly similar to cases, with the exception of outcome status. However, techniques such as matching controls to cases and stratified analysis can improve the precision of estimates from case-control studies.

Case-control studies are useful when attempting to isolate a cause or causes for an emerging disease condition. Most recently this approach was used in attempts to determine the factors contributing to Colony Collapse Disorder (CCD) (vanEnglesdorp *et al.*, 2009b, 2010; Dainat *et al.*, 2012)

2.1.3.1. Data analysis and interpretation

The data analysis is similar to that presented in cross-sectional study designs above. However, the results from case-control studies have more importance in determination etiology because exposure status is ascertained at a time prior to case and control status are determined.

2.2. Experimental study designs

In contrast to observational studies, an experimental study assigns subjects to different treatment or exposure levels. This type of study design can be used to investigate the change in health status due to disease screening programs, prevention plans, interventions, diagnostic techniques or treatment procedures. Ultimately, a research team decides who will be treated or exposed, which consequently results in experimental intervention, not just observation of natural events.

Randomized studies are very powerful for investigating cause and effect because of the random assignment of study subjects to two or more intervention strategies, which leads to a compelling test of causality. The most simple randomized trial design consists of participants being randomly assigned to one of two treatment arms, the experimental arm (receive treatment of interest) or the control arm (receive no, placebo or standard treatment). Data from randomized trials can be utilized to calculate incidence of outcomes per treatment arm and then compare the incidence using the relative risk or risk differences. Randomization helps protect against bias, because it is likely that potential confounders are equally distributed across the treatment and control study groups. The scope of randomized studies is limited because these studies aim to confirm or disprove a specific hypothesis. Additionally, the cost and time needed to conduct trials are two primary disadvantages of this study design. A third concern is that the results from a controlled randomized trial may not be generalizable to uncontrolled real-world settings. There are many different variations on the simple randomized study design in which randomization schemes are modified and researchers are blinded to study conditions.

3. Economic considerations

Understanding those factors that are associated with a lower rate of loss may provide potential treatment options for beekeepers. However, just because a practice appears to be effective in reducing loss does not mean that it is necessarily in the beekeeper's best interest to adopt it. An additional piece of information for apiary managers is how much the treatment will cost and how much money the producer will likely save with its application.

Calculating the costs of practices in beekeeping is relatively straight forward, in that it includes the purchase cost of treatment and any labour or materials costs associated with its application. While each producer can calculate their costs, accurate aggregate data are more difficult to obtain, particularly for labour costs, or for applications where producers use their own recipe. Thus, the true costs of treatment may vary from producer to producer, and individual managers can be guided to compare their own costs to the average for a better cost estimate.

Calculating the benefit from reducing disease is more nuanced. One simple approach is to use the replacement cost of a hive as an estimate for the benefit of losing one less colony. To be as close as possible to the actual cost, one would like to find the replacement process that most closely replicates the scenario of having not lost the colony in the first place, such as a nuclear colony. Thus, one would not simply want to use the cost of splitting a hive, but would want a replacement that would be as productive as quickly as an existing colony while not reducing the productivity of surviving colonies. The true replacement costs would include extra feeding and labour costs associated with getting that colony to productivity (Equations 3.0).

Equation 3.0.a

Benefit of saving one colony = Replacement cost

Where replacement cost = cost of nuclear colony + cost of feed + cost of labour

Once one has a measure of the benefit of saving one colony, one can determine the expected net benefit of treatment for a disease.

Equation 3.0.b

Expected net benefit of treatment =
Replacement cost x (mean survival of untreated colonies – mean survival of treated colonies)

Where mean survival = 100 - Average Loss

If the cost of treatment exceeds the expected benefit, generating a negative expected net benefit, then despite the fact that the treatment may reduce colony loss, it may not be in the producer's best interest to use that treatment.

Note that the above calculation, even if all treatment and replacement costs are included, will tend to underestimate the benefits associated with treatment. Disease not only affects mortality, it also affects productivity, which is not captured in the above calculation. Thus, the above calculation should be thought of as generating a lower bound on expected net benefit. A more nuanced approach would be to estimate the effect of treatment on disease load, and the effect of disease load on productivity of honey production, pollination or other revenue-generating activities. Further, some beekeepers may place personal value on not losing a colony, and for them, their expected benefit of treatment may be higher still. These data are more difficult to collect, and will likely vary greatly from producer to producer. Nonetheless, giving beekeepers an estimate of the net benefit of treatment should allow them to compare the pure monetary costs and benefits to any other idiosyncratic costs of colony loss and help them in their management decisions.

4. Inferring causal relationships using Hill's Criteria

To diagnose the cause of a disease in honey bees, scientists typically compare observed symptoms with a list of exposures in colonies that implicate a particular pathogen, toxin or other detrimental aspect of the environment. Confirming the cause of the particular instance of these symptoms is relatively straightforward – the scientist either tests for the presence of the diagnosed causal agent itself or removes it and checks for amelioration of the symptoms. These approaches are feasible when the symptoms occur at the level of the individual or colony, because effects on growth, short-term survival or reproduction are readily measured (see the BEEBOOK paper on measuring colony strength parameters (Delaplane *et al.*, 2013)). In principle, it is possible to estimate the impact of the disease on the population's dynamics by using demographic models that quantify the effect on population growth (Varley *et al.*, 1973).

There are some cases, however, that are problematic for two reasons. First, the symptom is itself a population-level attribute; for instance, a general population decline. Second, the normal procedure is reversed because the causal agent is already identified, albeit as a hypothesis. An example is the supposed role of trace dietary pesticides in causing honey bee declines. In this case, scientists are asked whether dietary exposure to the pesticide is capable of causing the observed population decline. Studying impacts at the population level by experiments with replicated comparisons presents a severe logistical challenge because the required manipulations are at the landscape scale. Some alternative tools are available, such as the classic 'life table' method of insect population ecology (Varley *et al.*, 1973), but these can be applied only if detailed census data are

---

Box 8.

Using the numbers from the average winter loss determined by a management survey given in Box 6, we observed that beekeepers that used a known varroa mite control product lost 7.2 percentage points (or 20%) fewer colonies than beekeepers that did not use a product. To calculate the 95% CI for the difference in the mean, we need to add and subtract $1.96 \times se_d$, where se_d is the standard error of the difference in means.

The standard error of the difference, se_d is defined as $\sqrt{se_1^2 + se_2^2}$ where se_1 is the standard error of the mean for sample 1, and se_2 is the standard error of the mean for sample 2. (The standard error calculations come from the confidence interval calculations in box 6 above.) The standard error for the sample using treatment is 1.02 and the standard error for the control sample (or no-treatment sample) is 0.92. Thus, the standard error of the difference in means is $\sqrt{1.02^2 + 0.92^2} = 1.37$. Thus, we get a 95% confidence interval of the difference in means of 7.2 plus or minus 1.96×1.37, or 4.51 to 9.89.

If the replacement costs of a hive, including labor and feeding are $150, then the expected benefit of the treatment is the change in probability of loss times the replacement costs, or $0.07 \times \$150 = \$10.80$ (with a 95% CI of $6.77 to $14.83). Assume the cost of treatment is $7.50 per colony. Thus the expected net benefits would be $10.80 - $7.50 = $3.30 (with a 95% CI of -$0.78 to $7.33) per hive.

Thus, on average the producer is expected to benefit from the treatment, but could in fact lose from treatment. Net gains are expected to range from a loss of $0.78 per colony to a gain of $7.33 per colony, 95 times out of 100.

---

available that precisely identify causes of death over extended time periods. Where such resorts are stymied, scientists must use the available circumstantial evidence to pass an expert judgement. Hill's criteria (Hill, 1965) provide a valuable framework that supports a repeatable and quantitative evaluation process.

Sir Austin Bradford Hill, a leading 20th century epidemiologist, identified nine types of information that provide 'viewpoints' from which to judge a proposed cause-effect relationship (Hill, 1965). The nine criteria include not only experimental evidence, but also eight kinds of circumstantial evidence that fall into two categories (Table 4).

For each criterion, scientists survey the available evidence and then formally describe the level of conviction with which they subsequently hold the proposed cause-effect hypothesis to be true: slight; reasonable; substantial; clear; and certain (Weiss, 2006). The descriptors are then associated with numerical values to produce a quantitative score of certainty (Cresswell *et al.*, 2012). Specifically, an eleven-point scale for each criterion returns a positive value

Table 4. The nine criteria established by Hill (1965), each with a brief description.

Criterion	Brief description
1. Experimental evidence	
2. Coherence	Fails to contradict established knowledge
3. Plausibility	Probable given established knowledge
4. Analogy	Similar examples known
5. Temporality	Cause precedes effect
6. Consistency	Cause is widely associated with effect
7. Specificity	Cause is uniquely associated with effect
8. Biological gradient	Monotonic dose-response relationship
9. Strength	Cause is associated with a substantive effect

(maximum five) if the evidence suggests that the agent certainly causes population decline, a negative value (maximum minus five) if the factor certainly does not and a zero if the evidence is equivocal or lacking. For example, if the evidence for a criterion gives a reasonable indication that an agent does not cause the symptom, the score for that criterion would be -2, etc.

One major value of the criteria is that they disaggregate the different kinds of evidence, requiring the scientist to consider each kind carefully, separately and explicitly. Once the scores are given, there is no *a priori* reason either to give equal weight to the nine criteria or to calculate an average score. It is important, moreover, to consider whether any large scores have arisen principally on the theoretical criteria, because it is conventional in science to favour material evidence (i.e. associational criteria) over conjecture. For example, an evaluation by Hill's criteria (Cresswell *et al.*, 2012) revealed that the proposition that dietary pesticides cause honey bee declines was a substantially justified conjecture in the context of current knowledge (positive scores on the theoretical criteria), but was substantially contraindicated by a wide variety of circumstantial evidence (negative scores on the associational criteria). The disparity in the scores on the two categories of criteria explains in part the controversy over this question, because different constituencies make differential use of the two kinds of evidence. Hill (1965) himself refused to weight the criteria because the evaluation of circumstantial evidence cannot be made algorithmic.

The use of Hill's criteria formalizes the evaluation of cause-consequence associations and applies a quantitative scoring method which makes the conclusions both apparent and repeatable. Since their inception over 40 years ago and subsequent widespread use, no criterion has been abandoned and none added, which means that they provide a stable and well-established infrastructure in which to process scientific evidence.

5. Conclusions

The general aim of all scientists studying honey bee health is the same; preservation of the bees. However, without common methods and shared terminology, it is difficult to confidently compare reported results. In an effort to standardize the efforts of those interested in improving honey bee health and make studies comparable, we have introduced epidemiological terminology, experimental design, and methods of calculation that are often different enough to facilitate comparisons between studies.

6. Acknowledgements

The COLOSS (Prevention of honey bee COlony LOSSes) network aims to explain and prevent massive honey bee colony losses. It was funded through the COST Action FA0803. COST (European Cooperation in Science and Technology) is a unique means for European researchers to jointly develop their own ideas and new initiatives across all scientific disciplines through trans-European networking of nationally funded research activities. Based on a pan-European intergovernmental framework for cooperation in science and technology, COST has contributed since its creation more than 40 years ago to closing the gap between science, policy makers and society throughout Europe and beyond. COST is supported by the EU Seventh Framework Programme for research, technological development and demonstration activities (Official Journal L 412, 30 December 2006). The European Science Foundation as implementing agent of COST provides the COST Office through an EC Grant Agreement. The Council of the European Union provides the COST Secretariat. The COLOSS network is now supported by the Ricola Foundation - Nature & Culture.

7. References

ANDERSON, D (2013) Standard methods for *Tropilaelaps* mites research. In *V Dietemann; J D Ellis; P Neumann (Eds) The COLOSS BEEBOOK, Volume II: standard methods for* Apis mellifera *pest and pathogen research. Journal of Apicultural Research* 52(4): http://dx.doi.org/10.3896/IBRA.1.52.4.21

BREIMAN, I; FRIEDMAN, J H; OLSEN, R A; STONE, C J (1984) *Classification and regression trees.* Wadsworth; Pacific Growe, CA, USA. 358 pp.

CRESSWELL, J E; DESNEUX, N; VANENGELSDORP, D (2012) Dietary traces of neonicotinoid pesticides as a cause of population declines in honey bees: an evaluation by Hill's epidemiological criteria. *Pest Management Science*: 68: 819-827. http://dx.doi.org/10.1002/ps.3290

DAINAT, B; VANENGELSDORP, D; NEUMANN, P (2012) Colony Collapse Disorder in Europe. *Environmental Microbiology Reports* 4: 123-125. http://dx.doi.org/10.1111/j.1758-2229.2011.00312.x

DE GRAAF, D C; ALIPPI, A M; ANTÚNEZ, K; ARONSTEIN, K A; BUDGE, G; DE KOKER, D; DE SMET, L; DINGMAN, D W; EVANS, J D; FOSTER, L J; FÜNFHAUS, A; GARCIA-GONZALEZ, E; GREGORC, A; HUMAN, H; MURRAY, K D; NGUYEN, B K; POPPINGA, L; SPIVAK, M; VANENGELSDORP, D; WILKINS, S; GENERSCH, E (2013) Standard methods for American foulbrood research. In *V Dietemann; J D Ellis; P Neumann (Eds) The COLOSS* BEEBOOK, *Volume II: standard methods for* Apis mellifera *pest and pathogen research. Journal of Apicultural Research* 52(1): http://dx.doi.org/10.3896/IBRA.1.52.1.11

DELAPLANE, K S; VAN DER STEEN, J; GUZMAN, E (2013) Standard methods for estimating strength parameters of *Apis mellifera* colonies. In *V Dietemann; J D Ellis; P Neumann (Eds) The COLOSS* BEEBOOK, *Volume I: standard methods for* Apis mellifera *research. Journal of Apicultural Research* 52(1): http://dx.doi.org/10.3896/IBRA.1.52.1.03

ESTOUP, A; SOLIGNAC, M; CORNUET, J-M (1994) Precise assessment of the number of patrilines and of genetic relatedness in honey bee colonies. *Proceedings of the Royal Society of London B, Biological Sciences* 258: 1-7.

FOSGATE, G T (2009) Practical sample size calculations for surveillance and diagnostic investigations. *Journal of Veterinary Diagnostic Investigation* 21: 3-14. http://dx.doi.org/10.1177/104063870902100102

GARDNER, M J; ALTMAN, D G (1986) Confidence intervals rather than P values: estimation rather than hypothesis testing. *British Medical Journal* 292: 746-750.

GENERSCH, E; VON DER OHE, W; KAATZ, H H; SCHROEDER, A; OTTEN, C; BERG, S; RITTER, W; GISDER, S; MEIXNER, M; LIEBIG, G; ROSENKRANZ, P (2010) The German bee monitoring project: a long term study to understand periodically high winter losses of honey bee colonies. *Apidologie* 41: 332-352. http://dx.doi.org/10.1051/apido/2010014

GIRAY, T; GUZMAN NOVOA, E; ARON, C W; ZELINSKY, B; FAHRBACH, S E; ROBINSON, G E (2000) Genetic variation in worker temporal polyethism and colony defensiveness in the honey bee, *Apis mellifera. Behavioural Ecology* 11(1): 44-55. http://dx.doi.org/10.1093/beheco/11.1.44

GISDER, S; HEDTKE, K; MÖCKEL, N; FRIELITZ, M C; LINDE, A; GENERSCH, E (2010) Five-year cohort study of *Nosema* spp. in Germany: does climate shape virulence and assertiveness of *Nosema ceranae*? *Applied and Environmental Microbiology* 76: 3032-3038. http://dx.doi.org/10.1128/AEM.03097-09

HENDRIKX, P; CHAUZAT, M-P; DEBIN, M; NEUMAN, P; FRIES, I; RITTER, W; BROWN, M; MUTINELLI, F; LE CONTE, Y; GREGORC, A (2009) Bee mortality and bee surveillance in Europe. Scientific report, European Food Safety Authority, Parma, Italy, pp. 217 (available at http://www.efsa.europa.eu/en/scdocs/doc/27e.pdf)

HILL, A B (1965) The environment and disease: association or causation? *Proceedings of the Royal Society of Medicine* 58: 295-300.

HOLST, E C (1945) A simple field test for American foulbrood. *American Bee Journal* 14: 34.

KOEPSELL, T D; WEISS, N S (2003) *Epidemiologic methods: studying the occurrence of illness*. Oxford University Press; New York, USA.

MAUSNER, J S; KRAMER, S (1985) *Epidemiology: an introductory text.* W B Saunders Company; Philadelphia, USA. 361

NUTTER, F W Jr (1999) Understanding the interrelationships between botanical, human, and veterinary epidemiology: the Ys and Rs of it all. *Ecosystems Health* 5: 131-140.

PAOLI, B; HAGGARD, L; SHAH, G (2002.) *Confidence intervals in public health*. Office of Public Health Assessment, Utah Department of Health, USA. p 8.

SAEGERMAN, C; SPEYBROECK, N; ROELS, S; VANOPDENBOSCH, E; THIRY, E; BERKVENS, D (2004) Decision support tools for clinical diagnosis of disease in cows with suspected bovine spongiform encephalopathy. *Journal of Clinical Microbiology* 42: 172-178. http://dx.doi.org/10.1128/JCM.42.1.172-178.2004

SAEGERMAN, C; PORTER, S R; HUMBLET, M F (2011) The use of modelling to evaluate and adapt strategies for animal disease control. *Revue Scientifique et Technique - Office International des Épizooties* 30: 555-569.

SPEYBROECK, N; BERKVENS, D; MFOUKOU-NTSAKALA, A; AERTS, M; HENS, N; HUYLENBROECK, G V; THYS, E (2004) Classification trees versus multinomial models in the analysis of urban farming systems in Central Africa. *Agricultural Systems* 80: 133-149. http://dx.doi.org/10.1016/j.agsy.2003.06.006

TOMA, B; BENET, J-J; DUFOUR, B; ELOIT, M; MOUTOU, F; SANAA, M (1991) *Glossaire d'épidémiologie animale*. Editions du Point Vétérinaire; Maisons-Alfort, France. 365 pp.

VANENGELSDORP, D; CARON, D; HAYES, J Jr; UNDERWOOD, R; HENSON, K R M; SPLEEN, A; ANDREE, M; SNYDER, R; LEE, K; ROCCASECCA, K; WILSON, M; WILKES, J; LENGERICH, E; PETTIS, J (2012) A national survey of managed honey bee 2010-11 winter colony losses in the USA: results from the Bee Informed Partnership. *Journal of Apicultural Research* 51: 115-124. http://dx.doi.org/10.3896/IBRA.1.51.1.14

VANENGELSDORP, D; EVANS, J D; SAEGERMAN, C; MULLIN, C; HAUBRUGE, E; NGUYEN, B K; FRAZIER, M; FRAZIER, J; COX-FOSTER, D; CHEN, Y; UNDERWOOD, R; TARPY, D R; PETTIS, J S (2009b) Colony Collapse Disorder: A descriptive study. *PloS ONE* 4: e6481. (available at: http://www.plosone.org/article/info:doi/10.1371/journal.pone.0006481)

VANENGELSDORP, D; EVANS, J D; DONOVALL, L; MULLIN, C; FRAZIER, M; FRAZIER, J; TARPY, D R; HAYES, J Jr; PETTIS, J S (2009a). "Entombed Pollen": a new condition in honey bee colonies associated with increased risk of colony mortality. *Journal of Invertebrate Pathology* 101: 147-149. http://dx.doi.org/10.1016/j.jip.2009.03.008

VANENGELSDORP, D; SPEYBROECK, N; EVANS, J D; NGUYEN, B K; MULLIN, C; FRAZIER, M; FRAZIER, J; COX-FOSTER, D; CHEN, Y; TARPY, D R; HAUBRUGE, E; PETTIS, J S; SAEGERMAN, C (2010) Weighing risk factors associated with bee Colony Collapse Disorder by classification and regression tree analysis. *Journal of Economic Entomology* 103: 1517-1523. (available at: http://www.bioone.org/doi/pdf/10.1603/EC09429)

VANENGELSDORP, D; BRODSCHNEIDER, R; BROSTAUX, Y; VAN DER ZEE, R; PISA, L; UNDERWOOD, R; LENGERICH, E J; SPLEEN, A; NEUMANN, P; WILKINS, S; BUDGE, G E; PIETRAVALLE, S; ALLIER, F; VALLON, J; HUMAN, H; MUZ, M; LE CONTE, Y; CARON, D; BAYLIS, K; HAUBRUGE, E; PERNAL, S; MELATHOPOULOS, A; SAEGERMAN, C; PETTIS, J S; NGUYEN, B K (2011) Calculating and reporting managed honey bee colony losses. In *D Sammataro; J A Yoder (Eds). Honey bee colony health: challenges and sustainable solutions*. CRC Press; FL, USA. pp. 237-244

VANENGELSDORP, D; SAEGERMAN, C; NGUYEN, K B; PETTIS J S (2013) Honey bee health surveillance. In *W Ritter (Ed.). OIE Technical Series no 12: The vet and the bee*. OIE. (in press).

VANENGELSDORP, D; TARPY, D R; LENGERICH, E J; PETTIS, J S (2013) Idiopathic brood disease syndrome and queen events as precursors of colony mortality in migratory beekeeping operations in the eastern United States. *Preventive Veterinary Medicine*. (in press).

VARLEY, G C; GRADWELL, G R; HASSELL, M P (1973) *Insect population ecology*. Blackwell Scientific Publications; Oxford, UK.

WEISS, N S (2006) *Clinical epidemiology: the study of the outcome of illness (Third edition)*. Oxford University Press; UK. 178 pp

WORLD HEALTH ORGANISATION (WHO) (1999) Norms and standards in epidemiology: case definitions. *Epidemiological Bulletin* 20.

WOODWARD, M (2005) *Epidemiology. Study design and data analysis*. Chapman & Hall; New York, USA.

Journal of Apicultural Research 52(4): (2013)
DOI 10.3896/IBRA.1.52.4.18

REVIEW ARTICLE

INTERNATIONAL BEE
RESEARCH ASSOCIATION

Standard survey methods for estimating colony losses and explanatory risk factors in *Apis mellifera*

Romée van der Zee[1]*, Alison Gray[2], Céline Holzmann[3], Lennard Pisa[1], Robert Brodschneider[4], Róbert Chlebo[5], Mary F Coffey[6], Aykut Kence[7], Preben Kristiansen[8], Franco Mutinelli[9], Bach Kim Nguyen[10], Adjlane Noureddine[11], Magnus Peterson[12], Victoria Soroker[13], Grażyna Topolska[14], Flemming Vejsnæs[15] and Selwyn Wilkins[16]

[1]Netherlands Centre for Bee Research, Durk Dijkstrastr. 10, 9014 CC, Tersoal, Netherlands.
[2]University of Strathclyde, Department of Mathematics and Statistics, Glasgow, G1 1XH, UK.
[3]Formerly at Institut Technique et Scientifique de l' Apiculture et de la Pollinisation, Institut de l' abeille, 75595 Paris, France.
[4]Department of Zoology, Karl-Franzens University Graz, Universitätsplatz 2, A-8010 Graz, Austria.
[5]Slovak University of Agriculture, Department of Poultry Science and Small Animals Husbandry, Tr. A. Hlinku 2, 94976 Nitra, Slovakia.
[6]Department of Life Sciences, University of Limerick, Limerick, Ireland.
[7]Middle East Technical University, Department of Biology, 06800, Ankara, Turkey.
[8]Swedish Beekeepers Association, Trumpetarev 5, SE-59019 Mantorp, Sweden.
[9]Istituto Zooprofilattico Sperimentale delle Venezie, National Reference Laboratory for Beekeeping, Viale dell'Universita' 10,35020 Legnaro (PD), Italy.
[10]University of Liege, Gembloux Agro-Bio Tech, Department of Functional and Evolutionary Entomology, B-5030 Gembloux, Belgium.
[11]Departement of Biology, University of Boumerdes, Algeria.
[12]Scottish Beekeepers Association, 20, Lennox Road, Edinburgh, EH5 3JW, UK.
[13]Agricultural Research Organization The Volcani Center, 50250 PO Box 6, Bet Dagan, Israel.
[14]Warsaw University of Life Sciences, Faculty of Veterinary Medicine, Ciszewskiego 8, 02-786 Warsaw, Poland.
[15]Danish Beekeepers Association, Fulbyvej, DK-4140 Sorø, Denmark.
[16]National Bee Unit, Food and Environment Research Agency, Sand Hutton, York, YO41 1LZ, UK.

Received 15th June 2012, accepted subject to revision 18 December 2012, accepted for publication 3 June 2013.

*Corresponding author: Email: romee.van.der.zee@beemonitoring.org

Summary

This chapter addresses survey methodology and questionnaire design for the collection of data pertaining to estimation of honey bee colony loss rates and identification of risk factors for colony loss. Sources of error in surveys are described. Advantages and disadvantages of different random and non-random sampling strategies and different modes of data collection are presented to enable the researcher to make an informed choice. We discuss survey and questionnaire methodology in some detail, for the purpose of raising awareness of issues to be considered during the survey design stage in order to minimise error and bias in the results. Aspects of survey design are illustrated using surveys in Scotland. Part of a standardized questionnaire is given as a further example, developed by the COLOSS working group for Monitoring and Diagnosis. Approaches to data analysis are described, focussing on estimation of loss rates. Dutch monitoring data from 2012 were used for an example of a statistical analysis with the public domain R software. We demonstrate the estimation of the overall proportion of losses and corresponding confidence interval using a quasi-binomial model to account for extra-binomial variation. We also illustrate generalized linear model fitting when incorporating a single risk factor, and derivation of relevant confidence intervals.

Footnote: Please cite this paper as: VAN DER ZEE, R; GRAY, A; HOLZMANN, C; PISA, L; BRODSCHNEIDER, R; CHLEBO, R; COFFEY, M F; KENCE, A; KRISTIANSEN, P; MUTINELLI, F; NGUYEN, B K; ADJLANE, N; PETERSON, M; SOROKER, V; TOPOLSKA, G; VEJSNÆS, F; WILKINS, S (2012) Standard survey methods for estimating colony losses and explanatory risk factors in *Apis mellifera*. In *V Dietemann; J D Ellis; P Neumann (Eds) The COLOSS BEEBOOK, Volume I: Standard methods for Apis mellifera research. Journal of Apicultural Research* 52(4): http://dx.doi.org/10.3896/IBRA.1.52.4.18

Métodos estándar de encuestas para la estimación de la pérdida de colonias y los factores de riesgo que los explican en *Apis mellifera*

Resumen

Este capítulo trata sobre la metodología de encuestas y el diseño del cuestionario para la recogida de datos relativos a la estimación de las tasas de pérdida de colonias de abejas de la miel y la identificación de los factores de riesgo de la pérdida de colonias. Se describen las fuentes de error en las encuestas. Se presentan las ventajas y desventajas de las diferentes estrategias de muestreo aleatorio y no aleatorio y diferentes modos de recogida de datos que permitan al investigador tomar una decisión informada. Discutimos sobre la metodología de las encuestas y los cuestionarios con cierto detalle, con el propósito de dar a conocer las cuestiones a tener en cuenta durante la fase de diseño de la encuesta con el fin de minimizar el error y el sesgo en los resultados. Se ilustran aspectos de la encuesta que a través de encuestas realizadas en Escocia. Se da como ejemplo parte de un cuestionario estandarizado, desarrollado por el grupo de trabajo COLOSS de Monitoreo y Diagnóstico. Se describen enfoques para el análisis de datos, centrándose en la estimación de las tasas de pérdida. Se utilizaron datos de un monitoreo holandés de 2012 como ejemplo de análisis estadístico con el software de dominio público R. Demostramos la estimación de la proporción total de las pérdidas y el intervalo de confianza correspondiente usando un modelo cuasi-binomial para dar cuenta de la variación extra-binomial. También ilustramos ajustes del modelo lineal generalizado al incorporar un solo factor de riesgo, y la derivación de los intervalos de confianza correspondientes.

西方蜜蜂估测蜂群损失和风险因子的标准调查方法

摘要

本章针对为估测蜂群损失率和鉴定蜂群损失相关风险因子采集数据时所需的调查方法和问卷设计。描述了调查中的误差来源，介绍了不同的随机或非随机取样方法和不同数据采集模式的利弊，以供研究者做出合理选择。为提高对于在调查设计阶段需要考虑的问题的认识，使结果中的误差和偏差最小化，我们在一些细节上进一步讨论了调查和问卷的方法学。调查设计部分以在苏格兰的调查为例说明，并以由 COLOSS 工作组开发的用于监测和诊断的标准化问卷的一部分作为另一个例子进一步说明。描述了针对估测损失率的数据分析方法。以荷兰 2012 年的监测数据为例，使用公共领域 R 软件，介绍统计分析方法。我们使用类似二项式模型解释额外二项式变化，示范了总体损失比例及相应置信区间的估算方法。同时，我们举例说明了当合并单一风险因素时广义线性模型的拟合，以及相关置信区间的推导

Keywords: COLOSS *BEEBOOK*, estimating colony losses, *Apis mellifera*, surveys, questionnaire, random / randomized sampling, non-random sampling, generalized linear models (GZLMs)

Table of Contents

1. Introduction

Surveys on honey bee colony losses have been conducted by many researchers over the years to understand the factors that contribute to colony losses. Recognizing the importance of standard questionnaires for use in surveys, a network of honey bee specialists preceding the establishment of the COLOSS Action network, initiated by Cost Action FA0803, established at its first meeting a working group (Working Group 1 - WG1) whose aim was to develop and implement research surveys for the purpose of identifying such factors. The working group currently represents a global network of scientists who monitor colony losses. This group was conscious of the fragility of many survey results and addressed crucial issues to obtain a valid research framework (Van der Zee *et al.*, 2012). Using other literature sources, the group developed and/or recognized appropriate case definitions, statistics and relevant factors associated with honey bee colony losses. The present manuscript aims to make the results of these efforts available to all researchers working in this field and to provide guidelines for conducting effective surveys.

Conditions in which to perform surveys on honey bee colony losses and achieve results which meet methodological standards are very different between and within countries. The present chapter offers guidelines to attain good quality surveys, even under unfavourable conditions. The main objective of these surveys (section 2.1.) is the estimation of winter colony loss rates, identification of specific areas with a higher or lower risk of honey bee mortality and information on possible determinants such as the control of *Varroa destructor*. This will enable the provision of advice on loss prevention and control.

We are conscious that the case definitions we present (section 2.3.) may be refined or changed in the following years because not enough knowledge is yet available to resolve many important issues. However what we present here does, in our view, give a good set of standards to which all researchers in this field should aim to conform in order to produce robust and reliable results.

The target population of the surveys is usually the set of active beekeepers in a country or specific area. The possibilities for reaching the target population vary between and within countries. Sometimes registers of beekeepers can be used for collecting data; but more often, cooperation with beekeepers' associations is necessary (section 5.). In some situations, both are absent and the investigator has to develop other survey strategies. Suggestions are given for sampling frames in situations where cooperation with a beekeeper association is not possible or if a beekeeper infrastructure is absent (section 6.4.).

The sample selection method used is one of the main issues for obtaining reliable survey outcomes. Selecting a random sample of beekeepers gives results whose accuracy can be quantified (section 6.1.), if, as is usually the case, studying the whole target population instead of a sample is not feasible. At present most monitoring surveys with questionnaires will gain in quality if the shift is made from the present common practices of self-selected samples (samples in which participants volunteer to take part) towards at least simple random sampling. The same or better survey results may be achieved by using other more sophisticated forms of probability sampling, and relatively small sample sizes might then be sufficient.

Detailed consideration will also be given to the various sources of bias which may affect survey outcomes (De Leeuw *et al.*, 2008), whose effect will usually be to introduce errors into results whose effects are difficult or impossible to assess. In particular, it is good practice to strive for high response rates, although there is no empirical support for the notion that low response rates necessarily produce estimates with high nonresponse bias (Groves, 2006). However that risk is inevitably present if response rates are low.

Attention is given here to a variety of methods of statistical data analysis. These range from simple analyses to examine the effects of different Varroa controls or other individual risk factors on mortality, to more advanced methods involving the use of Generalized Linear Models (GZLMs) to investigate simultaneously the possible effects of multiple different factors on colony loss rates. We use the statistical program R to illustrate an analysis in section 10, using data from the Netherlands. Survey design and sampling methodology is illustrated throughout the manuscript using Scotland as a case study. An introduction to this is provided in Box 1.

Box 1. Introduction to the example of surveys of beekeepers in Scotland.

Throughout this chapter we illustrate some of the methodology and concepts using as an example surveys of beekeepers in Scotland. These surveys are not perfect in the way they have been conducted, but most of them have used random sampling and we describe how this was achieved. They provide a case study, for those who may be interested, of how random sampling can be done in a situation where reasonably good records of beekeepers are available for use as a sampling frame. It is recognised that this approach is not possible in all situations. We describe the sampling design, the use of records for sample selection, obtaining permission for use of records, maintaining anonymity etc.

These surveys have run since 2006. They have all used as the survey population members of the Scottish Beekeepers' Association (SBA), the national body for beekeepers in Scotland. Persons keeping any number of honey bee colonies, or with no bees but having an interest in bees, can belong to this organisation. Affiliated with the SBA are a large number of local associations for beekeepers. It is possible to belong to one or more local associations as well as, or instead of, the SBA.

The first survey used a quota sampling approach, as permission had not at that time been sought for using the membership records of the SBA for sampling purposes (see section 6. and box 5), and made use of SBA representatives to identify beekeepers to include in the sample. Subsequent surveys have used a stratified random sampling approach, by dividing the membership into those belonging to each of several large administrative areas and taking a separate random sample from each of those identified groups of SBA members using the membership records (see section 6.).

As these administrative areas are geographical, this approach was chosen to try to ensure coverage of the different geographical areas, since conditions for beekeeping vary across the country. In particular, the more remote parts of the north and west of Scotland are thought to be free of *Varroa destructor*. Weather patterns also vary geographically.

All of the surveys from 2006 to 2012 have been conducted using a postal questionnaire.

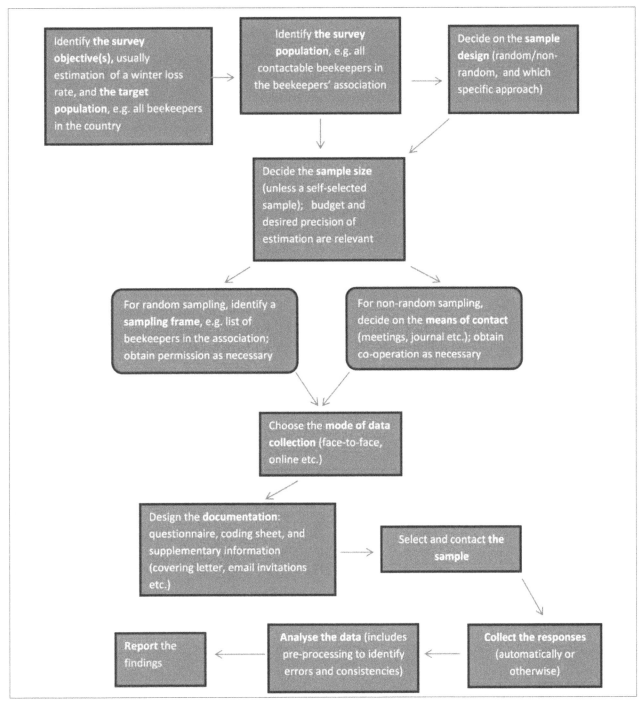

Fig. 1. A basic flowchart of the key steps in carrying out a survey.

2. Objectives and case definitions

Fig. 1 shows the steps to be addressed in designing a survey. We address each of these in the sections below.

2.1. Objective of epidemiological studies on honey bee colony losses

The epidemiological study of honey bee colony losses aims to determine explanatory factors for, and to monitor the magnitude and the spatial distribution of honey bee colony losses at the operation, apiary or colony level. This enables the formulation of good advice on prevention and control of colony stressors. In this section, we propose standardized case definitions to facilitate various objectives as follows:

1. Reporting and classification of cases of honey bee colony losses by national and international honey bee experts.
2. Standardization of language for communication purposes.
3. Comparability of data across time and geographical areas.

2.2. Application of definitions associated with honey bee colony loss

1. The case definitions for use in surveillance are based on available epidemiological data summarizing what is currently known about the magnitude and spatial distribution of honey bee colony losses. Countries may need to adapt case definitions depending on their own situation.
2. The case definitions have been developed to help national authorities classify and track cases.
3. The case definitions are not intended to provide complete descriptions of the symptomatology of lost colonies but rather to standardize reporting of these losses.
4. The case definitions will describe the symptoms of dwindling and lost colonies and the timeframe of observation during which honey bee colony losses occurred.

2.3. Case definitions

1. Lost honey bee colony is a honey bee colony that:
   a. is reduced to such a small number of bees that it cannot perform the normal biological activities needed for survival (brood rearing, resource gathering) or
   b. has queen problems, such as drone laying queens or drone laying worker bees in absence of a queen, which could not be solved or
   c. was missing due to burglary, or didn't survive fire, inundation, desert storms or similar causes unrelated to health problems.
   d. no longer has any living bees present.
2. Weak honey bee colony:
A honey bee colony that is not considered as lost, but in which the number of bees is less than would be expected from the colony size observed at an earlier inspection.
3. Colony Depopulation Syndrome (CDS):
This is observed if a honey bee colony shows the following conditions within a certain time-frame:
   a. reduced to no, or only a few remaining, living bees in the hive and
   b. no, or only a few dead bees in or in front of the hive or at the apiary while
   c. food is present in the hive (Van der Zee *et al.*, 2012).
4. Colony Collapse Disorder (CCD)
This is observed if the following conditions are present:
   a. a rapid loss of adult worker bees from affected honey bee colonies, as evidenced by weak or dead colonies with excess brood populations present relative to adult bee populations (vanEngelsdorp *et al.*, 2009);

   b. a noticeable lack of dead worker bees both within and surrounding the hive (vanEngelsdorp *et al.*, 2009);
   c. the delayed invasion of hive pests (e.g., small hive beetles (Neumann *et al.*, 2013) and wax moths (Ellis *et al.*, 2013)) into affected colonies and kleptoparasitism of affected colonies by neighbouring colonies (Cox-Foster *et al.*, 2007);
   d. the absence of *Varroa* and *Nosema* at levels thought to cause economic damage (vanEngelsdorp *et al.*, 2009).
5. Time frames during which honey bee colony losses occurred can be distinguished as:
   a. Time frames related to seasonal characteristics:
   For example winter: the period between the moment that a beekeeper finished pre-winter preparations for his/her honey bee colonies and the start of the new foraging season.
   b. Fixed time frames:
   For example: observations every half year.

It is difficult to come to conclusions on losses with a fixed timeframe approach since the outcome depends on beekeeper practices such as merging, splitting, buying and selling of colonies. Recalling the numbers of colonies involved in these practices later when a questionnaire is disseminated may easily lead to errors in the data (Van der Zee *et al.*, 2012). Another problem is that no information is collected on when these increases/reductions were made within the timeframe, with the effect that colonies bought at the start of a time frame have the same weight in the risk estimation as colonies bought later on. Not recognising these problems may severely bias the outcome.

3. Data collection methods

3.1. Choosing the method of data collection

When choosing the mode of data collection, a number of issues must be considered: effective coverage of target populations, data accuracy, potential bias of the survey sample, the survey mode(s), and the effort and cost of data collection (De Leeuw, 2008; Charrière and Neumann, 2010; Dahle, 2010; Hatjina *et al.*, 2010; Mutinelli *et al.*, 2010; Nguyen *et al.*, 2010). In a survey, the data can be collected by direct contact of an interviewer with the respondent (face-to-face interviews, telephone interviews) or the questions can be administered and answered by beekeepers without the assistance of an interviewer by means of a self-administered questionnaire (postal surveys, email surveys, internet surveys). The advantages and disadvantages of each mode are described in section 3.2. Further references are given in the online supplementary material.

3.2. Available data collection methods with advantages and disadvantages

3.2.1. Surveyor administered questionnaires

3.2.1.1. Face-to-face interviews

Pros:

- The interviewer can explain the importance of a survey and clarify questions if needed.
- The data can often be entered directly into the computer database and checked for a valid data entry, question by question.
- Answers can easily be corrected *in situ*. It may become apparent immediately that some answers are inconsistent or wrong, especially if suitable data checks are built into a computer questionnaire programme.

Cons:

- A representative list of beekeepers is needed; there are few countries in which it would be available through beekeeping associations, census registers or the veterinary services.
- The presence of an interviewer may influence the answers, unless the interviewer is well-trained.
- This method is time-consuming and costly because of: travel costs (unless some form of cluster sampling is used), the need for many highly trained interviewers, and the need for multiple call-backs to ensure a high response rate.

3.2.1.2. Telephone interviews

Pros:

- The interviewer can explain the importance of a survey and clarify the questions if needed.
- Beekeepers may feel obliged to participate in the survey, though others may simply say they do not have time to participate. The data can often be entered directly into the computer database, as above.
- Lower costs than face-to-face interviews and call-backs are much faster and easier.
- Many interviews can be completed in a relatively short time.

Cons:

- A representative list of telephone numbers of the target population of beekeepers is needed; there are very few countries where it would be available.
- Beekeepers under pressure may give the answer without careful consideration, and it is difficult to correct such answers later.
- Time necessary for some interviews may be longer than is necessary to answer the questions because some people tend to be garrulous, although other people will become impatient if the time required is long.

3.2.2. Self-Administered questionnaires

3.2.2.1. Postal or email survey

Beekeepers receive a questionnaire by mail (or email), answer the questions and return the questionnaire.

Pros:

- Beekeepers have time to check their apiary notes and to answer the questions fully.
- Quick distribution of questionnaires for survey organisers and quick return (if emailed).

Cons:

- A good (complete) list of addresses (email addresses) of the target population of beekeepers is needed. Limited access of beekeepers to the internet can very badly influence the survey coverage if only an email survey is performed.
- Especially clear questions and instructions are necessary.
- Beekeepers are not always actively involved and very often they do not respond. Free return postage raises the costs for postal surveys but also beekeepers' participation in the survey.
- Reminders are likely to be necessary for a good response rate.
- The questionnaires may be filled in carelessly with answers missing.
- The time for and cost of data entry can be high, unless the sample is very small.

3.2.2.2. Internet survey

Beekeepers complete a questionnaire hosted on the internet. The same questionnaire could be hosted or linked on different websites, e.g. research institutes, reference laboratories, beekeepers' associations or beekeeping journals. There are several approaches to the calculation of response rates for internet surveys (AAPOR, 2011). The response rate for an online survey is often comparable with the response to a questionnaire published in a beekeeper journal, if the invitation to participate in an online survey is published on the website of a beekeepers' association.

Pros:

- Large numbers of completed questionnaires can be collected in a very short time.
- There is no need for transferring data from paper questionnaires to an e-database.
- Low cost of data collection and reduced cost of data analysis.
- Beekeepers' associations or beekeeping journals could contribute to increasing the number of filled in questionnaires by advertising the survey.
- Beekeepers have enough time to check their apiary notes and to answer the questions fully.

Cons:

- A list of email addresses of the target population of beekeepers is needed to advise them of the survey or issue reminders.
- Beekeepers with ready computer access may not be wholly representative of the general beekeeping population of interest, which can adversely affect the survey coverage.
- In principle, a beekeeper could complete the survey more than once; however, there is a choice of a suitable survey software that makes it possible to prevent duplicate submissions.

3.2.2.3. Questionnaire published in beekeeping journals

Completed questionnaires are usually posted, faxed or emailed by beekeepers.

Pros:

- If a journal has a large circulation, the dissemination of a questionnaire is widespread.
- Cost of questionnaire dissemination is low.
- Cost of data collection is low. Free return shipment raises the costs but also beekeepers' participation in the survey.
- Beekeepers have enough time to check their apiary notes and to answer the questions fully.
- Non-response can be estimated, if all beekeepers receive the journal for free because they are a member of a beekeeper association.

Cons:

- Compared to disseminating copies of a questionnaire at beekeepers' association meetings and encouragement to respond by leaders of those associations, participation rates are lower.
- The survey will not cover non-readers of these journals, so its representativeness may be limited to the readers of the journal.
- The questionnaires may be filled in carelessly with answers missing.
- The time for and cost of data entry can be high.

3.2.2.4. Questionnaires disseminated during meetings

Completed questionnaires can be collected immediately by the survey organiser or an association representative, or posted, faxed or emailed by beekeepers.

Pros:

- An easy way to disseminate the questionnaires if co-operation with the beekeeping association hosting the meeting is good.
- The survey organiser can explain the importance of the survey and clarify the questions if needed.
- If beekeepers post the filled-in questionnaires by themselves they have enough time to check their apiary notes and to answer the questions fully.

Cons:

- Not all beekeepers attend beekeepers' association meetings, leading to coverage problems.
- The survey will cover beekeepers only from specific regions or associations.
- The questionnaires are often filled in carelessly with many answers missing.
- The time for and cost of data entry can be high.

3.3. Data validity and accuracy

The use of unambiguous questions is critical. However, the clarity of any international questionnaire may well be culturally dependent. In the absence of a face-to-face or telephone interviewer to conduct the survey, the questions could be misunderstood by a respondent though this may not be immediately obvious. This would reduce the validity or accuracy of the response to the question asked. In self-administered surveys, the respondent is the locus of control and can spend as much time as s/he wants to consult records to answer detailed questions. Especially in telephone interviews, but also face-to-face, one may feel pressured to answer and not let the interviewer wait and give an estimate instead of looking up the correct answer. Multi-stage manual data entry, such as is often involved in the collection of data in electronic form from paper questionnaires, completed by individual beekeepers, and then read and entered by another individual later, is error-prone and needs careful checking.

4. Quality issues in surveys

4.1. Errors

Any survey will be vulnerable to errors which may invalidate the extrapolation of the sample results to the target population. The most important sources of error are discussed in detail below. The aim in each investigation must be to minimise the non-sampling errors, and to quantify as far as possible the (unavoidable) sampling error. Therefore we describe these various sources of error in detail, as they should be borne in mind at the planning stage of the survey.

4.1.1. Coverage error

Coverage error arises when the survey population listed in the "sampling frame" – the list from which the sample selection is made (see section 6.2., usually of beekeepers in a particular country or region) – does not match well with the target population (the population about whom an inference is to be made; usually the set of all beekeepers in that country or region). The results of the survey will be seriously affected if those omitted from the sampling frame differ in some respect relevant to the aims of the investigation (in size of enterprise, for example, so that perhaps large commercial beekeepers are not represented).

4.1.2. Sampling error

Sampling error is the error that occurs because a sample is taken instead of examining the whole population (Lohr, 2010). A survey allows estimation of characteristics of variables (typically a population mean, sum, or proportion) concerning a whole population, on the basis of a sample. Such an estimate is inevitably not the exact value of the population quantity. The only way to obtain the exact value is to calculate it from the whole population, but this is rarely possible.

If the sample has been randomly selected from the population, this error can be quantified by calculating the standard error (standard deviation) of the estimate. This is an estimate of the variation of the estimator used between different samples of the same size selected from the same population. When a non-random sample is used, there is no appropriate analytical form for the standard error (see any elementary textbook on survey sampling, e.g. Schaeffer *et al.*, 1990) and therefore the results of any such calculation should be viewed with caution.

This variation from sample to sample is usually presented by quoting a confidence interval for the estimate obtained from the sample. A confidence interval can only be reported if the sample is representative for the population of interest. An example of the calculation is given in Box 2.

The sampling error can be reduced by increasing the size of the sample. It is possible to calculate the optimum size of the sample for a satisfactory estimate at a given cost, depending on the chosen confidence level or required precision or margin of error. Having some approximate knowledge of the population quantities is needed for such a calculation of the sample size (see section 9.).

4.1.3. Measurement errors

Measurement error occurs when the recorded answer to a question deviates from the true answer. The risk of such error is increased by an imprecise or wrongly formulated question. These errors can also occur in face to face surveys, if the interviewer influences the respondent (interviewer bias). Moreover, a survey is always declarative, so the respondent can voluntarily give a wrong answer if the question is in some way sensitive.

4.1.4. Non-Response errors

Non-response can be complete non-response (if a sampling unit, i.e. a beekeeper, did not answer at all) or partial non-response (some questions are not answered or only partially answered).

Keeping questions clear and simple can help to reduce the chance of missing data. Emphasising the importance of the survey and explaining how the results will be used may increase the response rate. If the participant can appreciate that there is some benefit to completing the survey, they will be more likely to take part. Use of rewards and incentives can be useful to increase participation of people selected already to take part in the survey.

Non-response bias occurs when there is something systematically different about participants who do not respond from those who do. Trying to minimise non-response is therefore very important. Reminders are useful in this regard. Non-response can in principle be estimated by randomly sampling some sample units after termination of the survey, approaching the non-respondents with another survey mode and comparing the main outcome with the survey response on these sample units. However such an effort is rarely regarded as a good use of scarce resources of time and money.

4.1.5. Errors caused by selection bias

Selection bias should also be considered as a source of error. Selecting an unrepresentative sample will lead to bias in the results. Actively selecting a random sample rather than a self-selecting sample who may well differ from those not included in the survey (volunteer bias), is the best way to avoid this. Using well-trained personnel will avoid any haphazard substitution of properly selected participants with more conveniently available participants who had not been selected, which is another possible source of selection bias.

4.1.6. Processing errors

These errors affect the data set. They can arise by errors caused by the person who records the data (mistyping, copy/paste errors, stretching cells in an Excel spread sheet etc.). If different people capture the data, harmonisation of notation and careful procedures for recording data should be in place, checks should be made, and personnel should be well-trained and informed to avoid introducing errors and biases (Schaeffer *et al.*, 1990).

Unlike the sampling error, the non-sampling errors are very difficult, if not impossible, to measure, and cannot be reduced by increasing the sample size. In a large sample, non-sampling errors are the more important source of errors, as the sampling error is reduced. The only way of controlling non-sampling error is to know what sorts of errors are possible, and to be very careful to avoid these as much as possible in the conduct of the survey.

Box 2. Example of confidence interval.

The overall proportion of colonies lost from those at risk is estimated as 19.5%, with a standard error (s.e.) of 1.5%. The corresponding 95% confidence interval is obtained in the usual way as the estimate +/- 1.96 (s.e.) giving 19.5% +/- 1.96 (1.5)%, which yields 16.6% to 22.4%. This means that in about 95% of cases when such a calculation is made, using a sample of the same size from the same population, the interval quoted will contain the true value of the overall proportion of colonies lost. This calculation assumes that the sample estimate is approximately normally distributed.

4.2. Effort and costs in data accumulation

Manual data entry is costly and time consuming. Although Optical Character Recognition systems for automatic data entry are available, the fastest mode of data collection and accumulation is when beekeepers directly answer the questionnaire using an internet database. Furthermore, after the end of the data collecting period, the respondent can receive feedback and evaluate his/her losses or other aspects of beekeeping experience relative to data accumulated from other participants in the survey. Such systems can encourage participation by other respondents and so achieve a higher response, and are available nowadays for affordable prices. However this approach will fail to achieve a representative sample and may be a source of selection bias if the availability of internet access is associated with questions of underlying interest.

As none of the discussed survey methods is flawless, nowadays survey organisers often use a combination of data collection modes (a mixed-mode survey) to offset the weaknesses of one mode with the strength of another (Brodschneider *et al.*, 2010; Topolska *et al.*, 2010; Soroker *et al.*, 2010, van der Zee *et al.*, 2012). Data validity and accuracy can be improved by interviewer-administered questionnaires of a selected group of beekeepers, ideally randomly selected, by following up a postal or email survey by a telephone interview or offering the opportunity to clarify any points of difficulty. Such a limited follow-up can also sometimes reveal the kinds of bias incurred by the more extensive survey. However, unless follow-up is so limited as to produce little information, it is a very costly option.

Repeating annual surveys among the same group of beekeepers will provide information on time trends, either by simply sampling the same population or possibly by following the same sample of beekeepers through time (i.e. using a panel design), when this is feasible. Use of panel surveys does require some replacement of panel of members who are no longer available to participate in the survey, while trying to keep the sample representative. Comparing the results of self-administered surveys which are widely distributed and interviewer-administered surveys of a selected group of beekeepers will enable evaluation of the extent of colony losses area/countrywide and indicate the reliability or otherwise of a non-randomly selected sample. It may also identify any special cases that require further study, such as extreme losses in specific geographical areas that were overlooked by random sampling and in the event that identification of such areas is the purpose of the study.

4.3. Issues of anonymity and ethical approval

Considering the issue of anonymity versus confidentiality is important before a questionnaire is disseminated. In an 'anonymous' type questionnaire, the subject is totally unknown to the survey organiser, while in a 'confidential questionnaire' all the data is known to the survey organiser, but kept confidential. Box 3 gives an example of how anonymity can be preserved when postal surveying is used, using experience from surveys in Scotland.

With the increase in use of email / webmail, using these means of communication make it virtually impossible to guarantee total anonymity, as the respondent's name – or at least the email address – is automatically included in their reply, although satisfactory survey software packages include the option of suppressing from the recorded responses all means of identifying individual respondents. While the perceived possibility of lack of anonymity may raise levels of non-response or compromise the validity of responses to any sensitive questions in an email questionnaire, the ease of access to a worldwide population of beekeepers, the low administration costs and its unobtrusiveness to respondents generally outweigh this negative effect. It is also a simple matter to issue reminders by email.

However, it is important that the level of confidentiality of the questionnaire is clearly outlined to participants. Hence, the covering

---

Box 3. Example: Preserving anonymity in a postal survey in Scotland.

The membership records of the Scottish Beekeepers' Association (SBA) provide a well-organised sampling frame of the target population. The help of the SBA Membership Convener was obtained. He is the only person with full access to those records. He was asked not to supply the survey organisers with the full records (which he would not have been permitted to do in any case), but only to supply the list of "Short Reference Numbers", each of which uniquely identifies one of the members, along with the associated postal code. Before supplying that list, he was asked to remove from the list those ineligible to participate in the survey, including for example members not resident in Scotland, institutional members (such as libraries), and those who had declared themselves unwilling to participate in surveys. (The opportunity to opt out of surveys is available to new members when joining the SBA, and the opportunity was given to all existing members of the SBA to opt out via a short article published in the SBA's regular publication for members, prior to the first survey using the SBA records for sample selection).

The postal codes were abbreviated in the list supplied, so that while preserving the broad geographical location of each potential survey participant, it was not possible to identify any particular address. The postal codes were used to assign each member on the list to a particular geographical region in Scotland, so that the sample selected could be stratified on a geographical basis. This was done by dividing the country into a number of regions related to the administrative areas used by the SBA, in order both to give greater precision in estimation and to ensure greater geographical coverage in sampling and therefore hopefully a more representative sample.

Then a stratified sample of the agreed size was selected from the list, using a sampling function available in the R software for objective random selection. Each questionnaire sent out was put into an envelope on which was written a questionnaire number provided by the survey organisers, also written on the questionnaire in the envelope. These envelopes were then sent to the Membership Convener along with a key file linking the questionnaire numbers to the corresponding "Short Reference Numbers". This enabled the Membership Convener to print the appropriate address on each envelope and to mail out the questionnaires, without the organisers knowing the identity of the selected members.

In fact, the majority of participants in the Scottish surveys have willingly provided personal contact details as part of their questionnaire return. (For example, in the survey in 2011, 85% out of 94 respondents did so; there was a 47% response rate).

letter with the original questionnaire should clearly state that a reminder will be forthcoming if no response is received. The availability of this option to issue reminders is important since research indicates that reminders increase the response rate (Campbell and Waters, 1990; see also the Scottish example below on reminders and incentives). Furthermore, email software allows the dispatcher of the questionnaire the option of notification when the recipient has opened the message.

The use of questionnaires raises the question of personal liberty and ethics and, because of this, many research institutes/universities require the survey organiser to acquire ethical approval prior to disseminating questionnaires. Since questionnaires related to honey bee research are primarily concerned with generic rather than personal information, it may be possible to acquire multi-annual ethical approval in advance, thus allowing the annual dissemination of the questionnaire. However, this will be specific to different institutes and thus clarification on the ethical requirements should be sought locally before a questionnaire is disseminated.

5. Coverage

5.1. Effective coverage

Collecting representative data on the extent of colony losses in any area or country depends on one's ability to identify and reach the target population (beekeepers in the country, commercial and/or non-commercial). This ability is affected by factors such as:

- the size of the country and beekeeper community
- the means of contacting the beekeeper community
- the degree of affiliation of beekeepers with beekeeping associations
- availability of professional magazines (including the possibility of publishing the questionnaire in a beekeeping journal)
- the holding of regular meetings of beekeepers
- the extent and accessibility of internet and telephone networks
- availability of addresses, e-mail addresses or telephone numbers
- willingness and ability of beekeepers' associations to cooperate in providing information
- the possibility of cooperation with beekeeping inspectors and veterinary services who may hold registers of beekeepers
- the number of staff engaged or available to conduct the survey and analyse the data
- and, of course, available time and funds.

Legal issues in relation to the preservation of the confidentiality of data also arise in some countries – for example, all organisations in the UK are constrained in what data they make available outside their own membership by the requirements of the Data Protection Act 1998.

If any method of probability sampling is to be used for a survey, access is also required to a sampling frame (see section 6.2.) which gives good coverage of the target population.

5.2. Potential bias of the survey sample

The responsiveness of the target population can be biased. For example, beekeepers suffering higher colony losses might be more likely to respond to surveys than those suffering fewer losses, although other biases are also possible. For example, Brodschneider *et al.* (2010) found a potential bias in reported colony losses from the same region collected by different media. Respondents who returned a postal questionnaire from a beekeeping journal reported higher proportions of losses than those responding online or at a convention in this particular region. This suggests that responses of some groups (such as from visitors of a convention) may not constitute a statistically representative sample, being from a different population than the target population. In such situations a mixed sampling approach must be considered. This enables comparison of the outcome of the different sampling methods, which should be reported in the final report. This could be due to different experiences of the target groups and hence a different level of motivation of the beekeeper to respond. A randomized sample may suffer from the same non-response problems, but maybe to a lesser extent because the approach is more focussed on the individual beekeeper. It may be possible to overcome this problem by increasing the response rate via encouragement of broad participation in the survey and the use of reminders. Using mixed media surveys or surveys using random surveying of the population of beekeepers to achieve a more representative sample may help, but the underlying problem remains that of the association of the response rate with the underlying questions of interest to the survey organisers. Response rates can be rather low. Dahle (2010) quotes a 15% response rate from surveys sent out in a beekeeping journal. Van der Zee (2010) quotes a 7.5% response rate from beekeepers who were invited in a national beekeeper journal to participate in an internet survey as well and they found differences between the results of these surveys and those from random surveys. In Denmark, response rates of up to 33% have been achieved (Vejsnæs *et al.*, 2010). In the Netherlands, an average of 22% of the beekeepers surveyed from 2006-2012 (van der Zee *et al.*, 2012) responded to a mixed mode approach of a questionnaire included in the 2 national beekeeper journals. The letters could be

Table 1. Some sampling methods with advantages and disadvantages.

Method groups	Method	Explanation	Advantages	Disadvantages
Complete	Census	Whole population selected.	With good response rate should give excellent information.	Potentially expensive and often infeasible. If attempted without a sampling frame, may be misleading.
Random	Stratified	Population divided into groups thought to be relevant to the survey objective, and each group sampled independently at random.	Gives well-targeted information where size of errors can be properly estimated if response rate is good. Has the potential for lower sampling error than simple random sampling.	Infeasible without a sampling frame.
	Simple	One random sample chosen from whole population.	Size of errors can be properly estimated if response rate is good.	Infeasible without a sampling frame. Less well-targeted than a well-stratified sample.
	Cluster	Population split into clusters (convenient subgroups), and a census is conducted within each of a randomly selected set of clusters.	Size of errors can be properly estimated if response rate is good and clusters are well chosen. Only requires good sampling frames within clusters selected. Potentially cheaper than stratified or simple random sample.	Requires sampling frame of clusters, and of individuals within selected clusters only. If clusters selected are not representative then may introduce bias to results. Sampling error is higher than with simple or stratified sampling.
Non-random	Quota	"Quotas" are set for researchers to fill of respondents matching given criteria.	Requires no sampling frame. Can work well, if quotas are based on sound underlying information.	Can fail very badly in achieving representative results. It is sometimes hard to fill quotas, and non-response is disguised.
	Purposive or judgemental	Researchers aim to select "representative" respondents.	Requires no sampling frame, and aims to achieve a good mix of respondents.	Personal selection is notoriously bad at choosing a genuinely representative sample.
	Convenience	Respondents are chosen because they provide a conveniently available sample.	Cheap and easy to implement. Useful for pilot studies. Requires no sampling frame.	Extremely unreliable and impossible to assess accuracy of results.

sent back without charges or through an email with a personalized link to the questionnaire on the internet.

5.3. Identifying the target population

A survey of beekeepers can be used to collect reliable data on beekeepers and beekeeping activity and/or practice in a certain area or territory of a country. The target population of such a survey should be defined according to the data that one aims to collect. It could be targeted to beekeepers' associations (local to national in scope) or the individual beekeeper. However, the target population is usually the set of all active beekeepers whose colonies are kept in the area of interest during the time period concerned. Consequently, coverage of individual beekeepers' operations via associations might be incomplete and variable. How easy it is to access such a population depends on whether beekeepers voluntarily or under legal compulsion are registered with some record-keeping organisation. Examples might be:

- A legally required register of all beekeepers within a given country;
- A voluntary register of beekeepers within a region to which most beekeepers subscribe;

- One or more regional or national beekeeping associations to which most beekeepers belong.

The coverage provided by such potential sources of data is clearly very variable, and the reliability of survey efforts will depend heavily on the adequacy of this coverage.

6. Sampling

6.1. Random and non-random sample selection methods

There are numerous sample selection methods for drawing the sample from the population, broadly classified into random or probability-based sampling schemes or survey design methods, and non-random or non-probability based sampling. An overview may be found for example in Schaeffer *et al.* (1990) or any such survey sampling textbook. A number of terms relating to surveys are introduced in the following sections, and may all also be found in such standard texts. The main methods which might be used for sampling beekeepers are summarised in Table 1. Some more detail is given about each method below. In general those methods higher up in the table will cost more, but will give more reliable results, provided that a good response rate can be achieved.

6.1.1. A census

It is possible to approach most of the beekeeper population in smaller countries such as the Netherlands, where questionnaires are included in both of the two beekeeper journals, which are sent to all Dutch beekeepers who are by their membership of a local organization also a member of a national association (>90%). The questionnaire could be returned without postal costs. Beekeepers who provided an email address (>85%) in the past received a personalised link to the online questionnaire. In 2012, about 70% of the data was submitted this way and processed immediately, which reduced costs substantially. This approach is in fact not sampling, but addressing the total population of organized beekeepers.

6.1.2. Random sampling

Survey designs based on random sampling are designed to select sampling units from the population with known probabilities. This means that the sampling properties of estimators of population quantities can be determined, such as whether or not the estimator is unbiased (i.e., does it on average give the right answer?) and what is its precision (i.e., how do we calculate its variance or its standard error).

This is the objective scientific approach to sampling and the only one for which sampling properties of estimators are known. Other methods may provide good information but there is no guarantee that they will, and their sampling properties are unknown. However even with random sampling, if response rates are poor then the possibility of non-response bias will compromise the estimation.

Implementation of random sampling methods requires a mechanism for random selection, usually accomplished by use of random number generators in computer software, e.g. the "sample" function in the public domain software R (downloadable from http://www.r-project.org/). It also usually requires a sampling frame, or list of sampling units in the population (section 6.2).

The simplest scheme is *simple random sampling*, which samples randomly without replacement from the sampling frame so that at every stage every sampling unit not already selected from the sampling frame is equally likely to be chosen. This results in all samples of a given size being equally likely to be selected.

Systematic sampling is sometimes used as a simple alternative to simple random sampling and works at least as well in situations where the population sampled from is "randomly ordered" with respect to the value of a quantity being measured or recorded, or is ordered in order of size of such a quantity. It does not always require a sampling frame. For example, if 1000 beekeepers attend a convention, to achieve a 10% sample of those attending, a participant may be selected at random from the first 10 beekeepers to arrive or register, and then every 10th person after that also selected.

Stratified sampling splits the population into subgroups or *strata*, using stratification factors such as geographical area or degree of experience of the beekeeper, or beekeepers/bee farmers, which are judged to be important in terms of coverage of the population and which are likely to be related to the response variable(s) or interest. Then a random sample, in the simplest case a *simple random sample*, is selected separately from each stratum, using predetermined sample sizes. This ensures representation of all these important groups in the sample (which might not be achieved by a single simple random sample), and the random sampling should compensate for any other relevant stratification factors which may have been overlooked in the survey design. It also allows comparison of the responses from each stratum, provided enough responses are achieved in each stratum.

If the average responses do differ between the strata, and/or the variation in recorded responses differs between strata, stratified sampling should provide estimates with a lower variance than simple random sampling. The lower variance is achieved because separate samples have been taken from populations with smaller variation within them compared to the population as a whole (Schaeffer *et al.*, 1990).

One basis for stratified sampling is operation size. The scale of beekeeping operations and management practices are very different for hobbyist beekeepers and professional/commercial operators (bee farmers). Due to the potential for different numbers of lost colonies and consequences of losses among these two groups, both should be included whenever possible in a survey. This allows the colony loss rates experienced by both groups to be compared and it is more representative of overall levels of loss. Box 4 gives an example.

The migration of colonies (the movement of colonies to/from nectar flows or for purposes of crop pollination) differs widely between beekeeping operations. Therefore, it is also desirable to consider different classes of migratory practice where possible when designing and analysing the survey. As migration may be a factor in loss rates (although see VanEngelsdorp *et al.*, 2010), comparing migratory and non-migratory beekeepers is important, if the sample

---

Box 4. Example: Case study of stratified random sample selection.

In Scotland, any beekeeper (or other person interested in bees) can choose to become a member of the Scottish Beekeepers' Association, while there is a separate Bee Farmers' Association for the UK, the qualification for the latter being that the beekeeper should keep at least 40 colonies of honey bees within the UK. There is known to be some overlap between the two membership lists, and care needs to be taken not to request survey participation of the same person twice for the same survey. Despite the fact that there are far fewer bee farmers than hobbyist beekeepers in Scotland, it is clear that they manage more than half the managed colonies, so that their contribution to the overall bee population is far greater than their numbers would suggest. Therefore in a recent survey it was decided to sample all of the bee farmers who could be identified, while selecting a random sample of non-commercial beekeepers (Gray and Peterson, 2012).

sizes permit valid comparisons. In places where there is widespread practice of migration on a large scale, this comparison becomes much more important. Identifying beekeepers practising migration of bees in advance of drawing a sample may be difficult, unless auxiliary sources like membership records include this information. If this information is available, then a stratified approach may be adopted to ensure coverage of both migratory and non-migratory beekeepers.

Geographical stratification may also be important, especially if different regions are subject to different weather conditions and differing exposure to bee diseases. However, combining multiple stratification factors with lower than ideal response rates can make the desired comparisons statistically invalid or impossible due to small samples.

Cluster sampling is the other main method of probability sampling. If the population can be divided into convenient groups of population elements rather than strata thought to differ in ways relevant to the response(s) of interest, then randomly selecting a few of the groups and including everyone in those selected groups as part of the sample will provide a representative sample from the whole population if the groups or *clusters* are representative of the population. For example, these clusters might consist of local beekeeping associations, which would be viewed as groups of beekeepers. This is a one-stage cluster sample design. There are other variants of this method, but they are unlikely to be of practical importance in this field of application.

6.1.3. Recommended approach for random sampling

In view of the discussion above, stratified sampling is recommended to achieve good spatial coverage of all main geographical areas in a country or region whose beekeepers are to be surveyed, and to give a more representative sample and more precise estimation. Using proportional allocation (see section 9.) is the simplest way to implement this. By ensuring that the chosen strata are represented in the sample, comparisons can be made, and these stratification factors can also be used as risk factors in modelling the risk of colony loss, for example. Achieving good spatial coverage of the population is also essential for spatial or spatial-temporal model fitting, which requires a high degree of data resolution.

6.1.4. Non-random methods

Non-random sampling, is any other kind of sampling. Such methods are often used for speed and convenience, and also they do not require a sampling frame. Their big disadvantage is that sampling error cannot reliably be quantified, as the sampling properties of any estimators used are not known (since the probability of choosing any one individual or sample cannot be determined).

Convenience or accessibility sampling involves asking a sample of people to respond to a survey. An example is distributing survey questionnaires at a meeting of a local beekeeping association or at a beekeepers' convention. However these people may not be representative of the whole target population of beekeepers, for

example due to local weather conditions in the first case, or the fact that attendees at a convention may be real enthusiasts whose bee husbandry practices are not typical of the general beekeeping population. A small convenience sample may be very useful for a pilot survey (see section 7.7) but is not recommended more generally.

An invitation to respond to a survey available on a web-site for example, is an example of taking a *self-selected sample* unless the people invited to respond to the survey have been selected already (as in Charrière and Neumann, 2010).

In some countries, such as Algeria, the most effective method in terms of response rates is a face-to-face survey in the beekeeper's home or at meetings of beekeepers' associations or co-operatives, as using mailed surveys produces an extremely low response. In Slovakia, it is also reported that the only method which works well is to disseminate questionnaires at meetings, as data collection via emails, web pages and journals has very low rates of return. For example, only 5 questionnaires were returned from a beekeeping journal with a circulation of 8 thousand copies (Chlebo, 2012; Pers. Comm.).

Given access to a population to be sampled, a survey organiser could try to take a "representative" sample, which is called *judgemental* or *purposive sampling*, to select what they think is a suitable mix of people to participate in the survey. The difficulty is that some important factors which have a bearing on the responses made to the questions may have been overlooked. Using judgemental sampling leads to a serious risk of badly biased samples.

Quota sampling is like stratified sampling in that stratification factors are identified which are thought to be relevant to the survey, but instead of sampling randomly the participants to come from each stratum, the survey samplers themselves choose the people subjectively from each stratum until sufficient people have been chosen and have responded. The main difficulty with this is the subjective choice of participants. Use of quota sampling also disguises non-response, as invited participants may decline to take part but the sampling will continue until the quotas are achieved. Quota sampling can work well, but can also fail spectacularly badly (as seen most notably in pre-election polling; see Schaeffer *et al.* (1990) for an overview of this and other methods). An example of using non-randomized quota sampling to survey American beekeepers is described in VanEngelsdorp *et al.* (2010), who recognised the dangers of using this approach but judged that it had given results consistent with the pattern of US beekeeping. Box 5 gives an example comparing quota and stratified random sampling.

A fundamental guiding principle in survey sampling is to use randomisation wherever possible in sample selection, to avoid subjective selection bias affecting survey results. Genuinely random samples are well-known to have the best chance of being representative of the survey population and should therefore be used unless it really is not possible (Schaeffer *et al.*, 1990) or will lead to such a low response rate that the results are of little use.

Box 5. Example: A case study: comparing quota and stratified random sampling.

In Scotland, the first survey conducted in 2006 for the Scottish Beekeepers' Association (SBA) used a form of quota sampling. This survey used strata which were broadly geographical, as has also been done subsequently. After the organisers decided on the split of the sample size between strata, they contacted the SBA Area Representatives, in order for them to choose the required number of participants from those known to them personally and known by the Secretaries of the Local Associations of beekeepers in that area. This allowed a known quota to be obtained from each geographical area. This was done purely because permission had not been gained at that stage to use the SBA membership records for sampling purposes and there was no other means of obtaining a list of beekeepers. The results (Peterson *et al.*, 2009) suggested to the organisers that the participants ran larger beekeeping operations than were typical of beekeepers in Scotland as a whole, and also that they were more conscientious and organised beekeepers than was typical. This is not entirely surprising, as the Area Representatives and Local Association Secretaries probably would have chosen people they thought were more organised and more likely to complete and return their questionnaire.

Subsequent surveys from 2008 onwards have used stratified random sampling. In the 2008 survey a modified Neyman allocation method (Schaeffer *et al.*, 1990; Särndal *et al.*, 1992; section 9.) was used to split the sample between the main SBA areas, and subdivided proportionally within these large areas to smaller geographical areas according to the number of SBA members (Gray *et al.*, 2010). In 2010 the simpler proportional allocation was used, as there was insufficient data from the 2008 survey on which to base Neyman allocation and the 2006 data was felt to be out of date. In 2011, Neyman allocation was used again, based on winter loss rates. The results were more in accord with what was expected, and therefore are probably more representative samples than the earlier one. The response rates however have been lower, and the higher response rate in the 2006 survey (of 77%, compared to 42% in the 2008 survey) almost certainly resulted from the element of personal contact.

Finally, it is essential in any reporting of survey results that the survey methodology and response rate should be clearly stated. This enables assessment of the reliability of the results, based on how representative the sample is likely to be. One way to assess whether or not the survey has been successful in achieving a representative sample is to check the responses to a standard question, to which the responses are not expected to change much from survey to survey, if past surveys have been carried out on the same population. If the results of this are different from what is expected this may indicate that the sample is not a representative one. The breakdown of the participants by key indicators such as geographical area or class of operation size can also be examined, although some of these factors will ideally have been controlled for in the sample design, by use of stratification.

6.2. Need for and use of a sampling frame in random sampling

Implementation of random sampling generally requires a list of sampling units, the *sampling frame*, from which to select the sample by random means, although systematic sampling can in some cases be carried out without one. This operates using a numbering system for each person listed. Random selection without replacement of the required sample size from the list of numbers of those having given permission for their records to be used for the purposes of sampling identifies the selected numbers and hence the sample selected, where a simple random sample is required.

For a stratified design, sub-samples are selected in the same way from each stratum, by selecting each sub-sample separately from the list of identifying numbers for those belonging to the relevant stratum. A cluster sample might select a few local beekeeping associations randomly from a centrally held list of associations, as for a simple random sample, and either target everyone in the selected associations as a survey participant, or take a further random sample from the membership list of each local association again using random selection of numbers as the means of sampling.

6.3. Availability of a sampling frame

In some countries, a substantial number of hobbyist beekeepers may choose not to belong to any kind of association of beekeepers or to be registered on an official list of beekeepers, meaning that there can never be 100% coverage in any list used as a sampling frame. Personal knowledge of some of these beekeepers may enable survey organisers to extend their sampling frame, however the possibilities for this are likely to be very limited. If such independent, unregistered beekeepers form a significant proportion of the beekeepers in a country, then it will be virtually impossible to obtain for that country a truly representative sample of beekeepers. To make matters worse, it is difficult to determine how many such unregistered beekeepers there are. This may be a cause of biased survey results, if the beekeeping experience and loss rates of non-registered beekeepers are likely to be systematically different from those that are registered.

6.4. Sources of sampling frames appropriate for different target populations

The ideal situation is one in which all beekeepers of a country (or other geographical unit) are equally represented in a sample. In some countries, beekeepers, usually with a minimum number of colonies, may be required to register on an official list, in which case gaining access to that list enables access to a selected part of the beekeeping population. In practice, not all beekeepers will register even if this is legally required. The level of compliance with registration requirements may vary greatly from one country to another. In the absence of a satisfactory list of registered beekeepers, other sources of sampling frames may be membership lists of beekeepers' associations or records held by veterinary services. In some countries (e.g. Norway, Sweden, Denmark and the Netherlands), beekeepers' associations may represent up to 90% of the beekeepers, so use of these association records seems to be the best approach currently.

Use of any of the above sampling frames for random sampling does require prior consent, by some means, of the beekeepers on the list sampled from, for their record to be used in the selection of a survey sample. Those who would not wish this must have the opportunity to opt out and, having done so, should be omitted from the list before random selection takes place. Ethical approval may also be required (section 4.3.). If cooperation with beekeeper associations which represent the majority of beekeepers in a country is not possible, or complicated because there are many small ones all with a limited number of members, another approach is advised.

Firstly, generally most of the bee stocks in the country are managed by large scale commercial beekeepers (even though there are also large numbers of small scale beekeepers). Often the commercial beekeepers have their own trade organisation which will list them all, as well as the approximate sizes of their operations, and if access can be obtained to them, estimation is possible.

If cooperation with a commercial beekeeper association is not possible, an approach using the fragmented smaller organisations may result at least in some kind of sampling frame from which a sample can be drawn with some hope of being representative of beekeepers who belong to these associations, at least in some local areas. How representative these associations are of all beekeepers is of course unknown.

If no beekeeper infra-structure is available, even if, say, in some parts of the world the post office is the only main central information hub, it may be possible to find for each post office district a nucleus of beekeepers. A representative sample of post offices with respect to climate and suitability of area for beekeeping could be drawn up. If such a sample were not too large, then putting out an enumerator for the survey into each of those post office areas might enable that enumerator to find within that area a fairly complete list of the beekeepers in the area. Then the cluster sampling approach would be sensible, where the post offices sampled were regarded as the clusters. A return would be made for each sampled post office area, and the usual techniques for cluster sampling could be used to analyse the results. These approaches may provide a way forward in situations where there is very little by way of an existing sampling frame and limited resources are available or only small scale surveys are possible. Even an imperfect investigation will yield some information. The important thing in considering the results of such work is to be open about the shortcomings of the results, and not to claim more for them than is justified.

7. Questionnaire design

7.1. Completeness of the questionnaire

In the framing of survey questions, the aims and objectives of the desired analysis, and the methods to be used in the analysis, should be borne in mind, in order to ask all of the questions needed to enable collection of the appropriate data. For example, for modelling the odds of colony loss through CDS, questions should be asked relating to any suspected risk factors as well as collecting data on numbers of colonies lost and what number of losses are attributed to each of a list of possible causes.

In a survey concerned with honey bee colony losses, migratory habits need to be stated as clearly as possible by the respondents in order to avoid misunderstanding about the place(s) where losses were/were not recorded, as well as possible causes of losses, since migratory habit could be one such cause or a contributing factor in honey bee loss.

It is also important to collect and record auxiliary information for statistical purposes such as weighting and multivariate analysis.

7.2. Appropriate designs for different sampling methods

A questionnaire for use by a trained interviewer conducting face-to-face or telephone interviews can be much more elaborately constructed than one used for general distribution (say at a conference) or for a postal or email-based survey or a web-based survey.

In the former situation, the time and cost of obtaining the interviews at all is so large that the additional expenditure for training specialist interviewers and possibly providing them with aids such as laptop computers with purpose-designed questionnaire software is often considered worthwhile. Many large scale government-funded surveys seeking national statistical data are conducted in this way. Questionnaires may then have complicated question routing for various alternative pathways through the questionnaire, as indicated by answers to key initial questions. Sophisticated questionnaire software will have these paths encoded within it. In other cases the interviewer will be familiar with the routes to take depending on the responses given. This is the more common situation in surveys with visiting inspectors of the extension service, or in general in projects which require advanced diagnosis of disease and/or sampling of colonies.

In the second situation, where respondents to the survey have full control of the progress of the response and interpretation of the questions, it is vital that shorter and simpler questionnaires are used, with clear instructions and clear questions, to avoid low response rates and inaccurate responses. This is almost always the situation relevant to surveys of beekeepers.

7.3. Common problems to avoid in questionnaire design

Some common problems to avoid in questionnaire design include *ambiguity of interpretation, loaded questions* and *questions on sensitive issues.*

7.3.1. Ambiguity of interpretation

If respondents can interpret a question in various ways, the returns made will not be easy to interpret and the analysis can become difficult or impossible to conduct. Box 6 gives an example. The way to minimise ambiguity is, first of all, to ensure that the early drafts of a questionnaire are always criticised by an independent evaluator before they are used, and once all obvious ambiguities have been removed, to pilot the questionnaire (see section 7.7.) in order to try to detect any remaining problems with the questions.

Box 6. Example: Colony management in Canada.

An example is provided by recent COLOSS surveys in which beekeepers were asked about increases and decreases during a certain timeframe. All Canadian respondents who reported increases or decreases during the defined wintering period were contacted to verify whether such changes truly reflected the dynamics of the wintering population. Invariably, these changes reflected spring-time activities (typically splitting colonies), where these activities could occur in warmer areas of the country prior to the defined end date of the wintering period. Moreover, these changes were not reflected in total colony counts at the end of the wintering period. The question was clear about the timeframe, but a substantial number of beekeepers ignored this information (van der Zee *et al.*, 2012).

7.3.2. Loaded questions

Questions can often be framed in such a way that the respondent is guided towards selecting a particular response, even when that response does not reflect the true state of affairs of interest to the investigator. Box 7 gives an example.

Critical analysis of the original questions for possible loading, and careful analysis of pilot survey results, with subsequent revision of the questionnaire where necessary, are essential.

Box 7. Example: Case study: Experience in a Scottish survey.

An example is provided by a question used in a recent Scottish survey in which respondents were asked in what year they had first become aware of varroa infestation of their colonies. The question was intended to discover how far in the past it was when this parasite had first been detected in that area of the country, since there are still remote areas of Scotland where it has not yet been found. However some newly established beekeepers interpreted this as meaning that they were expected to have personally observed the parasite, and so were inclined to respond that the parasite had "not yet been found" - a biased answer leading to an over-optimistic interpretation of the extent of the parts of the country which were still free of varroa.

7.3.3. Questions on sensitive issues

Even in surveys of beekeepers, some issues can be sensitive. Matters such as financial returns, incidence of disease, location of beehives, and methods for treating diseases and parasites may be sensitive topics for some beekeepers. Taxation, personal and commercial confidentiality issues may be important in financial questions. Beekeepers may feel sensitive about exposure to criticism for poor management if they report disease. If unorthodox treatments for disease have been used, then exposure to the risk of prosecution may make respondents reluctant to respond. Concern about safety of

beehives, or any of these other issues may mean that some beekeepers will not answer those questions at all, will provide incomplete information, or will supply wrong information. There is no doubt that seeking too much sensitive information will seriously reduce response rates and also lead in many cases to incomplete survey returns, thus defeating the object of asking the questions.

It is hard to know how best to address this problem. Firstly, investigators should be aware of what may be sensitive points in their target population, and if necessary, avoid directly asking about them. Sometimes a suitable non-sensitive substitute question may be available for some of these issues, but that too can be problematic. Perhaps seeking information about the mean honey yield per production colony may be felt to be less threatening to a respondent than asking about the financial return for their honey harvest in a particular season, for example. Participants should be assured that all information held will be treated in confidence and only used for the stated purposes of the survey, and that permission would be sought for any subsequent use of the data for other purposes. Any wider data sharing should only be undertaken with extreme caution and great care taken to remove any information which could lead to the identification of the individual beekeeper. There are limits on how successful this can be, in a small area, for example, where beekeepers may be well-known. Information on exact hive location is probably best not shared at all.

7.4. Questionnaire design for minimisation of measurement error and ease of analysis

It is important to word questions carefully, using neutral language, clearly and unambiguously to avoid misunderstanding and consequent errors in data supplied. Carrying out a pilot survey of a questionnaire (see section 7.7), or part of it, involving any new questions is essential in order to check that the questions are appropriately worded. Modifications of existing questions are best tested in the same way, unless the change is minor.

It is worth considering the order of questions asked. The usual guideline is to ask more general questions before more specific ones. This avoids attitudes/responses from becoming fixed early on and may encourage a more flexible way of thinking. It also allows for appropriate question routing, i.e. based on their response to certain questions the respondent is then directed to go to the next appropriate question for them to answer. For example, the questionnaire can state something like "If you responded "yes" to this question, please go next to question X". This is important in surveys drawn randomly from membership lists of beekeeping associations, where not all members may be active beekeepers at the time of the survey and it cannot be determined prior to carrying out the survey which members are active beekeepers. Asking whether or not the respondent is an active beekeeper early on in the survey allows asking such respondents any questions directed at them specifically, while directing the active beekeepers to the start of the main part of the questionnaire and the questions intended for them.

For ease of data coding and analysis it is best to use closed format questions where possible, with a fixed number of response options of most interest and/or thought to be the most common, and to provide an "other" or "not applicable" category to cater for responses that cannot be fitted into the supplied list of responses, with the means to provide further details of the answer to the question. Closed questions with a given format make it possible to compare responses from surveys of different populations, for example in different countries, or of the same or a similar population at different times. Completely open questions inviting a written response are much less easy to deal with in data coding and analysis and are best avoided. The number of response options is best not to be too long, to avoid confusion or error.

Questions asking for a numerical response are best worded and set out to allow the respondent to supply the exact number, of colonies managed, for example, as the answers can be categorised later if required but having the exact numbers provides more information for analysis.

It is well-known that asking sensitive questions in surveys (see section 7.3.3.) is less likely to elicit an accurate or complete response than less emotive questions. Some survey methods are more successful in this matter than others (Schaeffer *et al.*, 1990). Questions requiring more knowledge than a participant has are likely to be answered inaccurately. Either some background information should be provided, or a screening question(s) should be asked first to determine whether or not it is appropriate for the participant to be asked the particular question of interest.

While constructing the questionnaire and accompanying documentation, including a covering letter/invitation to participate and instructions to the survey participant, a coding sheet should also be prepared. In online surveys, responses will automatically be entered into a database, and the coding of them is part of the questionnaire design (0 = no, 1 = yes, N/A for missing, for example). In surveys requiring manual data entry, a coding sheet is important for consistent translation of survey responses into data entered in a spreadsheet. This is especially important in situations where there is more than one person involved in data entry.

7.5. Need to limit data sought, for a high response rate and accurate measurement

Constructing a long and detailed questionnaire offers the survey organisers the opportunity to collect a great deal of useful information from those survey participants who return their questionnaire, but is likely to result in a rather lower response rate than would be desirable, owing to the time and possible difficulty involved in completing it. Few respondents may return the questionnaire and it is likely that amongst those who do, some of the information will be missing. Asking very detailed questions is also likely to result in information being less reliable, as not all beekeepers will recall the details or will not have kept sufficiently detailed records to be able to provide correct information, or be unwilling to take time to find the information requested. Longer, more detailed questionnaires and more complex response options are more likely to be successful in face-to-face surveys, but less so in more modern forms of survey. (In telephone surveys, participants may become impatient with long and complicated surveys with many response options, and are likely to terminate the interview prematurely, so shorter and simpler is best.) Postal and self-administered surveys in general require especially clear questions and response options and should be kept to a manageable length (Schaeffer *et al.*, 1990; Brodschneider *et al.*, 2010). Balancing the desire for more information with simpler questions and a shorter questionnaire is likely to produce a higher response rate and more accurate data.

7.6. Problems of multi-lingual/multi-cultural questionnaires

Care should be taken in constructing a multi-lingual or multi-cultural questionnaire, to ensure that the questions and response options are relevant to those receiving them, to avoid needless complication and needless irritation of survey participants, with a view to securing the goodwill and co-operation of the questionnaire recipients and hopefully therefore a high response rate. Accurate translation of specialised concepts requires translation by those familiar with specialist terms in both languages involved, which can be hard to achieve.

Local modifications may be necessary, for example in specifying in relevant questions the month of the start of the winter/summer season for beekeeping, however care should be taken to preserve the meaning intended by the original question. Similarly, differing response options may be appropriate in different countries. For example, in a question about possible disturbances to bee colonies, bears are a possible hazard in some countries, but not in others. Bee races kept will also vary from country to country. Providing "Other" as a response option allows for any more unusual responses, while keeping the specific response options relevant to the participants. Even some questions may not be felt to be relevant ones for some countries. These local variations have implications for the return of the data for central processing and also for its interpretation. Data coding needs to allow for the different response options and care is required in returning accurate data to avoid introducing errors.

One difficulty when colony losses are being recorded is the time period of observation. Lost colonies are common within a period when colonies are not foraging. Depending on the climatic zone this may be winter or other periods. Such periods also differ in duration between years and areas. Using seasonal characteristics allows for comparing effects on honey bee survival of the length of the non-foraging periods between climate zones. However, in some parts of the world like the USA, Southern Europe and Asia, migration of colonies for pollination purposes to warmer zones during winter can be substantial. This suggests the use of fixed timeframes and determining how many colonies are present at some fixed moments in time.

7.7. Testing survey questions: importance of pilot studies

It is always sensible to test a new or modified questionnaire in a small scale pilot survey before circulating it more widely to a larger group of survey participants. Inevitably in the answering and reviewing of the pilot questionnaires, some unanticipated problems will be highlighted, from minor issues such as duplication of question numbers, to misinterpretation of question wording and issues requiring modification of question wording, new response options and/or additional questions. Box 8 gives an example.

If re-using a well-tried and tested questionnaire, clearly there is less need for a pilot run. However, if new questions are added a small pilot run is still advisable. Almost always some small point has been overlooked or can be improved upon, despite the most careful survey design. Even with an old questionnaire, piloting is often advisable to ensure that questions are still comprehensible and relevant.

Box 8. Case study: Pilot surveys in Scotland.

In recent surveys in Scotland, for example, about 6 people known to one of the survey organisers through his local beekeeping association were identified as suitable candidates who were readily contactable, covering a wide span of years of beekeeping experience from the beginner to the much more experienced. The questionnaire was delivered to them personally at a time when they able to deal with it immediately or an arrangement made to collect it shortly thereafter, so that no responses went missing. In the face to face situation, any immediate difficulties in understanding the questions are easily dealt with and explained, and a note made that these questions need to be re-worded. In all cases this exercise has suggested some points to be changed in the survey questionnaire, if only minor ones, and has been felt to be very useful.

7.8. Example of a standardized questionnaire on colony losses

An example of a standardized questionnaire, produced by the monitoring working group of the COLOSS network (section 1), is provided in Fig. 2. The questionnaire can be split into essential questions which should be implemented in all participating countries and optional questions which are left to the national survey organisers to use or not. Optional questions ask more information about the operation such as postal code and location of the apiary, migration, bee race, increases and decreases made by the beekeeper during winter, origin of queens, queen replacement, pollination services, honey and pollen sources, comb replacement and winter feeding. The survey organisers may replace the concept of winter by another seasonal concept suitable for the local situation.

8. Response rates

There are different ways to calculate response rates. The examples given here simply use the number of usable responses (complete or otherwise) divided by the size of the sample selected or number of participants approached. Variations on this as well as several other measures of outcome rate are discussed fully in the reference AAPOR (2011).

8.1. Use of incentives and reminders to improve response rates

As mentioned above, reminders are an important means of improving response rates in self-administered surveys. A personal reminder is likely to be more effective than a more general public one. Providing an incentive to participate in the survey to those already selected to participate can also encourage return of a questionnaire and hence may have some beneficial effect on response rates.

In telephone surveys call-backs are easy to arrange. Sending repeat emails is also straightforward. In an online web-based survey, as in self-selected survey samples generally, it is the more motivated who will respond to a general call for participation and these may well coincide with those who have more extreme opinions or experiences to report. Therefore reminders are important to try to overcome the bias which this creates, by involving some of those who are less inclined to participate but who may be more representative of the population as a whole. Box 9 gives an example.

Box 9. The Scottish surveys: use of reminders and incentives.

In the 2008 survey in Scotland, a short public reminder was published in "The Scottish Beekeeper", but the final response rate was only 42%. In that survey no personal reminder was possible as anonymity was built into the survey and questionnaire numbering was not used. Numbering of the questionnaires allows identification of the selected survey participants who have not responded. In recent Scottish surveys the numbers of questionnaires returned by the deadline were removed from the list of numbers of all the questionnaires sent out, the remaining numbers matched to the reference number of the person concerned and this list of numbers sorted into order and sent to the membership convenor for identification of the people in order for him to send a short reminder letter. The first time this was done, in 2010, the response rate was considerably improved, to 69%, although in 2011 it had little effect (response improved from 45% to 49%) and there was barely any effect in 2012. Nonetheless reminders are recommended.

In the last few annual surveys of beekeepers, a well-known commercial supplier of beekeeping equipment has willingly provided a generous voucher to be awarded to the winner of a prize draw at the end of the deadline specified for return of the questionnaire. The winner was randomly selected from the list of questionnaire numbers returned by that deadline. The winning number was matched to the identifying short reference number for that participant, and the details were sent to the SBA membership convenor. The convenor identified and contacted the winner, and contacted the commercial company to arrange for the sending of the prize to the winner. The winner was asked what details they would be willing to have published in the SBA's monthly publication for members, for example, information such as "The winner of the £50 voucher kindly offered by Company A as a prize to the successful participant in the SBA 2010 survey lives in Argyll"), hence giving some publicity to Company A.

Survey on Colony Losses 2012

In this questionnaire we try to gather information about production colonies. Please consider colonies which are queen-right and strong enough to provide a honey harvest as production colonies.

<u>1</u>   Country _____

<u>2</u>   Region/Province_____

In the next questions you are asked for numbers of colonies lost. Please consider a colony as lost if it is dead, or reduced to a few hundred bees, or alive but with queen problems, like drone laying queens or no queen at all, which you couldn't solve.

Please consider winter as the period between the moment that you finished the pre-winter preparations for your colonies and the start of the new foraging season.

<u>3</u>   How many production colonies did you have before winter 2011-2012?

<u>4</u>   How many of <u>these</u> colonies were lost during winter 2011-2012?

<u>5</u>   How many of the lost colonies did not have dead bees in the hive or in the apiary?

<u>6</u>   How many of the lost colonies had dead workers in cells;

/   and no food present in the hive?
         and food present in the hive?

<u>7</u>   How many of the lost colonies had queen problems, like drone laying queens or no queen at all?

8   How many production colonies did you have <u>after;</u>
        winter 2010-2011
        winter 2011-2012

<u>9</u>   Have you treated your colonies against Varroa during the period November 2010 - January 2012? | yes | no |

To learn about <u>the number</u> of Varroa treatments during the period November 2010-January 2012 we would like to get information on the months you <u>started</u> a Varroa treatment with a product.

<u>10</u>   Could you please indicate in what month and year you <u>started</u> every Varroa treatment of your colonies <u>with a product</u> during the period November 2010 - January 2012?

2010		2011												2012		
November	December	January	February	March	April	May	June	July	August	September	October	November	December	January	February	March

Fig. 2. An example of a standardized questionnaire, produced by the monitoring working group of the COLOSS network.

9. Choice of sample size

In a probability-based sample, the sample size can be calculated statistically in order to achieve a required level of precision of estimates from the data collected, where these estimates have been identified in advance as being of interest. The formulae required depend on the sampling scheme to be used. Schaeffer *et al.* (1990) give details.

For example in a simple random sample, to estimate a mean, e.g. average number of colonies kept per beekeeper, to within a distance or error bound B of the correct value with approximately 95% confidence, the formula for the sample size is $n = \frac{N\sigma^2}{(N-1)D + \sigma^2}$ where $D = \frac{B^2}{4}$ and σ^2 is the variance in the population of the quantity of interest, e.g. the number of colonies kept, and N is the population size. In the case of a very large population of beekeepers, where N is not known exactly, an approximation to this sample size is given by $n = \frac{\sigma^2}{D}$. The population variance may be estimated from the variance calculated from data in a previous survey of the same population, or from a pilot survey. To estimate a total (by the population size N times the sample average) with the same precision uses this same formula but with $D = \frac{B^2}{4N^2}$. Box 10 provides an example of the calculations.

Box 10. Sample size calculation for a survey to estimate a mean or a total.

For example, using a simple random sampling approach, to estimate the average number of colonies kept to within a margin of error of 10% ($B = 0.10$) of the true value with an approximate confidence level of 95%, the sample size is calculated as follows. We use the formula $n = \frac{N\sigma^2}{(N-1)D + \sigma^2}$ where $D = \frac{B^2}{4} = \frac{0.10^2}{4} = 0.0025$. Assuming that the total number of beekeepers in the population is 1500, and if we have recent information from a previous survey that the variance σ^2 of the number of colonies per beekeeper is about 4, then we should sample $n = \frac{1500(4)}{(1499)(0.0025)+4} = 775$ beekeepers, rounding up to the nearest integer. If we wished to estimate the total number of colonies kept, say to within 200 of the actual total with the same level of confidence, then making use of the same information, we calculate instead $D = \frac{B^2}{4N^2} = \frac{200^2}{4(1500^2)} = 0.00444$, which now gives $n = \frac{1500(4)}{(1499)0.00444+4} = 563$ beekeepers to be sampled.

To estimate a proportion p to within an error bound B of the true value with approximately 95% confidence, the same exact and approximate formulae are used as for estimating a mean, but with $\sigma^2 = p(1-p)$, so in the large population case $n = \frac{p(1-p)}{D}, D = \frac{B^2}{4}$. These formulae require an approximate value for p based on prior experience, or else substitution of a conservative value of $p = 0.5$ to maximise the required sample size. Box 11 shows the calculations.

Box 11. Sample size calculation for a survey to estimate a proportion.

For example, using a simple random sampling approach, to estimate an overall proportion of losses which was 20% last year (so $p = 0.20$ approximately), to within a margin of error of 5% ($B = 0.05$) of the true value with an approximate confidence level of 95%, the sample size is calculated as follows. The population size is assumed large, but is unknown. So we use the large population version of the sample size formula for estimation of a proportion given by $n = \frac{p(1-p)}{D}, D = \frac{B^2}{4}$. Here this gives $n = \frac{0.20(0.80)}{D}, D = \frac{0.05^2}{4}$, giving $n = 256$ exactly. So the sample should be composed of at least 256 individuals to achieve the required level of precision.

If there is more than one quantity to be estimated, as there will be in surveys of beekeepers, the larger of the relevant calculated sample sizes can be used, where this is feasible, or it can be decided to focus on one more important estimator, e.g. the proportion of beekeepers experiencing winter colony loss or the proportion experiencing CDS losses. It is then accepted that any other estimates requiring a larger sample size will be estimated with lower precision than is desirable.

For a stratified sample, which takes simple random samples from each stratum, similar calculations may be done to obtain the overall sample size required to estimate the mean or total or proportion to within an error bound B of the true value with approximately 95% confidence. See Schaeffer *et al.* (1990), for example, for details.

Various approaches are possible to divide the chosen sample size between the strata, including the *proportional* method which takes the sample size n_i in the ith stratum proportional to N_i/N, where N_i is the size of the ith stratum and N is the population size. This means taking $n_i = \left(\frac{N_i}{N}\right)n = W_i n$, where W_i is the ith stratum weight or the proportion of the population belonging to stratum i.

Neyman allocation is a more complex method which splits the sample between strata in order to minimise the variance of the unbiased estimator of the population mean (given by $\sum w_i \bar{x}_i$, where $k = N_i/N$, where k is the number of strata, $W_i = N_i/N$ and $\bar{x}_i$ is the mean of the sample from stratum i) or of the total (taken as N times the estimator for the mean) by taking the ith stratum sample size proportional to $N_i\sigma_i / N$ or $W_i\sigma_i$, where σ_i^2 is the variance within stratum i and σ_i is the standard deviation the variance within stratum i. So

$$n_i = \left(\frac{N_i\sigma_i}{\sum_{i=1}^{k} N_i\sigma_i}\right)n = \left(\frac{W_i\sigma_i}{\sum_{i=1}^{k} W_i\sigma_i}\right)n.$$

The within stratum variances may be estimated from previous experience or a pilot survey.

To estimate a proportion (by $\sum w_i \hat{p}_i$, where $\hat{p}_i$ is the sample proportion in stratum i), the same formula can be used for allocation as for estimating a mean, but σ_i is replaced by $\sqrt{p_i(1-p_i)}$ where p_i is the value of the population proportion in stratum i (and in practice an estimate of this is used).

The Neyman approach can also be modified, if required, to incorporate different sampling costs for each stratum. More complex modified Neyman allocation schemes are also possible (Särndal *et al.*, 1992).

More generally it may be decided, in order to achieve a suitable coverage of the population, that a fixed percentage of the population should be sampled. For some of the COLOSS surveys, a guideline for acceptable coverage has been that, where possible, at least 5% of beekeepers should be surveyed. This is a simple way to choose sample size, especially in a non-probability sample for which sample size calculations are not valid.

Another concern in a smaller population which may be surveyed repeatedly is not to overburden individuals, but to maintain goodwill.

This may mean taking a smaller sample than is ideal. Data processing concerns may also limit the sample size.

If the level of non-response can be anticipated, for example, from recent experience, the calculated or chosen sample size can be increased accordingly, in order still to give a sample of the required size, as $n_2 = n_1 / (1 - r)$, where n_1 is the original sample size, n_2 is the new size, and r is the expected non-response rate as a proportion, e.g., $r = 0.25$.

Obtaining standard errors of estimates, or confidence intervals, as part of the data analysis indicates how precisely the various quantities of interest have been estimated (see sections 4.1.2. and 10.).

10. Analysis of survey data

10.1 Assessing data quality

Prior to the analysis, some assessment and possible improvement of the quality of the data is essential. This is of utmost importance when these data are to be used in statistical models. Errors of different kinds can easily result in false inferences of general patterns, meaning that effort expended in complex modelling may be largely wasted if the data are unreliable.

As numerical data is not directly measured but derived from surveys, the means of data collection used in the surveys has to be designed in such a way that respondents have limited opportunity to generate extreme or erroneous responses. Thorough data validation must precede modelling procedures, i.e. checking for out-of-range data (invalid responses), and inconsistent responses. The proportion of missing values is also an indication of data quality. See De Leeuw *et al.* (2008), chapters 17-22, for an overview of quality control and data validation for survey data.

If the results of data checking suggest that the data are unreliable, then it may be sensible to limit analysis to simple procedures, or else interpret the results of model fitting with some caution. This is also true for small data sets where complex model fitting may not be feasible.

If the selected sample size is known, as for example in a randomized sample, the overall non-response rate can be calculated as a first indicator of quality, as a survey with a high non-response rate (a low achieved sample size) may be unrepresentative of the population of interest. Assessing non-response involves comparison of the actual sample size and the planned sample size (see § 9 on choice of sample size).

Examining responses to individual questions is also necessary. For each question, several simple quality measures may be calculated:

1. The missing data rate can be checked (for partial non-response).

A high proportion of missing data may indicate inappropriate or sensitive questions, for example those which will be important to reconsider for the question design of future surveys. Missing data may be left as missing for the purposes of the analysis, or a data imputation method may be used to replace the missing data with a plausible value (De Leeuw *et al.*, 2008, chapter 19).

2. The proportion of invalid values can be checked. The size of any deviations from what is a valid response is also of interest.

The response may be a value outside the valid range of responses for that variable, such as a percentage above 100 or a negative number of colonies lost, or it may be a suspiciously extreme value. This problem occurs when the question was not correctly answered by the respondent, or when the data was not correctly captured at the point of data collection or data entry. A question with many invalid answers should probably be reformulated. If there is no way of checking what is the correct answer, the response should be considered as missing data and should be omitted from the analysis.

3. The proportion of inconsistent values must be checked. It may be clear from examination of the data that the responses to some questions are inconsistent with the responses to some other questions. For example, the calculation of the number of colonies lost in periods when bee management is practised may give a different answer from the number of colonies stated as having been lost. A variable recording the difference in these two quantities may be used as a filter to remove cases with inconsistent data from analysis.

These data quality descriptors can be obtained from descriptive analysis, for example using summary statistics including the range of a variable, tabulations, cross-tabulations and histograms.

10.1.1. Dealing with missing data

The treatment of missing data is a rather specialised statistical topic. Missing data is difficult to deal with adequately in the analysis of questionnaire data, especially if it is not "missing at random". Data which is missing at random is such that the responses that would have been given are not related to the probability of non-response. If data is missing at random, then the data that is available can be analysed and the results should still be representative of the population, provided that the selected sample was representative. If it is not missing at random, then the results of analysing the available data are likely to be badly biased. Missing data reduces the number of responses available to analyse and hence reduces the precision of any estimates made. The best approach therefore is to try to minimise the chances of data being missing, by careful questionnaire design and by choosing a survey mode which gives respondents time to complete all the questions and secures their co-operation to do so.

10.2. The use of weighting in statistical analysis

In an analysis of a survey based on random selection, if the survey does not use simple random sampling or sampling with replacement, then participants are not all equally likely to be selected from the population. In this case, to achieve unbiased estimation, a case weight should be assigned to each of the participants returning data. These sampling weights should be inversely proportional to the probability of selection of each participant and should sum, over all the participants, to the sample size. Sampling weights can only be calculated if probability sampling is used. The software package SPSS, for example, allows a weighted analysis to be carried out.

For example, if a stratified sample is used, based on geographical area, for a case sampled from stratum i the sampling case weight is given by $(N_i/N)/(n_i/n) = (N_i n)/(n_i N)$ where N is the population size, N_i is the size of stratum i in the population, n_i is the number of people sampled from stratum i and n is the total sample size. This requires knowledge of which area or stratum a respondent comes from. Numbering questionnaires and recording in the data spreadsheet the area in a column beside the questionnaire number is probably the best way to ensure that the required information is available. Inclusion of appropriate questions can make it feasible to set up weights to be used in a weighted analysis of the data, however not all participants may respond to these questions, so it is safer to record the information in advance.

Weights can also be used to allow for *unit non-response*, i.e. where some people do not respond at all. These weights are inversely proportional to the probability of responding. So in stratum i, each person would have a non-response weight of n_i/n_{ir} where n_i is the number of people selected from stratum i and n_{ir} is the number of people responding from that stratum. This and more sophisticated methods are discussed in Lehtonen and Pahkinen (2004).

The weights for sample design and the weights for non-response can both be used at once, by multiplying the two columns of weights together and rescaling so that the new weights add to the sample size.

In a multi-cultural survey, in which different sampling designs may have been used to select participants in different countries, different weight calculations will be needed for respondents from each constituent country, and this requires detailed knowledge of the different survey designs which were used to select the samples. In practice this information may not be readily available.

10.3. Elementary analysis

10.3.1. Descriptive analysis

Any statistical analysis of data should begin with simple data description and presentation using summary statistics, tables and plots. While the main interest may well be in modelling, the initial analysis is still an essential first step. The results of such analysis with well-designed graphs and/or tables can reveal unsuspected patterns in the data, and will ensure that the obvious characteristics of the data are clearly understood by readers of any resulting report.

Responses to any simple survey question clearly require this approach, e.g. to determine the proportion of respondents in a postal survey who are currently beekeepers, where this cannot be determined in advance of choosing the sample, or the proportion wishing to remain anonymous, or the proportion experiencing any colony losses over a specified period. The data on any categorical variable can also be presented in a bar chart and/or a contingency table, with frequencies and relative frequencies, for an overview of the responses, the range of values and the most common category. This will also help in identifying invalid responses.

Extending this analysis to more than one categorical variable, e.g. to compare the proportions of losses experienced by respondents in different countries, or by geographical area within a single country, or for different sizes of beekeeping operation, two-way tables are useful. Relevant follow-up tests include chi-squared tests of association or homogeneity, which will permit the statistical investigation of the possible significance of differences in sample proportions. Even if observations contributing to each cell in the table are not all independent, the results of this can inform any subsequent modelling, by identifying potential risk factors for colony loss, for example, to be included in the model.

For questions with a quantitative response, of most interest is some measure of a typical or central value. The most appropriate measure depends on the distribution of the numerical responses. Where these are fairly symmetrically distributed, and there are not many extreme atypical values, the best measure is the *mean* or arithmetical average of the observations. However if the distribution of the data is very skewed, and/or there is a fairly large proportion of extreme atypical values, then the mean can be seriously misleading. For example, in the distributions of number of colonies kept per beekeeper, or honey yield, the existence of a few very large numbers of colonies kept or correspondingly high honey yields has the effect that the mean will give a grossly inflated idea of what is a typical value. The number of lost colonies per beekeeper also tends to have a highly positively skewed distribution. For such cases the *median* is preferred. This is the central observation, or the mid-point between the two central observations, after the data have been arranged in increasing order of magnitude.

Almost as important is some measure of dispersion of the observations around the mean or median, whichever has been chosen as being most appropriate. The usual choices are either the *standard deviation* for variables for which the mean is used, or the *inter-quartile range* for situations where the median is the appropriate measure of a typical value. (Any first level statistics textbook, such as Ott and Longnecker (2009) or Samuels *et al.* (2010), will describe the computation of these quantities). Confidence intervals based on the mean are *z*-intervals in the case of a large sample, or *t*-intervals for smaller samples. Population means may be compared using Analysis of Variance (ANOVA; see Ott and Longnecker (2009), and Pirk *et al.* (2013)), assuming independent observations and independent samples from normal distributions. For medians, nonparametric confidence intervals and tests are available, including the Kruskal-Wallis test as a nonparametric equivalent of ANOVA. Nonparametric procedures generally are robust to data which does not conform exactly to the assumptions of the test procedure.

Histograms are essential graphical tools to study the nature of the probability distribution of a quantitative variable such as number of colonies kept or number of colonies lost or honey yield, and hence to determine whether this is symmetric or skewed. Boxplots can also be useful in this regard. Comparing these between countries for example can indicate differences.

Comparing a histogram visually with the theoretical density functions of a range of possible probability distributions is also a simple first step in selecting and justifying a plausible model for use in more advanced statistical modelling of a dataset. The most frequently used probability model for the distribution of continuous numerical data is the symmetric bell-shaped *Normal* or *Gaussian* distribution. However for data which are clearly asymmetrical and skewed, the choice is wider. For continuous positive data, the *Gamma* distribution provides a large family of shapes of probability distribution, or the *Beta* distribution can be used for positive data over a finite range between 0 and some given positive value *a*. For skewed count data, the *Negative Binomial* distribution may be appropriate. For example, data describing number of colony losses contains many zeroes, but may also have some rather high numbers lost. Various tests for goodness of fit can be used to see if any of these models can be clearly ruled out, but often the final choice is governed by considerations of convenience and mathematical tractability.

10.3.2. Loss calculations and Confidence Intervals

1. Regarding loss rates, rather than the raw numbers of colonies kept and number of colonies lost which are used in their calculation, different quantities are of interest. The overall loss rate is the proportion calculated as the total number of lost colonies in the sample of beekeepers divided by the total number of colonies at risk of loss in the sample. (VanEngelsdorp *et al.* (2013) refer to this as "total loss". As this suggests to us the total number of colonies lost rather than any kind of rate or proportion, we prefer the terms overall loss rate or overall proportion of colonies lost). Adjustments can be made to this calculation to take account of colony management (VanEngelsdorp *et al.*, 2012). The overall loss rate is influenced disproportionately by the larger beekeepers, who are fewer in number. Using this approach,

confidence intervals for proportions may be calculated. There are several ways to do this. Alternatively, the average loss rate is the average of the individual loss rates (number of colonies lost divided by number of colonies at risk) experienced by different beekeepers in the sample. Using this approach, confidence intervals should be those for an average, not a proportion. However, a difficulty of using the average loss rate is that the loss rates experienced by beekeepers with different sizes of operation are not equally variable, yet they are weighted equally in the calculation of this average. While the loss rates can only range between 0 and 1 (0 to 100%), larger scale beekeepers have many more colonies which can be lost, and can experience a much larger set of possible loss rates within this range; therefore, their loss rates are subject to greater variation. Also, there are many ties in the individual loss rates, for example due to the large number of beekeepers with no losses. The median individual loss rate could well be zero. Average individual loss rate is often higher than overall loss rate, owing to the larger number of small scale beekeepers present in many populations of beekeepers, who can suffer extreme individual loss rates. For this reason, the use of medians and Kruskal-Wallis tests to compare loss rates should be avoided. Owing to these various difficulties, we recommend use of the overall loss rate.

2. Another difficulty is that the usual procedure to calculate standard errors and confidence intervals for the overall loss rate (the proportion of colonies lost) is based on the binomial distribution, as the number of losses is limited by the number of colonies at risk. This assumes that each bee colony is lost or not independently of any other colony, and also that the probability of loss is the same for all colonies. Within apiaries, whether or not a colony is lost is likely dependent on whether or not neighbouring colonies are lost. Furthermore, the probabilities of losing a colony are likely to differ between beekeepers. One way to account for that extra source of variation in the data is to model the data using a generalisation of the binomial distribution. There are different ways to do this. One approach uses generalised linear modelling using a quasi-binomial distribution and a logit link function, and derives a confidence interval for the overall loss rate based on the standard error of the estimated intercept in an intercept-only model (see VanEngelsdorp *et al.* (2012) and below).

3. Another approach to calculating confidence intervals, when it is felt that formulae based on parametric models are not appropriate, is to use the nonparametric bootstrap approach, based on resampling the data (Efron and Tibshirani, 1994). This avoids the need to specify any particular model for the data. This is easy to implement in a software package such as R.

10.3.3. Loss rate per factor including stratification on the operation size

The loss calculations and confidence intervals described above can be used as a means to identify risk factors for colony loss, by looking for confidence intervals that do not overlap each other. Total loss of operations reporting or not reporting a particular management type (e.g. transport of colonies) can be compared using the chi-square test (as in VanEngelsdorp *et al.*, 2010, 2011, for example). The loss rates of operations grouped by factors presumed to be involved in colony mortality (starvation, high varroa infestation etc.) can also be compared. Of course, this analysis does not give any information on, or account for, interdependencies of different factors, for which model fitting is needed (as described below).

To account for known or obvious differences among beekeeping operations, a first stratification, for example on operation size, can be accomplished, by classifying operations as hobby, side-line or commercial. Alternatively the number of colonies per beekeeper can be used as a basis for stratification.

Depending on the size of the survey and cultural differences between the target populations, beekeeping operations can be split into three operation size classes, for example

- small operations (≤50 colonies),
- intermediate operations (51-500 colonies),
- large scale operations (> 500 colonies).

If the scale of beekeeping in the survey population is limited mainly to small and intermediate operations, the classes can be split further as:

- small hobbyist beekeepers (≤15 colonies),
- large hobbyist beekeepers (16-50 colonies),
- small-commercial beekeepers (51-150 colonies),
- larger-commercial beekeepers (150-500 colonies).

When comparing several operation size classes, a chi-square test can be used first to compare all size classes, and if the result of this is significant, it can be followed up by pairwise multiple comparisons, again using the chi -square test or a *z*-test of the difference in two proportions. In each such pairwise test, the significance level to reject the null hypothesis should be Bonferroni adjusted (i.e. divided by the number of tests being conducted) to reduce the rate of false rejections of the null hypothesis that operations of different sizes have equal rates of loss. It should be borne in mind that the chi-squared test and *z*-test assume independent observations and therefore have their limitations.

10.4. Advanced analysis; identification of risk factors by logistic regression

10.4.1. Logistic regression

Elementary analysis of the answers to the essential questions, regarding colony losses, can yield an estimate of the overall loss rate for the observations (beekeepers or operations) grouped together by a single factor (such as country, or involvement (or not) in commercial pollination). Comparing these loss rate estimates and confidence intervals for the loss rates can indicate differences between the groups and hence potential risk factors relating to the risk of colony loss. The overall loss rate is a problematic estimator when the contribution of multiple factors to the risk of loss has to be determined, since factor responses may be associated, not independent of each other. For example, commercial pollination is more common in certain countries than others. Larger scale beekeepers contribute more to the overall loss rate than smaller scale beekeepers.

A statistical approach that deals with the difficulties of overall loss rate and enables conclusions on how factors (bee race, pollination practices, size of operation, honey yield, location etc.) influence colony losses is regression analysis (see Zuur *et al.* (2009) and Pirk *et al.* (2013)). In regression analysis, the numerical outcome of the essential questions (number of colonies lost, number of colonies alive or the calculated population at risk) is linked to the factors through a linear model. In the analysis of bee colony losses, many of the response variables of interest are positively skewed (having a long tail to the right) and so generalized linear regression models (GZLMs) are appropriate. These models assume that the observations y_i arise independently from a specified family of probability distributions, and independent variables or factors $x_{j,i}$, $j=1,..., k$, are used to provide a set of linear predictors

$$\eta_i = \beta_0 + \beta_1 x_{1,i} + \cdots + \beta_k x_{k,i}$$

such that $g(\mu_i) = \eta_i$, where μ_i is the mean of y_i, and the β_i are model coefficients to be estimated. Using GZLMs requires the specification of an appropriate probability distribution for the response variable y and also an appropriate form for the link function g (Krzanowski, 1998; McCullagh and Nelder, 1983).

The dependent variable of interest, the loss rate, is binary in the nature of its components (the number of lost colonies divided by the number of colonies at risk makes up the loss rate). This property leads to models that use a binomial distribution for the dependent variable. Each colony can be regarded as a "Bernoulli trial" resulting in no loss or a loss (0 or 1 respectively), and the number of lost colonies for a beekeeper can be regarded as a "binomial trial" of a certain size n (total number of colonies at risk, or number alive before the winter rest period) with a certain probability (p) of any one colony being lost after winter (an "event") and probability $1-p$ of the colony being alive after winter (a "non-event"). If x is the number of events per beekeeper, then the binomial probability distribution describing the probability of x events has the formula

$$p(x, n, p) = \binom{n}{k} p^x (1 - p)^{n-x},$$

with the mean value of x given by np and variance of x by $np(1-p)$.

Groups of beekeepers or operations can be seen as series of binomial trials which vary in size, and also with different probabilities of an event, *p*. Hence it is of interest to model the probability of loss for (groups of) beekeepers or operations characterized by different values of the risk factors involved, such as country or operation size or migratory practice.

Probabilities cannot be used directly as a response variable in a classical linear regression model, as probabilities can only have values ranging from 0 to 1, whereas continuous response variables can have any value. The solution for this problem is moving from the probability to the "odds" (*p/(1-p)*) and calculating the logarithm of the odds, the "logit", to be used as the dependent variable. The first step, taking the odds, removes the boundary of 1 as the odds can have any positive value, while taking the logit in the second step removes the boundary of 0 as the logarithm can be negative (for odds less than 1). A probability of 50% has an odds of 1 and a logit of 0, with negative and positive logits corresponding to probabilities of less than and more than 50% respectively.

Generalized linear models of this nature are called *logistic regression models*, and can be expressed in the form

$$logit(p_i) = \ln\left(\frac{p_i}{1 - p_i}\right) = \beta_0 + \beta_1 x_{1,i} + \cdots + \beta_k x_{k,i},$$

where the β_i are model coefficients to be estimated and $x_{j,i}$, *j=1,..., k*, are the values of the *k* independent variables or factors used in the model for prediction of the log odds of loss for case *i*.

Substituting the values and the estimated parameters into the right hand side of the equation enables prediction of the log odds of an event for that beekeeper or operation or group of operations. If this gives a value *y*, then taking the inverse logit $e^y/(1+e^y)$ gives the prediction of the probability p_i itself.

Kleinbaum and Klein (2002), Hosmer and Lemeshow (2000) and Agresti (2002) give an in-depth explanation of the principles of logistic regression, their interpretation, and the construction of best fitting models.

When honey bee loss data are involved in the analysis, several specific characteristics of these data and their analysis have to be addressed, as are now described.

10.4.2. Dispersion in statistical models
For a binomial distribution, the variance *np(1-p)* depends on the mean *np*. When the variance in the observations is bigger or smaller than the expected variance, data are said to show over- or under-dispersion. Both types of dispersion are indicated by the goodness-of-fit tests of fitted models by the ratio of the residual deviance of the fitted model to the number of degrees of freedom, values appreciably larger than 1 indicating over-dispersion and values lower than1 indicating under-dispersion. Both types can strongly affect and invalidate model hypothesis testing (standard errors, confidence intervals and p-values). See Twisk (2010), Zuur *et al.* (2009), Hardin and Hilbe (2007) and Myers *et al.* (2002) for examples. Causes of under- or over-dispersion can be related to the frequency characteristics of the data, with relatively small and large beekeepers/operations present in

different numbers (heterogeneity of the sample population). An important assumption of a binomial distribution, namely independence of observations (independent Bernoulli trials), might be violated when losses are not independent (are clustered) through an unknown factor (i.e. effects of a certain location, incidence of pathogens) that cannot be used (properly) in the model.

When under- or over-dispersion are not reduced after using the most significant model factors derived from the data and/or stratifying available data according to binomial trial size, the solution is using a different distribution for the dependent variable. A suitable candidate is the quasi-binomial distribution, in which variance is characterised by adding an additional parameter to the binomial distribution, and hypothesis testing can be corrected for the extra-binomial variance. The form of the quasi-binomial probability distribution is:

$$p(x,p,\varphi) = \binom{n}{k} p(p + x\varphi)^{x-1}(1 - p - x\varphi)^{n-x}; \ x = 0,1,2,\dots,n \ ; \ 0 \leq p \leq 1; \ -\frac{p}{n} < \varphi < \frac{1-p}{n}$$

See the manual available online by Kindt and Coe (2005) for an excellent example of the use of a quasi-binomial distribution and its differences compared to the standard binomial distribution. An excess of zero values (no loss) can be a cause of over-dispersion. To investigate the relation between predictor variables and the presence of zero values (no loss), zero-inflation techniques can be used (for example, Hall (2000)).

10.4.3. Multilevel analysis
Clustering of losses results in over-dispersed data, but clustering might very well be a biologically relevant phenomenon. A method to investigate correlations between groups of observations is to perform multilevel analysis by means of fitting models that contain random effects (random effects models and mixed models). Classic examples of multilevel analysis include schools or hospitals as random factors in an analysis of dependent variables on the level of students or patients respectively. In the case of colony losses, suitable data levels for random effects are often spatial in nature, as colonies are clustered by beekeepers, beekeepers are clustered in regions or habitat types and the latter are clustered within countries.

See Twisk (2010) and Zuur *et al.* (2009) for practical application of multilevel analysis methods. Rodríguez (2008) is also useful. A good online resource for multilevel analysis can be found at the homepage of the University of Bristol Centre for Multilevel Modelling (at http://www.bristol.ac.uk/cmm/).

10.4.4. Software for logistic regression models
Logistic regression can be conducted using generally available statistical software. The software packages R, SAS, STATA and MLwiN are able to perform logistic multilevel regression. The latest version of SPSS (19) also has a mixed model procedure but has no option (at the time of writing) to use quasi-binomial distributions. For an evaluation of different software used for logistic multilevel analysis, see Twisk (2010).

10.4.5. Example of advanced analysis

The analysis below uses the Dutch data collected with the full 2011 COLOSS questionnaire, as an example of how to estimate overall loss rates, calculate confidence intervals and fit GZLMs. It uses the quasi-binomial family of GZLMs, to account for any extra-binomial variation in the data. It is a simple illustration of how model fitting can be done in R, with factors and covariates, rather than a procedure for determining a best fitting model. Guidance on model building may be found, for example, in Dobson (2002) and Zuur *et al.* (2009).

The data was "cleaned" prior to use to remove some inconsistent values. The "glm" procedure in R is sensitive to invalid values in the data, and will generate error messages rather than omit the cases with invalid data values, so it is best to deal with these before attempting model fitting (or any other kind of analysis). The analysis below uses the variables ColOct10 as the number of colonies kept at 1st October 2010, and Loss1011, the stated number of colonies lost over winter 2010/2011, rather than the calculated population at risk or calculated colonies lost. Even so, in one case Loss1011 was missing and in six other cases Loss1011 was greater than ColOct10, causing negative calculated values of a new variable, NotLost, the number of colonies surviving. In some cases, though not all, this was due to winter management (making in/decreases) of colonies. These few cases were also removed before carrying out the analysis shown below.

The analysis does not show all available options for the "glm" procedure. Several diagnostic plots are available, for example.

a. Calculation of overall loss rate and confidence interval from a null model (Boxes 12-14).

b. Fitting a GZLM with an explanatory term.

Box 12. Reading in and setting up the data for further analysis.

Read in the data, in this case in csv format:

>dutch<-read.csv("cleaner_dutchdata.csv", header=T,sep=",")

Check the first few rows and columns:

>dutch[1:5,1:6]

	Validity	COLOSSID2011	IDBeekeeper	Country	Region	City
1	0	6528	1426	167	NA	
2	0	6529	1607	167	NA	
3	0	6531	5048	167	NA	Den Hoorn
4	0	6532	5296	167	NA	Amsterdam
5	0	6533	5396	167	NA	Amsterdam

Load the data into the memory, so variables can be identified by name:

>attach(dutch)

Check the descriptive data for the variables:

>summary(ColOct10)

Min.	1st Qu.	Median	Mean	3rd Qu.	Max.
1.000	3.000	5.000	8.932	9.000	401.000

> summary(Loss1011)

Min.	1st Qu.	Median	Mean	3rd Qu.	Max.
0.000	0.000	1.000	1.905	2.000	67.000

Calculate a new variable, the number of colonies not lost (alive), combine this with the data set and check the descriptive statistics of this variable:

> NotLost<-ColOct10 - Loss1011

> dutch<-cbind(dutch,NotLost)

> summary(NotLost)

Min.	1st Qu.	Median	Mean	3rd Qu.	Max.
0.000	2.000	4.000	7.027	7.000	354.000

Box 13. Fitting a quasibinomial intercept-only model for estimation of the overall loss rate.

Estimate the overall loss rate by fitting the null (intercept only) model, omitting any missing values.
The overall loss rate is the predicted probability of loss.

```
> dutch.glm1<-glm(cbind(Loss1011,NotLost)~1,
+family=quasibinomial(link="logit"),data=dutch,na.action=na.omit)
```

```
> summary(dutch.glm1)
```

Call:

```
glm(formula = cbind(Loss1011, NotLost) ~ 1,
+ family = quasibinomial(link = "logit"),data = dutch, na.action = na.omit)
```

Deviance Residuals:

Min	1Q	Median	3Q	Max
-9.0730	-1.3851	-0.6480	0.8932	10.7613

Coefficients:

	Estimate	Std. Error	t value	Pr(>|t|)
(Intercept)	-1.30553	0.03732	-34.98	<2e-16 ***

Signif. codes: 0 '***' 0.001 '**' 0.01 '*' 0.05 '.' 0.1 ' ' 1

(Dispersion parameter for quasibinomial family taken to be 3.196163)

```
    Null deviance: 4979  on 1530  degrees of freedom
Residual deviance: 4979  on 1530  degrees of freedom
AIC: NA
```

Number of Fisher Scoring iterations: 4

```
> dutch.glm1$fitted.values[1]
        1
0.2132358
```

The overall loss rate 0.213 can also be calculated directly:

```
> overall_loss<-sum(Loss1011)/sum(ColOct10)
> overall_loss
[1] 0.2132358
```

Or it can be calculated as the inverse logit of the estimated coefficient (the intercept) of the model.
For this the inverse logit function of the bootstrap library is used:

```
> library(boot)
```

```
> inv.logit(coef(dutch.glm1))
(Intercept)
0.2132358
```

Box 14. Calculating a confidence interval for the overall loss rate, using results from Box 12.

To calculate the 95% confidence interval (CI) for the overall loss rate, the standard error of the null model
intercept is stored and used in the formula for the normal approximation interval or t-interval, and then the
inverse logit of the result is calculated. If using the t-interval, the value of df below is the residual degrees of
freedom from the model fitting above. The t-interval is recommended for smaller samples. While it makes
little difference for this data, it is used in the further analysis for greater generality.

```
> se.glm1<-0.03732
```

```
> inv.logit(coef(dutch.glm1)+c(-1,1)*1.96*se.glm1)
```

```
[1]    0.2012216    0.2257647
```

```
> inv.logit(coef(dutch.glm1)+c(-1,1)*qt(0.975, df=1530)*se.glm1)
```

```
[1]    0.2012125    0.2257746
```

The second step in model building is the use of explanatory variables. Explanation of the methods for evaluating model fit and determining optimal models is outside the scope of this document. For this example analysis, the variable Region is used. The region variable is one that is largely outside of the beekeeper's control, rather like pesticide use by farmers, yet for various reasons may be associated with the loss rate. In some countries, region may be a substitute for meteorological variables. Boxes 15 to 18 and Fig. 3 show the analysis.

Box 15. Fitting a quasibinomial GZLM with one explanatory factor.

The categorical predictor variable "Region" is added to the model by means of the as.factor command. A continuous predictor variable would be added in the same way, but omitting the use of as.factor(). Significant effects of several regions are found. The intercept corresponds to the first level of the factor, i.e. region 2073.
Note that there are large differences in the number of observations between regions (shown by the tabulation of Region below, giving frequencies for each of regions 2073 to 2086) so differences in loss have to be interpreted cautiously.

```
> region.glm1<-glm(cbind(Loss1011,NotLost)~ as.factor(Region),
+ family=quasibinomial(link="logit"),data=dutch,na.action=na.omit)

> region.summary<-summary(region.glm1)
>region.summary

Call:
glm(formula = cbind(Loss1011, NotLost) ~ as.factor(Region),
+ family = quasibinomial(link = "logit"), data = dutch, na.action = na.omit)
```

Deviance Residuals:

Min	1Q	Median	3Q	Max
-8.0276	-1.2511	-0.6123	0.8868	9.3824

Coefficients:

	Estimate	Std. Error	T value	Pr(>|t|)	
(Intercept)	-1.36639	0.12719	-10.743	< 2e-16	***
as.factor(Region)2074	0.41251	0.32822	1.257	0.20902	
as.factor(Region)2075	0.49789	0.18588	2.679	0.00747	**
as.factor(Region)2076	-0.16554	0.15226	-1.087	0.27713	
as.factor(Region)2077	0.52147	0.20211	2.580	0.00997	**
as.factor(Region)2079	-0.11394	0.18460	-0.617	0.53720	
as.factor(Region)2080	0.36701	0.15172	2.419	0.01568	*
as.factor(Region)2081	0.01721	0.18930	0.091	0.92759	
as.factor(Region)2082	0.08012	0.18101	0.443	0.65809	
as.factor(Region)2083	0.05483	0.19392	0.283	0.77742	
as.factor(Region)2084	-0.78537	0.39354	-1.996	0.04615	*
as.factor(Region)2086	-0.21667	0.19578	-1.107	0.26861	

Signif. codes: 0 '***' 0.001 '**' 0.01 '*' 0.05 '.' 0.1 ' ' 1

(Dispersion parameter for quasibinomial family taken to be 3.106299)

    Null deviance: 4895.9 on 1509 degrees of freedom
Residual deviance: 4734.8 on 1498 degrees of freedom
(21 observations deleted due to missingness)
AIC: NA

Number of Fisher Scoring iterations: 4
>table(Region)
Region

2073	2074	2075	2076	2077	2079	2080	2081	2082	2083	2084	2086
116	25	108	327	76	95	208	118	151	115	37	134

Box 16. Testing factor significance, and obtaining confidence intervals for log odds of loss per region.

To determine if the fitted model gives better prediction than the null model, the models
are compared by means of an ANOVA. In this case the model with the factor region is a significantly
better predictor of loss than the null model:

\>anova(region.glm1,test="F")

Analysis of Deviance Table
Model: quasibinomial, link: logit
Response: cbind(Loss1011, NotLost)
Terms added sequentially (first to last)

	Df	Deviance	Resid. Df	Resid. Dev	F	Pr(>F)
NULL			1509	4895.9		
as.factor(Region)	11	161.09	1498	4734.8	4.7143	3.912e-07***

Signif. codes: 0 '***' 0.001 '**' 0.01 '*' 0.05 '.' 0.1 ' ' 1

Odds, probabilities and corresponding CIs can be calculated for the factor levels. For example, to get
just the predicted loss rate for the region coded 2074:

\> predict(region.glm1, data.frame(Region=2074),type="response")

              1
    0.2781065

Or for all the regions, and requesting standard errors for calculation of confidence intervals:

\> values<-predict(region.glm1,data.frame(Region=levels(as.factor(Region)))),
+ type="link",se.fit=T)
\> logodds<-values$fit
\> lowerlim<-values$fit-qt(0.975, df= 1498)*values$se.fit
\> upperlim<-values$fit+qt(0.975, df= 1498)*values$se.fit

Approximate 95% CIs for the log odds of loss per region, given as the lower limit, log odds and
upper limit respectively:

\> cbind(lowerlim, logodds, upperlim)

	lowerlim	logodds	upperlim
1	-1.615869	-1.3663880	-1.1169065
2	-1.547393	-0.9538734	-0.3603536
3	-1.134385	-0.8685001	-0.6026148
4	-1.696136	-1.5319275	-1.3677194
5	-1.153013	-0.8449141	-0.5368147
6	-1.742775	-1.4803233	-1.2178717
7	-1.161645	-0.9993760	-0.8371073
8	-1.624211	-1.3491820	-1.0741526
9	-1.538904	-1.2862660	-1.0336283
10	-1.598711	-1.3115591	-1.0244077
11	-2.882287	-2.1517622	-1.4212369
12	-1.875024	-1.5830575	-1.2910912

Box 17. Obtaining confidence intervals for the odds of loss and the model coefficients.

Approximate 95% CIs for the odds of loss per region, given as the lower limit, odds and upper limit respectively:

```
> odds<-exp(logodds)
> cbind(exp(lowerlim),odds, exp(upperlim))
```

		odds	
1	0.1987178	0.2550265	0.3272907
2	0.2128020	0.3852459	0.6974296
3	0.3216197	0.4195804	0.5473785
4	0.1833908	0.2161187	0.2546871
5	0.3156840	0.4295943	0.5846074
6	0.1750340	0.2275641	0.2958592
7	0.3129710	0.3681091	0.4329611
8	0.1970670	0.2594524	0.3415871
9	0.2146163	0.2763006	0.3557140
10	0.2021570	0.2693997	0.3590091
11	0.0560065	0.1162791	0.2414152
12	0.1533513	0.2053463	0.2749706

Approximate 95% CIs for the odds ratios per region, relative to the reference region, can be obtained from 95% CIs for the coefficients in the model, which we find first:

```
>coeffs<-region.summary$coef[,1]
>se.coeffs<-region.summary$coef[,2]
>coeffs.lowerlim<-coeffs-qt(0.975, df=1498)*se.coeffs
>coeffs.upperlim<-coeffs+qt(0.975, df=1498)*se.coeffs
>coeffs.CIs<-cbind(coeffs.lowerlim, coeffs, coeffs.upperlim)
```

These are the CIs for the model coefficients:

```
>coeffs.CIs
```

	coeffs.lowerlim	coeffs	coeffs.upperlim
(Intercept)	-1.61586947	-1.36638799	-1.1169065
as.factor(Region)2074	-0.23130745	0.41251455	1.0563366
as.factor(Region)2075	0.13328402	0.49788793	0.8624918
as.factor(Region)2076	-0.46421217	-0.16553956	0.1331331
as.factor(Region)2077	0.12503197	0.52147392	0.9179159
as.factor(Region)2079	-0.47604280	-0.11393532	0.2481722
as.factor(Region)2080	0.06940126	0.36701199	0.6646227
as.factor(Region)2081	-0.35411884	0.01720602	0.3885309
as.factor(Region)2082	-0.27493680	0.08012204	0.4351809
as.factor(Region)2083	-0.32556165	0.05482889	0.4352194
as.factor(Region)2084	-1.55732512	-0.78537421	-0.0134233
as.factor(Region)2086	-0.60070769	-0.21666949	0.1673687

Box 18. Obtaining confidence intervals for odds ratios and probability of loss, per region.

```
>odds.ratios<-exp(coeffs)
>odds.ratios.CIs<-cbind(exp(coeffs.lowerlim), odds.ratios, exp(coeffs.upperlim))
```

The CIs for the odds ratios (excluding the baseline category) are as follows, given as the lower limit, odds ratio and upper limit respectively :

```
>odds.ratios.CIs[-1,]
```

		odds.ratios	
as.factor(Region)2074	0.7934955	1.5106115	2.8758163
as.factor(Region)2075	1.1425745	1.6452427	2.3690566
as.factor(Region)2076	0.6286302	0.8474363	1.1424020
as.factor(Region)2077	1.1331847	1.6845087	2.5040662
as.factor(Region)2079	0.6212369	0.8923157	1.2816806
as.factor(Region)2080	1.0718662	1.4434152	1.9437570
as.factor(Region)2081	0.7017916	1.0173549	1.4748125
as.factor(Region)2082	0.7596201	1.0834193	1.5452425
as.factor(Region)2083	0.7221217	1.0563598	1.5453021
as.factor(Region)2084	0.2106989	0.4559490	0.9866664
as.factor(Region)2086	0.5484234	0.8051960	1.1821901

Note that the CIs excluding 1 correspond to the significant regions in the summary.region model output.
95% confidence intervals and point estimates of the probability of loss for each region, given as the
lower limit, estimated probability and upper limit respectively:

```
> library(boot)

> prob<-inv.logit(logodds)

> cbind(inv.logit(lowerlim), prob, inv.logit(upperlim))
```

		prob	
1	0.16577531	0.2032040	0.2465855
2	0.17546308	0.2781065	0.4108740
3	0.24335270	0.2955665	0.3537457
4	0.15497064	0.1777118	0.2029886
5	0.23993910	0.3005008	0.3689289
6	0.14896082	0.1853786	0.2283112
7	0.23836856	0.2690641	0.3021444
8	0.16462490	0.2060041	0.2546142
9	0.17669472	0.2164855	0.2623813
10	0.16816191	0.2122261	0.2641697
11	0.05303613	0.1041667	0.1944677
12	0.13296150	0.1703629	0.2156682

The row numbers 1-12 of the above output correspond with the order of the region names in the Figure below, which shows the estimated probability and its 95% CI for each region. Some of the confidence intervals overlap each other, indicating that there is no significant difference between these pairs of regions in terms of probability of loss. However, significant differences between some groups of regions can be seen. Gelderland, Limburg, Zeeland and Zuid-Holland have a lower probability than Friesland and Groningen and Noord-Brabant.

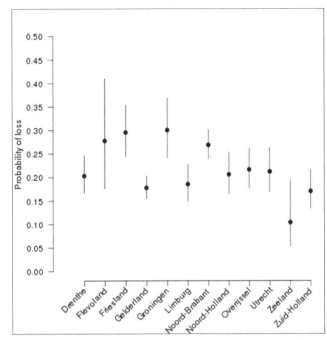

Fig. 3. Estimated probability of loss and 95% confidence interval per region.

11. Conclusions

Estimating colony loss rates reliably depends both on selecting representative samples of beekeepers and also on using suitable methods of estimation. Examining potential risk factors for losses also requires use of suitable statistical methodology. Standardisation of methodology will enable valid comparisons of loss rates to be made across time and between/within countries. Standardisation of terminology avoids confusion and facilitates the required comparisons.

In this manuscript, we have defined terminology associated with colony losses, and have presented the concepts involved in conducting a survey. The latter ranges from choosing the method of data collection to designing the questionnaire and how to select a representative sample and guidelines for choosing the sample size. Practical suggestions and examples are given.

We have examined many of the difficulties of conducting surveys and the important or most likely sources of error in surveys. Being aware of the potential for error makes it more likely that the survey organiser will be careful to avoid practices which are likely to introduce error into a survey, and therefore should achieve a more reliable result. We have also reviewed relevant methods for assessing the quality of the data and for statistical analysis, and have illustrated the more advanced techniques proposed for analysis.

12. Acknowledgements

The COLOSS (Prevention of honey bee COlony LOSSes) network aims to explain and prevent massive honey bee colony losses. It was funded through the COST Action FA0803. COST (European Cooperation in Science and Technology) is a unique means for European researchers to jointly develop their own ideas and new initiatives across all scientific disciplines through trans-European networking of nationally funded research activities. Based on a pan-European intergovernmental framework for cooperation in science and technology, COST has contributed since its creation more than 40 years ago to closing the gap between science, policy makers and society throughout Europe and beyond. COST is supported by the EU Seventh Framework Programme for research, technological development and demonstration activities *(Official Journal L 412, 30 December 2006)*. The European Science Foundation as implementing agent of COST provides the COST Office through an EC Grant Agreement. The Council of the European Union provides the COST Secretariat. The COLOSS network is now supported by the Ricola Foundation - Nature & Culture.

13. References

AAPOR (2011) *Standard definitions: final dispositions of case codes and outcome rates for surveys*. The American Association for Public Opinion Research. Available at: http://www.esomar.org/uploads/public/knowledge-and-standards/codes-and-guidelines/ESOMAR_Standard-Definitions-Final-Dispositions-of-Case-Codes-and-Outcome-Rates-for-Surveys.pdf

AGRESTI, A (2002) *Categorical data analysis (2nd Ed.)*. John Wiley and Sons, Inc., Hoboken, NJ, USA.

BRODSCHNEIDER, R; MOOSBECKERHOFER, R; CRAILSHEIM, K (2010) Surveys as a tool to record winter losses of honey bee colonies: a two year case study in Austria and South Tyrol. *Journal of Apicultural Research* 49(1): 23-30. http://dx.doi.org/10.3896/IBRA.1.49.1.04

CAMPBELL, M J; WATERS, W E (1990) Does anonymity increase the response rate in postal questionnaire surveys about sensitive subjects? A randomised trial. *Journal of Epidemiology and Community Health* 44: 75-76. http://dx.doi.org/10.1136/jech.44.1.75

CHARRIERE, J D; NEUMANN, P (2010) Surveys to estimate winter losses in Switzerland. *Journal of Apicultural Research* 49(1): 132-133. http://dx.doi.org/10.3896/IBRA.1.49.1.29

COX-FOSTER, D L; CONLAN, S; HOLMES, E C; PALACIOS, G; EVANS, J D; MORAN, N A; QUAN, P; BRIESE, T; HORNIG, M; GEISER, D M; MARTINSON, V; VANENGELSDORP, D; KALKSTEIN, A L; DRYSDALE, A; HUI, J; ZHAI, J; CUI, L; HUTCHISON, S K; SIMONS, J F; EGHOLM, M; PETTIS, J S; LIPKIN, W I (2007) A metagenomic survey of microbes in honey bee colony collapse disorder. *Science* 318: 283-287. http://dx.doi.org/10.1126/science.1146498

DAHLE, B (2010) The role of *Varroa destructor* for honey bee colony losses in Norway. *Journal of Apicultural Research* 49(1): 124-125. http://dx.doi.org/10.3896/IBRA.1.49.1.26

DE LEEUW, E D (2008) Choosing the method of data collection. In *E D De Leeuw, J J Hox, and D A Dillman (Eds). International handbook of survey methodology.* Lawrence Erlbaum Associates, Taylor & Francis Group; New York, USA. pp 113-135.

DE LEEUW, E D; HOX, J J; DILLMAN, D A (Eds) (2008) *International handbook of survey methodology.* Lawrence Erlbaum Associates, Taylor & Francis Group; New York, USA.

DOBSON, A J (2002) *An introduction to Generalized Linear Models (2nd Ed.).* Chapman & Hall / CRC; Boca Raton, Florida, USA.

EFRON, B; TIBSHIRANI, R J (1994) *An introduction to the bootstrap.* Chapman and Hall / CRC; New York, USA.

ELLIS, J D; GRAHAM, J R; MORTENSEN, A (2013) Standard methods for wax moth research. In *V Dietemann; J D Ellis; P Neumann (Eds) The COLOSS BEEBOOK, Volume II: standard methods for* Apis mellifera *pest and pathogen research. Journal of Apicultural Research* 52(1): http://dx.doi.org/10.3896/IBRA.1.52.1.10

GRAY, A; PETERSON, M; TEALE, A (2010) An update on colony losses in Scotland from a sample survey covering 2006-2008. *Journal of Apicultural Research* 49(1): 129-131. http://dx.doi.org/10.3896/IBRA.1.49.1.28

GRAY A; PETERSON, M (2012) SBA member surveys 2010/2011 (main results). *The Scottish Beekeeper* 89(12): 312-316.

GROVES, R M (2006) Nonresponse rates and nonresponse bias in household surveys. *Public Opinion Quarterly* 70(5): 646-675. http://dx.doi.org/10.1093/poq/nfl033

HALL, D B (2000) Zero-inflated Poisson and Binomial regression with random effects: a case study. *Biometrics* 56: 1030-1039.

HARDIN, J W; HILBE, J M (2007) *Generalized Linear Models and extensions (2nd Ed.).* Stata Press; Texas, USA.

HATJINA, F; BOUGA, M; KARATASOU, A; KONTOTHANASI, A; CHARISTOS, L; EMMANOUIL, C; EMMANOUIL, N; MAISTROS, A-D (2010) Data on honey bee losses in Greece: a preliminary note. *Journal of Apicultural Research* 49(1): 116-118. http://dx.doi.org/10.3896/IBRA.1.49.1.23

HOSMER, D W; LEMESHOW, S (2000) *Applied Logistic Regression (2nd Ed.).* John Wiley and Sons, Inc.; New York, USA.

KINDT, R; COE, R (2005) Tree diversity analysis. A manual and software for common statistical methods for ecological and biodiversity studies. *Nairobi: World Agroforestry Centre (ICRAF).* Available at http://www.worldagroforestry.org/downloads/publications/PDFs/B13695.pdf.

KLEINBAUM, D G; KLEIN, M (2002) *Logistic Regression: a self-learning text (2nd Ed.).* Springer-Verlag; New York, USA.

KRZANOWSKI, W J (1998) *An introduction to statistical modelling.* Arnold Publishers; London, UK.

LEHTONEN, R; PAHKINEN, E (2004) *Practical methods for design and analysis of complex surveys (2nd Ed.).* John Wiley and Sons Ltd; Chichester, UK.

LOHR, S L (2010) *Sampling: design and analysis (2nd Ed.).* Brooks / Cole, Cengage Learning; Boston, USA.

MUTINELLI, F; COSTA, C; LODESANI, M; BAGGIO, A; FORMATO, G; MEDRZYCKI, P; PORRINI, C (2010). Honey bee colony losses recorded in Italy. *Journal of Apicultural Research* 49(1): 119-120. http://dx.doi.org/10.3896/IBRA.1.49.1.24

MYERS, R H; MONTGOMERY, D C; VINING, G G (2002) *Generalized Linear Models: with applications in engineering and the sciences.* John Wiley and Sons, Inc.; New York, USA.

MCCULLAGH, P; NELDER, J A (1983) *Generalized Linear Models.* Chapman and Hall Ltd; London, UK.

NEUMANN, P; PIRK, C W W; SCHÄFER, M O; EVANS, J D; PETTIS, J S; TANNER, G; WILLIAMS, G R; ELLIS, J D (2013) Standard methods for small hive beetle research. In *V Dietemann; J D Ellis, P Neumann (Eds) The COLOSS BEEBOOK: Volume II: Standard methods for* Apis mellifera *pest and pathogen research. Journal of Apicultural Research* 52(4): http://dx.doi.org/10.3896/IBRA.1.52.4.19

NGUYEN, B K; MIGNON, J; LAGET, D; DE GRAAF, D C; JACOBS, F J; VANENGELSDORP, D; BROSTAUX, Y; SAEGERMAN, C; HAUBRUGE, E (2010) Honey bee colony losses in Belgium during the 2008-2009 winter. *Journal of Apicultural Research* 49(4): 337-339. http://dx.doi.org/10.3896/IBRA.1.49.4.07.

OTT, R L; LONGNECKER, M (2009) *An introduction to statistical methods and data analysis. (6th Ed.).* Brooks / Cole, Cengage Learning; Belmont, CA, USA.

PETERSON, M; GRAY, A; TEALE, A (2009) Colony losses in Scotland in 2004-2006 from a sample survey. *Journal of Apicultural Research* 48(2): 145-146. http://dx.doi.org/10.3896/IBRA.1.48.2.11

PIRK, C W W; DE MIRANDA, J R; FRIES, I; KRAMER, M; PAXTON, R; MURRAY, T; NAZZI, F; SHUTLER, D; VAN DER STEEN, J J M; VAN DOOREMALEN, C (2013) Statistical guidelines for *Apis mellifera* research. In *V Dietemann; J D Ellis; P Neumann (Eds) The COLOSS BEEBOOK, Volume I: standard methods for* Apis mellifera *research. Journal of Apicultural Research* 52(4): http://dx.doi.org/10.3896/IBRA.1.52.4.13

RODRIGUEZ, G (2008) Multilevel Generalized Linear Models. In *J De Leeuw; E Meijer (Eds). Handbook of multilevel analysis.* Springer; New York, USA. Chapter 9.

SAMUELS, M L; WITMER, J A; SCHAFFNER, A (2010) *Statistics for the life sciences (4th Ed.).* Pearson; USA.

SÄRNDAL, C-E; SWENSSON, B; WRETMAN, J (1992) *Model-assisted survey sampling.* Springer-Verlag; New York, USA.

SCHAEFFER, R L; MENDENHALL, W; OTT, L (1990) *Elementary survey sampling (4th Ed.).* PWS-Kent Publishing Company; Boston, USA.

SOROKER, V; HETZRONI, A; YAKOBSON, B; DAVID, D; DAVID, A; VOET, H; SLABETZKI, Y; EFRAT, H; LEVSKI, S; KAMER, Y; KLINBERG, E; ZIONI, N; INBAR, S; CHEJANOVSKY, N (2010) Evaluation of colony losses in Israel in relation to the incidence of pathogens and pests *Apidologie* 42: 192-199. http://dx.doi.org/10.1051/Apido/2010047

TOPOLSKA, G; GAJDA, A; POHORECKA, K; BOBER, A; KASPRZAK, S; SKUBIDA, M; SEMKIW, P (2010) Winter colony losses in Poland. *Journal of Apicultural Research* 49(1): 126-128. http://dx.doi.org/10.3896/IBRA.1.49.1.27

TWISK, J W R (2010) *Applied multilevel analysis. A practical guide (4th Ed.)*. Cambridge University Press; Cambridge, UK.

VAN DER ZEE, R (2010) Colony losses in the Netherlands. *Journal of Apicultural Research* 49(1): 121-123. http://dx.doi.org/10.3896/IBRA.1.49.1.25

VAN DER ZEE, R; PISA, L; ANDONOV, S; BRODSCHNEIDER, R; CHARRIERE, J-D; CHLEBO, R; COFFEY, M F; CRAILSHEIM, K; DAHLE, B; GAJDA, A; GRAY, A; DRAZIC, M; HIGES, M; KAUKO, L; KENCE, A; KENCE, M; KEZIC, N; KIPRIJANOVSKA, H; KRALJ, J; KRISTIANSEN, P; MARTIN-HERNANDEZ, R; MUTINELLI, F; NGUYEN, B K; OTTEN, C; ÖZKIRIM, A; PERNAL, S F; PETERSON, M; RAMSAY, G; SANTRAC, V; SOROKER, V; TOPLOSKA, G; UZUNOV, A; VEJSNÆS, F; WEI, S; WILKINS, S (2012) Managed honey bee colony losses in Canada, China, Europe, Israel and Turkey, for the winters of 2008–2009 and 2009–2010. *Journal of Apicultural Research* 51(1): 100-114. http://dx.doi.org/10.3896/IBRA.1.51.1.12

VANENGELSDORP, D; EVANS, J D; SAEGERMAN, C; MULLIN, C; HAUBRUGE, E; NGUYEN, B K; FRAZIER, M; FRAZIER, J; COX-FOSTER, D; CHEN, Y; UNDERWOOD, R; TARPY, D; PETTIS, J S (2009) Colony Collapse Disorder: A Descriptive Study. *PLoS ONE* 4 (8): e6481. http://dx.doi.org/10.1371/journal.pone.0006481

VANENGELSDORP, D; HAYES, J Jr; UNDERWOOD, R M; PETTIS, J (2010) A survey of honey bee colony losses in the United States, fall 2008 to spring 2009. *Journal of Apicultural Research* 49(1): 7-14. http://dx.doi.org/10.3896/IBRA.1.49.1.03

VANENGELSDORP, D; UNDERWOOD, R M; CARON, D; HAYES, J Jr; PETTIS, J S (2011) A survey of managed honey bee colony losses in the USA, fall 2009 to spring 2010. *Journal of Apicultural Research* 50(1): 1-10. http://dx.doi.org/10.3896/IBRA.1.50.1.01

VANENGELSDORP, D; BRODSCHNEIDER, R; BROSTAUX, Y; VAN DER ZEE, R; PISA, L; UNDERWOOD, R; LENGERICH, E J; SPLEEN, A; NEUMANN, P; WILKINS, S; BUDGE, G E; PIETRAVALLE, S; ALLIER, F; VALLON J; HUMAN, H; MUZ, M; LE CONTE, Y; CARON, D; BAYLIS, K; HAUBRUGE, E; PERNAL, S; MELATHOPOULOS, A; SAEGERMAN, C; PETTIS, J S; NGUYEN, B K (2012) Calculating and reporting managed honey bee colony losses. In *D Sammatro; J A Yoder (Eds). Honey bee colony health: challenges and sustainable solutions.* CRC Press / Taylor and Francis Group; Boca Raton, Florida, USA.

VANENGELSDORP, D; LENGERICH, E; SPLEEN, A; DAINAT, B; CRESSWELL, J; BAYLISS, K; NGUYEN, K B; SOROKER, V; UNDERWOOD, R; HUMAN, H; LE CONTE, Y; SAEGERMAN, C (2013) Standard epidemiological methods to understand and improve *Apis mellifera* health. In *V Dietemann; J D Ellis, P Neumann (Eds) The COLOSS BEEBOOK: Volume II: Standard methods for Apis mellifera pest and pathogen research. Journal of Apicultural Research* 52(4): http://dx.doi.org/10.3896/IBRA.1.52.4.15

VEJSNÆS, F; NIELSEN, S L; KRYGER, P (2010) Factors involved in the recent increase in colony losses in Denmark. *Journal of Apicultural Research* 49(1): 109-110. http://dx.doi.org/10.3896/IBRA.1.49.1.20

ZUUR, A F; IENO, E N; WALKER, N J; SAVELIEV, A A; SMITH, G M (2009) *Mixed Effects Models and Extensions in Ecology with R.* Springer; New York, USA.

Online Supplementary Material: Useful resources for survey design. (http://www.ibra.org.uk/downloads/20130805_1/download)

Journal of Apicultural Research 52(4)

Journal of Apicultural Research 52(4): (2013)
DOI 10.3896/IBRA.1.52.4.19

REVIEW ARTICLE

Standard methods for small hive beetle research

Peter Neumann[1,2]*, Jay D Evans[3], Jeff S Pettis[3], Christian W W Pirk[2], Marc O Schäfer[4], Gina Tanner[1] and James D Ellis[5]

[1]Institute of Bee Health, Vetsuisse Faculty, University of Bern, Bern, Switzerland.
[2]Social Insect Research Group, Department of Zoology & Entomology, University of Pretoria, Pretoria, South Africa.
[3]USDA-ARS Bee Research Laboratory, Beltsville, Maryland, USA.
[4]National Reference Laboratory for Bee Diseases, Friedrich-Loeffler-Institute (FLI), Federal Research Institute for Animal Health, Greifswald Insel-Riems, Germany.
[5]Honey Bee Research and Extension Laboratory, Department of Entomology and Nematology, University of Florida, Gainesville, Florida, USA.

Received 28 March 2013, accepted subject to revision 20 June 2013, accepted for publication 15 July 2013.

*Corresponding author: Email: peter.neumann@vetsuisse.unibe.ch

Summary

Small hive beetles, *Aethina tumida*, are parasites and scavengers of honey bee and other social bee colonies native to sub-Saharan Africa, where they are a minor pest only. In contrast, the beetles can be harmful parasites of European honey bee subspecies. Very rapidly after *A. tumida* established populations outside of its endemic range, the devastating effect of this beetle under suitable climatic conditions prompted an active research effort to better understand and control this invasive species. Over a decade, *A. tumida* has spread almost over the entire USA and across the east coast of Australia. Although comparatively few researchers have worked with this organism, a diversity of research methods emerged using sets of diverse techniques to achieve the same goal. The diversity of methods made the results difficult to compare, thus hindering our understanding of this parasite. Here, we provide easy-to-use protocols for the collection, identification, diagnosis, rearing, and for experimental essays of *A. tumida*. We hope that these methods will be embraced as standards by the community when designing and performing research on *A. tumida*.

Métodos estandar para la investigación del pequeño escarabajo de las colmenas

Resumen

Los pequeños escarabajos de la colmena, *Aethina tumida*, son parásitos y carroñeros de la abeja de la miel y otras colonias de abejas sociales nativas de África subsahariana, donde sólo son una plaga menor. Sin embargo, los escarabajos pueden ser parásitos dañinos de las subespecies de abejas europeas. Muy rápidamente después de que *A. tumida* estableciera poblaciones fuera de su área endémica, el efecto devastador de este escarabajo en condiciones climáticas adecuadas impulsó un activo esfuerzo de investigación para comprender y controlar

Footnote: Please cite this paper as: NEUMANN, P; EVANS, J D; PETTIS, J S; PIRK, C W W; SCHÄFER, M O; TANNER, G; ELLIS, J D (2013) Standard methods for small hive beetle research. In *V Dietemann; J D Ellis, P Neumann (Eds) The COLOSS BEEBOOK: Volume II: Standard methods for* Apis mellifera *pest and pathogen research. Journal of Apicultural Research* 52(4): http://dx.doi.org/10.3896/IBRA.1.52.4.19

mejor esta especie invasora. En más de una década, *A. tumida* se ha extendido prácticamente por todos los EE.UU. y en toda la costa este de Australia. Aunque comparativamente son pocos los investigadores que han trabajado con este organismo, surgió una diversidad de métodos de investigación utilizando diversas técnicas para lograr el mismo objetivo. La diversidad de métodos complica la comparación de los resultados, lo que dificulta la comprensión de este parásito. Aquí, ofrecemos protocolos fáciles de utilizar para la recolección, la identificación, el diagnóstico, la crianza, y para ensayos experimentales con *A. tumida*. Esperamos que estos métodos se adopten como normas estándar por la comunidad para el diseño y la realización de investigaciones sobre *A. tumida*.

蜂箱小甲虫研究的标准方法

摘要

蜂箱小甲虫（*Aethina tumida*）是源于撒哈拉以南非洲地区土著蜜蜂和其它社会性蜂群的一种寄生虫和食腐动物。在当地，它们危害极小。但是对于欧洲蜜蜂亚种，这种甲虫却是有害寄生虫。小甲虫在其原产地以外建立种群后，在适宜的气候条件下造成了破坏性的影响，迅速促进了针对了解和有效控制这一入侵物种的热点研究。十多年来，*A. tumida*几乎已经扩散至整个美国以及澳大利亚的东海岸。虽然针对这一物种的研究者相对较少，但是出现了针对同一研究目标利用不同技术手段的多样化研究方法。研究方法的多样性导致了研究结果难以进行比较，妨碍了我们对这一寄生虫的进一步认识。本章我们提供了有关*A. tumida*的样品采集、鉴定、诊断和实验研究的简易方法。我们希望这些方法今后能被业内当作针对*A. tumida*的研究设计和实施的标准。

Keywords: COLOSS, *BEEBOOK*, *Aethina tumida*, *Apis mellifera*, research method, protocol, small hive beetle, honey bee

Table of Contents

Table of Contents cont'd

1. Introduction

Aethina tumida (Coleoptera: Nitidulidae) was named and described in 1867 by Andrew Dickson Murray in the Annals and Magazine of Natural History London (Murray, 1867). He received two specimens originating from Old Calabar, on the west coast of Africa. Small hive beetles (SHB) are parasites and scavengers of honey bee and other social bee colonies native to sub-Saharan Africa, where they are considered to usually be a minor pest only (Lundie, 1940; Schmolke, 1974; Hepburn and Radloff, 1998; Neumann and Elzen, 2004; Neumann and Ellis, 2008). In 1996, SHB were discovered outside of their native range in colonies of European subspecies of honey bees in the southeastern USA (Elzen *et al.*, 1999; Hood, 2004). Since then, SHB introductions have been reported from a number of other countries (Neumann and Elzen, 2004; Ellis and Munn, 2005; Neumann and Ellis, 2008). In these new ranges, the beetles can be harmful parasites of colonies of European honey bee subspecies (Elzen *et al.*, 1999; Hood, 2004) and may also damage colonies of non-*Apis* bees such as bumble bees and stingless bees (Spiewok and Neumann, 2006a; Hoffmann *et al.*, 2008; Greco *et al.*, 2010; Halcroft *et al.*, 2011). Very rapidly after *A. tumida* established populations outside of its endemic range, the devastating effects of this beetle on honey bee colonies (given suitable climatic and soil conditions, Ellis *et al.*, 2004c) resulted in an active research effort to better understand and control this pest. Over a decade, *A. tumida* has spread and established almost over the entire USA and across the east coast of Australia (Neumann

and Ellis, 2008). Although comparatively few laboratories have worked with this organism, a range of different research methods emerged. Here we provide an overview on methods in the field and in the laboratory for experimental essays of *A. tumida*, which we hope will be embraced as standards by the community when designing and performing research on SHB.

2. Diagnostics and sampling

Small hive beetle research requires diagnostic as well as experimental approaches and involves a wide range of different techniques for laboratory and field trials. For both, diagnostics as well as experiments it is crucial to find and identify the SHB. Knowledge about beetle appearance and location in the colony during the different developmental stages is necessary to work with this pest.

2.1. Beetle appearance during the different developmental stages
2.1.1. Eggs

A. tumida eggs are pearly-white and about 2/3 the size of a honey bee egg (Fig. 1) and can usually be found in typical egg-clutches (Lundie 1940). Normally, eggs hatch in about three days, occasionally in up to six days (Lundie, 1940). In low relative humidity, egg hatching may be reduced (Somerville, 2003; Stedman, 2006).

Fig. 1. Small hive beetle eggs oviposited in clutches in a honey bee mating nucleus colony. They are ~1.4 mm in length and ~0.26 mm in width (Lundie, 1940). Photo: Peter Neumann.

2.1.2. Larvae

SHB larvae are creamy-white and emerge from the egg shell through a longitudinal slit at the anterior end (Lundie, 1940). After hatching, the larvae are about 1.3 mm long and grow to a final size of about 8.6 to 10.5 mm in length (Schmolke, 1974; Fig. 2). Larval development takes 10-14 days, but may extend to 30 days, depending on food resources and temperature (Stedman, 2006). They have three pairs of legs close to the head, a characteristic row of paired dorsal spines on each segment and two larger paired spines protruding sub terminal on the rear end of the dorsum which distinguishes it from the greater wax moth larvae (*Galleria mellonella*).

Fig. 2. Small hive beetle larvae. The post-feeding wandering stage is shown. Photo: Marc Schäfer.

2.1.3. Pupae

A. tumida pupae are free pupae (*pupa libera*), which means that their extremities such as legs and wing sheaths are, in contrast to *pupa obtecta*, not glued to the body (Fig. 3). Pupae in early stages (Fig. 3a) are pearly-white with a series of characteristic projections on the thorax and the abdomen, but darken as they develop (Fig. 3b) and their exoskeleton hardens, with pigmentation first appearing in the eyes and then the under wings (Lundie, 1940). Depending on soil temperature pupation time varies between 15 – 74 days (Neumann *et al.*, 2001a), but may take up to 100 days in cold periods (Stedman, 2006). During eclosion, the puparium bursts and the adult beetle emerges (Fig. 3c).

2.1.4. Adults

Adult *A. tumida* beetles (Fig. 3d) are about 1/3 the size of a honey bee worker. They are oval-shaped and vary in size ranging from 5 to 7 mm in length and 3 to 4.5 mm in width. The adults' bodies are broad and dorsoventral flattened. For a short period after emergence, they are coloured reddish-brown (Fig. 3c), but during maturation they darken to dark brown or even black (Fig. 3d). The head, thorax and abdomen are well separated. Their elytra are short, so that a few segments of their abdomens are visible, and their antennae are distinctively club shaped. The entire body is covered with short, fine hair.

Note: Caution must be taken not to confuse SHB with closely related Nitidulidae beetles. This is especially true for *Cychramus luteus*, which might be easily confused with not fully mature adult SHB (Neumann and Ritter, 2004, Fig. 4). In cases of doubt, species status should be confirmed based on definitive morphological characteristics (e.g. shape of ovipositor and antennae, coloration, shape of pronotum, length of elytra, and other defined attributes, Freude *et al.*, 1967).

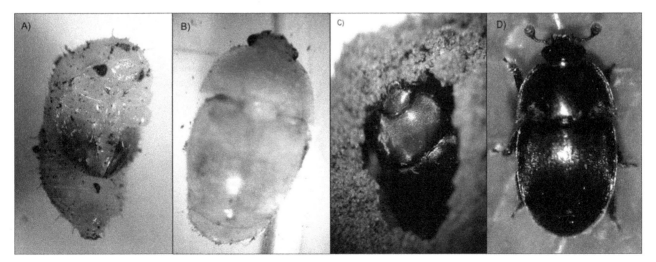

Fig. 3. Small hive beetle pupae and adults: **A.** early pupal stage; **B.** late pupal stage; **C.** adult ready to emerge from pupation chamber; **D.** adult on honey bee comb (dorsal view) with typical antenna and short elytra. Photos A-C: Anna Röttger; D: Nelles Ruppert.

Fig. 4. Adult beetles of the family Nitidulidae, which are associated with honey bee colonies: **a.** *Aethina tumida* ready to emerge from the pupation chamber. Photo: Anna Röttger;   **b.** *Cychramus luteus.* Photo: Frank Koehler.

The following SHB characteristics are most useful to distinguish them from other nitidulids in the field:

- Broadened ends of club shaped antennae are as long as wide
- Short wing cases: end of abdomen is visible
- Spiky edges on the lateral margins of the pronotum.

2.2. Beetle locations during the different development stages

2.2.1. Eggs

A. tumida eggs are mostly deposited in clutches (Fig. 1) that can be found anywhere within the hive. Females appear to prefer crevices or cavities for oviposition (Neumann and Härtel, 2004), however, eggs can also be found on combs or underneath sealed honey combs (Neumann and Hoffmann, 2008) or directly with the honey bee pupae underneath the cell cappings (Ellis *et al.*, 2003f; Ellis and Delaplane, 2008).

2.2.2. Larvae

All stages of larvae may be found crawling on and in the combs where they especially pierce the cappings and side walls of rather fresh combs. By contrast, older combs with several layers of cocoons seem to withstand penetration better. Larvae can also be found underneath sealed honey combs (Neumann and Hoffmann, 2008) and on the bottom boards provided that sufficient food can be found in the debris (e.g. dead adult bees, Spiewok and Neumann, 2006b). Fully grown, so-called wandering larvae (Fig. 5; Lundie, 1940) may also be found on the ground surrounding the infested colony, where they search for suitable earth to pupate, e.g. underneath hives. Larvae can also be spotted in apicultural facilities (e.g. honey houses) given low sanitary

standards or unsuitable storage conditions for (honey) combs or other material. Please note that SHB wandering larvae can crawl considerable distances to find suitable soil (> 200 m, Stedman, 2006). SHB larvae have rarely been reported outside of hives, apiaries and other apicultural facilities (e.g. occasionally in fruit buckets, Buchholz *et al.*, 2008).

Fig. 5. Small hive beetle wandering larvae underneath hives.

Photo: Peter Neumann.

2.2.3. Pupae

Pupation takes place in the soil surrounding the colony in individual chambers at shallow depth (seldom deeper than 20 cm, de Guzman *et al.*, 2009). Pupae can be carefully sieved out of the soil of a cavity (sieve: 20 x 20 cm; mesh width: 1 mm) and then be collected with tweezers or an aspirator (Pettis and Shimanuki, 2000).

2.2.4. Adults

In the hive, SHB are regularly found everywhere, where they can hide from bees or in areas of the hive with low bee densities, but seem to prefer bottom boards (up to 40% of the hive SHB population, Neumann and Hoffmann, 2008). Most SHB are prevented from freely roaming the hive because honey bees defend against SHB by chasing and corralling them into confinement sites (Neumann *et al.*, 2001c; Ellis, 2005). Beekeepers working a colony often open these confinement sites, which sets the beetles free and running over the hive material. Nevertheless, adult SHB are often notoriously difficult to spot during colony surveys due to their dark coloration and their tendencies to avoid sunlight and hide in corners or underneath material. Outside of the hives, adult SHB can often be spotted in honey houses and underneath hives, pending local beetle population density and sanitation efficacy of the respective beekeeper(s) (Spiewok *et al.*, 2007; 2008). Besides hives and apicultural facilities adult SHB have rarely been reported (e.g. in fruit buckets, Buchholz *et al.*, 2008).

2.3. Shipped bee samples

Rapid detection and identification of SHB is crucial to limit further distribution of this species. To prevent the beetle from being introduced to other areas or countries, imported queens, worker bees or honey bee colonies should be inspected carefully. In any case, suspect honey bee sample shipments from countries where SHB is known to be present should be immediately sent to the respective national reference laboratory or competent authority for examination and identification.

2.3.1. Inspection of honey bee queen shipments for small hive beetle

When receiving a shipment of honey bee queens that are to be inspected for presence of small hive beetle, the following methods can be used:

1. Remove the queen from queen cage and visually inspect her for signs of SHB (adult, larval, or egg stages; Fig. 1-3) or any other potentially introduced species, such as the mites *Tropilaelaps* spp. (see the *BEEBOOK* paper on *Tropilaelaps* (Anderson *et al.*, 2013)) or *Varroa destructor* (see the *BEEBOOK* paper on *Varroa* (Dietemann *et al.*, 2013)).
2. Store queen appropriately (e.g. in a colony), according to local procedures and regulations.
3. Immediately place the shipping container and queen cage, including the worker attendants, in a sealable bag and store in a freezer for at least 24 hours to kill attendants and any introduced species such as SHB or *Tropilaelaps* spp.
4. Perform a second visual inspection on the queen cage, the dead worker attendants, the shipping container and other accompanying materials as quickly as possible after the freeze period.

Note: A similar protocol can be applied if inspecting shipped worker honey bees.

2.4. Sampling of adult small hive beetles

For many experimental studies it is necessary to collect small hive beetles in the field. *Aethina tumida* can be sampled not only directly from honey bee colonies (Neumann and Hoffmann, 2008), but also from their surroundings (Pettis and Shimanuki, 2000). Several techniques can be used for the sampling of SHB.

2.4.1. Manual collection

For the manual collection of adult SHB, many devices are feasible to use (e.g. battery operated vacuums) and can be modified for SHB collection. However, the easiest and most common ones are aspirator devices and suction devices (e.g. Spiewok *et al.*, 2007; 2008; Neumann and Hoffmann, 2008, Fig. 6), which are widely used for collection of insects in general. Such aspirators consist of a container for the specimen and two tubes connected to the container. A longer tube allows suction by connecting the container to the collector's mouth, while a shorter tube guides a targeted insect into the container. The longer tube is covered with cloth mesh to prevent accidental aspiration of the specimen. As this method is very easy and requires no prearrangements, we recommend this method for any type of collection of SHB, but especially for collection at apiaries which are visited only once and where none of the following methods can be applied.

Note: It is highly recommended to clearly label the two tubes differently to avoid accidental inhalation of SHB and bees during collection.

2.4.2. Traps in colonies

Several devices to trap adult SHB inside colonies are available. All of them have entrances which allow SHB to enter but exclude honey bees. Most of these traps use mineral or vegetable oil as a killing agent. There are traps available for all positions in the hive: under the bottom board, on the bottom board, in the frame, as a replacement for a frame, between frame top bars, and at the entrance of the hive. Two very similar devices that are widely used are the "Beetle Eater®" and the "Beetle Blaster®". Both are designed to be placed between two frame top bars either in the bottom chamber or in supers. Most of these in-hive trapping-devices are designed to kill SHB and only allow an approximately measurement of the SHB infestation rate of colonies. For collection of live SHB for laboratory studies, we recommend a very simple method to catch adult SHB inside of the colonies using diagnostic strips (Schäfer *et al.*, 2008; 2010a). These devices are made of corrugated plastic strips (75 × 500 × 4 mm) which create rows of narrow tunnels for SHB to hide in, but prevent access of bees. Diagnostic strips have a mean capture efficacy of 30% of the colony SHB population, and provide a fast, cheap, and easy quantitative diagnosis for estimating SHB population sizes in the field (Schäfer *et al.*, 2008). Furthermore, in light of the very efficient detection rate (96.3%), these diagnostic strips are also a suitable tool

Fig. 7. Usage of diagnostic strips to detect adult small hive beetles. The strips are placed on the bottom board of a hive by sliding it through the flight entrance. Special care must be taken that the bottom boards are clean and even to avoid SHB hiding underneath the strips and not inside.   Photo: USDA.

Fig. 6. Manual collection of small hive beetle adults in a hive using an aspirator (for details please refer to Neumann and Hoffmann (2008)).
Photo: Dorothee Hoffmann.

to screen colonies in areas, where SHB have not been reported yet (Schäfer *et al.*, 2010a). This will foster a fast sanitation response to prevent SHB spread given accidental introductions. Please refer to the Portugal case for successful prevention of SHB spread (Murilhas, 2005) and to Australia/USA for consequences of too late detections (Neumann and Ellis, 2008). Usage is as follows (Schäfer *et al.*, 2008; 2010a):

1. Place one diagnostic strip on the bottom board of a colony by sliding it through the flight entrance (Fig. 7).
2. Leave the strips in the hives for two nights to provide SHB time to find the refuges.
3. Remove the strips by pulling them out quickly and shake the enclosed SHB into bright trays for counting or for collection with an aspirator or other suction devices (see section 2.4.1.).

Note: Caution must be taken with respect to both bottom board debris (boards should be clean to allow no space below the trap in which SHB can crawl) and ambient weather conditions. For example, under cold conditions sample SHB from bee clusters, where beetles cluster with the resident bees for warmth (Pettis and Shimanuki, 2000; Schäfer *et al.*, 2011).

2.4.3. Traps outside of colonies
2.4.3.1. Trapping of adult small hive beetles
A variety of different traps and baits have been used to capture adult SHB outside the hive (Elzen *et al.*, 1999; Arbogast *et al.*, 2007; Benda *et al.*, 2008; Buchholz *et al.*, 2008; Neumann and Hoffmann, 2008; de Guzman *et al.*, 2011). Plastic bucket traps with 8-mesh hardware cloth glued across 7 cm diameter holes (large enough for adult beetles to enter) have been used to trap SHB outside of colonies. Selection of an attractive bait is critical to capture of SHB outside the hive environment. While a mixture of honey, pollen, and adult bees showed the highest catch numbers in the field (Elzen *et al.*, 1999), fruits were not efficient (Buchholz *et al.*, 2008). Arbogast *et al.* (2007) used pollen dough inoculated with the yeast *Kodamaea ohmeri* (Benda *et al.*, 2008) as bait in traps made of 25.5 cm PVC pipe sections with a removable cap at each end. Two openings covered with 4-mesh screen allowed beetles to enter the trap.

An 18-mesh screen inverted cone, located just below the openings, funnelled SHB into the bottom cap through a small hole in the cone apex. Traps of this type positioned at 46 cm (same height as colony entrances) showed the highest catch numbers in the field. Although these traps generally do not catch high numbers of beetles, these traps allow for continuous observations that provide a relative measure of *A. tumida* migration (de Guzman *et al.*, 2011).

2.4.3.2. Trapping of small hive beetle larvae
To our knowledge, the wandering larvae trap (mounted on the entrance of the colony) of Arbogast *et al.* (2012) is the only published trap that captures wandering larvae at a colony scale. The trap

consists of two parts constructed of three-eighths inch (0.95 cm) acrylic plastic held together by catches. The lower part of the trap is water tight and half filled with a solution of detergent and water. The upper part intercepts the larvae and is covered, except for a 3 mm gap at the level of the bottom board. Larvae enter through this gap and fall through a screen (18 gauge stainless steel wire with 2 mm openings). The screen prevents bees from falling into the detergent solution. The trap is attached to the bottom board by two 18 cm extensions on the upper part. We recommend the trap primarily as a research tool for colony-wide SHB population dynamics.

Note: Please see the original publications for a drawing of the trap.

2.4.4. Storage and shipment of SHB samples

Storage and shipping conditions obviously depend on the planned subsequent analyses (e.g. for viruses see the *BEEBOOK* paper by de Miranda *et al.*, 2013). For morphometric analysis, the samples should be immediately preserved in 70-95% EtOH. This ensures that the specimens are suitable for this kind of analyses for at least a few months, and often considerably longer. Alternatively, for later usage in DNA analysis, the samples should be stored in a freezer at ≤ -20°C to slow down the degeneration of DNA in tissues. DNA from specimens frozen at -20°C remains viable for several years, but to remain viable for longer, samples should be stored at -70°C (for details please refer to section 9.3. in 'Storing dead adults' in the *BEEBOOK* paper on miscellaneous methods by Human *et al.*, 2013). If SHB samples are shipped for scientific purposes, it is important to note that no living material should be sent to avoid the further distribution of this honey bee pest. Adult beetles or larvae should be first killed by over-night freezing or by soaking the specimen in 70-95% EtOH, and sent in a sealed container.

2.4.5. Molecular diagnostics

Below we describe and reference tools for using modern molecular-biology techniques to diagnose SHB. They can also be used to gain knowledge about the introduction(s) of this beetle from Africa into and among the current ranges (Lounsberry *et al.*, 2010). This will elucidate pest populations and invasion pathways and contribute to knowledge of how this parasite expands in new populations (Lounsberry *et al.*, 2010). These tools will also enable future studies to better understand SHB behaviour, health, and other aspects of their biology. We here cover only DNA techniques.

2.4.5.1. DNA extraction

For rapid DNA analyses, including genotyping with the markers described below, an extraction step is necessary:

1. Place one leg from an adult SHB into a 1.7 ml centrifuge tube.
2. Grind with a disposable pestle.
3. Add 200 µl of 5% Chelex®-100 solution (Bio-Rad Corp.; Walsh *et al.*, 1991).

4. Vortex the solution.
5. Incubate for 15 min. at 95°C and 30 min. at 55°C before cooling.
6. Remove aliquots of 1 µl from this solution, which will suffice for genetic tests.

When removing these aliquots, care must be taken to avoid including the Chelex® resin, as this can inhibit the next steps.

Alternate extraction methods for long-term use of DNA (e.g. the 'CTAB' method and the commercial DNEasy kit from Qiagen) are also effective and are described in detail in the *BEEBOOK* paper on molecular methods by Evans *et al.*, (2013). Extractions of legs are advised since this avoids the heavily pigmented elytra, which can lead to inhibition of enzymatic assays and procedures.

2.4.5.2. Mitochondrial DNA analyses

The mitochondrion, a maternally inherited organelle, provides numerous genetic polymorphisms that can be useful for inferring sources for new SHB populations and for some small-scale studies of populations. Mitochondrial DNA (mtDNA) sequences for this species have been analysed for the gene encoding the Cytochrome oxidase I enzyme, a gene often used for DNA 'barcoding' (species-level and population-level identification) in insects. Evans *et al.* (2000) described two oligonucleotide primers, AT1904S (5'-GGTGGATCTTCAGTTGATTTAGC-3') and AT2953A (5'-TCAGCTGGGGGATAAAATTG-3') that amplify a ca. 1,000 base-pair region of the SHB mitochondrion. These can be amplified and analysed as below:

1. Amplify using a thermal cycling protocol of
   - 93°C (1 min.),
   - 54°C (1 min.),
   - 72°C (2 min.) for 35 cycles,
   - followed by a 5 min. elongation step at 72°C.
2. Prepare a 1% agarose gel in 1X TAE buffer by mixing:
   2.1. 10X solution of 48.4 g of Tris base (tris (hydroxymethyl) aminomethane),
   2.2. 11.4 ml of glacial acetic acid (17.4 M),
   2.3. 3.7 g of EDTA, disodium salt.
   2.4. Bring to a final volume of 1 l with deionized water.
   2.5. Mix 1 g agarose/100 ml gel buffer.
   2.6. Microwave in an Erlenmeyer flask for ca. 1 min.
   2.7. Swirl the liquids.
   2.8. Microwave again for *ca.* 1 min.
   Solution should reach a rapid boil and be fully dissolved.
   2.9. Let cool for 10 min.
   2.10. Add 1 µl of 10 mg/ml Ethidium bromide per 100 ml of gel volume.
3. Prepare sucrose loading buffer by mixing:
   - 4 g sucrose,
   - 25 mg bromophenol blue or xylene cyanol (0.25%),
   - dH$_2$O to 10 ml.

4. Mix 3 µl of each sample with 2 µl sucrose loading buffer.
5. Load each sample in an individual well.
6. Include a DNA size standard in a well alongside the products (e.g. the 100 base-pair ladder from Invitrogen).
7. Electrophorese the products at ca. 100 V/25 amps (depending on gel apparatus) in a horizontal gel rig.
8. Visualize under UV light.

There should be a single strong band at ca. 1,000 base pairs.

9. Purify the resulting products (e.g. PCR Purification Kit, Qiagen).
10. Sequence from each direction via standard Sanger dideoxy sequencing (see the *BEEBOOK* paper on molecular methods, Evans *et al.*, 2013) and the amplification primers.

Sequences can be aligned against numerous SHB haplotypes in public sequence databases (e.g., http://www.ncbi.nlm.nih.gov/gquery/?term=aethina+tumida) to determine novelty or placement into a described haplotype.

2.4.5.3. Microsatellite analyses

Microsatellites are regions containing tandem repeats of 1-six-nucleotide repeats that are widespread in eukaryotic genomes including that of the SHB (please refer to Evans *et al.*, 2013 for details). For example a CA_{12} microsatellite has 12 adjoining 'CA' nucleotides in the middle of non-repetitive DNA. These regions are inherently unstable, adding and clipping repeat units during cell replication and recombination. As a result, individuals often differ from each other in their genotypes at these loci, and they offer a powerful tool for describing populations and gene flow. Several microsatellite loci have been described for the SHB (Evans *et al.*, 2008) and these loci have proven useful in mapping the movement of SHB in the Americas (Lounsberry *et al.*, 2010). Below is a protocol for genotyping using dinucleotide (CA) microsatellites.

1. Prepare 5 µl reaction mixes for each sample and primer containing:
   - 1 Unit *Taq* DNA polymerase with appropriate 1X buffer (Invitrogen),
   - 1 mM dNTP mix,
   - 2 mM added $MgCl_2$,
   - 0.2 µM of each forward and reverse primer,
   - 0.08 µM of the Forward primer end-labelled on the 5' end with FAM or HEX fluorophores (ordered from Invitrogen, or another supplier).
2. Carry out the polymerase chain reaction (PCR) with a cycling program of
   - 96°C for 2 min., then
   - 3 cycles of 96°C for 30 sec.,
   - 60°C for 30 sec. (-1°C/Cycle),
   - 65°C for 1 min.,
     followed by 35 cycles of
     - 96°C for 30 sec.,

   - 56°C for 30 sec.,
   - 65°C for 1 min., and
   - a final extension at 65°C for 2 min.
3. Add 1 µl each PCR product (diluted 1:20) to 10 µl Formamide containing 1X LIZ size standard (Applied Biosystems, ABI).
4. Determine labelled product size via one capillary gel run using an ABI 3730 DNA machine and GeneMapper software (e.g. version 3.7, ABI)
5. Carry out population-genetic analyses with the programs GenAlex (Peakall and Smouse, 2006), STRUCTURE (Falush *et al.*, 2007) or other software programs designed to assess multi-allelic data, see section 3.3.2. on DNA microsatellites of the *BEEBOOK* paper on methods for characterising subspecies and ecotypes of *Apis mellifera* (Meixner *et al.*, 2013).

2.4.5.4. Screening hive debris for SHB

Since adult SHB show cryptic behaviour, they are notoriously difficult to spot in hives. Moreover, beetles are highly migratory (Spiewok *et al.*, 2008; Neumann *et al.*, 2012) and may have left the hive prior to inspection. We therefore recommend the following molecular method to screen imported hives for SHB (modified after Ward *et al.*, 2007):

1. Extract DNA from hive debris samples:
   1.1. Place ~10 g samples of debris into grinding mill canisters (Kleco, Visalia, California, USA).
   1.2. Add CTAB (Hexadecyltrimethylammoniumbromide, (Sigma)) lysis buffer (12% Sodium phosphate buffer pH 8.0, 2% CTAB, 1.5 M NaCl), (20 ml) containing 1% antifoam B (Sigma) to each canister.
2. Seal canisters.
3. Load onto Kleco grinding mill.
4. Ground for 2 min. at top speed.
5. Pour the lysate from each canister into a 50 ml Falcon tube.
6. Spin the tubes at 4,000 g for 5 min.
7. Remove 2 ml of the cleared lysate.
8. Place into a 2 ml micro-centrifuge tube.
9. Spin for a further 3 min. at 10,000 g.
10. Transfer cleared lysate (1 ml) to a fresh 2 ml micro-centrifuge tube.
11. Add 250 µl of lysis buffer B (Promega) and 750 µl of precipitation buffer (Promega).
12. Vortex.
13. Spin tubes at 10,000 g for 10 min.
14. Transfer clarified extract (750 µl) to a fresh micro-centrifuge tube.
15. Add 50 µl of kit MagneSil™ beads and 600 µl of isopropanol.
16. Incubate tubes at RT for 10 min.
17. Extract DNA from each sample using the robotic magnetic particle processor (Kingfisher mL, Thermo Labsystems) in conjunction with the Promega DNA purification system for food kit following the instructions of the supplier.

2.4.5.5. Future genetic work with SHB

The SHB is well poised for genomic studies of olfaction (finding colonies to exploit or finding other SHB for mating), reproduction, and symbioses with fungi and other microbes. Hence, analyses of genomic and transcriptomic (high-throughput sequencing of expressed genes) traits in the SHB are sure to follow. The methods for these analyses will be identical to those used previously for honey bees and other insects, and these methods are described in the *BEEBOOK* paper on molecular methods (Evans *et al.*, 2013).

3. Techniques for experimental investigations

The experimental work with SHB can basically be divided into two main categories; laboratory and field experiments. The following part of this manual includes recommendations and protocols of previously applied methods for various investigations.

3.1. Laboratory techniques

3.1.1. Maintaining adult SHB in the laboratory and transport

3.1.1.1. SHB diet

SHB feed and reproduce on honey, pollen, and most rapidly, on bee brood (Lundie, 1940; Ellis *et al.*, 2002b). A diet of bee brood alone is impractical, however, because (1) many colonies are needed to produce enough brood to sustain a SHB rearing program, especially a large one, (2) of the destructive nature of removing brood from a colony and (3) using brood is not economical (a full brood frame can be easily consumed by the offspring of 10-15 breeding pairs; Neumann *et al.*, 2001a).

Although other Nitidulidae are often reared on fruits (see Peng and Williams, 1990b), rearing SHB on fruits alone is impractical because of the beetle's low fecundity on such diet (Ellis *et al.*, 2002b; Buchholz *et al.*, 2008). We also recommend avoiding artificial diets (like that proposed for rearing multiple species of Nitidulidae, Peng and Williams, 1990a), because of the general expense of artificial diets, the difficulty one has in obtaining the ingredients, and to keep the diet of SHB as natural as possible. One successful semi-defined SHB diet consists of dry granulated pollen, honey, and a honey bee protein supplement (e.g. Brood Builder™, Dadant and Sons, Inc.; Hamilton, IL, USA or Booster bee Protein Feed™, Beequipment South Africa, Mike Miles) mixed together in a 1:1:2 volume ratio. The exact ratio varies depending on how moist the honey is:
To make the diet,

1. Add in a large stand mixer the three ingredients
   - 2000 ml of pollen,
   - 2000 ml honey,
   - 4000 ml protein supplement
2. Mix for about 20 min. until the mixture has a firm but pliable consistency.

This recipe makes enough for 20, 400 g sections of diet.

3. If sticky, add protein supplement to the diet mixture incrementally until it is no longer sticky.

The diet should not be sticky to the touch because larvae feeding on a sticky substrate are able to crawl up vertical surfaces (their bodies become sticky) and are difficult to collect..

4. Provide water *ad libitum*.

3.1.1.2. SHB maintenance and transport conditions

Whenever maintaining or transporting adult SHB individually or *en masse* for experimental purposes, it is recommended besides *ad libitum* water supply to avoid direct sunlight exposure, to keep temperatures in general low (RT) and to allow for sufficient air circulation, since unsuitable storage conditions can quickly result in death (Peter Neumann, unpublished data).

3.1.1.2.1. Mass maintenance

1. Provide autoclaved glass containers with food (see 3.1.1.1.) and water *ad libitum* (or not in case starving is required).
2. Place adult SHB in the glass containers.
3. Adjust SHB numbers according to experimental needs (single, pairs, groups).
4. Store containers at RT in darkness.

3.1.1.2.2. Individual maintenance

1. Provide 0.5 g food (see section 3.1.1.1.) in standard Eppendorf® 1.5 ml reaction tubes.
2. Place them in laboratory trays and puncture the lids with a needle (3-4 small holes) to allow air circulation and feeding.
3. Place individual adult SHB in the Eppendorf® tubes using tweezers and seal them.
4. Store trays at RT in darkness.
5. Provide to each tube honey : water in a 1:1 ratio daily by carefully pipetting via the small holes.

Notes: The SHB can be maintained for up to 8 weeks in darkness at RT. Take care to seal containers quickly to limit SHB escape. To enable ovary activation in female SHB, use protein-rich food (see section 3.1.1.1.) instead of honey : water in a 1:1 ratio. Ensure that water is always provided *ad libitum* to prevent SHB dehydration.

3.1.3. Rearing techniques

3.1.3.1. Rearing Procedure

The rearing procedure is summarized in Table 1.

1. Conduct all laboratory manipulations with adult SHB under a screened insect cage to prevent adult escape.
2. Place 25 sexually mature adult pairs (see section 3.1.4.) into a 3 l plastic mating container with ~400 g of SHB diet. No other

Table 1. Summary of SHB mass rearing procedure.

Number	Description	Maintained at:	Duration
1	50 adult SHB with 400 g diet in a plastic container (~3 l volume)	25°C, no light, > 80% humidity	2 weeks
2	Remove adults from container; SHB eggs, larvae, and food present	25°C, no light, > 80% humidity	2 weeks
3	Mature larvae + food placed on ridged tray (e.g.: 45 × 35 × 6 cm, l × w × h with 4 ridges 8 cm apart running the width of the tray)	25°C, no light, humidity needs to be high enough to keep larvae from desiccating	> 99% larvae will leave the food < 5 days and can be collected
4	80 ml mature larvae added on top of ~1.75 l sandy soil in a plastic container (25.5 cm height, 15 cm diameter, 3.7 l capacity); add another ~1.75 l sand to burry larvae	25°C, constant light for 1 week, then no light	20 days
5	Place funnel trap on pupation container	25°C, > 80% humidity, full light	In 2-5 days, > 80% of adults will be in top container
6	Remove funnel from trap and store adult SHB in top container	25°C, > 80% humidity, light optional	Adults will survive 8+ weeks. Adult mortality will increase over time

substrate, including wax comb or water should be placed in a mating container, because the inclusion of additional materials makes future larval collection difficult. Each container should be fitted with a lid ventilated with tiny holes.

3. The mating containers should be kept at 25°C, > 80% humidity with no light in incubators for 14 days to promote SHB reproduction and oviposition.

A high relative humidity is needed because SHB egg hatch rate is positively correlated with humidity (Jeff Pettis, unpublished data cited in Somerville, 2003).

4. Following the 2 week oviposition period, remove the adult beetles from the old diet with an aspirator, leaving the SHB eggs and larvae behind.

5. Transfer the adults in a new mating container with 400 g of beetle diet to continue the rearing program.

SHB diet should not be reused because old diet supports mould and fungus growth and may promote increased mortality in SHB larvae and pupae.

6. Move the mating containers from which the adult SHB were removed (they contain old diet, eggs, and developing larvae) to a second incubator for 2 weeks under similar conditions.

Larvae are thus allowed additional time to develop in the absence of adult beetles.

7. Following the 2 week larval growth period, place the larvae and food opposite of a hole (~5 cm) cut into the bottom at the far end of a tray that has raised (2 cm) ridges spaced 5-8 cm apart that run the width of the tray.

Larvae developing on the recommended beetle diet are not sticky and are unable to crawl up the tray walls (= dry rearing approach;

Neumann and Härtel, 2004). Any moisture added at this part of the procedure permits larvae to escape the tray.

8. Post-feeding mature larvae (wandering stage, Lundie, 1940) become positively phototactic, wander away from the diet in search of soil and fall through the hole on the opposite end of the tray. The ridges in the tray prevent the beetle diet from spreading over the tray and falling down the hole.

9. Place a second walled tray (same dimensions as aluminium tray) under the hole of the first to collect the falling wandering larvae.

10. Maintain the trays at 25°C, ~40% humidity, and no light for 5 days.

Placement of the larvae trays in a dark room encourages the larvae to crawl away from the diet when finished feeding. This takes advantage of a biological characteristic of the larvae, which normally leave bee colonies during the night. The larvae will remain hidden in the diet if maintained under constant light.

11. Collect the wandering larvae daily from the bottom tray as long as they are observed wandering from the food (usually up to 4 days).

Collected larvae are ready to pupate in the soil.

12. To facilitate pupation, half fill cylindrical plastic containers (pupation containers, 25.5 cm height, 15 cm diameter, 3.7 l capacity) with ~1.75 l sandy autoclaved soil (see section 3.1.6.2.) that is ≥ 10% moisture by mass.

13. Add about 2,000 wandering larvae (~80 ml larvae at 25 larvae/ml) to a pupation container and bury the larvae with ~1.75 l additional moist sand.

This number of individuals added to the chamber seems to create an optimal density of pupae in the volume of sand used (approximately 1.9 ml soil/pupa). If too many larvae are added, those that cannot create pupation chambers will crawl back to the surface and wander the soil in an attempt to leave the container.

14. The pupation containers should be maintained at 25°C, > 80% humidity and constant light for 1 week, followed by constant darkness for 13 days, until the adults begin emerging.

Notes for pupation containers:

- Use sandy soils when rearing SHB because sand is easier to sift through to expose and recover buried adult SHB if necessary. Most soil types will work; sand is easier to use. See Ellis *et al.* (2004c) for details.
- Soil should be discarded after one use or sterilized since there often is an increase in pupal mortality when unsterilized soil is reused.
- Never rely on wandering larvae to bury themselves naturally. If put on top of the soil, some larvae may not burrow into the ground for more than 2 weeks, thus widening the range of adult emergence. Manually-buried larvae emerge in a much narrower time period. Larvae will remain buried if the pupation containers are exposed to constant light. Wandering larvae may crawl to the soil surface if the containers are put in a dark room, thus widening the time range of adult emergence.
- Container depth (rather than width) appears to be important to pupation success.

15. Twenty days after burying the larvae in the pupation containers, fit the pupation containers with an inverted, similar container (adult beetle funnel trap) with a funnel pointing up into the mouth of the top container (Fig. 8).

The adult trap should be ventilated.

16. Loosely cover the funnel hole with a small strip of nylon mesh (1 × 3 cm) that can be secured with tape.

Adult beetles emerging from the soil crawl through the funnel into the top container where they are unable to go back down. The nylon mesh discourages the adults from returning to the pupation container once they have entered the top container.

17. Spray a 1:1 mixture (by volume) of honey/water with a hand-held pump sprayer through the ventilation holes into the container to feed the emerging adults.

Spray enough honey water to wet the side walls of the containers but not too much to promote pooling of the honey water on the bottom of the container.

18. Maintain the adults at 25°C and > 80% humidity under full light. The light encourages adults to move into the top container.

19. After 2-5 days, more than 80% of the emerging adult SHB will

Fig. 8. Small hive beetle pupation chamber with an adult beetle funnel trap fitted on top. There is an inverted funnel in the middle that beetles from the pupation chamber (container with soil) crawl through to access the adult chamber (top). The adults can be fed with honey water that is sprayed through the ventilation holes in the upper container. Photo: James Ellis.

be present in the top container. At this time, remove the funnel from the top container, invert the adult trap, and securely fasten its lid.

Since adult SHB often remain underneath the soil surface without emerging (up to 35 days, Muerrle and Neumann, 2004), they can be sifted out of the sand and collected using an aspirator. Adult SHB can be kept alive in these containers for more than 8 weeks by feeding them 1:1 honey water with a sprayer as previously described (see step 17).

Note: Since buried adult SHB can sexually mature before emergence, adult SHB may mate before emergence from the soil or before they are collected. Thus, rearing programs aimed at unmated adults should use the individual rearing approaches described below.

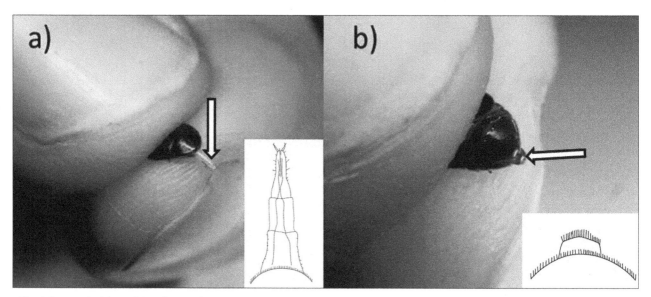

Fig. 9. Sexing of adult small hive beetles. Squeezing the abdomen will cause the female to extend her ovipositor (arrowed in *a.*) and the male to protrude his 8th tergite (arrowed in *b.*). The line sketches inset in both pictures are redrawn from Schmolke, 1974 and show the female ovipositor and male 8th tergite in detail.   Sketches: Kay Weigel (University of Florida); photos: Marc Schäfer.

3.1.3.2. Individual rearing approach

1. Fill standard 1.5 ml Eppendorf® reaction tubes 1/3 full with autoclaved pupation soil (see section 3.1.3.1.). Seal the tubes, place them in laboratory trays, and puncture the lids with a needle (3-4 small holes) to allow air circulation.
2. Collect wandering larvae from the rearing program (see section 3.1.3.1.).
3. Gently introduce an individual larva onto the soil of each tube using soft tweezers.
4. Cover the larvae with another 1/3 full with autoclaved pupation soil (see section 3.1.3.1.).
5. Seal the tubes.

Follow instructions for storage conditions of trays in incubators until adult emergence (see section 3.1.3.1.).

6. Transfer adults, which have finished pupation (Fig. 3c), into new individual tubes.

7. Provide with 1:1 honey water *ad libitum* using a pipette.
The SHB can be maintained for up to 8 weeks in darkness at RT.

Notes:

If any other diet is used to rear or maintain SHB (e.g. to test reproductive success (Ellis *et al.*, 2002b; Buchholz *et al.*, 2008)), the food should be frozen first, especially if field collected (e.g. fruit) to kill any other insect eggs that may be present.

3.1.4. Sex determination

It is often required to determine the sex of individual SHB for experimental or monitoring purposes. SHB seem to have a female-biased adult sex ratio both in the field and in laboratory rearing approaches (**field:** USA:

Ellis *et al.*, 2002a, Africa: Schmolke, 1974; Spiewok and Neumann, 2012; Australia: Spiewok and Neumann 2012; **laboratory:** Neumann *et al.*, 2001a; Ellis *et al.*, 2002b; Muerrle and Neumann, 2004).

3.1.4.1. Sex determination of adults

The easiest approach for adults has been developed by Schmolke (1974; Fig. 9).

1. Grasp the adult beetle so that the ventral side of the tip of the abdomen can be viewed.
2. Gently squeeze the beetle abdomen.
3. Squeezing the abdomen will cause the female to extend her ovipositor and the male to protrude his 8th tergite (Fig. 9).

Females readily extend their ovipositor without much squeezing. If you are squeezing "hard" with little result, the beetle likely is a male. It is more difficult to cause male beetles to protrude their 8th tergite so care must be taken to avoid harming the beetle.
If done gently, the sexed adults will not be affected adversely.

Note: When sexing many adult beetles (Ellis *et al.*, 2002b; Spiewok and Neumann, 2012), it is recommended to slow them down by placing them in a fridge or in a vial on ice for at least 5 min. (up to 10 min.).

3.1.4.2. Sex determination of pupae

The sex of pupal SHB can be determined visually without the need to manipulate the individual (Fig. 10).

1. Observe the ventral side of the distal section of the abdomen.
2. Look for two bulbous projections on this location.

If present (Fig. 10a), it is a female pupa. These projections are absent in male pupae (Fig. 10b).

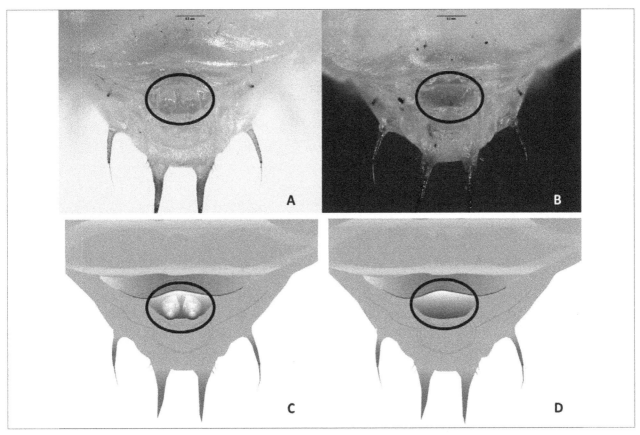

Fig. 10. Sexing pupal small hive beetles. **A.** and **C.** show the tip of the ventral side of the abdomen of a female small hive beetle. The two bulbous protrusions are circled. **B.** and **D.** show the tip of the ventral abdomen of a male small hive beetle. The bulbous protrusions are absent from the circled area. **A.** and **B.** are shown with enhanced contrast to highlight the genital area.

Sketches: Kay Weigel, University of Florida; photos Lyle Buss, University of Florida

3.1.5. Marking techniques

Experimental purposes often require cohort-specific or individual marking of SHB, e.g. for behavioural studies (see the *BEEBOOK* paper on behavioural methods by Scheiner *et al.*, 2013). Therefore various marking techniques employing dyes, food colouring, dusting, and thoracic notching have been attempted in SHB larvae and adults with different levels of persistence and mortality.

3.1.5.1. Marking SHB larvae

Coloured food is a good method to mark SHB larvae.

1. Dissolve 1 g of the dye (Sudan Blue 670, Sudan M Red, Rhodamine B; S. Spiewok, unpublished data) in 9 g olive oil to obtain a stock solution.
2. Stir 3 g stock solution into 57 g honey.
3. Mix the honey with 90 g powdered pollen to obtain dye-concentrations of 2,000 ppm.
4. Fill 10 g of the coloured food into small containers with lids ventilated with tiny holes that allow airflow but prevent escape of SHB larvae.

5. Place 10 young larvae (see section 3.1.3.1.) into each container and incubate for 8 days at 30°C till they reach the wandering stage.
6. Wandering larvae fed with Sudan Blue 670 will show green colouration. Sudan M Red results in reddish and Rhodamine B in violet coloured larvae.

Note: Feeding SHB larvae with food colours showed no decent results in the adults after metamorphosis (S. Spiewok, unpublished data).

3.1.5.2. Marking adult SHB
3.1.5.2.1. Feeding adult SHB with coloured food

1. Dissolve the dye (Rhodamine B, green and blue food colouring; S. Spiewok, unpublished data) in sugar water (40%).
2. Allow recently emerged, unfed SHB adults to feed for 3 days *ad libitum*.

Note: After two weeks in field colonies, SHB fed with green and blue food colouring were distinguishable from other SHB, but both colours came out turquois (S. Spiewok, unpublished data). Rose Bengal, fed to bees, was used successfully to identify the feeding interactions between SHB and honey bees (Ellis *et al.*, 2002c) but resulted in high mortality when fed to SHB directly (S. Spiewok, unpublished data).

3.1.5.2.2. Thoracic notching of adult SHB

For capture-mark-recapture studies of adult we recommend thoracic notching, as notched beetles (Fig. 11) survived and resulted in a high rate of recovery (de Guzman *et al.*, 2012).

1. Hold the SHB between two fingers under a stereo microscope.
2. Carefully notch the edge of either the right or left margin of the SHB's pronotum (Fig. 11) using the tip of an iris scissor (8 mm).

Note: With double notches or different cutting angles, individual marking may also be possible. De Guzman *et al.* (2012) also tested the use of blue and red chalk dusts to mark SHB, but this was not persistent and caused high mortality. For specific details please refer to the original publication.

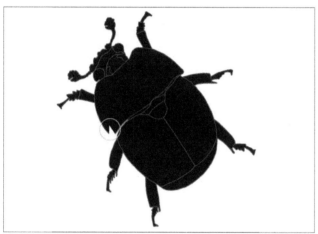

Fig. 11. Adult SHB marked by thoracic notching.   Drawing: FLI.

3.1.6. Effect of external conditions on SHB in the laboratory
3.1.6.1. Investigating diet effects on SHB reproduction

SHB feed on honey, pollen, and bee brood in honey bee colonies. However, they have been shown to reproduce on other diets as well (Ellis *et al.*, 2002b; Buchholz *et al.*, 2008). Below, we outline a method for investigating diet effects on SHB reproduction modified after Ellis *et al.* (2002b).

1. Laboratory rear (see section 3.1.3.2.) and sex (3.1.4.) adult male and female beetles individually to ensure that mating occurs only in the food containers. In reproduction studies, it is essential to know the ratio of males and females. Only use > 1 week old adults to ensure their reproductive maturity.
2. Place a SHB couple (one female and one male) into a plastic container (Fig. 12).

The container lids should contain small holes to allow sufficient ventilation.

3. Prepare at least 10 containers of beetles (10 replicates) per diet tested.
4. Add an appropriate amount of diet per adult container to accommodate the amount of offspring that will be produced.

Unfortunately, there is no established formula for the volume of food that should be added given an established number of adult beetles. The investigator will have to experiment with this since this is an important consideration. For example, SHB adults reproduce explosively on bee brood but considerably less so on fruit. Therefore, larvae can quickly exhaust food resources in some circumstances. The diet should be frozen first, especially if field collected (e.g. fruit) to kill any other insect eggs that may be present.

5. Move adult beetles from the food chambers to new ones every 3-4 days to alleviate problems associated with explosive population growth on limited food resources, if necessary.

Limiting the amount of time adults spend in a given container equally limits the number of eggs female beetles will be able to oviposit in the container, thus lessening the likelihood that food will be rapidly exhausted by larvae. This way, adults can be moved from container to container throughout their reproductive lifetime.

6. Monitor the containers daily to ensure that the beetles always have food available (only in case of *ad libitum* studies, otherwise adjust accordingly, e.g. let SHB starve).

In reproduction studies, larvae should be able to feed *ad libitum* if one is determining how diet impacts the number of larvae produced and amount of time it takes the larvae to reach the wandering phase.

7. Once larvae in the diet container reach the wandering phase, empty the diet and larvae into a metal tray to facilitate capture (see section 3.1.3.1.).
8. Place wandering larvae in soil pupation containers (see section 3.1.3.1.).
9. Keep all rearing containers at 25°C and > 80% humidity for one week under light and then until adult emergence under dark conditions (see section 3.1.3.1.).
10. Determine reproductive success.

Reproductive success is defined by Ellis *et al.* (2002b) collectively as the total number of offspring produced per female on a given diet (determined by counting the number of larvae produced in all of the food containers assigned to the adult pairs) and the percentage of those offspring that pupate successfully (= live adult SHB). One may also look at the reproductive capacity of the F1 generation for further evidence of reproductive success.

Notes:

- Allowing SHB adults to reproduce on wet or sticky diets is of special concern because the adults can drown in the food if it becomes fermented and begins to pool in the container. In these instances, it is advisable to place a piece of paper towel in the bottom of the container to absorb moisture.

- Larvae maturing on wet or sticky diets cause the container environment to become inhospitable in many instances. So, these containers should be well ventilated (with holes that larvae cannot traverse) to facilitate airflow and checked 2-3 times daily for any problems or inconsistencies.
- Field-collected adult beetles are **not** suited for determining maximum reproductive capacity because their age and mating status at the time of collection is not known and they may have already reproduced prior to the experiment.
- Manipulations with adult beetles should be done under insect nets to minimize beetle escape (see section 3.1.).
- The rearing containers should be provisioned with water (moistened cotton balls (Fig. 12) similar to maintain adult bees in cages see respective *BEEBOOK* paper Williams *et al.*, 2013), especially when the adults are attempting to reproduce on a dry food (such as pollen).
- Freezing the diet to kill eggs of other insects potentially present will alter its microbial balance. This could alter the diet's nutrition or attractiveness to SHB.

Fig. 12. Plastic container for maintenance of adult SHB (Photo: Elise Jeanerat) with standard food *ad libitum* (see section 3.1.1.1. pollen, honey, and a honey bee protein supplement in a 1:1:2 volume ratio), tap water in a small glass vial sealed with a piece of cotton wool to prevent draining, and two microscope slides as oviposition site (see section 3.2.4.). Food and equipment can be adjusted according to experimental needs (e.g. fruits instead of standard food and two or no oviposition sites).

3.1.6.2. Determining soil effects on SHB pupation
Ellis *et al.* (2004c) determined the impact of soil type, moisture, and density of SHB pupation success see section 3.1.3.1.). The methods they used are applicable to any study focusing on pupation habits of SHB.

1. Autoclave test soils prior to use to kill pathogens.
2. The soil moisture can be manipulated by first drying the soil in an oven.
   2.1. Spread the soil in a shallow, metal tray to facilitate drying. The drying oven can be set to 85°C. The soil is "dry" when the tray holding the soil no longer loses weight over a 24 h period. This takes 6 weeks or longer at 85°C. If decomposition of organic soil compounds is of concern to the research, the soils can be dried at lower temperatures for longer periods of time.
   2.2. Measure soil moisture as water by weight (either a dry mass or wet mass basis). For the dry mass gravimetric method, soil moisture = grams of water/grams of dry soil.
   2.3. Using this formula, create soils of various moistures by first drying the soil as outlined above, weighing it, and then adding an appropriate amount of distilled water to create the desired water by weight moisture level.
3. For each soil type of interest, place a desired amount of loose soil into plastic pupation containers (pupation chambers are better if they are tall rather than wide).

Ellis *et al.* (2004c) used cylindrical containers that were ~22 cm in height and 1,000 ml in volume. The containers should have small holes drilled in the bottom to allow water to exit.
4. Compact the soil while in the pupation chamber if soil density is an important component of the study.
   4.1. Run 1 l of distilled water through the soil-filled pupation container.
   The holes in the bottom of the container will allow the water to drain. The water compacts the soil naturally.
   4.2. If a predetermined amount of water by weight is desired in the packed soil, dry the packed soil while in the pupation container.
   The oven must be set at a temperature that does not melt or distort the plastic container, or decompose soil compounds/ nutrients (for the microbes, not the beetles themselves) if that is of interest. The container of soil must remain in the oven until it no longer loses weight over a 24 h period.
   4.3. Add the appropriate amount of distilled water to the dried, packed soil to achieve the desired water by weight moisture.
5. Place wandering SHB larvae produced according to standard rearing procedures (see section 3.1.3.1.) in the pupation containers once the soils are conditioned to need.

3.1.7. Transmission of bee pathogens by small hive beetles
Similar to other pests, SHB may act as vectors of honey bee pathogens (AFB: Schäfer *et al.*, 2010b; viruses: Eyer *et al.*, 2009a, b). This section describes experimental methods to investigate such pathogen transmission by SHB.

3.1.7.1. Investigating potential interactions between A. tumida *and* Paenibacillus larvae, *the causative organism of American foulbrood*

Brood combs with clinical American foulbrood (AFB) symptoms can be used to contaminate larval and adult SHB in the laboratory (Schäfer *et al.*, 2010b). This contamination was shown to persist in pupae and newly emerged adults. Contaminated adult SHB can be used to expose honey bee field colonies with *P. larvae* spores. The corresponding methods are outlined below:

1. Take clinical combs with sealed and unsealed brood from infected AFB colonies and arrange into plastic containers with an equal amount of infected brood cells each.
2. Collect adult SHB from rearing programs or from infested field colonies and introduce into the containers (N = 20).
3. Keep the containers in darkness at 30°C and high relative humidity (> 50%).
4. After seven days, collect contaminated adult small hive beetles.
5. Three days later, collect wandering larvae.

Both adults and larvae can be used depending on the needs of the experiment. If needed, larvae can be moved into sand containers placed in darkness in an incubator to allow further development (see section 3.1.3.).

6. To quantify the number of *P. larvae* spores per specimen (see respective *BEEBOOK* paper by de Graaf *et al.*, 2013), immediately freeze the samples after collection.
7. To expose field colonies to contaminated adult SHB, introduce the collected beetles into the experimental field colonies, which should be free of *P. larvae* spores.

Note: Please always consider the biosafety risks when manipulating *P. larvae* for research! (see the *BEEBOOK* paper on American foulbrood by de Graaf *et al.*, 2013). This method might also be practicable for European foulbrood and other brood diseases.

3.1.7.2. Investigating the potential of SHB to vector honey bee viruses

To investigate the potential of SHB as a vector of honey bee viruses, SHB can be fed with virus-infected workers or brood or with virus-contaminated wax or pollen. Eyer *et al.* (2009a, b) found no virus infections through feeding of contaminated pollen, but SHB became infected through feeding on infected worker or brood and on contaminated wax:

1. Maintain laboratory-reared adult SHB (see section 3.1.1. and Fig. 8 & 12) and provide virus contaminated food or other material.
2. Collect adult SHB after 6 days and store adequately until virus analysis (see respective *BEEBOOK* paper by de Miranda *et al.*, 2013).

3.1.8. Investigations of SHB pathogens and parasites

Knowledge about pathogens and parasites of SHB may offer alternative avenues for pest control and will in general contribute to our understanding of this species.

3.1.8.1. Inoculating SHB larvae with entomopathogenic fungi

Fungal infected cadavers (Ellis *et al.*, 2004e) or conidial suspensions (Muerrle *et al.*, 2006) can be used to infect SHB. Ellis *et al.* (2004e) achieved about 30% mortality among pupating beetles exposed as wandering larvae to beetle cadavers from which *Aspergillus flavus* and *A. niger* were isolated (see *BEEBOOK* paper by Jensen *et al.*, 2013):

1. Collect healthy-looking wandering SHB larvae produced using autoclaved soil and the rearing method (see section 3.1.3.).
2. Place desired amount of wandering larvae in a small plastic container (11 x 11 x 9 cm) that has SHB larvae/pupae that show signs of being infected with a fungal pathogen (Ellis *et al.*, 2004e). It is important to confirm that the cadavers are carrying the pathogen of interest using adequate microbiological techniques (Mürrle *et al.*, 2006 and the respective BEEBOOK paper (Jensen *et al.*, 2013)).
3. Allow the healthy larvae to wander in the container among the cadavers for 24 hours.
4. After 24 hours exposure to infected cadavers, place the wandering larvae in soil pupation chambers and allow to pupate (see section 3.1.3.).

3.1.8.2. Infesting SHB with nematodes

This method to infest SHB with nematodes was developed by Ellis *et al.* (2010). It allows for the testing of a variety of nematode species aiming at alternative control of SHB.

3.1.8.2.1. In vitro *infestation of wandering larvae with an aqueous suspension of nematodes*

Nematode infective juveniles (IJs) often are suspended in water by the company from which the nematodes are obtained. It is such suspension that the following method is based on to infest SHB larvae.

1. Place a piece of filter paper in the bottom of a petri dish. The filter paper should cover the bottom of the dish entirely.
2. Determine the density of IJs suspended in the water solution.
3. Add the desired number of IJs to the filter paper using a pipette. One should gently invert the container of IJs to ensure even dispersal of nematodes throughout the liquid. The total volume of water including the suspended IJs added to the filter paper should be 1.5 ml.

   For example, suppose that you are trying to infect wandering beetle larvae with 200 IJs per larvae. If you plan to add 10 beetle larvae to the petri dish, you will need to add 2,000 IJs to the dish. If you determine that there are 2,000 IJs per 0.5 ml of water, pipette 0.5 ml of the water/nematode suspension and 1 ml of distilled water to the paper for a total of 1.5 ml solution added to the paper.

4. Add desired number of wandering SHB larvae to the petri dish.

5. Replace the lid to the petri dish and secure the lid to the bottom using a tight rubber band. SHB larvae can push up the lid of a petri dish and escape so a rubber band is necessary to keep this from occurring.

6. Place the dishes in an incubator at 25°C and no light.

7. Assess mortality in the SHB larvae at any time period though resolution is greater the more often one views the dishes.

8. Experimental controls include adding nothing to the dish (no paper, water or nematodes), adding only filter paper, adding only 1.5 ml of distilled water, or adding filter paper and 1.5 ml of distilled water.

3.1.8.2.2. In vitro *soil infestation of larvae using a sand bioassay*

This method is similar to that presented above.

1. Rather than using filter paper in the petri dish, use 30 g of dried autoclaved sand.

2. Pipette the aqueous nematode suspension directly onto the sand, which is then wetted to about 10% water by weight.

3. Add wandering SHB larvae to the petri dish.

4. Replace and secure the top.

5. Place the dishes in an incubator at 25°C and no light.

6. Assess mortality in the SHB larvae at any time period though resolution is greater the more often one views the dishes.

7. Control petri dishes include adding nothing to the dish (no sand, water or nematodes), adding only sand, adding only 1.5 ml of distilled water, or adding sand and 1.5 ml of distilled water.

3.1.8.2.3. In vitro *soil infestation of pupae with an aqueous suspension of nematodes*

1. Autoclave soil.

2. Moisten soil to 10% water by weight.

3. Place autoclaved soil in small plastic cups.

Ellis *et al.* (2010) used 118 ml cups. A similar size is appropriate.

4. Place 5 wandering SHB larvae in the soil cups and allow them to bury themselves and begin the process of pupating.

5. Two days after the addition of larvae, pipette the desired amount of IJs suspended in an aqueous solution onto the soil in the cup.

6. Add 5 ml of water to the soil cups every 3-4 days as needed to maintain adequate soil moisture.

7. Place the cup lids on the cups to enclose the SHB.

One can expect the SHB in the control cups (no nematodes) to begin to emerge as early as day 17 post addition to the soil.

3.1.8.2.4. In vitro *soil infestation of pupae with nematode-infected cadavers*

Follow the method outlined in section 3.1.8.2.3. with one modification at step 5.

5. Rather than applying nematodes via aqueous suspension, add a SHB cadaver infected with the target nematode to the cup. Bury the cadaver about 0.5 cm below the soil surface. The cadaver can be produced according to the method outlined in section 3.1.8.1.

Notes:

- The methods outlined can be scaled up and modified for field use.

- Ellis *et al.* (2010) did this by burying the soil cups outside. The cups had screened lids and bottoms to allow rain drainage through the cups. Otherwise, nematodes and SHB larvae were added to the cups as in the *in vitro* assays. The methods also can be used to determine the generational persistence of the nematodes. This is accomplished by adding 5 additional SHB larvae to the soil cups weekly. If the nematodes are reproducing and moving into the newly-burrowing larvae, adult SHB emergence rates in the soil cups will be perpetually low.

3.2. Field techniques

The described field techniques involve a wide range of different protocols for investigations including interactions between honey bees and SHB, behavioural studies, quantification of SHB in colonies and more general techniques such as the introduction of SHB into honey bee colonies.

3.2.1. Investigating intra-colonial interactions between adult bees and adult SHB

SHB and adult worker honey bees interact with one another regularly within a colony (see Elzen *et al.*, 2001; Neumann *et al.*, 2001c; Ellis 2005; Ellis and Hepburn, 2006; Ellis *et al.*, 2003c, d; 2004a, b; Pirk and Neumann, 2013). Most of the studies on bee/beetle behavioural interactions have been conducted using honey bee colonies hosted in observation hives (see the *BEEBOOK* paper on behavioural methods, Scheiner *et al.*, 2013). To that end, we discuss initial experimental establishment and then various methods that can be used to investigate a suite of beetle/bee interactions. The series of procedures outlined below likely are amendable for use by those investigating the interactions between adult bees and larval SHB.

3.2.1.1. General experimental establishment

1. Establish equalized 3-frame observation hives (see the respective *BEEBOOK* paper Delaplane *et al.*, 2013) according to the parameters set for the study (one brood frame and two

honey frames; queenright or queenless, etc., see the *BEEBOOK* paper on behavioural methods by Scheiner *et al.*, 2013 for details on establishing and maintaining observation hives) and remove any beetles currently living in the colony using aspirators (see Fig. 6).

2. The observation hives should be kept in a dark room, with no outside or artificial light shining on the hive.

3. According to the needs of the experiment, collect adult SHB from rearing programs (see section 3.1.3.) or from infested field colonies. The beetles can be all males, all females, age-cohort specific, etc. per the needs of the experiment. Adult beetles should be sexed if required (see section 3.1.4.) and introduced into the observation hives after dusk, the preferred time window for SHB flights (Neumann *et al.*, 2012).

4. Hives should be monitored at least twice daily (Neumann *et al.*, 2001c; Ellis *et al.*, 2003c, d, 2004a, b) at time windows suitable for the experimental needs (morning and afternoon or only after local dusk for nocturnal behaviour). All night observations of the hives should be done using red lights to minimize behaviour disturbance to colonies. It may be necessary to use a small flashlight to find beetles confined in particularly hard-to-view areas. The order of observed hives should be altered at a daily basis at each time window.

5. During behavioural studies, it is essential to precisely define the behavioural categories/pattern. They have to be clearly defined and distinct; under no circumstance should the observer come into the situation that a certain observed behavioural pattern could be either behaviour 1 or behaviour 2. As a hypothetical example let's assume that two categories are labelled "walking" and "running". These categories would be examples of poorly defined behaviours, since they do not give a clear objective definition. The categories "walking, speed less than 5 mm/sec" and "running, speed greater than 5 mm/sec" would be clearly defined and are distinct. As another hypothetical example, the categories "feeding" and "trophallactic interactions" would be not helpful category choices since they overlap.

6. In most behavioural studies, it is important to know where the interaction took place, especially in a social insect colony like a honey bee colony and especially in the case of SHB/honey bee interactions. Therefore, it is useful to superimpose a transparent grid of squares onto both side windows of an observation hive to fix the location of the interaction. Obviously, the size of the square determines the spatial resolution. The general consensus is to use 5 x 5 cm squares (Moritz *et al.*, 2001; Neumann *et al.*, 2001b, 2003; Ellis *et al.*, 2003d), which then could be even further subdivided with slightly drawn internal 1 x 1 cm squares (Ellis *et al.*, 2003d). One has to avoid overloading the side windows with markings

to ensure that one can actually observe the behaviour. The labelling of the rows and columns should be included on the edge of the grid. To avoid confusion, the rows should be labelled with numbers and the columns with letters (or *vice versa*). It is highly recommended when preforming observations on both sides of an observation frame to continuously label the columns around both sides. For example with a 6 x 9 grid on each side, the rows are labelled 1-9 and the columns are labelled A-F on one side and G-L on the opposite side.

7. At each observation period, the observer should screen the colony in a left-to-right pattern, following the uppermost 5 × 5 cm bold square. This pattern should be followed with the second row of 5 × 5 cm bold squares, continuing to the bottom of the grid. This observation pattern minimizes the chance that the same area will be viewed more than one time and that behaviour will be double counted. Both sides of the observation hive should be monitored this way.

3.2.1.2. Behavioural categories

The following behavioural categories are a synthesis of the numerous studies (Elzen *et al.*, 2001; Neumann *et al.*, 2001c; Solbrig, 2001; Ellis 2005; Ellis *et al.*, 2002c; 2003c; 2004a, b; Pirk and Neumann 2013) and can be used to investigate interactions. The following behavioural categories were defined:

SHB:

Resting/Ignore (within 5 mm of a honey bee) — not moving at all, none of the other categories can be observed.

Walking — moving around (< 5 mm/sec. without being chased by a SHB or worker).

Running — moving fast (> 5-10 mm/sec.), without being chased by a SHB or worker.

Flee — moving fast (> 5-10 mm/sec.), while being chased by a SHB or worker.

Being chased — a) by fellow SHB, b) by a worker c) by more than one worker.

Mating — male is mounting the female

NOTE: in this definition of the behavioural category mounting is used as a proxy for mating.

Feeding on — a) pollen store, b) nectar store, c) larvae, d) a dead honey bee worker, e) a dead SHB, f) a live honey bee worker.

Antennating with a SHB — antennal contact with one or more a SHB.

Antennating with a honey bee worker — antennal contact with any bee e.g. with a guard bee (guarding the entrance of the hive (Rösch, 1925) or at a confinement site or prison (Neumann *et al.*, 2001c; Ellis, 2005)).

Trophallactic contact with worker bee — obtaining a drop of food from a bee that is presenting food between its mandibles.

Interfering with other SHB in trophallactic contact with worker bee — obtaining food while another SHB gets fed.

Turtle-defence posture — stays motionless and tucks its head underneath the pronotum with the legs and antennae pressed tightly to the body.

Mount a bee — mounting the workers' abdomen and cutting with the mandibles through the tissue between the tergites.

Honey bees:

There are more than 50 behavioural categories described for honey bees (see for example Kolmes, 1985; Neumann *et al.*, 2001b; Pirk, 2002). The following state only the basic ones related to SHB:

Resting/Ignore (within 5 mm of a SHB) — not moving at all, none of the other categories can be observed.

Walking — moves around at < 1 cm/sec.

Running — moves around at > 1 cm/sec.

Chasing — running and following an intruder (SHB).

Biting — mandibles are open and the bee tries to bit and grab the SHB.

Stinging — bends the abdomen to bring the stinger into position and/or sting is moving in and out of the sting sheath .

Trophallactic interaction with/feeding — a) another worker, b) queen, c) confined or SHB, d) one free running SHB – the worker regurgitates a droplet from its honey stomach, which can be observed between the mandibles and from which one or more food receivers will feed (SHB or bees).

Guarding — a worker prevents a SHB from moving freely by keeping it at bay in a crack or a prison/confinement. This kind of pattern is often followed by feeding and/or biting.

Building prison (confinement) — chewing or/moulding the wax-propolis mixture constituting the barrier of the prison (confinement).

Final note: All categories can be further subdivided or merged depending on the needs of the experiment.

3.2.1.3. Investigating behavioural interactions of adult honey bees and adult SHB at the colony entrance – two alternative options

Option 1 – at the entrance:
Two methods have been developed to investigate adult bee and SHB (or other nest invader) interactions at the hive entrance (Halcroft *et al.*, 2011; Atkinson and Ellis, 2011a):

1. Build a modified observation hive to facilitate nest entrance observation.

The entrance corridor to the hives should contain a test arena (Atkinson and Ellis, 2011a, used a test arena of 10 x 20 cm, L x W), with a floor marked with a 1 cm^2 grid system (Fig. 13 and 14).

2. Build a closable partition at both ends of the test arena using

acrylic glass or another material. Pierce the acrylic glass with holes to accommodate normal airflow into and out of the hive.

3. Build a side entrance to the test arena through which beetles can be introduced (Fig. 13 and 14).

4. Conduct the trials under red light conditions to take into account usual SHB flight activity after dusk (Neumann *et al.*, 2012).

5. Close both doors to the test arena to "trap" guard bees in place in the nest entrance. The doors should be closed slowly and with minimum disturbance in order to not excite the guard bees.

6. Introduce, for each trial, one adult SHB through the side entrance of the test arena.

7. Insert likewise a glass bead into the test arena as a control. The bead should be roughly the same size and colour as the invader. Atkinson and Ellis (2011a) tethered a 60 mg black bead to a 15 cm piece of monofilament fishing line so that the bead could be retracted from the test arena after the observation period.

8. Once introduced, record the responses of guard bees to the beetle or glass bead for any length of time though Atkinson and Ellis (2011a) recorded responses for 60 sec.

9. Three potential guard bee responses can be recognized (perhaps more can be discovered using this method):
   - Ignore (a bee's head comes within 5 mm of the subject but without making contact),
   - Contact/interacting (the bee physically contacts the subject in a non-defensive manner, which involves licking of the beetle and antennating), and
   - Defend (the bee attempts to sting and/or remove the subject from the nest, Elzen *et al.*, 2001, see section 3.2.1.2. on behaviour)

In the event that the test arena is not sealed well enough to prevent beetle escape, only trials in which the beetles remain in the arena for ≥ 30 sec. should be counted.

10. After the observation period, open the acrylic glass doors on either side of the test arenas for > 10 sec. to allow movement of honey bees into/out of the central nest area and to reduce guard bee agitation.

11. Allow time between trials for the beetles to naturally exit the test arena or for the bead to be withdrawn. In Atkinson and Ellis (2011a), the average time between trials was > 1 min.

12. Use test beetles and beads only once.

13. It is common to trap different numbers of guard bees in the test arena for each trial and colony when following this method. Consequently, convert bee responses to beetles or beads to proportional data to facilitate fair data comparison across trials. Atkinson and Ellis (2011a) state that this is not the proportion of bees performing a given response but rather the proportion of all responses that were Ignore, Contact, and

Defend responses. This way, a single bee may demonstrate these behaviours multiple times throughout the observation period and the behaviours be counted.

14. Analyse data as outlined in the statistical guidelines of the *BEEBOOK* (Pirk *et al.*, 2013).

Note: Trials should be conducted using multiple observation hives and simultaneously if resources permit. The latter allows one to minimize observation period and weather impacts on behaviour at the nest entrance.

Option 2: T-shaped arena entrances in an observation hive
This option was developed in the Social Insect Research Group at the University of Pretoria (Strauss, 2009). The advantage is that it does not interfere with the ongoing foraging activity of the colony and utilizes the observation, that any given natural colony often has more than one entrance. It further allows the manipulation of the intruder, in this case SHB, without having the risks of releasing guard bees. The T-shaped form (Fig. 13) offers two chambers to introduce intruders from more than one position and prevents any line of sight between the guards and the intruder before the experiments starts.

1. Attach the T-shape container to the hive.
2. Give at least three days for the guard bees to recognize the additional hive entrance.
3. Cover the arena with a glass lid to allow for observations and to prevent SHB and honey bees from escaping.

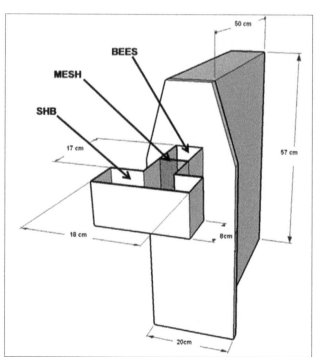

Fig. 13. Observation hive with t-shaped container as a testing area attached to it. Figure with permission from the author.

Drawing: Ursula Strauss.

4. Superimpose a grid on the glass if spatial information has to be recorded (see section 3.2.1.1.).
5. Insert a piece of metal mesh (± 9.5 x 4.9 cm) into the wooden box as a barrier to separate the intruder/SHB and honey bees. Instead of wood also polycarbonate could be used (Köhler *et al.*, 2013). Holes in the mesh are only large enough for SHB to move through, thereby preventing honey bees from moving into the part of the container where the SHB were released. In this design, SHB have access to both the hive and container whereas honey bees only have access to the hive and the part of the container closest to the hive entrance.
6. To optimize the observations and to reduce the influence of the observer, record the interactions using a remotely controlled CCTV system.

The recorded footage can be analysed using software designed for behavioural studies.

Advantage of option 2 is that the setup is not interfering with the gas exchange of the colony, normal activity is unaffected, and therefore observations during the day are possible. In addition, ample of space for the observer or additional equipment is available, so one does not have to squeeze between the exit hole and the observation hive.

Note: Depending on the questions asked both options have their advantages or disadvantages. If one needs constant flow of foragers coming into contact with the SHB option 1 might be better. If the experiment should not interfere with the normal activities of the colony option 2 might be more suitable.

3.2.2. Investigating SHB oviposition behaviour and bee hygienic responses to SHB eggs and young larvae

Adult SHB females will oviposit in cracks and crevices around the honey bee nest (Fig. 1, Lundie, 1940; Neumann and Härtel, 2004). Occasionally, the females will bite holes in the capping and/or side of brood cells, insert their ovipositor into the hole, and oviposit on the bee developing in the cell (Ellis *et al.*, 2003a; 2003c; 2004d; Ellis and Delaplane, 2008). In response, adult worker honey bees can detect SHB eggs/young larvae present in capped brood cells, uncap the cell, and remove the brood and/or SHB eggs/larvae (termed hygienic behaviour). A number of methods have been developed to facilitate studying SHB oviposition behaviour and the resulting honey bee hygienic responses to the behaviour.

3.2.2.1. Promoting SHB oviposition behaviour in capped bee brood

The idea of this method is that SHB adults are trapped and allowed to reproduce on a section of brood.

1. SHB used in the study can be laboratory-reared, field-collected, all males, all females, age-cohort specific, etc., per

the needs of the experiment. They should be at least 1 week old to ensure that they have reached sexual maturity.

2. Remove a frame of capped brood (~60-90% capped) from a colony.

The capped brood should not contain any uncapped larvae as these will crawl out of the cells and be macerated by the SHB in this procedure. Any uncapped larvae in the brood patch should be removed with forceps. It is best to use a section of brood that does not contain any honey stores. Furthermore, the selected brood should be > 6 days from enclosing (determined by uncapping and examining select brood cells in the test area, see the section on obtaining workers and brood of know age in the *BEEBOOK* paper on miscellaneous methods, Human *et al.*, 2013) so that no bees will emerge during the test period.

3. Prepare a sheet-metal, push-in cage (10 x 10 x 2.5 cm cage, L x W x H).

The face of the cage should be screen mesh to allow for ventilation. The screen mesh should be too small for SHB or bees to traverse.

4. Trap the required number of SHB in small vials.

5. Place the vials on ice 4-5 min.

6. Once anesthetized, dump the beetles onto the brood section. Trap the beetles on the section by pushing the cage onto the section and into the wax to the comb midrib.

7. Return the frame containing the beetles caged on a section of brood to the centre of the bee cluster in the colony from which the brood was removed.

8. Allow SHB to mate and the females to oviposit for 24 hours.

Though the length of time for this can vary, the beetles may not oviposit in the brood cells sufficiently if allowed less time. In contrast, they decimate the brood section beyond use if allowed access longer than 24 hours. This is especially true if > 20 adult SHB are added to the comb section.

9. After the oviposition period concludes, remove the combs from the colony

10. Collect the cages and beetles.

Note: It is important to know that some adult SHB may exit the cage during the oviposition period unless the cage is pushed firmly into the comb. Likewise, SHB already present in the colony at the time of the study can migrate into the cage. These situations usually only present a problem if one is trying to investigate the oviposition behaviour of an exact number of beetles.

3.2.2.2. Identifying and marking brood cells in which SHB have oviposited

1. Capped brood cells in which SHB have oviposited naturally or per the methods outlined in section 3.2.2.1. contain perforated cappings or side walls of the cell. The former is easy to note though the latter often requires a keen eye and

the use of a flashlight.

2. Mark capped brood cells exposed to SHB and containing perforated cappings and/or side walls using a transparent sheet of plastic (acetate).

3. Cut the acetate to about the size of the face of the frame.

4. Mark the sheet with a permanent marker in one of the corners of the frame to identify its correct placement on the comb in subsequent observations.

5. Label the sheet according to frame and colony if several are used for replication.

6. Once the acetate is stable on the frame, use a permanent marker to place a small dot on top over all brood cells containing perforated cappings and/or side walls.

7. Once marked and labelled, remove the acetate sheet from the frame and store until needed.

3.2.2.3. Determining honey bee hygienic responses to SHB eggs and young larvae.

1. Following the methods outlined in sections 3.2.2.1. and 3.2.2.2., expose reproductive SHB adults to brood comb. They will breed, and the females might oviposit in capped brood cells.

2. Mark the cells (see section 3.2.2.2.).

3. Use a pin to put small holes in brood cell cappings as a positive control (Ellis *et al.,* 2004d).

When pin-pricking, the holes should be made around the cap perimeter to avoid damaging the developing bee within the cell.

4. Mark capped brood cells with no perforations in the cappings and/or side walls as negative controls (Ellis and Delaplane, 2008).

5. Return the frame containing the perforated brood cells to the centre of the nest cluster.

Consequently, the brood section returned to the colony in step 3 will be investigated and attended to by worker bees. Honey bee workers that possess heightened hygienic responses can detect SHB eggs/ young larvae present in capped brood cells, uncap the cell, and remove the brood and/or SHB eggs/larvae (Ellis *et al.*, 2003a, c; 2004d; Ellis and Delaplane, 2008).

6. Leave the frame containing the test section of brood in the nest for a predetermined amount of time.

Ellis *et al.* (2003c) left the brood in the colonies post exposure to SHB for 48 hours. However, worker bee hygienic response to the capped brood neared 100% in all test colonies so Ellis and Delaplane (2008) lessened the brood time in the colony to 24 hours in subsequent tests to look for colony differences in hygienic responses.

7. After the 24 hour period, remove the frame from the colony.

8. Replace the marked acetate (see section 3.2.2.2.) on the corresponding frame and align to the section of brood.

9. Consider hygienically removed the marked brood cells having

perforated cappings/side walls the day before that no longer contained a developing bee.

10. Calculate the level of hygienic behaviour (or proportion of brood removed) as

$$\frac{\text{\# marked brood cells containing no developing bee after 24 hours exposure to adult bees}}{\text{total number of marked cells containing perforated side walls and/or cappings}}$$

The higher the proportion, the more hygienic the colony.

3.2.2.4. Determining proportion of perforated capped brood cells containing SHB eggs (oviposition rate)

1. Use the method outlined in section 3.2.2.3. to determine the oviposition rate of SHB females in capped brood.
2. A female SHB will not oviposit in all cells in which she perforates the capping and/or side wall. Consequently calculate the "oviposition rate" as:

$$\frac{\text{\# marked capped brood cells containing SHB eggs}}{\text{total number of marked capped brood cells containing perforated cappings and/or side walls}}$$

3. Open the cells marked according to section 3.2.2.3. with forceps.
4. Determine the number and presence of SHB eggs.
5. Remove the developing pupa/prepupa to facilitate egg quantification.

Potential additional uses:

The procedures outlined in sections 3.2.2.1. – 3.2.2.4. can be used to:

- investigate beetle density effects on oviposition (simply vary the number of SHB in the cages during the oviposition period),
- screen and possibly select for the level of hygienic expression of honey bees (within and between subspecies) toward SHB eggs and young larvae,
- determine time, environmental, bee subspecies, etc. impacts on SHB oviposition behaviour,
- determine colony strength impacts on hygienic behaviour (Ellis and Delaplane, 2008),
- and investigate other similar areas.

3.2.3. Determining the number and distribution of adult SHB inside a field colony and winter clusters and starting colonies without SHB

To estimate infestation loads of colonies and preferential SHB locations, it is essential to quantify and locate the beetles adequately. Some experiments also require beetle-free colonies. Here we provide any overview on the respective methods.

3.2.3.1. Visually screening the number of beetles in live colonies and/or removing beetles to start beetle-free colonies

These methods are modified from Ellis et al., 2002a; Ellis and Delaplane, 2006; Spiewok et al., 2007; 2008; Neumann and Hoffmann, 2008. The procedure is best accomplished with two people, one to work the colony and the second to collect the beetles.

1. Place a sheet of opaque plastic (~2 x 2 m, preferably white or light in colour) or plywood in front of the colony in which you want to count the number of beetles.
2. Lightly smoke the colony.
3. Remove the lid from the colony.
4. Bounce the lid on the plywood. This dislodges all bees and beetles adhering to the lid.
5. A second individual (the beetle collector) combs through the bees by hand or with a small stick and collects all visible adult beetles with an aspirator. All bees on the plywood should be inspected since beetles can easily be concealed in clusters of bees.
6. Remove the outermost frame in the uppermost super.
7. Shake the bees from the frame onto the plywood.
8. The beetle collector repeats step 5.
9. Once the bees have been shaken from the frame, turn the frame onto its face.
10. Bounce the frame against the plywood to dislodge adult beetles from the comb.

This step should be repeated 2-3 times for both sides of the frame.

11. The beetle collector repeats step 5.
12. After all frames in a box have been examined, the individual working the colony bounces the empty box on the plywood to remove the remaining SHB. This step should be repeated for all supers, all frames, and the bottom board of the colony.
13. The bees accumulated on the plywood can be bounced off the board in front of the reassembled colony. The bees will return to the hive.

Notes:

A certain proportion of the adult SHB will remain undetected during such visual inspections (Neumann and Hoffmann, 2008). This procedure can be repeated 2-3 times in a 24 hour period in order to create beetle-free colonies for experiments.

3.2.3.2. Counting the number and distribution of beetles in freshly-killed colonies

A more accurate approach for counting SHB is to examine beetles in freshly-killed colonies (modified from Ellis et al., 2003a; Neumann and Hoffmann, 2008; Schäfer et al., 2011). One can, presumably, find 100% of the beetles inhabiting the nest if all the beetles and bees are dead. However, these methods are fatal to the colony and are useful only under certain circumstances. One of the following methods can be used to conduct absolute counts of SHB in colonies depending on what data are required and what facilities are available.

3.2.3.2.1. Killing with liquid nitrogen

It is possible to kill whole colonies by dipping them into liquid nitrogen. The hives have to be manipulated to allow quick air displacement while keeping all bees and SHB inside. This can be achieved with a screened lid (mesh width, 1 mm). As killing with liquid nitrogen is fast, the spatial distribution of the bees and the beetles inside or on the combs will remain stable.

1. Place an adequate container, filled with liquid nitrogen, next to the colony.
2. Install the screened lid.
3. Dip the colony into the liquid nitrogen.
4. Store the colonies in a cold room until they are carefully dissected.

Note: it is very important to follow the safety regulations for handling liquid nitrogen.

3.2.3.2.2. Killing with petrol fuel

Colonies can also be killed with petrol fuel. This method doesn't require hive manipulation, but one has to make sure, that no petrol fuel leaks out of the colony for environmental safety reasons.

1. Seal the complete hive with masking tape, except the lid.
2. Quickly open the top lid.
3. Pour 500 ml standard petrol fuel into the colony.
4. Close the colony.
5. Store in a cold room until dissection.

Note: Petrol is poisonous for humans. Inspections should be done open air and gas masks or similar protection is highly recommended. Furthermore it is very important to be aware of the great flammability of petrol and its fumes!

3.2.3.2.3. Freeze killing

If large freezers are available, colonies can be killed by freezing them, but the placement of colonies into a cool room or freezer will change the distribution of SHB, because of the resulting clustering behaviour of the bees.

1. Close all colony entrances with tape, grass, or other similar material.
2. Place the colonies into a freezer room (< -20°C) for 2 weeks to ensure that all the bees in the colonies are dead. It is important to note that honey bees are able to thermoregulate, so strong colonies with honey reserves (the fuel for thermoregulation) may die slowly. Wherever possible, colonies should be placed into very cold freezers and left for at least 2 weeks. Colonies kept at temperatures > -20°C may die too slowly.

Regardless of which way of killing, the SHB in the stored colonies are counted as follows:

1. Thaw the colonies at RT for 24 hours prior to inspection for beetles.

2. Once thawed, remove the lid to the colony.
3. Carefully inspect for beetles. Bees should be removed to facilitate beetle visualization and cracks/crevices examined carefully.
4. Remove all frames from the colony
5. Inspect each frame for beetles.

This includes removing bees from the combs, tapping the combs on their sides to dislodge beetles hiding in the wax cells and uncapping of sealed honey combs to detect mining larvae (Neumann and Hoffmann, 2008). It is important to note that bees cluster in cold temperatures, with many bees clustering head first into empty cells. Beetles often can be found at the bottom of cells that bees are in (Ellis *et al.*, 2003a) so all clustering bees should be removed from all cells (this can be done using forceps) in order to find every beetle present in the nest.

6. Inspect all supers and the bottom board for hiding beetles.

3.2.3.3. Counting the number of beetles and defining their spatial distribution in winter clusters killed by dipping in liquid nitrogen

This method is modified from Schäfer *et al.* (2011). It is an accurate method for counting the number and determining the spatial distribution of adult SHB inside honey bee winter clusters. One can, presumably, find 100% of the beetles inhabiting the colony. However, the method is fatal to the colony and is useful only under certain circumstances. Furthermore, it requires manipulation of the hives and the availability of liquid nitrogen and a cool room that can accommodate the clustering colonies.

1. Place the colonies in hives without entrances and with screened bottom and lid (mesh width, 1 mm).
2. Transfer the colonies into a cold room at -5°C and constant darkness, to allow the formation of the clusters.
3. Kill the colonies by dipping the whole hives into liquid nitrogen (see section 3.2.3.2.1.), immediately after taking them out of the cool room.

This will fix the spatial distribution of adult *A. tumida* inside the clusters and keep the structure of the cluster.

4. Leave the colonies inside the liquid nitrogen long enough to kill all the bees and SHB (1 min. for a cluster of ~4,000 bees).
5. Store the colonies in a cold room until they are carefully dissected.
6. Record the position of the cluster (e.g. by taking pictures) and of each SHB (all bees inside cells have to be removed to investigate the cell bottoms).

The positions of SHB can be recorded as follows: core (= central area of the cluster), periphery (= inside cluster, except core area) and outside clusters or elsewhere inside the colony.

Note: it is very important to follow the safety regulations for handling liquid nitrogen.

3.2.4. Collecting SHB eggs

This method has been designed by Ellis and Delaplane (2007)

1. Lay a microscope slide on a flat surface.
2. Place ½ of a cover slip on both ends of the slide.
3. Place another microscope slide on top of the cover slip halves.

In effect, you are separating two microscope slides with a cover slip.

4. Tape both ends of the slides so that they will remain together.
5. Place the slides in a petri dish that contains a small amount (~5 g) of SHB food (see section 3.1.1.1. for food recipe).
6. Add mated females or adult beetles (see section 3.1.3.) to the petri dish and replace the petri dish top.
7. Secure the top and bottom of the petri dish to one another with a rubber band.

SHB females will oviposit in the space between the two slides created by the cover slip (see Fig. 12).

8. Handle SHB eggs with a small paint brush.

This procedure can be modified by using a small piece of wax paper (~10 cm x 10 cm) that has been folded back and forth 10 times (making ~1 cm folds). The folds should be tight and the ends of the paper should be paper clipped to keep it from unfolding. The SHB females will oviposit between the folds in the paper.

3.2.5. Introducing adult SHB into colonies

Ellis *et al.* (2003b) investigated the impacts of adult SHB on nests and flight activity of Cape and European subspecies of honey bees. They did this by adding SHB to nucleus colonies nightly for 15 nights to simulate a large-scale, chronic invasion of SHB into colonies. They developed the method outlined below for introducing adult SHB into colonies daily.

1. Rear adult SHB according to the methods outlined in section 3.1.3.
2. Prepare/equalize colonies prior to the experiments according to the methods outlined in the *BEEBOOK* paper on estimating colony strength (Delaplane *et al.*, 2013).
3. Collect adult SHB from the rearing program with an aspirator (see Fig. 6).
4. Place the exact number and sex (if needed see Fig. 9) of adult SHB intended for each colony into small vials before their release into the colonies. Adult SHB are much easier to release from a vial into a colony than by other methods.
5. Colonies can be artificially-infested with SHB on any time schedule (daily, weekly, etc.) provided the introductions are done during evening hours. SHB adults preferentially invade colonies during evening hours (Neumann *et al.*, 2012) so adults should be introduced during regionally-appropriate hours, between 1-2 hours before sunset and 1-2 hours after sunset. Ellis *et al.* (2003b) introduced SHB between 17:00 – 21:00 h.
5. To introduce SHB into the colony, lightly tap the vial containing the beetles to cause the beetles to fall to the bottom of the vial.

SHB are quick crawlers and can easily escape the vial once the lid is removed if they are not tapped to the bottom of the vial first.

6. Open the lid to the colony just enough to allow room to add the beetles.
7. Add the adults to the uppermost super of the hive and close to the nest periphery to avoid overreaction by the host bees ("beetle shock": bees being exposed to SHB abnormally and immediately removing the beetles from the hive). If the beetles are dumped into the centre of the bee cluster, the bees will attack the beetles immediately and many beetles may exit the colony within minutes of introduction. Beetles should not be anesthetized prior to introduction into the nest because anesthetized beetles are easily removed from colonies by adult bees.
8. Replace the lid to the colony immediately after beetle introduction.

Modifications, additional uses, research on future improvements:

- The same method can be used to add adult SHB to full-size or nucleus colonies and to observation hives. Modifications can be made to the observation hive to accommodate SHB introduction since observation hive lids often cannot be removed (depending on hive design) (Fig. 14).
- Ellis *et al.* (2003b) added 100 SHB/night for 15 nights (totalling 1,500 SHB). Delaplane *et al.* (2010) added from 75 – 1,200 SHB/introduction every two months for a beetle threshold study. Consequently, the method is useful to simulate chronic, small scale beetle invasions or large, acute beetle invasions into the bee nest.
- For possible future improvement, one should investigate how SHB can be "control-released" into the nest over a longer period of time. The method outlined above involves the sudden addition of SHB adults to the nest, heightening the likelihood of "beetle shock". Controlled introductions should be a point of future investigation.
- This procedure can be used to investigate SHB impacts on colony absconding behaviour, honey and pollen stores, amount of bees/brood in the nest, average colony flight activity, colony weight gain, SHB reproduction, SHB migration between colonies, etc. The beetles can be laboratory-reared, field-collected, all males, all females, age-cohort specific, etc. according to the needs of the experiment.
- Neumann and Härtel (2004) investigated the removal of SHB larvae by honey bee colonies. They introduced larvae using petri-dishes. Since SHB larvae are neither quick crawlers nor able to show the turtle defence posture of adult SHB (Neumann *et al.*, 2001c), worker bees quickly remove them (Neumann and Härtel, 2004).

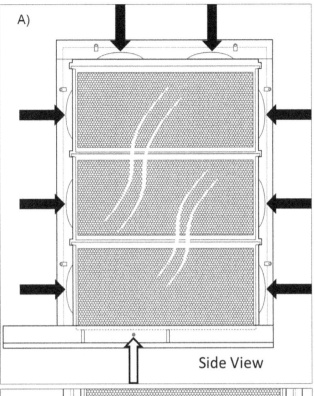

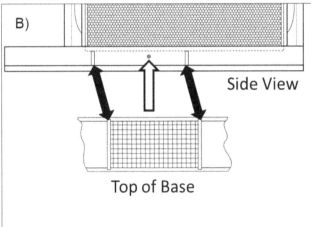

SHB have now well established populations in North America and Australia (Neumann and Ellis, 2008), and are likely to spread into more areas (Asia, South America, Europe) with potentially devastating effects on local managed honey bees and possibly other bees under suitable climatic conditions. This calls for concerted efforts of the community to better control this invasive species. However, despite this *BEEBOOK* paper there are still significant gaps in SHB methodology, thereby limiting its further study and control. This calls for more research in this rather small field. Development of alternative treatments with natural enemies or an optimized trapping of adult/ larval SHB in the field might constitute promising future avenues. On the other hand, this beetle has a truly fascinating biology and there is considerable potential to shed light on numerous fundamental questions in ecology and evolution. We therefore hope that this set of standard methods will attract more researchers to join the SHB research field, thereby stimulating exciting future research on this species.

4. Acknowledgements

Appreciation is addressed to Geoff Williams and an anonymous referee for constructive comments on an earlier draft of the manuscript and to Sebastian Spiewok for sharing his unpublished data. The COLOSS (Prevention of honey bee COlony LOSSes) network aims to explain and prevent massive honey bee colony losses. It was funded through the COST Action FA0803. COST (European Cooperation in Science and Technology) is a unique means for European researchers to jointly develop their own ideas and new initiatives across all scientific disciplines through trans-European networking of nationally funded research activities. Based on a pan-European intergovernmental framework for cooperation in science and technology, COST has contributed since its creation more than 40 years ago to closing the gap between science, policy makers and society throughout Europe and beyond. COST is supported by the EU Seventh Framework Programme for research, technological development and demonstration activities (Official Journal L 412, 30 December 2006). The European Science Foundation as implementing agent of COST provides the COST Office through an EC Grant Agreement. The Council of the European Union provides the COST Secretariat. The COLOSS network is now supported by the Ricola Foundation - Nature & Culture.

Fig. 14. Diagram of the modified observation hive used by Atkinson and Ellis (2011a, b). **A.** shows a side view of the observation hive. The white arrow indicates the location where beetles are introduced into the observation hive, while the black arrows indicate the location of the eight grooves (confinement sites) located on the periphery of the observation hive. The confinement sites are present on both sides, totalling 16 sites. Invading beetles are more likely to be confined in these sites (facilitating their observation) than in other locations in the nest. **B.** shows the bottom board of the observation hive (top picture) and a top view of the gridded base (test arena) of the observation hive. The white arrow indicates the location where beetles or control beads can be introduced into the test arena. The black arrows indicate the location of the acrylic glass doors that, when slid in place, capture guard bees in the test arena.

5. References

ANDERSON, D L; ROBERTS, J M K (2013) Standard methods for *Tropilaelaps* mites research. In *V Dietemann; J D Ellis; P Neumann (Eds) The COLOSS BEEBOOK, Volume II: standard methods for* Apis mellifera *pest and pathogen research. Journal of Apicultural Research* 52(4): http://dx.doi.org/10.3896/IBRA.1.52.4.21

ARBOGAST, R T; TORTO, B; VAN ENGELSDORP, D; TEAL, P E A (2007) An effective trap and bait combination for monitoring the small hive beetle, *Aethina tumida* (Coleoptera: Nitidulidae). *Florida Entomologist* 90(2): 404-406.

ARBOGAST, R T; TORTO, B; WILLMS, S; FOMBONG, A T; DUEHL, A; TEAL, P E A (2012) Estimating reproductive success of *Aethina tumida* (Coleoptera: Nitidulidae) in honey bee colonies by trapping emigrating larvae. *Environmental Entomology* 41(1): 152-158. http://dx.doi.org/10.1603/EN11186

ATKINSON, E; ELLIS, J D (2011a) Adaptive behaviour of honey bees (*Apis mellifera*) toward beetle invaders exhibiting various levels of colony integration. *Physiological Entomology* 36: 282-289. http://dx.doi.org/10.1111/j.1365-3032.2010.00774.x

ATKINSON, E; ELLIS, J D (2011b) Honey bee, *Apis mellifera* L., confinement behaviour toward beetle invaders. *Insectes Sociaux* 58: 495-503. http://dx.doi.org/10.1007/s00040-011-0169-7

BENDA, N D; BOUCIAS, D; TORTO, B; TEAL P (2008) Detection and characterization of *Kodamaea ohmeri* associated with small hive beetle *Aethina tumida* infesting honey bee hives. *Journal of Apicultural Research* 47(3): 194-201. http://dx.doi.org/10.3827/IBRA.1.47.3.07

BUCHHOLZ, S; SCHÄFER, M O; SPIEWOK, S, PETTIS J S, DUNCAN M, RITTER W, SPOONER-HART R, NEUMANN P (2008) Alternative food sources of *Aethina tumida* (Coleoptera: Nitidulidae). *Journal of Apicultural Research* 47(3): 201-208. http://dx.doi.org/10.3827/IBRA.1.47.3.08

DE GRAAF, D C; ALIPPI, A M; ANTÚNEZ, K; ARONSTEIN, K A; BUDGE, G; DE KOKER, D; DE SMET, L; DINGMAN, D W; EVANS, J D; FOSTER, L J; FÜNFHAUS, A; GARCIA-GONZALEZ, E; GREGORC, A; HUMAN, H; MURRAY, K D; NGUYEN, B K; POPPINGA, L; SPIVAK, M; VANENGELSDORP, D; WILKINS, S; GENERSCH, E (2013) Standard methods for American foulbrood research. In *V Dietemann; J D Ellis; P Neumann (Eds) The COLOSS BEEBOOK, Volume II: standard methods for* Apis mellifera *pest and pathogen research. Journal of Apicultural Research* 52(1): http://dx.doi.org/10.3896/IBRA.1.52.1.11

DE GUZMAN, L I; PRUDENTE, J A; RINDERER, T E; FRAKE, A M; TUBBS, H (2009) Population of small hive beetles (*Aethina tumida* Murray) in two apiaries having different soil textures in Mississippi. *Science of Bee Culture*, 1: 4-8.

DE GUZMAN, L I; FRAKE, A M; RINDERER, T E; ARBOGAST R T (2011) Effect of height and colour on the efficiency of pole traps for *Aethina tumida* (Coleoptera: Nitidulidae). *Journal of Economic Entomology* 104(1): 26-31. http://dx.doi.org/10.1603/EC10300

DE GUZMAN, L I; FRAKE, A M; RINDERER, T E (2012) Marking small hive beetles with thoracic notching: effects on longevity, flight ability and fecundity. *Apidologie* 43: 425-431. http://dx.doi.org/10.1007/s13592-011-0107-8

DELAPLANE, K S; VAN DER STEEN, J; GUZMAN, E (2013) Standard methods for estimating strength parameters of *Apis mellifera* colonies. In *V Dietemann; J D Ellis; P Neumann (Eds) The COLOSS* BEEBOOK, *Volume I: standard methods for* Apis mellifera *research. Journal of Apicultural Research* 52(1): http://dx.doi.org/10.3896/IBRA.1.52.1.03

DE MIRANDA, J R; BAILEY, L; BALL, B V; BLANCHARD, P; BUDGE, G; CHEJANOVSKY, N; CHEN, Y-P; GAUTHIER, L; GENERSCH, E; DE GRAAF, D; RIBIÈRE, M; RYABOV, E; DE SMET, L VAN DER STEEN, J J M (2013) Standard methods for virus research in *Apis mellifera*. In *V Dietemann; J D Ellis; P Neumann (Eds) The COLOSS* BEEBOOK, *Volume II: standard methods for* Apis mellifera *pest and pathogen research. Journal of Apicultural Research* 52(4): http://dx.doi.org/10.3896/IBRA.1.52.4.22

DELAPLANE, K S; ELLIS, J D; HOOD, W M (2010) A test for interactions between *Varroa destructor* (Acari: Varroidae) and *Aethina tumida* (Coleoptera: Nitidulidae) in colonies of honey bees (Hymenoptera: Apidae). *Annals of the Entomological Society of America* 103(5): 711-715. http://dx.doi.org/10.1603/AN09169

DIETEMANN, V; NAZZI, F; MARTIN, S J; ANDERSON, D; LOCKE, B; DELAPLANE, K S; WAUQUIEZ, Q; TANNAHILL, C; FREY, E; ZIEGELMANN, B; ROSENKRANZ, P; ELLIS, J D (2013) Standard methods for varroa research. In *V Dietemann; J D Ellis; P Neumann (Eds) The COLOSS* BEEBOOK, *Volume II: standard methods for* Apis mellifera *pest and pathogen research. Journal of Apicultural Research* 52(1): http://dx.doi.org/10.3896/IBRA.1.52.1.09

ELLIS, J D (2005) Reviewing the confinement of small hive beetles (*Aethina tumida*) by western honey bees (*Apis mellifera*). *Bee World* 86(3): 56-62.

ELLIS, J D; DELAPLANE, K S (2006) The effects of habitat type, ApilifeVAR™, and screened bottom boards on small hive beetle (*Aethina tumida*) entry into honey bee (*Apis mellifera*) colonies. *American Bee Journal* 146(5): 537-539.

ELLIS, J D; DELAPLANE, K S (2007) The effects of three acaricides on the developmental biology of small hive beetles (*Aethina tumida*). *Journal of Apicultural Research* 46(4): 256-259. http://dx.doi.org/10.3896/IBRA.1.46.4.08

ELLIS, J D; DELAPLANE, K S (2008) Small hive beetle (*Aethina tumida*) oviposition behaviour in sealed brood cells with notes on the removal of the cell contents by European honey bees (*Apis mellifera*). *Journal of Apicultural Research* 47(3): 210-215. http://dx.doi.org/10.3896/IBRA.1.47.3.09

ELLIS, J D; DELAPLANE, K S; HEPBURN, H R; ELZEN, P J (2002a) Controlling small hive beetles (*Aethina tumida* Murray) in honey bee (*Apis mellifera*) colonies using a modified hive entrance. *American Bee Journal* 142(4): 288-290.

ELLIS J D; DELAPLANE K S, HOOD W M (2002b) Small hive beetle (*Aethina tumida* Murray) weight, gross biometry, and sex proportion at three locations in the southeastern United States. *American Bee Journal* 142(7):520–522.

ELLIS, J D; HEPBURN, H R (2006) An ecological digest of the small hive beetle (*Aethina tumida*), a symbiont in honey bee colonies (*Apis mellifera*). *Insectes Sociaux* 53: 8-19. http://dx.doi.org/10.1007/s00040-005-0851-8

ELLIS, J D; MUNN, P A (2005) The worldwide health status of honey bees. *Bee World* 86(4): 88–101.

ELLIS, J D; HEPBURN, H R; DELAPLANE, K S; ELZEN, P J (2003a) A scientific note on small hive beetle (*Aethina tumida*) oviposition and behaviour during European (*Apis mellifera*) honey bee clustering and absconding events. *Journal of Apicultural Research* 42(3): 47-48.

ELLIS, J D; HEPBURN, H R; DELAPLANE, K S; NEUMANN, P; ELZEN, P J (2003b) The effects of adult small hive beetles, *Aethina tumida* (Coleoptera: Nitidulidae), on nests and flight activity of Cape and European honey bees (*Apis mellifera*). *Apidologie* 34: 399-408. http://dx.doi.org/10.1051/apido:2003038

ELLIS, J D; HEPBURN, H R; ELLIS, A M; ELZEN, P J (2003c) Prison construction and guarding behaviour by European honey bees is dependent on inmate small hive beetle density. *Naturwissenschaften* 90: 382-384. http://dx.doi.org/10.1007/s00114-003-0447-y

ELLIS, J D; HEPBURN, H R; ELLIS, A M; ELZEN, P J (2003d) Social encapsulation of the small hive beetle (*Aethina tumida* Murray) by European honey bees (*Apis mellifera* L.). *Insectes Sociaux* 50: 286-291. http://dx.doi.org/10.1007/s00040-003-0671-7

ELLIS, J D; HOLLAND, A J; HEPBURN, H R; NEUMANN, P; ELZEN, P J (2003e) Cape (*Apis mellifera capensis*) and European (*Apis mellifera*) honey bee guard age and duration of guarding small hive beetles (*Aethina tumida*). *Journal of Apicultural Research* 42 (3): 32-34.

ELLIS, J D; HEPBURN, H R; ELZEN, P J (2004a) Confinement behaviour of cape honey bees (*Apis mellifera capensis* Esch.) in relation to population densities of small hive beetles (*Aethina tumida* Murray). *Journal of Insect Behaviour* 17(6): 835-842. http://dx.doi.org/10.1023/B:JOIR.0000048992.26016.7f

ELLIS, J D; HEPBURN, H R; ELZEN P J (2004b) Confinement of small hive beetles (*Aethina tumida*) by Cape honey bees (*Apis mellifera capensis*). *Apidologie* 35(4): 389-396. http://dx.doi.org/10.1051/apido:2004030

ELLIS, J D; HEPBURN, H R, LUCKMANN, B; ELZEN, P J (2004c) The effects of soil type, moisture, and density on pupation success of *Aethina tumida* (Coleoptera: Nitidulidae). *Environmental Entomology* 33(4): 794-798. http://dx.doi.org/10.1603/0046-225X-33.4.794

ELLIS, J D; DELAPLANE, K S; HOOD, W M (2002a) Small hive beetle (*Aethina tumida* Murray) weight, gross biometry, and sex proportion at three locations in the south-eastern United States. American Bee Journal 142: 520–522.

ELLIS, J D; NEUMANN, P, HEPBURN, H R; ELZEN, P J (2002b) Longevity and reproductive success of *Aethina tumida* (Coleoptera: Nitidulidae) fed different natural diets. *Journal of Economic Entomology* 95(5): 902-907. http://dx.doi.org/10.1603/0022-0493-95.5.902

ELLIS, J D; PIRK, C W W; HEPBURN, H R; KASTBERGER, G; ELZEN, P J (2002c) Small hive beetles survive in honey bee prisons by behavioural mimicry. *Naturwissenschaften* 89: 326-328. http://dx.doi.org/10.1007/s00114-002-0326-y

ELLIS, J D; RICHARDS, C S; HEPBURN, H R; ELZEN, P J (2003f) Oviposition by small hive beetles elicits hygienic responses from Cape honey bees. *Naturwissenschaften* 90(11): 532-535. http://dx.doi.org/10.1007/s00114-003-0476-6

ELLIS, J D; RICHARDS, C S; HEPBURN, H R; ELZEN, P J (2004d) Hygienic behaviour of Cape and European *Apis mellifera* (Hymenoptera: Apidae) toward *Aethina tumida* (Coleoptera: Nitidulidae) eggs oviposited in sealed bee brood. *Annals of the Entomological Society of America* 97(4): 860-864. http://dx.doi.org/10.1603/0013-8746(2004)097[0860:HBOCAE]2.0.CO;2

ELLIS, J D; RONG, I H; HILL, M P; HEPBURN, H R; ELZEN, P J (2004e) The susceptibility of small hive beetle (*Aethina tumida* Murray) pupae to fungal pathogens. *American Bee Journal* 144 (6): 486-488.

ELLIS, J D; SPIEWOK, S; DELAPLANE, K S; BUCHHOLZ, S; NEUMANN, P; TEDDERS, L (2010) Susceptibility of *Aethina tumida* (Coleoptera: Nitidulidae) larvae and pupae to entomopathogenic nematodes. *Journal of Economic Entomology* 103(1): 1-9. http://dx.doi.org/10.1603/EC08384

ELZEN, P J; BAXTER, J R; WESTERVELT, D; RANDALL, C; DELAPLANE, K S; CUTTS, L; WILSON, W T (1999) Field control and biology studies of a new pest species, *Aethina tumida* Murray (Coleoptera: Nitidulidae), attacking European honey bees in the Western Hemisphere. *Apidologie* 30: 361-366. http://dx.doi.org/10.1051/apido:19990501

ELZEN, P J; BAXTER, J R; NEUMANN, P; SOLBRIG, A J; PIRK, C W W; HEPBURN, H R; WESTERVELT, D; RANDALL, C (2001) Behaviour of African and European subspecies of *Apis mellifera* toward the small hive beetle, *Aethina tumida*. *Journal of Apicultural Research* 40: 40-41.

EYER, M ; CHEN, Y P ; SCHÄFER, M O; PETTIS, J; NEUMANN, P (2009a) Small hive beetle, *Aethina tumida*, as a potential biological vector of honey bee viruses. *Apidologie* 40: 419-428. http://dx.doi.org/10.1051/apido:2008051

EYER, M ; CHEN, Y P ; SCHÄFER, M O; PETTIS, J; NEUMANN, P (2009b) Honey bee sacbrood virus infects adult small hive beetles, *Aethina tumida* (Coleoptera: Nitidulidae). *Journal of Apicultural Research* 48(4): 296-297.
http://dx.doi.org/10.3896/IBRA.1.48.4.11

EVANS, J D; PETTIS, J S; SHIMANUKI, H (2000). Mitochondrial DNA relationships in an emergent pest of honey bees: *Aethina tumida* (Coleoptera: Nitidulidae) from the United States and Africa. *Annals of the Entomological Society of America* 93: 415-420.
http://dx.doi.org/10.1603/0013-8746(2000)093[0415:MDRIAE] 2.0.CO;2

EVANS, J D; PETTIS, J S; HOOD, H; SHIMANUKI, H (2003). Tracking an invasive honey bee pest: Mitochondrial DNA variation in North American small hive beetles. *Apidologie*. 34: 103-109.
http://dx.doi.org/10.1051/apido:2003004

EVANS, J D; SPIEWOK, S; TEIXEIRA, E W; Neumann, P (2008) Microsatellite loci for the small hive beetle, *Aethina tumida*, a nest parasite of honey bees. *Molecular Ecology Resources* 8: 698–700.
http://dx.doi.org/10.1111/j.1471-8286.2007.02052.x

EVANS, J D; SCHWARZ, R S; CHEN, Y P; BUDGE, G; CORNMAN, R S; DE LA RUA, P; DE MIRANDA, J R; FORET, S; FOSTER, L; GAUTHIER, L; GENERSCH, E; GISDER, S; JAROSCH, A; KUCHARSKI, R; LOPEZ, D; LUN, C M; MORITZ, R F A; MALESZKA, R; MUÑOZ, I; PINTO, M A (2013) Standard methodologies for molecular research in *Apis mellifera*. In *V Dietemann; J D Ellis; P Neumann (Eds) The COLOSS BEEBOOK, Volume I: standard methods for* Apis mellifera *research. Journal of Apicultural Research* 52(4): http://dx.doi.org/10.3896/IBRA.1.52.4.11

FALUSH, D; STEPHENS, M; PRITCHARD, J K (2007) Inference of population structure using multilocus genotype data: dominant markers and null alleles. *Molecular Ecology Notes* 7 (4): 574-578.
http://dx.doi.org/10.1111/j.1471-8286.2007.01758.x

FREUDE, H; HARDE, K W; LOHSE, G A (1967) Die Käfer Mitteleuropas, Goecke und Evers, Krefeld, 72 pp.

GRECO, M K; HOFFMANN, D; DOLLIN, A; DUNCAN, M; SPOONER-HART, R; NEUMANN, P (2010) The alternative Pharaoh approach: stingless bees mummify beetle parasites alive. *Naturwissenschaften* 97: 319-323.
http://dx.doi.org/10.1007/s00114-009-0631-9

HALCROFT, M; SPOONER-HART, R; NEUMANN, P (2011) Behavioral defense strategies of the stingless bee, *Austroplebeia australis*, against the small hive beetle, *Aethina tumida*. *Insectes Sociaux* 58: 245-253. http://dx.doi.org/10.1007/s00040-010-0142-x

HEPBURN, H R; RADLOFF, S E (1998) *Honey bees of Africa.* Springer Verlag; Berlin, Germany.

HOFFMANN, D; PETTIS, J S; NEUMANN, P (2008) Potential host shift of the small hive beetle (*Aethina tumida*) to bumble bee colonies (*Bombus impatiens*). *Insectes Sociaux* 55: 153–162
http://dx.doi.org/10.1007/s00040-008-0982-9

HOOD, W M (2004) The small hive beetle, *Aethina tumida*: a review. *Bee World* 85: 51–59.

HUMAN, H; BRODSCHNEIDER, R; DIETEMANN, V; DIVELY, G; ELLIS, J; FORSGREN, E; FRIES, I; HATJINA, F; HU, F-L; JAFFÉ, R; JENSEN, A B; KÖHLER, A; MAGYAR, J; ÖZIKRIM, A; PIRK, C W W; ROSE, R; STRAUSS, U; TANNER, G; TARPY, D R; VAN DER STEEN, J J M; VAUDO, A; VEJSNÆS, F; WILDE, J; WILLIAMS, G R; ZHENG, H-Q (2013) Miscellaneous standard methods for *Apis mellifera* research. In *V Dietemann; J D Ellis; P Neumann (Eds) The COLOSS* BEEBOOK, *Volume I: standard methods for* Apis mellifera *research. Journal of Apicultural Research* 52(4): http://dx.doi.org/10.3896/IBRA.1.52.4.10

JENSEN, A B; ARONSTEIN, K; FLORES, J M; VOJVODIC, S; PALACIO, M A; SPIVAK, M (2013) Standard methods for fungal brood disease research. In *V Dietemann; J D Ellis, P Neumann (Eds) The COLOSS* BEEBOOK: *Volume II: Standard methods for* Apis mellifera *pest and pathogen research. Journal of Apicultural Research* 52(1): http://dx.doi.org/10.3896/IBRA.1.52.1.13

KÖHLER, A; NICOLSON, S W; PIRK, C W W (2013) A new design for honey bee hoarding cages for laboratory experiments. *Journal of Apicultural Research* 52(2): 12-14.
http://dx.doi.org/10.3896/IBRA.1.52.2.03

KOLMES, S A (1985) An information-theory analysis of task specialization among worker honey bees performing hive duties. *Animal Behaviour* 33: 181-187.

LOUNSBERRY, Z; SPIEWOK, S; PERNAL, S F; SONSTEGARD, T S; HOOD, W M; PETTIS, J S; NEUMANN, P; EVANS, J D (2010) Worldwide diaspora of the small hive beetle, *Aethina tumida*, a nest parasite of honey bees, *Apis mellifera*. Annals of the Entomological Society of America 103: 671-677.
http://dx.doi.org/10.1603/AN10027

LUNDIE, A E (1940) The small hive beetle *Aethina tumida*. *Science Bulletin* 220: Department of Agriculture and Forestry, Government Printer; Pretoria, South Africa. 30 pp.

MEIXNER, M D; PINTO, M A; BOUGA, M; KRYGER, P; IVANOVA, E; FUCHS, S (2013) Standard methods for characterising subspecies and ecotypes of *Apis mellifera*. In *V Dietemann; J D Ellis; P Neumann (Eds) The COLOSS* BEEBOOK, *Volume I: standard methods for* Apis mellifera *research. Journal of Apicultural Research* 52(4): http://dx.doi.org/10.3896/IBRA.1.52.4.05

MORITZ, R F A; CREWE, R M; HEPBURN, H R (2001) Attraction and repellence of workers by the honey bee queen (*Apis mellifera* L.). *Ethology* 107: 465-477.
http://dx.doi.org/10.1046/j.1439-0310.2001.00681.x

MURILHAS, A M (2005) *Aethina tumida* arrives in Portugal. Will it be eradicated? *EurBee Newsletter* 2: 7–9.

MURRAY, A (1867) List of Coleoptera received from Old Calabar, on the west coast of Africa. *The Annals and Magazine of Natural History* 19: 167.

MUERRLE, T M; NEUMANN, P (2004) Mass production of small hive beetles (*Aethina tumida* Murray, Coleoptera: Nitidulidae). *Journal of Apicultural Research* 43(3): 144-145.

MUERRLE, T M; NEUMANN, P; DAMES, J F; HEPBURN, H R; HILL, M P (2006) Susceptibility of adult small hive beetle to entomopathogenic fungi. *Journal of Economic Entomology* 99: 1-6. http://dx.doi.org/10.1603/0022-0493(2006)099[0001:SOAATC]2.0.CO;2

NEUMANN, P; HÄRTEL, S (2004) Removal of small hive beetle (*Aethina tumida* Murray) eggs and larvae by African honey bee colonies (*Apis mellifera scutellata* Lepeletier). *Apidologie* 35: 31-36. http://dx.doi.org/10.1051/apido:2003058

NEUMANN, P; ELLIS, J D (2008) The small hive beetle (*Aethina tumida* Murray, Coleoptera: Nitidulidae): distribution, biology and control of an invasive species. *Journal of Apicultural Research* 47 (3): 180-183. http://dx.doi.org/10.3896/IBRA.1.47.3.01

NEUMANN, P; ELZEN, P J (2004) The biology of the small hive beetle (*Aethina tumida*, Coleoptera: Nitidulidae): Gaps in our knowledge of an invasive species. *Apidologie* 35: 229-247. http://dx.doi.org/10.1051/apido:2004010

NEUMANN, P; RITTER, W (2004) A scientific note on the association of *Cychramus luteus* (Coleoptera: Nitidulidae) with honey bee (*Apis mellifera*) colonies. *Apidologie* 35: 665-666. http://dx.doi.org/10.1051/apido:2004051

NEUMANN, P; HOFFMANN, D (2008) Small hive beetle diagnosis and control in naturally infested honey bee colonies using bottom board traps and CheckMite+ strips. *Journal of Pest Science* 81: 43–48. http://dx.doi.org/10.1007/s10340-007-0183-8

NEUMANN, P; PIRK, C W W; HEPBURN, H R; EELZEN, P J; BAXTER, J R (2001a) Laboratory rearing of small hive beetle, *Aethina tumida* (Coleoptera: Nitidulidae). *Journal of Apicultural Research* 40: 111-112.

NEUMANN, P; PIRK, C W W; HEPBURN, H R; RADLOFF, S E (2001b) A scientific note on the natural merger of two honey bee colonies (*Apis mellifera capensis*). *Apidologie* 32: 113-114. http://dx.doi.org/10.1051/apido:2001116

NEUMANN, P; PIRK, C W W; HEPBURN, H R; SOLBRIG, A J; RATNIEKS, F L W; ELZEN, P J; BAXTER, J R (2001c) Social encapsulation of beetle parasites by Cape honey bee colonies (*Apis mellifera capensis* Esch.). *Naturwissenschaften* 88: 214–216. http://dx.doi.org/10.1007/s001140100224

NEUMANN, P; RADLOFF, S E; PIRK, C W W; HEPBURN, H R (2003) The behaviour of drifted Cape honey bee workers (*Apis mellifera capensis*): predisposition for social parasitism? *Apidologie* 34: 585-590. http://dx.doi.org/10.1051/apido:2003048

NEUMANN, P; HOFFMANN, D; DUNCAN, M; SPOONER-HART, R; PETTIS, J S (2012) Long-range dispersal of small hive beetles. *Journal of Apicultural Research* 51(2): 214-215. http://dx.doi.org/10.3896/IBRA.1.51.2.11

PEAKALL, R; SMOUSE, P E (2006) Genalex 6: genetic analysis in Excel. Population genetic software for teaching and research. *Molecular Ecology Notes* 6 (1): 288-295. http://dx.doi.org/10.1111/j.1471-8286.2005.01155.x

PENG, C; WILLIAMS, R N (1990a) Multiple-species rearing diet for sap beetles (Coleoptera: Nitidulidae). *Annals of the Entomological Society of America* 83(6): 1155-1157.

PENG, C; WILLIAMS, R N (1990b) Pre-oviposition period, egg production and mortality of six species of hibernating sap beetles (Coleoptera: Nitidulidae). *Journal of Entomological Science* 25(3): 453-457.

PETTIS, J S; SHIMANUKI, H (2000) Observations on the small hive beetle, *Aethina tumida*, Murray, in the United States, *American Bee Journal* 140, 152-155.

PIRK, C W W (2002) Reproductive conflicts in honey bee colonies. PhD thesis, Department of Zoology & Entomology, Rhodes University, South Africa.

PIRK, C W W; NEUMANN, P (2013) Small hive beetles are facultative predators of adult honey bees. *Journal of Insect Behaviour* 26: http://dx.doi.org/10.1007/s10905-013-9392-6

PIRK, C W W; DE MIRANDA, J R; FRIES, I; KRAMER, M; PAXTON, R; MURRAY, T; NAZZI, F; SHUTLER, D; VAN DER STEEN, J J M; VAN DOOREMALEN, C (2013) Statistical guidelines for *Apis mellifera* research. In *V Dietemann; J D Ellis; P Neumann (Eds) The COLOSS BEEBOOK, Volume I: standard methods for Apis mellifera research. Journal of Apicultural Research* 52(4): http://dx.doi.org/10.3896/IBRA.1.52.4.13

RÖSCH, G A (1925) Untersuchungen über die Arbeitsteilung im Bienenstaat. 1. Teil: Die Tätigkeiten im normalen Bienenstaate und ihre Beziehungen zum Alter der Arbeitsbienen. *Zeitschrift für vergleichende Physiologie* 2: 571-631.

SCHÄFER, M O; PETTIS, J S; RITTER, W; NEUMANN, P (2008) A scientific note on quantitative diagnosis of small hive beetles, *Aethina tumida*, in the field. *Apidologie* 39: 564-565. http://dx.doi.org/10.1051/apido:2008038

SCHÄFER, M O; PETTIS, J S; RITTER, W; NEUMANN, P (2010a) Simple small hive beetle diagnosis. *American Bee Journal* 150: 371-372.

SCHÄFER, M O; RITTER, W; PETTIS, J S; NEUMANN, P (2010b) Small hive beetles, *Aethina tumida*, are vectors of *Paenibacillus larvae*. *Apidologie* 41: 14-20. http://dx.doi.org/10.1051/apido/2009037

SCHÄFER, M O; RITTER, W; PETTIS, J S; NEUMANN, P (2011) Concurrent parasitism alters thermoregulation in honey bee (Hymenoptera: Apidae) winter clusters. *Annals of the Entomological Society of America* 104(3): 476-482 http://dx.doi.org/10.1603/AN10142

SCHEINER, R; ABRAMSON, C I; BRODSCHNEIDER, R; CRAILSHEIM, K; FARINA, W; FUCHS, S; GRÜNEWALD, B; HAHSHOLD, S; KARRER, M; KOENIGER, G; KOENIGER, N; MENZEL, R; MUJAGIC, S; RADSPIELER, G; SCHMICKLI, T; SCHNEIDER, C; SIEGEL, A J; SZOPEK, M; THENIUS, R (2013) Standard methods for behavioural studies of *Apis mellifera*. In *V Dietemann; J D Ellis; P Neumann (Eds) The COLOSS BEEBOOK, Volume I: standard methods for* Apis mellifera *research. Journal of Apicultural Research* 52(4): http://dx.doi.org/10.3896/IBRA.1.52.4.04

SCHMOLKE, M D (1974) A study of *Aethina tumida:* the small hive beetle. University of Rhodesia Certificate in Field Ecology Project Report, 178 pp.

SOLBRIG, A J (2001) Interaction between the South African honey bee, *Apis mellifera capensis* Esch., and the small hive beetle, *Aethina tumida* Murray. Diplomarbeit, Freie Universität Berlin, Institut für Zoologie, Berlin, Germany.

SOMERVILLE, D (2003) *Study of the small hive beetle in the USA.* Rural Industries Research and Development Corporation, Barton, Australian Capital Territory. 57 pp.

SPIEWOK, S; NEUMANN, P (2006a) Infestation of commercial bumble bee (*Bombus impatiens*) field colonies by small hive beetles (*Aethina tumida*). *Ecological Entomology* 31: 623-628. http://dx.doi.org/10.1111/j.1365-2311.2006.00827.x

SPIEWOK, S; NEUMANN, P (2006b) Cryptic low-level reproduction of small hive beetles in honey bee colonies. *Journal of Apicultural Research* 45(1): 47-48 http://dx.doi.org/10.3896/IBRA.1.45.1.11

SPIEWOK S; NEUMANN, P (2012) Sex ratio and dispersal of small hive beetles. *Journal of Apicultural Research* 51(2): 216-217. http://dx.doi.org/10.3896/IBRA.1.51.2.12

SPIEWOK, S; PETTIS, J, DUNCAN, M; SPOONER-HART, R; WESTERVELT, D; NEUMANN, P (2007) Small hive beetle, *Aethina tumida*, populations I: Infestation levels of honey bee colonies, apiaries and regions. *Apidologie* 38: 595–605. http://dx.doi.org/10.1051/apido:2007042

SPIEWOK, S; DUNCAN, M, SPOONER-HART, R; PETTIS, J S; NEUMANN P (2008) Small hive beetle, *Aethina tumida*, populations II: dispersal of small hive beetles. *Apidologie* 39: 683–693. http://dx.doi.org/10.1051/apido:2008054

STEDMAN, M (2006) Small hive beetle (SHB): *Aethina tumida* Murray (Coleoptera: Nitidulidae). Government of South Australia. Primary Industries and Resources for South Australia. *Factsheet* 03/06: 13 pp.

STRAUSS, U (2009) Small hive beetle *(Aethina tumida)* behaviour in honey bee (*Apis mellifera*) colonies. Honours thesis, Social Insect Research Group, Department of Zoology and Entomology, University of Pretoria, South Africa.

WALSH, P S; METZGER, D A; HIGUCHI, R (1991) Chelex 100 as a medium for simple extraction of DNA for PCR-based typing from forensic material. *BioTechniques* 10(4): 506-513.

WARD, L; BROWN, M; NEUMANN, P; WILKINS, S; PETTIS, J S; BOONHAM, N (2007) A DNA method for screening hive debris for the presence of small hive beetles (*Aethina tumida*). *Apidologie* 38: 289-295. http://dx.doi.org/10.1051/apido:2007004

WILLIAMS, G R; ALAUX, C; COSTA, C; CSÁKI, T; DOUBLET, V; EISENHARDT, D; FRIES, I; KUHN, R; MCMAHON, D P; MEDRZYCKI, P; MURRAY, T E; NATSOPOULOU, M E; NEUMANN, P; OLIVER, R; PAXTON, R J; PERNAL, S F; SHUTLER, D; TANNER, G; VAN DER STEEN, J J M; BRODSCHNEIDER, R (2013) Standard methods for maintaining adult *Apis mellifera* in cages under *in vitro* laboratory conditions. In *V Dietemann; J D Ellis; P Neumann (Eds) The COLOSS BEEBOOK, Volume I: standard methods for* Apis mellifera *research. Journal of Apicultural Research* 52(1): http://dx.doi.org/10.3896/IBRA.1.52.1.04

Journal of Apicultural Research 52(4): (2013)
DOI 10.3896/IBRA.1.52.4.20

© IBRA 2013

REVIEW ARTICLE

Standard methods for tracheal mite research

IBRA
INTERNATIONAL BEE
RESEARCH ASSOCIATION

Diana Sammataro[1*], Lilia de Guzman[2], Sherly George[3], Ron Ochoa[4] and Gard Otis[5]

[1]USDA-ARS Carl Hayden Honey Bee Research Center, Tucson, Arizona, USA.
[2]USDA-ARS Honey Bee Breeding, Genetics & Physiology Lab., Baton Rouge, Louisiana, USA.
[3]Plant Health & Environment Lab., Investigation & Diagnostic Centres & Response, Ministry for Primary Industries, New Zealand.
[4]USDA-ARS Systematic Entomology Lab., Beltsville, Maryland, USA.
[5]School of Environmental Sciences, University of Guelph, Guelph, Ontario, Canada.

Received 24 April 2012, accepted subject to revision 2 October 2012, accepted for publication 21 June 2013.

*Corresponding author: Email: Diana.Sammataro@ars.usda.gov

Summary

The honey bee tracheal mite (HBTM) *Acarapis woodi* (Rennie) (Acari: Tarsonemidae) is an obligate endoparasite of honey bees. First described from the Western (European) honey bee *Apis mellifera* L., this mite species was initially observed when honey bee colonies on the Isle of Wight, UK were dying between 1904 and 1919 (Rennie, 1921). Since then, this mite has been found in Europe, North and South America and parts of Asia, but its global distribution is not well understood. In this chapter, we outline protocols for collecting, detecting, identifying, diagnosing and measuring the infestation rates of *A. woodi*. We also describe methods to determine the damage threshold, outline several control measures, and describe methods for studying live mites.

Métodos estándar para el estudio del ácaro traqueal

Resumen

El ácaro traqueal de la abeja de la miel (HBTM son sus siglas en inglés) *Acarapis woodi* (Rennie) (Acari: Tarsonemidae) es un endoparásito obligado de la abeja. Fue descrito por primera vez en la abeja occidental europea *Apis mellifera* L. Esta especie se describió inicialmente en la mortandad de las colonias de abejas de la isla de Wight, Reino Unido, entre 1904 y 1919 (Rennie, 1921). Desde entonces, este ácaro ha sido encontrado en Europa, en el Sur y Norte de América y en zonas de Asia, aunque su distribución global aún no se comprende bien. En este capítulo, se describen los protocolos para la recolección, detección, identificación, diagnóstico y medición de las tasas de infestación de *A. woodi*. También se describen métodos para determinar el umbral de daño, se esbozan una serie de medidas de control, y se describen métodos para el estudio de ácaros vivos.

气管螨研究的标准化方法

摘要

蜜蜂气管螨（*Acarapis woodi*）(Rennie) (Acari: Tarsonemidae)是一种蜜蜂的专一性体内寄生虫。该螨最早发现于西方蜜蜂（*Apis mellifera* L）上，当时正值1904至1919年间英国怀特岛上蜂群死亡期（Rennie, 1921）。此后，该螨发现于欧洲、南北美洲和部分亚洲地区，但它在全球的分布情况尚未充分探明。本章我们概述了气管螨研究中样品采集、检测、鉴定、诊断和感染率测定的实验方法。同时，我们介绍了测定损伤阈值的方法，概述了多种防治措施，并介绍了研究活体螨的方法。

Keywords: COLOSS *BEEBOOK*, *Acarapis woodi*, *Apis mellifera*, honey bee, research methods, protocol

Footnote: Please cite this paper as: SAMMATARO, D; DE GUZMAN, L; GEORGE, S; OCHOA, R; OTIS, G (2013) Standard methods for tracheal mites research. In V Dietemann; J D Ellis, P Neumann (Eds) The COLOSS BEEBOOK: Volume II: Standard methods for Apis mellifera pest and pathogen research. Journal of Apicultural Research 52(4): http://dx.doi.org/10.3896/IBRA.1.52.4.20

Table of Contents

1. Introduction

Countries that export bees and bee products are required to conduct apiculture surveillance programmes to meet disease reporting and sanitary control requirements of the OIE (Office International des Epizooties) to facilitate international trade. A surveillance programme also aids in early detection of honey bee pests and diseases including any new introductions. One pest is the Honey Bee Tracheal Mite (HBTM) *Acarapis woodi* (Acari: Tarsonemidae), an obligate endoparasite of honey bees. This microscopic mite was discovered in 1919 in the UK (Rennie *et al.*, 1921). The identification and detection of the mite led to a law from the US Department of Agriculture restricting all live honey bee imports into the USA in 1922 (Phillips, 1923). Despite this restriction, HBTM was first seen in the USA by beekeepers in Texas in 1984 and the prairie provinces of Canada in 1985. Thereafter, *A. woodi* spread throughout the USA and most Canadian provinces, facilitated by commercial beekeepers transporting bees for pollination, and the sale of mite-infested package bees.

In addition, infested swarms, drifting bees, and the worldwide distribution of *A. mellifera* have contributed to the spread of this mite. Although its current range is not fully known, the HBTM has successfully been established in many countries in most continents, including Europe, Asia, parts of Africa, and North and South America (Ellis and Munn, 2005). To date, it is not known to occur in Australia, New Zealand or Scandinavia (Denmark *et al.*, 2000; Hoy, 2011). Recent work by Kojima *et al.* (2011a) reported *A. woodi* was found on Asian honey bees, *A. cerana japonica*, in Japan.

From what we currently know, *A. mellifera* is the original host of HBTM, as this mite has only been recorded on other *Apis* species following the introduction of *A. mellifera* to Asia. The exact causes of the loss of colonies infested with HBTM are still unknown. This problem is exacerbated by the lack of unique symptoms associated with tracheal mite.

2. Taxonomy and systematics

2.1. Taxonomy

Honey bee tracheal mites were first described by Rennie as *Tarsonemus woodi* in 1921, as parasites of the Western (European) honey bee *Apis mellifera* L. This discovery followed extensive colony mortality on the Isle of Wight and elsewhere in the UK between 1904 and 1919 when bee colonies began to die from unknown causes (Rennie, 1921). The initial suspicion that the "Isle of Wight Disease" was caused by tracheal mites (Hirst, 1921; Rennie, 1921) was never confirmed (Bailey, 1964). Hirst (1921) reclassified the species as *Acarapis woodi*, the name by which it is known today (Lindquist, 1986).

Table 1. Differential diagnosis of *Acarapis* species (Ritter, 1996).

Character	A. dorsalis	A. externus	A. woodi
Notch of the coxal plate	Deep	Flat to Short	Short
Space between stigmata	16.7 µm	16.8 µm	13.9 µm
Length of tarsal limb	7.6 µm	11.4 µm	7.5 µm

Its detection led to the restriction of all live honey bee imports into the USA in 1922 (Phillips, 1923). Despite this, the first report of colony losses from HBTM in the USA came from beekeepers in Texas in 1984. Thereafter, *A. woodi* spread to all states of the USA and most Canadian provinces. Their range expansion was facilitated by commercial beekeepers transporting bees for pollination, and from the sale of mite-infested package bees.

In addition to *A. woodi*, there are two external mite species in the genus *Acarapis* that infest honey bees. *A. externus* Morgenthaler is found on the cervix (the neck region) and *A. dorsalis* Morgenthaler is found on the dorsal groove of the thorax (Ibay and Burgett, 1989; Fig. 1. A, B; Table 1). They were considered to be harmless by Eckert (1961) and Delfinado-Baker (1984), but that may reflect a lack of information on these two *Acarapis* species. A third external species, *A. vagans* (Schneider), described from central Europe and New Zealand, was found principally on drones, on the basal part of the hind wing (Lindquist, 1986). However, other researchers considered this species to be *nomen dubium* and to date this issue has not been resolved (Delfinado-Baker and Baker, 1982; Lindquist, 1986).

Unfortunately, HBTM is now overshadowed by the ectoparasitic mite, *Varroa destructor* Anderson and Trueman. As a result, the presence of HBTM in some instances is not regularly investigated.

When found, they are often at very low levels, perhaps due to the treatments used to control *V. destructor* (see Dietemann *et al.*, 2013).

The current taxonomy of HBTM, based on Krantz *et al.* (2009) is:

Kingdom: Animalia
Phylum: Arthropoda
Class: Arachnida
Subclass: Acari
Superorder: Acariformes
Suborder: Prostigmata
Order: Trombidiformes
Cohort: Heterostigmatina
Superfamily: Tarsonemoidea
Family: Tarsonemidae
Genus: *Acarapis*
Species: *woodi*

2.2. Identifying tracheal mites

Also see Section 3 for methods of collecting and identifying mites.

2.2.1. Mite appearance

Adult female tracheal mites have a pyriform (pear-shaped) body (Fig. 2), measuring 120 to190 µm long by 77 to 80 µm wide. Female mites can be distinguished by their stubby form and the presence of five setae on leg IV (Fig. 1C). Adult males are 125 to 136 µm long by 60 to 77 µm wide (Fig. 2). Leg IV of the males lacks all the tarsal structures. Both

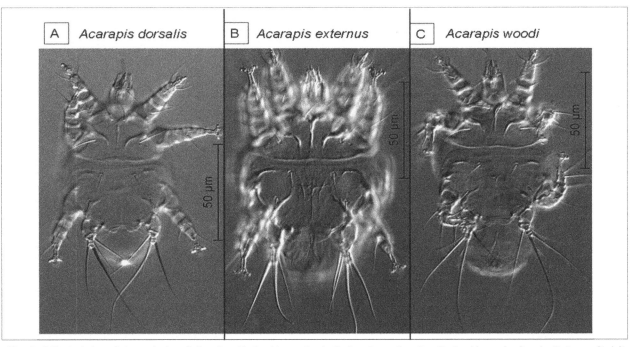

Fig. 1. A. Ventral view of *Acarapis dorsalis* (ex. New Zealand honey bee); **B.** *A. externus* (ex. New Zealand honey bee) and; **C.** *A. woodi* adult female (ex. Canadian honey bee) taken at a 400x magnification under light microscopy.   Photos: Dr Qing-Hai Fan.

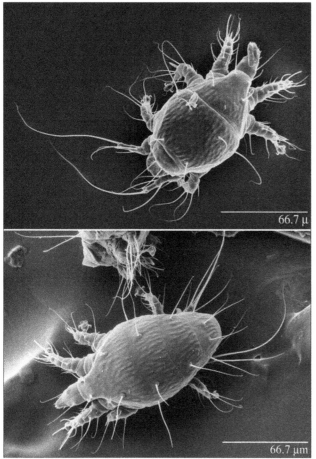

Fig. 2. LT-SEM micrographs of the dorsal view of a male (top) and female (bottom) of *A. woodi.*   Photos: E Erbe and R Ochoa.

sexes have the dorsum covered with round to elongated bumps and an absence of bothridial setae on the prodorsal area (Fig. 2). The gnathosoma contains the mouth parts, palps and the chelicerae. Ventrally, the adults have an anterior apodeme forming a Y-shaped juncture with the prosternal apodeme (Fig. 1C). All tracheal mites are a semi glossy white colour. Immature mites have 3 pairs of legs and are bigger than the adult female (Figs 3 and 6C). Delfinado-Baker and Baker (1982), Lindquist (1986) and Ochoa *et al.* (2005) present detailed descriptions of the genus *Acarapis,* and in particular of *A. woodi.*

2.2.2. New tarsonemid mite associations

Other mites in the family Tarsonemidae have been reported in association with *Apis* bees. *Tarsonemus apis* was reported on *Apis* spp. by Rennie (1921) and *Pseudacarapis indoapis* was reported on *Apis cerana* Fabricius by Lindquist (1986). In 2003, Ochoa *et al.* reported a new mite, *Pseudacarapis trispicula* (Fig. 4), collected from the comb of live *A. mellifera* in Mexico. Females of *P. trispicula* are similar in size (length 173-214 μm, width 85-93 μm) to *Acarapis* mites. Studies in India found *P. indoapis* was only associated with colonies of *A. cerana* in areas where colonies of *A. mellifera* were also present (Sumangala, 1999; Sumangala and Haq, 2001). Abou-Senna (1997)

reported the presence of *P. indoapis* on *A. mellifera* in Egypt. However, based on the information and drawing by Abou-Senna (1997), Ochoa *et al.* (2003) considered this association and the identification of the mite incorrect. The males, immatures and the feeding habits of *P. indoapis* have been described by Sumangala and Haq (2002), while males, immatures, and the feeding behaviour of *P. trispicula* remain unknown. Recently, *P. indoapis* were reported from both *A. cerana* and *A. mellifera* from Vanuatu and from China (Q-H Fan and S George, pers. comm. 2013; see Fig. 4).

2.2.3. Where to find mites

Acarapis woodi can hide under the flat lobe that covers the bee's first thoracic spiracle, accessing the main pro-thoracic tracheal trunk (Fig. 5). To look for or collect mites, bees must be dissected, see sections 3 and 4.

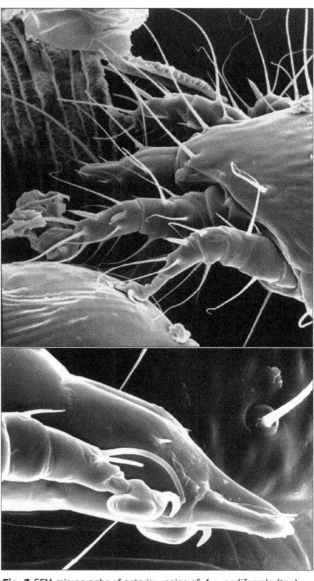

Fig. 3. SEM micrographs of anterior region of *A. woodi* female (top) and close-up of the larva (bottom).   Photos: E Erbe and R Ochoa.

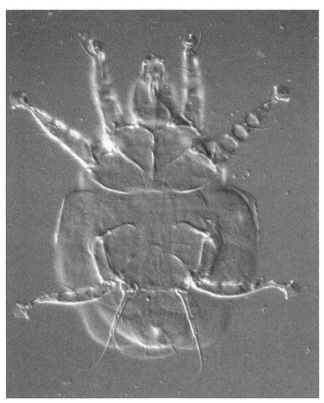

Fig. 4. *Pseudoacarapis trispicula* on *Apis mellifera*.

Photo: R Ochoa (400×).

2.2.4. Life cycle

The life stages are, egg, larva, and adult. The nymphal instar remains inside the larval skin (Fig. 5). Males complete their development in 11 to 12 days, females in 14 to 15 days; therefore, a new generation of mites can emerge in two weeks (Pettis and Wilson, 1996). All larval and adult stages of HBTM feed on bee haemolymph, which they obtain by piercing the walls of the trachea and air sacs with their sharply pointed stylets that move by internal chitinous levers (Hirschfelder and Sachs, 1952). Once the bee trachea is pierced, the mites' mouth presses close to the wound and bee haemolymph is sucked through the short tube into the pharynx.

All mite instars live within the tracheae and associated air sacs (Figs. 5, 6, 7, 8), except during a brief period when adult, mated females disperse to search for young (generally less than four days old) bee hosts. Mites are attracted to the outflowing air from the prothoracic spiracle and to specific hydrocarbons from the bees' cuticle (Phelan *et al.*, 1991; McMullan *et al.*, 2010) and immature stages may move into the trachea via air currents during bee respiration (Ochoa *et al.*, 2005). HBTM females are less attracted to older bees, which during the summer will usually not live long enough for the mites to complete their life cycle.

2.2.5. Mite dispersal

Once a suitable host is found (queens, workers or drones), the female mite enters a trachea via the spiracle to lay eggs. Drones have been found to have more mites than workers (Royce and Rossignol, 1991; Dawicke *et al.*, 1992), perhaps due to their larger tracheal trunks. Workers, however, which are more abundant through the year, are the prime host and reservoir for HBTM in bee colonies. Queens, even those commercially reared, often have HBTM. Camazine *et al.* (1998) found that infested queens weighed less, however, queens with completely black thoracic tracheae have been observed laying eggs and otherwise acting normally (D Sammataro, pers. obs.). Mites will also infest the air sacs of the bees' abdomen and head (Giordani, 1965), and can be found externally at the base of the bee's wings (Royce and Rossignol, 1991); the fate of the mites found in these areas and their effect on the host is unknown.

Female mites can disperse when the host bee is more than 12 days old, peaking at 15 to 25 days by questing on bee setae (Pettis and Wilson, 1996; Fig. 8); mites have a higher dispersal rate at night (Pettis *et al.*, 1992). Eggs and female HBTM have been found at the wingbase of the bee's thorax (Royce *et al.*, 1988). During this questing period, mites are vulnerable to desiccation and starvation, and their survival depends on the ambient temperature and humidity

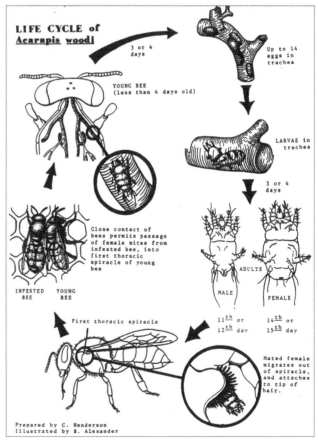

Fig. 5. Life cycle of tracheal mites. (Morse, 1991)

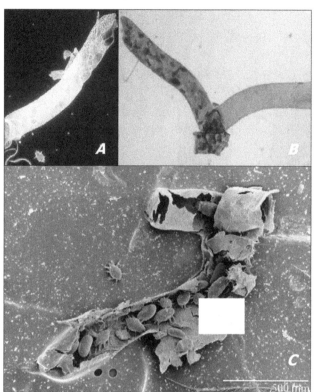

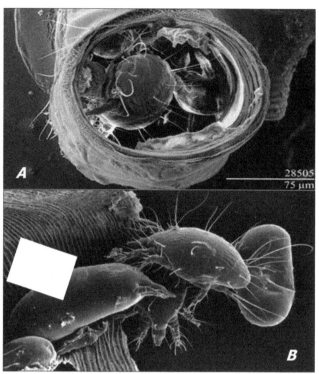

Fig. 6. A. Pro-thoracic trachea of a honey bee, filled with HBTM;
B. Micrograph of stained tracheal tubes with mites (left) and clean, no mites (right) (Photos: D Sammataro, Light Stereoscope 400×);
C. LT-SEM micrograph of an open tracheal tube showing female, male and immature mites and eggs. Note the smaller males further inside the trachea, arrowed.   Photo: E Erbe and R Ochoa.

Fig. 7. A. View of interior of tracheal tube containing a female HBTM; the long setae on Leg IV are thought to be used to help measure the interior walls of tracheae (Ochoa *et al.*, 2005);
B. Larval mite (arrow), adults and egg (far right).

LT-SEM photos: E Erbe and R Ochoa.

(Giordani, 1962). An exposed mite will die after a few hours unless it enters a host; they are also at risk of being dislodged during bee flight and grooming (Sammataro and Needham, 1996; Sammataro *et al.*, 2000). In infested and crowded tracheal tubes, males move about and locate pharate nymphal females that are about to moult to adulthood and guard them in advance of mating (Ochoa *et al.*, 2005). The males do not attach to the immature stages as is common in other genera in the family Tarsonemidae (Ochoa *et al.*, 2005). Only the female HBTM go deep into the tracheal system, measuring the walls of the tracheal branches with their dorsal and ventral setae and the leg IV seta; see Fig. 7A (Ochoa *et al.*, 2005). The eggs are 5 to 15 μm longer than the adult females (see Fig. 7B).

The genotype of honey bees and the location of the colonies influence the levels of HBTM infestations. Buckfast, ARS-Y-C-1 (Yugoslavian) and Russian honey bees are known to be resistant to HBTM (Lin *et al.*, 1996; Danka *et al.*, 1995; de Guzman *et al.*, 2002, 2005). Heat is also associated with mite mortality (Harbo, 1993). Exposing hives to direct sun impedes HBTM mite population growth and shading them tends to accelerate it (L. de Guzman, unpub. data).

Fig. 8. A female tracheal mite questing on bee seta. Drawing: D Sammataro (Sammataro and Avitabile, 2011).

3. Collecting mites

Beekeepers often use unreliable bee and colony symptoms, such as dwindling populations, abandoned overwintered hives full of honey, or weak bees crawling on the ground, as symptoms of HBTM infestation. These symptoms are not dependable and are not recommended.

3.1. Field methods

In general, bees do not show symptoms that are reliable indicators of mite infestation. Tracheal mites affect the overwintering capability of bee colonies and have been associated with bees displaying disjointed wings (called 'K-wing') or crawling on the ground near hives. Crawling bees are only apparent during winter or early spring months, particularly when HBTM infestations are very high. With the widespread distribution of nosemosis (see Fries *et al.*, 2012) and some viral infections which may cause similar symptoms (de Miranda *et al.*, 2013), the presence of crawlers in front of colonies should not be used as a reliable indicator of HBTM infestation (Bailey, 1961; Bailey & Ball, 1991).

The only way to identify HBTM is to physically examine honey bees for mites. This section describes several sampling techniques to look for mites inside bees.

3.2. Sampling colonies

3.2.1. When to sample

When trying to detect tracheal mites, sampling time is crucial. Infestation by tracheal mites varies through time. For detection of HBTM in colonies, bees should be collected in winter or early spring when HBTM populations are highest because of the reduced bee brood production. During this time, a high proportion of older, overwintering bees is present in the colonies, and the mites have had a long time to reproduce. The large number of actively feeding mites can cause the tracheae to turn black. Infestation of HBTM decreases in summer due to the dilution of mite populations as they enter the large population of emerging host bees.

3.2.2. Collecting bee samples

Because HBTM infestations are influenced by the age of bees, the location within the hive from which bees are sampled should be considered. Since queens can be found on honey frames, it is recommended to examine frames of the entire colony to find the queen before taking any samples; this will ensure that the queen will not end up in the sample jar. Collect adult drones for sampling as well, as they tend to have higher mite abundance than worker bees (Royce and Rossignol, 1991; Dawicke *et al.*, 1992). However, because drones are seasonal, adult worker bees are most often sampled for detection or surveillance purposes.

Procedure:

1. Collect about 50 bees from frames in the honey super or from the inner covers where older bees congregate. Highly-infested, older bees will have darkened trachea and many stages of mites, although some of the younger female mites may have migrated out of the tubes.

2. If young nurse bees (found in the broodnest) are present, these young bees may only have foundress mites that just started reproducing. The presence of one foundress or a foundress and an egg near the opening of the trachea may be difficult to detect. Thus, to determine mite abundance (number of mites per bee) or mite intensity (number of mites per infested bee), it is best to sample bees from the honey and pollen regions of brood frames, where a good mixture of young and old bees is generally found.

3. Collect bees by using portable insect vacuums (Fig. 9) or by scooping bees with a plastic cup directly from the frames or inner cover.

4. Place samples into vials or plastic bags or sample directly into a wide-mouth jar containing 70% ethanol. Label each container or plastic bag with location, colony number and the date the samples were collected.

Fig. 9. Sampling bees for HBTM using a modified portable car vacuum, which collects bees directly into a plastic vial.

Photo of S Cobey by D Sammataro.

5. Although bees can be preserved in 70% alcohol, fresh or frozen bees are easier to dissect. Examination of tracheae is easier when no alcohol is inside them. Bees stored for a long time in alcohol will have darker muscle and tracheal tissues, making the mites harder to detect.

6. If molecular techniques are used for mite detection, bee samples should be stored in a refrigerator at 4°C or frozen at -80°C. Storage containers are similar to those used for varroa (see Dietemann *et al.*, 2013).

3.2.3. Shipping samples

If bees are to be used for molecular analysis, samples should be shipped overnight on dry ice. Otherwise, alcohol wet samples can be shipped through postal services (see similar for varroa: Dietemann *et al.*, 2013).

3.2.4. Number of bees to be examined to detect the presence of HBTM

About 30-50 bees are examined per colony in most studies. Because the tracheal networks on the two sides of the bees do not interconnect, this represents independent samples of 60-100 tracheae. There are different ways of determining the sample size needed to accurately detect tracheal mite infestation of a colony. Frazier *et al.* (2000) developed a sequential sampling technique which they validated twice by using two levels of significance (α = 0.10 and 0.20), and a precision level of β = 0.05 and 0.10 (Table 2). This technique allows one to classify low-infested (< 10%) and highly infested (> 10%) colonies. Such information is needed before deciding when to treat or not to treat a colony, and also if further sampling is necessary. This improved technique can save time and money since it requires fewer than 50 bees to reach a decision.

For example, when 3 bees are found infested after examining 3-7 bees, stop sampling and decide to treat the colony because it is highly infested. However, if only 1 or 2 bees are infested with the first 7 examined, continue dissecting samples. If after dissecting 17 bees only 1 bee is found infested, also stop sampling and declare the colony to be low-infested and therefore, no treatment is required. However, if 5 bees are infested out of 17 bees examined, then the colony is highly-infested and needs to be treated. If only 2, 3 or 4 bees are infested out of 17 bees, continue sampling (see Table 2).

The following equation developed by Cochran (1963) is another way of finding the number of bees that need to be sampled for each colony in order to get an accurate number of mites per bee and therefore if the colony needs to be treated:

$$n_0 = \frac{Z^2 pq}{e^2}$$

Where:

n_0 is the sample size needed,

Z^2 is the abscissa of the normal curve that cuts off an area at the tails (1 equals the desired confidence level). The value for Z is found in statistical tables which contain the area under the normal curve.

e is the desired level of precision (for example, setting it at 0.05 means that the sample size provides 95% certainty of detecting a 5% tracheal mite infestation level),

p is the estimated proportion of bees infested with tracheal mites,

q is 1-p.

Example: A colony has an expected infestation of about 5%. Using this equation to determine a sample size, we will have:

Z = 1.96; α (Alpha) = 0.05 (significance level)

p = 0.05 (5%, estimated proportion of bees that are infested)

q = 0.95 (1-0.05)

e (Beta, β) = 0.05 (95% precision level)

Substituting the values:

$$n_0 = \frac{1.96^2 \cdot 0.05 \, (0.95)}{0.05^2} = \frac{3.8416 \times 0.0475}{0.0025} = 72.99 \text{ or } 73 \text{ bees}$$

If, on the other hand, infestation is estimated to be 10%, about 17 bees should be examined; an estimated 20% infestation only requires about 4 bees to be examined, since there is a higher percentage of bees infestated. This method as well as the sequential sampling technique may be useful for detection purposes (to determine when to apply treatments or for regulatory purposes) and is not recommended for scientific reporting. In that case, a full sample should be analysed (e.g. 50 or 100 bees) to determine mite prevalence (percentage of hosts infested) and/or mite abundance (number of mites per host bee). In general, tracheal mite infestations lower than 20% do not require treatment, but this depends on the severity and length of the winter months (when bees are confined in their hives).

3.3. Detection methods

Since these parasitic mites reside inside the tracheae, their detection requires specialized techniques, such as thoracic disc preparation and examination under a dissecting microscope. This is a laborious procedure. Molecular techniques are currently being developed (see section 3.4.3.) for processing bees in bulk, and should provide increased sensitivity, specificity and speed of screening bees for tracheal mites.

3.3.1. Laboratory detection: microscopic detection of *Acarapis woodi*

The morphological technique most frequently used involves examining the prothoracic tracheae under a microscope. Detection of low level infestation by *A. woodi* requires careful microscopic examination of tracheae. When the infestation is heavy, the trachea will turn opaque and discoloured and mites can be noticed without the aid of a microscope. One method is to pull off the head and collar of a bee and examine the trachea (Sammataro, 2006; see video of bee dissection at: http://www.ars.usda.gov/pandp/docs.htm?docid=14370).

Table 2. How to make decisions using the sequential sampling technique (modified from Frazier *et al.,* 2000); Calderone and Shimanuki 1992; see text 3.2.4. for explanation.

No. of bees examined	Number of infested bees		
	Low infestation (stop, don't treat)	High infestation (stop, treatment)	Moderate infestation (continue sampling)
1	-	-	0,1
2	-	-	0,1,2
3	-	3	0,1,2
4	-	3	0,1,2
5	-	3	0,1,2
6	-	3	0,1,2
7	-	3	0,1,2
8	-	4	0,1,2,3
9	-	4	0,1,2,3
10	-	4	0,1,2,3
11	0	4	0,1,2,3
12	0	4	1,2,3
13	0	5	1,2,3,4
14	0	5	1,2,3,4
15	0	5	1,2,3,4
16	1	5	1,2,3,4
17	1	5	2,3,4
18	1	6	2,3,4,5
19	1	6	2,3,4,5
20	1	6	2,3,4,5
21	1	6	2,3,4,5
22	1	6	2,3,4,5
23	2	6	3,4,5
24	2	7	3,4,5,6
25	2	7	3,4,5,6
26	2	7	3,4,5,6
27	2	7	3,4,5,6
28	3	7	4,5,6
29	3	8	4,5,6,7
30	3	8	4,5,6,7
31	3	8	4,5,6,7
32	3	8	4,5,6,7
33	4	8	5,6,7
34	4	8	5,6,7,8
35	4	9	5,6,7,8
36	4	9	5,6,7,8
37	4	9	5,6,7,8
38	4	9	5,6,7,8
39	5	9	6,7,8
40	5	10	6,7,8,9
41	5	10	6,7,8,9
42	5	10	6,7,8,9
43	5	10	6,7,8,9
44	6	10	7,8,9
45	6	11	7,8,9,10
46	6	11	7,8,9,10
47	6	11	7,8,9,10
48	6	11	7,8,9,10
49	6	11	7,8,9,10
50	7	11	8,9,10

3.3.2. Screening individual bees

When the level of infestation is low, tracheae from an individual bee need to be examined. Bees may be anesthetized or killed by freezing before examination. Milne (1948) developed a technique to locate the internal mites on individual bees (see Figs. 10 and 11 for details).

1. The bee is placed under a dissecting microscope, held prone with forceps (across abdomen) and the head and the first pair of legs is scraped off using a scalpel or razor blade.
2. The ring of prothoracic sclerite (collar) is also removed using a fine forceps.
3. The exposed tracheae of both sides are removed after carefully detaching them from the thoracic wall.
4. The tracheae are removed and placed on a glass slide and examined under a microscope for mites; this technique is very time-consuming and also has the possibility to lose mites while separating tracheae from the thoracic wall and transferring them to the slide.
5. Lorenzen and Gary (1986) modified this technique where the thoracic tergite was removed as a flap to look at mites *in situ* (see also Ritter *et al.,* 2013).

Liu (1995) developed a rapid technique to distinguish live mites from dead by staining with thiazolyl blue tetrazolium, which stains the live mites purple. The tracheae, after mounting on glass slides, are perfused with thiazolyl blue tetrazolium solution (5 mg stain in 5 ml distilled water). The cuticle of live mites picks up the stain immediately and turns purple, dead mites turn greenish yellow.

3.3.3. Sample preparation: morphological analysis and slide preparation

HBTM specimens should be preserved carefully for microscopic examination. Since these mites are weakly sclerotized, no clearing agents are need. The mounting medium listed below is recommended for mounting specimens.

Mounting medium (Hoyer's medium from Kranz and Walter, 2009)

    50 ml of distilled water

    30 g gum Arabic

    200 g chloral hydrate

    20 ml glycerol

Note: The ingredients should be mixed in the sequence listed above. Allow each solid ingredient to dissolve before adding the next one. The mixture needs to be warmed and stirred gently so the gum arabic can melt. Filter and store in airtight containers with rubber stoppers. Do not use a screw top container. ***Caution:*** chloral hydrate is a toxic chemical, a mutagen and a chromosome-damaging agent.

3.3.4. How to mount specimens

1. Place a drop of Hoyer's medium (section 3.3.3.) in the centre of a clean microscope slide.
2. Using the tip of a fine needle, minute pin or a wire loop, pick up a mite.

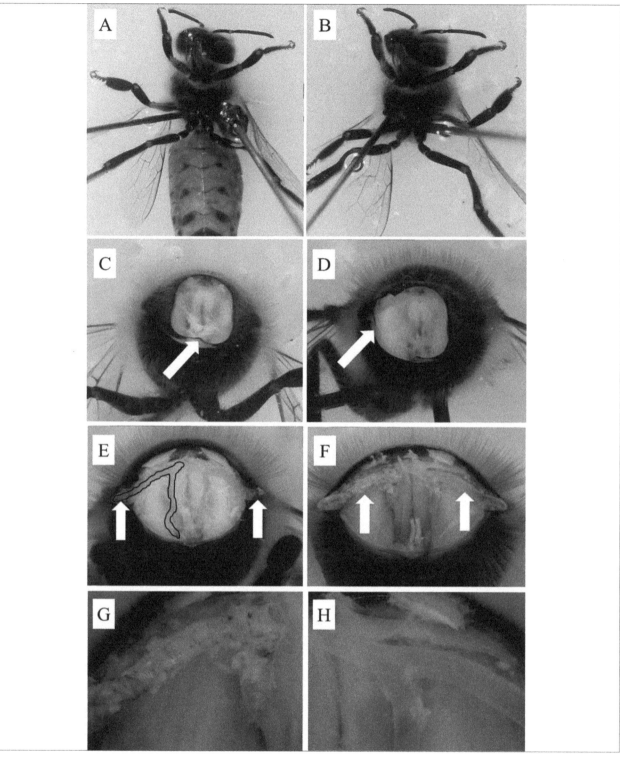

Fig. 10. Dissecting a bee to determine tracheal mite infestation. ***A.*** Pin a worker bee through the thorax using two insect pins (the bee's body will pivot in the dish if pinned with only 1 pin). The bee in this figure is pinned into a petri dish of hardened beeswax. The bee is covered in 70% ethanol to facilitate dissection. ***B.*** Remove the abdomen. This is a helpful technique to limit contents from the bee's digestive systems from emptying into the field of view. ***C.*** Remove the head and front pair of legs. The "collar" junction is arrowed. ***D.*** The beginning of collar removal. The area where the collar has been removed is arrowed. ***E.*** The collar has been removed, including the part that covers the spiracles (arrowed). The bee's right trachea (the one on the left in the figure) is outlined to show shape and position. ***F.*** Trachea infested with mites (on left, arrowed) and not infested with mites (on right, arrowed). ***G.*** Close-up of infested trachea and ***H.*** close-up of uninfested trachea.

Photos: Lyle Buss and Tricia Toth, University of Florida, USA.

3. Touch the tip of the tool to which the mite adhered to the droplet of Hoyer's medium.

4. Gently press the mite to the bottom of the droplet and position the mite on a vertical axis.

5. Using a pair of forceps, pick up a cover slip and gently place it on top of the droplet.

6. Mark the mite's location by drawing a ring around the specimen using a permanent marker, to facilitate locating the specimen later.

7. Label slides with date of collection, place of collection, species name, host species and name of collector.

8. Dry slides at 45ºC for 48 h to one week.

9. Slides can be sealed with a ring of glyptal (Glyptal, Inc.; Chelsea, MA, USA)

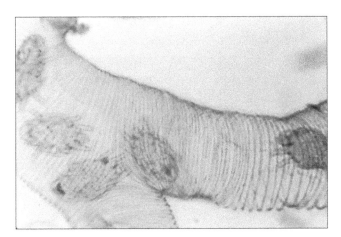

Fig. 11. Mites seen through tracheal tube.
Light Microscope photo: L de Guzman (140x).

3.3.5. Screening a large number of bees

For screening tracheae of many bees together to look for HBTM, a number of methods have been developed. Colin *et al.* (1979) developed the following technique.

1. Place bee thoraces in a blender with water

2. Grind for several seconds at 10,000 rpm 3 times to suspend the mites.

3. Strain the liquid through a screen mesh to remove larger particles and then centrifuge it to deposit the suspended particles at the bottom of the tube, which is then examined for mites.

The advantage of this technique is that a large number of bees (100-200) can be processed together. However, other *Acarapis* species such as *A. dorsalis* and *A. externus* that reside on the neck, thorax and wing bases will also be extracted with this method; see Section 2.2.

The morphological separation of these species is very time consuming. Washing bees prior to grinding was not found to be effective in removing *A. externus* or *A. dorsalis* (Lorenzen and Gary,

1986; S George, pers. obs. in New Zealand). A 'tracheal flotation technique' developed by Camazine (1985) reduced this risk somewhat (of mixed *Acarapis* species) by grinding the bee thoraces in water; however, this method would not be suitable to detect low levels of infestation.

3.3.6. Thoracic disc method (TDM)

TDM is a technique developed for detailed assessment of mite infestations. TDM involves cutting a thoracic disc that contains the prothoracic tracheae.

1. The bee is placed, dorsal side down and pinned in place and a razor blade or scalpel is used to cut off the head; then a thin transverse section is cut from the anterior face of the thorax, resulting in a 1 to 1.5 mm section, which includes the trachea (Fig. 12).

2. The discs are then heated on a hot plate (approximately 60ºC for a minimum of 2 h) in 5-10% potassium hydroxide (KOH) to dissolve the surrounding tissues.

3. The contents then are passed through a fine strainer over a sink and rinsed with cold water to remove dissolved matter.

4. The samples are returned to a hot plate to digest further for another hour after adding fresh KOH, if the muscle tissue has not been completely cleared.

5. When the thoracic discs become transparent in the middle, leaving only the sclerotized tergites around the outside, they are sieved and gently rinsed with cold tap water.

6. The discs are returned to the Petri dish and suspended in distilled water containing a few drops of aqueous methylene blue (1%) (Peng and Nasr, 1985).

7. Tracheae are then examined for tracheal mites under magnification (*ca.* 20-40×) using a dissecting microscope with lighting from below. Even a small number of mites can be detected through this method.

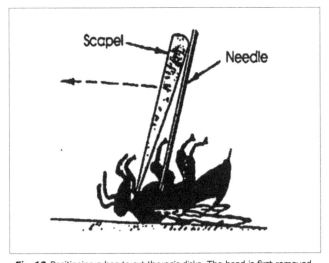

Fig. 12. Positioning a bee to cut thoracic disks. The head is first removed, then a thin section of the thorax is sliced. From Shimanuki and Knox, 1991.

8. Once cleared, the tracheae are then individually mounted on slides and examined under a microscope (Shimanuki and Cantwell, 1978; Delfinado-Baker, 1984; Shimanuki and Knox, 1991). A modified version of the thoracic disc method is used for detection of HBTM in New Zealand. Sampled bees are frozen for at least 24 h to facilitate cutting the thorax.

3.4. Serological and molecular detection of *Acarapis woodi*

3.4.1. Enzyme-linked Immunosorbent Assay (ELISA)

Ragsdale and Furgala (1987) developed antiserum against *A. woodi* where infested tracheae were detected using a direct enzyme-linked immunosorbent assay (ELISA); this method was further modified by Ragsdale and Kjer (1989). This assay was sensitive enough to detect a very low level of tracheal mite infestation but was found to cross-react with other proteins present in the haemolymph and thoracic muscles. The lack of specificity limits the application of this test to tracheal preparations. A practical ELISA test was developed by Grant *et al.* (1993) where whole bee samples could be analysed for HBTM detection, but the sensitivity of the test was reduced when mite prevalence fell below 5%.

3.4.2. Guanine visualization

This is an indirect method based on detecting Guanine (2-amino-6-oxypurine), which is the main end product of nitrogen metabolism in mites and other arachnids. It is present only in a negligible amount in bee excretions. In this method, bee tracheae are individually homogenized and their guanine content is visualized on thin layer chromatography (TLC) plates. Since bees need to be individually tested and low level infestations may go undetected (Mozes-Koch and Gerson, 1997), this method is not usually recommended.

3.4.3. Molecular detection of *Acarapis woodi* in *Apis mellifera*

Because morphological detection is time-consuming and requires detailed and sustained attention by the screener, detection of *A. woodi* using a molecular technique is currently being developed by various laboratories for routine screening and quarantine checking. A real-time PCR assay for *A. woodi* was designed by Giles Budge at the UK Food and Environment Research Agency (Fera) which amplifies a section of the internal transcribed spacer region 2 (ITS2); however when tested, it was found to also amplify the ITS sequence from other *Acarapis* species. Evans *et al.* (2007) developed a nested PCR to amplify part of the cytochrome oxidase1 gene (CO1) of *A. woodi*. The PCR was designed to detect a low level of infestation of *A. woodi* from the entire thorax of a bee. At the time of development, the assay was not tested against other *Acarapis* species, but subsequent testing has shown that these primers amplify sequences from them also (Delmiglio *et al.*, in prep.).

Kojima *et al.* (2011b) published a conventional PCR method shown to amplify *A. woodi* DNA and not *A. externus,* but this method has not been tested on *A. dorsalis* DNA. Although their test was able to detect *A. woodi* when four mites were present in a single bee, it was not able to detect the presence of a single mite unless a nested-PCR was performed. Furthermore, Kojima *et al.* (2011b) did not test the sensitivity of detection at different bulking rates for extractions and the need for post amplification handling (i.e. electrophoresis) increases the length of the detection procedure.

Delmiglio and Ward (unpublished) obtained sequences from the CO1 region for *A. woodi*, *A. externus* and *A. dorsalis* and designed real-time PCR primers and a TaqMan probe for *A. woodi* within a single variable region of the CO1 gene. The assay was able to amplify *A. woodi* DNA from a single mite (specimens obtained from Canada and UK).

Tests showed that the assay was able to detect *A. woodi* down to 1% and 2% prevalence in bees, and 200 copies of the target DNA when using plasmid standards. The real-time assay was not found to cross-react with mites of other genera associated with bees; however a very low level of cross-reaction occurs with the other two *Acarapis* species when sequence from these species are present at high concentrations in the form of plasmid DNA, or when there are a very high number *Acarapis* mites. To counter this, bees are externally washed before testing and a lower C_t cut-off of 35 cycles is used.

Nucleic acid extraction:

Note: the extraction method is based on the semi-automated system Thermo Kingfisher & InviMag® DNA Mini Kit (Invitek GmbH; Germany):

1. Using this semi-automated extraction method, 15 bees maximum can be sampled per extraction to allow reliable detection of *A. woodi* from a single infected bee.
2. Prepare bees by removing the abdomens using a clean scalpel, and place bee heads and thoraces into a filter grinding bag (Bioreba; Switzerland) of suitable size.
3. Add lysis Buffer P (Invitek GmbH; Germany) to the grinding bag at a rate of 0.5 ml buffer per bee (i.e. use 7.5 ml for 15 bees), and using a Homex grinder (Bioreba, Switzerland) grind the bees to form a lysate.
4. Decant 600 µl of the lysate into a clean micro-centrifuge tube, add 30 µl of proteinase K (Invitek GmbH; Germany) and incubate at 65°C for 30 min in a thermomixer (Eppendorf; Germany).
5. Centrifuge the resulting lysate at 8000 *g* for 1 minute to pellet debris.
6. Extract DNA from the cleared lysate using a KingFisher® ml Magnetic Particle Processor system (Thermo Fisher Scientific Inc.; USA) using the protocol and program recommended by the manufacturer, and using buffers and magnetic beads

supplied in an InviMag® DNA Mini Kit (Invitek GmbH; Germany).

7. The resulting DNA is eluted in 260 ml of elution buffer D (Invitek), transferred to a fresh 1.5 ml tube and stored at -80°C prior to use.

3.4.4. Real-time PCR

1. qPCR reactions are run in final volume of 10 µl using Platinum® qPCR SuperMix-UDG (Invitrogen-Life Technologies)

2. Each reaction contains a final concentration of:
   - 1x qPCR SuperMix
   - 3.5 mM of Mg ions
   - 0.3 µg of Bovine Serum Albumin (BSA)
   - 300 nM each of forward and reverse primer (Table 3) and
   - 100 nM of LNA probe (Table 3)

3. The cycling conditions used are (as optimized in a BioRad CFX -BioRad Laboratories) - 50°C for 2 min (UDG incubation-single hold)
   - 95°C for 2 min (initial denaturation-single hold)
   And 35 cycles of:
      - 95°C for 10 sec (denaturation)
      - 59°C for 45 sec (anneal and extension)
      - Reading of signal at end of each cycle.

Note: The PCR competency and success of nucleic acid extracted from bees is assessed using the internal control real-time PCR assay which amplifies part of the 18s rRNA gene of *A. mellifera* (Ward *et al.*, 2007). The 18s real-time reactions were set-up and cycled as described for the *A. woodi* assay.

4. Experimentation with live mites

Mite populations tend to fluctuate seasonally, in patterns that are moderately consistent at the regional level. However, individual colonies frequently experience unexpected increases or decreases in mite abundance, thereby complicating experimentation with them. There is no record of experimentation with individual mites until after they were discovered in North America in the 1990s. Because little has been published on techniques for experimenting with mites, most of the researchers who have studied mites in recent years provided information that is synthesized below.

4.1. Rearing HBTM in colonies

Tracheal mite prevalence (the proportion of infested bees in a sample; Margolis *et al.*, 1982) generally decreases during periods of the year when large numbers of young bees are emerging (due to extensive brood rearing and relatively short-lived adult bees). Prevalence increases during fall (when brood rearing is declining) or periods of confinement (e.g. rainy weather), presumably because there is extensive contact between older infested bees and young susceptible bees (Bailey & Ball, 1991). In cold temperate climates, mite prevalence usually increases rapidly from late summer until early winter (Otis *et al.*, 1988; de Guzman *et al.*, 2002). However, HBTM prevalence in individual colonies in summer is not correlated with mite prevalence in late fall; only when bee brood has largely disappeared can mite prevalence in wintering colonies be predicted with any assurance (r = ~0.8, Dawicke *et al.*, 1989). Due to variable bee mortality during the winter as well as brood production in infested colonies in early spring, mite prevalence in the fall is uncorrelated with mite prevalence the following spring (Dawicke *et al.*, 1989). This makes experimentation difficult because heavily infested colonies cannot be identified until shortly before they are needed for experiments.

Once the wintering population of bees has developed (Mattila *et al.*, 2001), mite prevalence tends to remain relatively constant over the winter months in the absence of newly emerging worker bees (Bailey, 1958); however, in some situations mite prevalence has increased over winter (Otis *et al.*, 1988; McMullan, 2011). It is possible that mite emigration from tracheae is stimulated by high or increasing titers of juvenile hormone (JH). This speculation would explain observations that although mite abundance (i.e. the mean number of mites per bee) continues to increase in bee tracheae during the winter months due to continuing mite reproduction, mite prevalence (percent of bees infested of the total number of bees in the samples) generally does not. J McMullan (pers. comm.) indicated that when there is little or no brood present in autumn and winter, it is difficult to influence mite infestations experimentally.

4.1.1. Removing sealed brood

During spring and summer, when brood rearing is continuous and extensive, it is possible to increase mite infestations in colonies by removing sealed bee brood (i.e. pupae). This has been done regularly

---

Table 3. Real-time PCR (qPCR). Sequence of primers and probe used for the detection of *Acarapis woodi*. Areas of sequences in bold (column 2) are non-complementary flaps; [+] Locked nucleic acid bases.

Primer/Probe name	Sequence 5'-3'	Amplicon length
aw_F1-flap	**AATAAATCATAA**TGATATCCCAATTATCTGAGTAATG	
aw_R3	AATATCTGTCATGAAGAATAATGTC	113 bp
aw-LNAProbe	6FAM-ACC[+T]GT[+C]AA[+T]CC[+A]CCTAC-BHQ1	

by numerous researchers (McMullan and Brown, 2005; G W Otis, pers. obs.; J Villa, pers. comm.). Confining bees in hives may also increase infestation of young workers, but this has not been tested. One challenge is to increase mite prevalence, yet not so much that winter survival is jeopardized. Because many heavily infested colonies die over winter, studying live mites often requires sampling hives in spring to find heavily infested colonies for experimentation. Several researchers have moved heavily infested colonies, only to subsequently find that the mite infestations declined drastically (J Pettis, pers. comm.; J McMullan, pers. comm.; G W Otis, pers. obs.). It is not known why this occurs, but suggested causes are overheating or chilling during transport, loss of infested bees due to their failure to return to their hives after orientation flights in new locations (J Pettis, pers. comm.), or emigration of infested bees from the nest (J McMullan, pers. comm.).

4.1.2. Artificial rearing of HBTM

Bruce *et al.* (1991) experimented with *in vitro* rearing of *A. woodi.* They encased modified insect tissue culture medium, Medium MD1 (Whitcomb, 1983) developed for rearing *Pyemotes tritici* mites (Bruce, 1989) in tiny parafilm® tubes with a thickness of approximately 10 μm, made by stretching the film several times. Live mites were placed on these tubes of medium within a closed Petri dish. Although only a small number of mites were studied, all of them successfully pierced the artificial membrane and fed on the medium. There has been no further research on artificial rearing methods for HBTM.

4.2. Infesting bees with HBTM

1. Allow worker bees to emerge in incubators.
2. Mark bees with paint on their abdomens or numbered tags glued to their thoraces shortly after emergence. Paint applied to the bees' thoraces may interfere with migration of mites into their tracheae.
3. Introduce them to infested colonies, preferably at dusk. Otherwise, screen the hive entrances for 1-2 h to prevent their rejection. Giordani (1962) referred to this method as *infestation expérimentale.* It has served as the method of obtaining infested bees to study the life history of the mite (Pettis and Wilson, 1996), host age-preference (Morgenthaler, 1931; Lee, 1963; Gary *et al.*, 1989), caste preference (Dawicke *et al.*, 1992), relative resistance of bees from different genetic sources (Page and Gary, 1990; Danka and Villa, 1999; Nasr *et al.*, 2001), and the effects of mites on adult bees (Bailey and Lee, 1959; Gary *et al.*, 1989).

4.3. Collecting live mites

1. Anesthetize a bee with CO_2 (or place bees in refrigerator until immobile).
2. Grab the paranotal lobe that covers the metathoracic spiracle with fine forceps and pull the lobe posteriorly at a modest

angle to the bee's surface (Smith *et al.*, 1987)

3. The lobe, spiracle, and most of the main tracheal trunk will be removed together. This technique was subsequently used by Phelan *et al.* (1991), Bruce *et al.* (1991), and Sammataro and Needham (1996) to obtain living mites.
4. Other researchers have removed the bee's head and pronotum to expose the large thoracic tracheal trunks. These are either grasped at the spiracle with fine forceps (J Pettis, pers. comm.; vanEngelsdorp and Otis, 2001b) or a cut is made around the spiracle before removing the tracheal trunk (McMullan *et al.*, 2010).
5. Allow the section of the trachea to dry onto a glass slide for ~ 2 min.
6. Tear open trachea with a minuten pin and pick up the mites with an eyelash attached to a wooden dowel (Phelan *et al.*, 1991; vanEngelsdorp and Otis, 2001b; McMullan *et al.*, 2010).
7. Young, lightly bronzed female mites can be selected; males, darker (older) females and those obviously gravid with an egg can be avoided because they have reduced mobility (McMullan *et al.*, 2010; J Pettis, pers. comm.).
8. A female mite placed externally near the paranotal flap of a bee will crawl towards and enter the spiracle, (D Sammataro, pers. obs.). When drones are present, they are preferred because mite intensity is generally higher in drones than in workers (Dawicke *et al.*, 1992).
9. Exposed mites are sensitive to desiccation, so they should be kept at high humidity.
10. vanEngelsdorp and Otis (2001b) could perform bioassays with living mites only at night, perhaps because of a circadian rhythm in mite activity that influences their nocturnal dispersal to new hosts (Pettis *et al.*, 1992).

5. Controlling tracheal mites

5.1. Effects on bees

HBTM can cause diminished brood area, smaller bee populations, looser winter clusters, increased honey consumption, lower honey yields and frequently, colony demise (Komejli *et al.,* 1989). In temperate regions, mite populations increase during the winter, when bees are confined to the hive in the winter cluster. Heavy mite infestation affects bee metabolism and the ability of colonies to regulate the cluster temperature (Skinner, 2000); chilling may be a significant cause of colony death. In North America, colony losses increased shortly after first exposure to HBTM (Wilson *et al.*, 1997). Colonies with 40-50% tracheal mite infestation or higher frequently die during the winter in northern USA and Canada (Furgala *et al.* 1989; Otis and Scott-Dupree, 1992). The colder the winter temperatures, the greater the probability of mortality at any mite prevalence value.

Treatments for HBTM include using vapours from menthol crystals, synthetic acaricides and oil or grease patties made from vegetable shortening and sugar. Additionally, resistant lines of bees have been developed.

5.2. Resistant bees

Some races of bees are less susceptible to tracheal mite infestations. Resistance is accomplished in part by the increased autogrooming behaviour of bees (Lin *et al.*, 1996; Pettis and Pankiw, 1998; Danka and Villa, 2005; Villa, 2006; de Guzman *et al.*, 2002, 2005). Russian and Buckfast queens are among the more resistant lines of bees. Fig. 13 illustrates steps needed to select and propagate bees that show resistance to HBTM. Colonies with 15-25% of workers infested experience increased winter mortality. Bee stock can be screened for HBTM resistance to develop resistant queen lines. In general, untreated colonies that survive the winter with low mite numbers should be selected for a breeding programme.

1. First, select colonies to be tested for mite resistance and collect sealed brood frames from them.
2. Brush off the bees clinging to the frame and place it into a frame cage (Fig. 13, insert) and then into an incubator set at 35ºC and 50% RH.
3. Mark each frame to identify the test colony it was removed from
4. After 12-24 hours in the incubator, remove and mark any emerging bees. Bees identified by coloured plastic tags or model paint can be separated by colony of origin and placed into an inoculation colony, where HBTM prevalence is high (> 50%). (see section 4.2)
5. After a week or more, the marked bees that were released into host colonies can be retrieved and dissected

to determine mite prevalences (percent of mite-infested bees in the samples) or mite abundances (average number of mites per bee/total number of individuals of the host bees in the sample) for the bees from each test colony source. Original colonies that tested lowest for HBTM can then be selected for a breeding program. For more information on breeding and selecting traits in queen bees, see Büchler *et al.*, 2013.

5.3. Chemical control

The overriding constraints for chemical control of mites are that the chemicals must be effective against the target (i.e. mites) and harmless to bees, and they must not accumulate in hive products. Because bees and mites are both arthropods, many of their basic physiological processes are similar, narrowing the possibilities for finding suitable toxicants. To control HBTM, the material must be volatile to reach the bee, inhaled, and lethal only to the parasite.

Based on early research by Giordani (1977), menthol was widely tested and registered for treatment of hives in the United States and Canada. Menthol is extracted from the mint plant *Mentha arvensis* and is sold as crystals, which can be applied inside hives (usually 50 g of crystals in a nylon mesh sac are placed on the top bars of frames). In cold conditions, menthol is ineffective for mite control because the rate of vaporization is too low to provide concentrations lethal to the mites. Some Canadian beekeepers obtained good results under cool conditions by applying pieces of cardboard saturated with a mixture of menthol (dissolved in alcohol) and vegetable shortening to hives. If the air temperatures outside are too high, menthol vapours may drive bees out of their hives. If bees are hanging outside the colony entrance in treated colonies, crystals should be removed until temperatures drop.

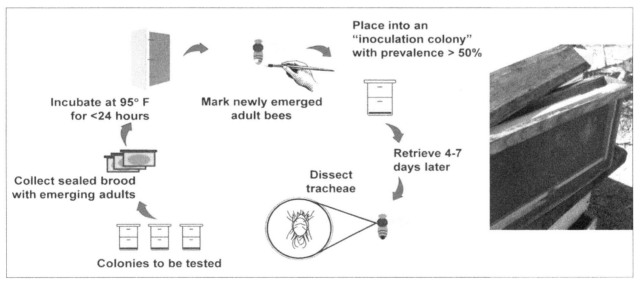

Fig. 13. Schematic diagram outlining a procedure to determine the relative susceptibility of honey bee colonies to HBTM infestation. Brood combs with emerging bees are brushed free of adult bees and then placed into frame cages (insert) or other containers to isolate bees; frames are placed into an incubator. After a week, emerging bees are marked and then placed into inoculation colonies. The marked bees are retrieved after a week and dissected to determine mite prevalences or mite abundances for the bees from each test colony source (from Danka, 2001). See text for explanation.

The synthetic pesticide amitraz, (sold as Apivar®) has been used for HBTM control. Check current regulations to determine its registration status. Formic acid has also been used against *A. woodi* (Hoppe *et al.*, 1989; Amrine and Noel, 2006; Underwood and Currie, 2009; Hood and McCreadie 2001). Widely used as a treatment for varroa mites, formic acid (along with other acaricides used against varroa) may be contributing to current lower populations of HBTM in North America. As many of these chemicals could have a negative impact on both the bees and the honey, they are not recommended as both non-toxic vegetable shortening and resistant queen stock are better alternatives.

5.4. Cultural control

An alternate, environmentally safe control is to use a 'grease' patty. More accurately, this is a vegetable shortening and sugar patty (2:1 sugar: shortening by volume, or sufficient shortening that the mixture does not break apart). Vegetable shortening (solid, hydrogenated vegetable oil, not animal fat) and white granulated sugar will keep mite prevalence to 10%, well below typical economic thresholds (< 20%). Animal fat will go rancid over time and is not recommended. Liquid vegetable oil can be used, but the patty will be looser and may not hold up in the hive, unless mixed with ample amount of sugar.

A 113 g patty, placed on the top bars at the centre of the broodnest where it comes in contact with the most bees, will protect young bees (which are most at risk) from becoming infested. Patties can be prepared ahead of time if they are wrapped in waxed paper and stored in the freezer. The shortening appears to disrupt the questing female mites as they search for new hosts (young bees) (Sammataro and Needham, 1996; Sammataro *et al.*, 1994). Because young bees are emerging continuously, the patty must be present for an extended period. The optimal application seasons are fall and early spring, when mite levels are increasing.

Requeening colonies is likely to alter HBTM infestations once offspring of the new queens begin to emerge; those new offspring may be more or less resistant to mites than the worker offspring of the original queens. Apiary location may also affect tracheal mite infestations (vanEngelsdorp and Otis, 2001a; L de Guzman, unpubl. data), perhaps due to temperature or humidity effects. Harbo (1993) demonstrated that hives exposed to the sun in Louisiana had reduced mite infestations. Integrated Pest Management (IPM) techniques have been successful in keeping HBTM under control; however, no biological controls currently exist (Fig. 14).

A cautionary note should be added. Many beekeepers are opting not to treat for mites or diseases, allowing survivor stock to become established. HBTM may reappear if treatments for varroa mites are suspended; sampling for HBTM is therefore recommended, especially if colony symptoms consistent with tracheal mite infestations are observed.

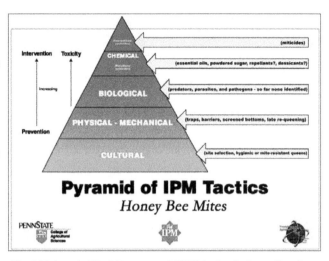

Fig. 14. Integrated Pest Management (IPM) tactics for bee mites. For HBTM, 'grease' patty should be listed with essential oils and powdered sugar (D Sammataro).

6. Acknowledgements

We are grateful to the authors and the reviewers who gave helpful suggestions and especially to Jamie Ellis for his patience. We wish to acknowledge the COLOSS network for making the BEEBOOK possible. The COLOSS (Prevention of honey bee COlony LOSSes) network aims to explain and prevent massive honey bee colony losses. It was funded through the COST Action FA0803. COST (European Cooperation in Science and Technology) is a unique means for European researchers to jointly develop their own ideas and new initiatives across all scientific disciplines through trans-European networking of nationally funded research activities. Based on a pan-European intergovernmental framework for cooperation in science and technology, COST has contributed since its creation more than 40 years ago to closing the gap between science, policy makers and society throughout Europe and beyond. COST is supported by the EU Seventh Framework Programme for research, technological development and demonstration activities (Official Journal L 412, 30 December 2006). The European Science Foundation as implementing agent of COST provides the COST Office through an EC Grant Agreement. The Council of the European Union provides the COST Secretariat. The COLOSS network is now supported by the Ricola Foundation - Nature & Culture.

7. References

ABOU-SENNA, F M (1997) A new record of phoretic mites on honey bee *Apis mellifera* L. in Egypt. *Social Parasitology* 27(3): 667-680.

AMRINE, J W Jr; NOEL, R (2006) Formic acid fumigator for controlling honey bee mites in bee hives. *International Journal of Acarology* 32: 115-124.

BAILEY, L (1958) The epidemiology of the infestation of the honey bee, *Apis mellifera* L., by the mite *Acarapis woodi* Rennie and the mortality of infested bees. *Parasitology,* 48: 493-506.

BAILEY, L (1961) The natural incidence of *Acarapis woodi* (Rennie) and the winter mortality of honey bee colonies. *Bee World* 42: 96-100.

BAILEY, L (1964) The 'Isle of Wight disease': the origin and significance of the myth. *Bee World* 45: 32-37, 18.

BAILEY, L; BALL, B V (1991) *Honey bee pathology.* Academic Press; London, UK.

BAILEY, L; LEE, D C (1959) The effect of infestation with *Acarapis woodi* (Rennie) on the mortality of honey bees. *Journal of Insect Pathology* 1: 15-24.

BRUCE, W A (1989) Artificial diet for the parasitic mite *Pyemotes tritici* (Acari: Pyemotidae). *Experimental and Applied Acarology* 6: 11-18.

BRUCE, W A; HENEGAR R B; HACKETT, K J (1991) An artificial membrane for *in vitro* feeding of *Varroa jacobsoni* and *Acarapis woodii,* mite parasites of honey bees. *Apidologie* 22: 503-507.

BÜCHLER, R; ANDONOV, S; BIENEFELD, K; COSTA, C; HATJINA, F; KEZIC, N; KRYGER, P; SPIVAK, M; UZUNOV, A; WILDE, J (2013) Standard methods for rearing and selection of *Apis mellifera* queens. In *V Dietemann; J D Ellis; P Neumann (Eds) The COLOSS* BEEBOOK, *Volume I: standard methods for* Apis mellifera *research. Journal of Apicultural Research* 52(1): http://dx.doi.org/10.3896/IBRA.1.52.1.07

CALDERONE, N W; SHIMANUKI, H (1992) Evaluation of sampling methods for determining infestation rates of the tracheal mite (*Acarapis woodi* R.) in colonies of the honey bee (*Apis mellifera*): Spatial, temporal and spatio-temporal effects. *Experimental and Applied Acarology* 15: 285-298.

CAMAZINE, S (1985) Tracheal flotation: a rapid method for the detection of honey bee acarine disease. *American Bee Journal* 125 (2): 104-105.

CAMAZINE, S; ÇAKMAK, I; CRAMP, K; FINLEY, J; FISHER, J; FRAZIER, M (1998) How healthy are commercially-produced US honey bee queens? *American Bee Journal* 138: 677-80.

COCHRAN, W G (1963) Sampling techniques *(2nd Ed.).* John Wiley and Sons Inc.; New York, USA.

COLIN, M E; FAUCON, J P; GIAUFFRET, A; SARRAZIN, C (1979) A new technique for the diagnosis of acarine infestation in honey bees. *Journal of Apicultural Research* 18: 222-224.

DANKA, R G (2001) Resistance of honey bees to tracheal mites. In *T Webster; K S Delaplane (Eds). Mites of the honey bee.* Dadant and Sons; Hamilton, IL, USA. pp 117-129.

DANKA, R G; VILLA, J D (1996) Influence of resistant honey bee hosts on the life history of the parasite *Acarapis woodi. Experimental and Applied Acarology* 20 (6): 313-322.

DANKA, R G; VILLA, J D (2005) An association in honey bees between autogrooming and the presence of migrating tracheal mites. *Apidologie* 36(3): 331-333.

DANKA, R G; VILLA, J D; RINDERER, T E; DELATTE, F T (1995) Field test of resistance to *Acarapis woodi* (Acari: Tarsonemidae) and of colony production by four stocks of honey bees (Hymenoptera: Apidae). *Journal of Economic Entomology* 88: 584–91.

DAWICKE, B L; OTIS, G W; SCOTT-DUPREE, C; NASR, M E (1992) Host preference of the honey-bee tracheal mite (*Acarapis woodi* (Rennie)). *Experimental and Applied Acarology* 15: 83-98.

DAWICKE, B; OTIS, G W; SCOTT-DUPREE, C D (1989) Predicting tracheal mite infestations and effects on colonies. *American Bee Journal* 129: 814.

DE GUZMAN, L I; RINDERER, T E; BIGALK, M; TUBBS, H; BERNARD, S J (2005) Russian honey bee (Hymenoptera: Apidae) colonies: *Acarapis woodi* (Acari: Tarsonemidae) infestations and overwintering survival. *Journal of Economic Entomology* 98 (6):1796-1801.

DE GUZMAN, L I; RINDERER, T E; DELATTE, G T; STELZER, J A; BEAMAN, G; KUZNETSOV, V (2002) Resistance to *Acarapis woodi* by honey bees from Far-eastern Russia. *Apidologie* 33(4): 411-415.

DE MIRANDA, J R; BAILEY, L; BALL, B V; BLANCHARD, P; BUDGE, G; CHEJANOVSKY, N; CHEN, Y-P; GAUTHIER, L; GENERSCH, E; DE GRAAF, D; RIBIÈRE, M; RYABOV, E; DE SMET, L; VAN DER STEEN, J J M (2013) Standard methods for virus research in *Apis mellifera.* In *V Dietemann; J D Ellis; P Neumann (Eds) The COLOSS* BEEBOOK, *Volume II: standard methods for* Apis mellifera *pest and pathogen research. Journal of Apicultural Research* 52(4): http://dx.doi.org/10.3896/IBRA.1.52.4.22

DELFINADO-BAKER, M (1984) *Acarapis woodi* in the United States. *American Bee Journal* 124: 805–806.

DELFINADO-BAKER, M; BAKER, E W (1982) Notes on honey bee mites of the genus *Acarapis* Hirst (Acari: Tarsonemidae). *International Journal of Acarology* 8: 211–26.

DELMIGLIO, C; WARD, L I; GEORGE, S; KUMARASINGHE, L; FLYNN, A; CLOVER, G R G (2013) Development and evaluation of a Real-Time PCR assay for the detection of *Acarapis woodi* (tracheal mites) in *Apis mellifera.* (in prep.).

DENMARK, H A; CROMROY, H L; SANFORD, M T (2000) Honey bee tracheal mite, *Acarapis woodi* (Rennie) (Arachnida: Acari: Tarsonemidae). DPI Entomology Circular, 267. Un. Florida, IFAS Extension.

DIETEMANN, V; NAZZI, F; MARTIN, S J; ANDERSON, D; LOCKE, B;
    DELAPLANE, K S; WAUQUIEZ, Q; TANNAHILL, C; FREY, E;
    ZIEGELMANN, B; ROSENKRANZ, P; ELLIS, J D (2013) Standard
    methods for varroa research. In *V Dietemann; J D Ellis;
    P Neumann (Eds) The COLOSS BEEBOOK, Volume II: standard
    methods for* Apis mellifera *pest and pathogen research. Journal of
    Apicultural Research* 52(1):
    http://dx.doi.org/10.3896/IBRA.1.52.1.09

ECKERT, J E (1961) Acarapis mites of the honey bee, *Apis mellifera*
    Linnaeus. *Journal of Insect Pathology* 3:409–25.

ELLIS, J D; MUNN, P A (2005) The worldwide health status of honey
    bees. *Bee World* 86(4): 88-101.

EVANS, J D; PETTIS, J S; SMITH, I B (2007) A diagnostic genetic test
    for the honey bee tracheal mite, *Acarapis woodi. Journal of
    Apicultural Research* 46(3): 1-5.

FRAZIER, M T; FINLEY, J; HARKNESS, W; RAJOTTE, E G (2000)
    A sequential sampling scheme for detecting infestation levels of
    tracheal mites (Heterostigmata: Tarsonemidae) in honey bee
    (Hymenoptera: Apidae) colonies. *Journal of Economic Entomology*
    93(3): 551-558.

FRIES, I; CHAUZAT, M-P; CHEN, Y-P; DOUBLET, V; GENERSCH, E;
    GISDER, S; HIGES, M; MCMAHON, D P; MARTÍN-HERNÁNDEZ, R;
    NATSOPOULOU, M; PAXTON, R J; TANNER, G; WEBSTER, T C;
    WILLIAMS, G R (2013) Standard methods for nosema research. In
    *V Dietemann; J D Ellis, P Neumann (Eds) The COLOSS BEEBOOK:
    Volume II: Standard methods for* Apis mellifera *pest and
    pathogen research. Journal of Apicultural Research* 52(1):
    http://dx.doi.org/10.3896/IBRA.1.52.1.14

FURGALA, B; DUFF, S; ABOULFARAJ, S; RAGSDALE, D; HYSER, R
    (1989) Some effects of the honey bee tracheal mite (*Acarapis
    woodi* Rennie) on non-migratory, wintering honey bee (*Apis
    mellifera* L.) colonies in east central Minnesota. *American Bee
    Journal* 129: 195-197.

GARY, N E; PAGE, R E JR (1989) Tracheal mite (Acari, Tarsonemidae)
    infestation effects on foraging and survivorship of honey bees
    (Hymenoptera, Apidae). *Journal of Economic Entomology* 82: 734-739.

GARY, N E; PAGE, R E JR; LORENZEN, K (1989) Effect of age of
    worker honey bees (*Apis mellifera* L.) on tracheal mite (*Acarapis
    woodi* (Rennie)) infestation. *Experimental and Applied Acarology*
    7: 153-160.

GIORDANI, G (1962) Recherches au laboratoire suyr *Acarapis woodi*
    Rennie, agent de l'acariose des abeilles (*Apis mellifica* L) Note 1.
    Bulletin Apicole, V: 33-48.

GIORDANI, G (1965) Richerche di laboratorio su *Acarapis woodi*
    Rennie agente dell'acariosi delle api (*Apis mellifica* L.) Nota 4.
    Bulletin Apicole, VIII: 159-175.

GIORDANI, G (1977) Facts about acarine mites. In *Proceedings of the
    XXVIth International Congress of Apiculture, Adelaide. Apimondia
    Publishing House;* Bucharest, Romania. pp 459-467.

GRANT, G A; NELSON, D L; OLSEN, P E; RICE, W A (1993) The ELISA
    detection of tracheal mites in whole honey bee samples. *American
    Bee Journal* 133: 652-655.

HARBO, J (1993) Field and laboratory tests that associate heat with
    mortality of tracheal mites. *Journal of Apicultural Research* 32(3-
    4): 159-165.

HIRSCHFELDER, H; SACHS, H (1952) Recent research on the acarine
    mite. *Bee World* 33: 201–9.

HIRST, S (1921) On the mite (*Acarapis woodi*, Rennie) associated
    with Isle of Wight bee disease. *Annals and Magazine of Natural
    History,* Ser.9, 7: 509-519.

HOOD, W M; MCCREADIE, J W (2001) Field tests of the Varroa
    Treatment Device using formic acid to control *Varroa destructor*
    and *Acarapis woodi. Journal of Agriculture and Urban Entomology*
    18(2): 87.

HOPPE, H; RITTER, W; STEPHEN, E W-C (1989) The control of
    parasitic bee mites: *Varroa jacobsoni, Acarapis woodi*, and
    Tropilaelaps clareae with formic acid. *American Bee Journal* 129:
    739-742.

HOY, M A (2011) *Agricultural acarology: introduction to integrated
    mite management.* CRC Press; Florida, USA. pp 303–308.

IBAY, L A; BURGETT, D M (1989) Biology of the two external *Acarapis*
    species of honey bees: *Acarapis dorsalis* Morgenthaler, and
    Acarapis externus Morgenthaler. *American Bee Journal* 129: 816.

KOJIMA, Y; TOKI, T; MORIMOTO, T; YOSHIYAMA, M; KIMURA, K;
    KADOWAKI, T (2011a) Infestation of Japanese native honey bees
    by tracheal mite and virus from non-native European honey bees
    in Japan. *Microbial Ecology,* 62: 895-906.
    http://dx.doi.org/10.1007/s00248-011-9947-z

KOJIMA, Y; YOSHIYAMA, M; KIMURA, K; KADOWAKI, T (2011b) PCR-
    based detection of a tracheal mite of the honey bee *Acarapis
    woodi. Journal of Invertebrate Pathology* 108: 135-137.

KRANTZ, G W; WALTER, D E (2009) *A manual of acarology (3rd Ed.).*
    Texas Tech University Press; Lubbock, TX, USA. 807 pp.

LEE, D C (1963) The susceptibility of honey bees of different ages to
    infestation by *Acarapis woodi* (Rennie). *Journal of Invertebrate
    Pathology* 6: 11-15.

LIN, H; OTIS, G W; SCOTT-DUPREE, C (1996) Comparative resistance
    in Buckfast and Canadian stocks of honey bees (*Apis mellifera* L.)
    to infestation by honey bee tracheal mites (*Acarapis woodi*
    (Rennie). *Experimental and Applied Acarology* 20: 87-101.

LINDQUIST, E E (1986) The world genera of Tarsonemidae (Acari:
    Hetersotigmata): A morphological, phylogenetic and systematic
    revision with a reclassification of family-group taxa in the
    Heterostigmata. *Memoirs of the Entomological Society of Canada*
    No. 136, 517 pp.

LIU, T P (1995) A rapid differential staining technique for live and
    dead tracheal mites. *Canadian Beekeeping* 18:155.

LORENZEN, K; GARY, N E (1986) Modified dissection technique for diagnosis of tracheal mites (Acari: Tarsonemidae) in honey bees (Hymenoptera: Apidae). *Journal of Economic Entomology* 79: 1401-1403.

MARGOLIS, L; ESCH, G W; HOLMES, J C; KURIS, A M; SCHAD, G A (1982) The use of ecological terms in parasitology (Report of an *ad hoc* committee of the American Society of Parasitologists). *Journal of Parasitology* 68: 131-133.

MATTILA, H R; HARRIS, J L; OTIS, G W (2001) Timing of production of winter bees in honey bee (*Apis mellifera*) colonies. *Insectes Sociaux* 48: 88-93.

MCMULLAN, J B (2011) A scientific note: migration of tracheal mites (*Acarapis woodi*) to old winter honey bees. *Apidologie* 42: 577-578.

MCMULLAN, J B; BROWN, M J F (2005) Brood pupation temperature affects the susceptibility of honey bees (*Apis mellifera*) to infestation by tracheal mites (*Acarapis woodi*). *Apidologie* 36: 97-105.

MCMULLAN, J B; D'ETTORRE, P; BROWN, M J F (2010) Chemical cues in the host-seeking behaviour of tracheal mites (*Acarapis woodi*) in honey bees (*Apis mellifera mellifera*). *Apidologie* 41: 568-578. http://dx.doi.org/10.1051/apido/2010004

MILNE, P S (1948) Acarine disease of bees. *Journal of the UK Ministry of Agriculture and Fisheries* 54: 473-477.

MORGENTHALER, O (1931) An Acarine disease experimental apiary in the Bernese Lake District and some of the results obtained there. *Bee World* 12: 8-10.

MORSE, R A (1991) Biology and control of tracheal mites of honey bees. *Cooperative Extension Bulletin, Cornell University; Ithaca, NY, USA.* 4 pp.

MOZES-KOCH, R; GERSON, U (1997) Guanine visualization: a new method for diagnosing tracheal mite infestation of honey bees. *Apidologie* 28: 3-9.

NASR, M E; OTIS, G W; SCOTT-DUPREE, C D (2001) Resistance to *Acarapis woodi* by honey bees (Hymenoptera: Apidae): divergent selection and evaluation of selection progress. *Journal of Economic Entomology* 94: 332-338.

OCHOA, R; PETTIS, J S; ERBE, E; WERGIN, W P (2005) Observations on the honey bee tracheal mite *Acarapis woodi* (Acari: Tarsonemidae) using low-temperature scanning electron microscopy. *Experimental and Applied Acarology* 35: 239-249.

OCHOA, R; PETTIS, J S; MIRELES, O M (2003) A new bee mite of the genus *Pseudacarapis* (Acari: Tarsonemidae) from Mexico. *International Journal of Acarology* 29(4): 299-305

OTIS, G W; BATH, J B; RANDALL, D L; GRANT, G M (1988) Studies of the honey bee tracheal mite (*Acarapis woodi*) (Acari: Tarsonemidae) during winter. *Canadian Journal of Zoology* 66: 2122-2127.

OTIS, S W; SCOTT-DUPREE, C D (1992) Effects of *Acarapis woodi* on overwintered colonies of honey bees (Hymenoptera: Apidae) in New York. *Journal of Economic Entomology* 85: 40-48.

PAGE, R E, JR; GARY, N E (1990) Genotypic variation in susceptibility of honey bees (*Apis mellifera*) to infestation by tracheal mites (*Acarapis woodi*). *Experimental and Applied Acarology* 8: 275-283.

PENG, Y S; NASR, M E (1985) Detection of honey bee tracheal mites (*Acarapis woodi*) by simple staining techniques. *Journal of Invertebrate Pathology* 46: 325-331.

PETTIS, J S; PANKIW, T (1998) Grooming behaviour by *Apis mellifera* L. in the presence of *Acarapis woodi* (Rennie) (Acari: Tarsonemidae). *Apidologie* 29: 241-253.

PETTIS, J S; WILSON, W T (1996) Life history of the honey bee tracheal mite (Acari: Tarsonemidae). *Arthropod Biology* 89(3): 368-374.

PETTIS, J S; WILSON, W T; EISCHEN, F A (1992) Nocturnal dispersal by female *Acarapis woodi* in honey bee (*Apis mellifera*) colonies. *Experimental and Applied Acarology* 15: 99-108.

PHELAN, P L; SMITH, A W; NEEDHAM, G R (1991) Mediation of host selection by cuticular hydrocarbons in the honey bee tracheal mite *Acarapis woodi* (Rennie). *Journal of Chemical Ecology* 17: 463–473.

PHILLIPS, E F (1923) *The occurrence of diseases of adult bees. II.* US Department of Agriculture Circular #287.

RAGSDALE, D W; FURGALA, B (1987) A serological approach to the detection of *Acarapis woodi* parasitism of *Apis mellifera*. *Apidologie* 18: 1-10.

RAGSDALE, D W; KJER, K M (1989) Diagnosis of tracheal mite (*Acarapis woodi* Rennie) parasitism of honey bees using a monoclonal based enzyme-linked immunosorbent assay. *American Bee Journal* 129: 550-553.

RENNIE, J (1921) Notes on acarine disease. VI. *Bee World* 13: 115-117.

RENNIE, J; WHITE, P B; HARVEY, E J (1921) XXIX. Isle of Wight disease in hive bees. 1. The etiology of the disease. *Transactions of the Royal Society of Edinburgh* 52(4): 737-754.

RITTER, W (1996) *Diagnostik und Bekämpfung der Bienenkrankheiten [Diagnosis and control of bee diseases]*. Gustav Fischer Verlag; Jena, Stuttgart, Germany.

RITTER, W; SAMMATARO, D; DE GUZMAN, L (2013) Acarapiosis of honey bees, Chap. 2.2.1. In *OIE, World Organisation for Animal Health, Manual of Diagnostic Tests and Vaccines for Terrestrial Animals 2012*. www.oie.int/en/international-standard-setting/terrestrial-manual/

ROYCE, L A; ROSSIGNOL, P A (1991) Sex bias in tracheal mite [*Acarapis woodi* (Rennie)] infestation of honey bees (*Apis mellifera* L.). *Bee Science* 1: 159–61.

ROYCE, L A; KRANTZ, G W; IBAY, L A; BURGETT, D M (1988) Some observations on the biology and behaviour of *Acarapis woodi* and *Acarapis dorsalis* in Oregon. In *G R Needham; R E Page, Jr; M Delifnado-Baker; C E Bowman (Eds). Africanized honey bees and bee mites*. Ellis Horwood; Chichester, UK. pp 498-505.

SAMMATARO, D (2006) An easy dissection technique for finding the tracheal mite, *Acarapis woodi* (Rennie) (Acari: Tarsonemidae), in honey bees, with video link. *International Journal of Acarology* 32: 339-343.

SAMMATARO, D; AVITABILE, A (2011) *Beekeeper's handbook (4th Ed.).* Comstock Publishing; Ithaca, NY, USA. 308 pp.

SAMMATARO, D; COBEY, S; SMITH, B H; NEEDHAM, G R (1994) Controlling tracheal mites (Acari: Tarsonemidae) in honey bees (Hymenoptera: Apidae) with vegetable oil. *Journal of Economic Entomology* 87: 910–16.

SAMMATARO, D; GERSON, U; NEEDHAM, G R (2000) Parasitic mite of honey bees: life history, implications and impact. *Annual Review of Entomology* 45: 519-548.

SAMMATARO, D; NEEDHAM, G R (1996) Host-seeking behaviour of tracheal mites (Acari: Tarsonemidae) on honey bees (Hymenoptera: Apidae). *Experimental and Applied Acarology* 20: 121–36.

SHIMANUKI, H; CANTWELL, G E (1978) *Diagnosis of honey bee diseases, parasites, and pests.* USDA Manual ARS-NE-87. 18 pp.

SHIMANUKI, H; KNOX, D A (1991) *Diagnosis of honey bee diseases.* US Department of Agriculture, Agriculture Handbook No. AH-690. 53 pp.

SKINNER, A J (2000) Impacts of tracheal mites (*Acarapis woodi* (Rennie)) on the respiration and thermoregulation of overwintering honey bees in a temperate climate. MSc. thesis, University of Guelph; Guelph, Ontario, Canada. 186 pp.

SMITH, A W; NEEDHAM, G W; PAGE, R E JR (1987) A method for the detection and study of live honey-bee tracheal mites (*Acarapis woodi* Rennie). *American Bee Journal* 127: 433-434.

SUMANGALA, K (1999) Ecobiology of *Pseudacarapis indoapis* (Acari: Tarsonemidae): Nutrition, dispersal and host range. *Entomon* 24: 235-239.

SUMANGALA, K; HAQ, M A (2001) Survey of the mite fauna associated with *Apis* spp. in Kerala, Southern India. In *R B Halliday et al. (Eds). Acarology: proceedings of the 10th International Congress.* CSIRO Publishing; Australia. pp 565-567.

SUMANGALA, K; HAQ, M A (2002) Ecobiology of *Pseudacarapis indoapis* (Acari: Tarsonemidae): Ontogeny and breeding behaviour. *Journal of Entomological Research* 26: 83-88.

UNDERWOOD, R M; R W CURRIE (2009) Indoor winter fumigation with formic acid for control of *Acarapis woodi* (Acari: Tarsonemidae) and nosema disease (*Nosema* sp.). *Journal of Economic Entomology* 102: 1729-1736.

VANENGELSDORP, D; OTIS, G W (2001a) Field evaluation of nine families of honey bees for resistance to tracheal mites. *Canadian Entomologist* 133: 793-803.

VANENGELSDORP, D; OTIS, G W (2001b) The role of cuticular hydrocarbons in the resistance of honey bees (*Apis mellifera*) to tracheal mites (*Acarapis woodi*). *Experimental and Applied Acarology* 25: 593-603.

VILLA, J D (2006) Autogrooming and bee age influence migration of tracheal mites to Russian and susceptible worker honey bees (*Apis mellifera* L). *Journal of Apicultural Research* 45: 28-31.

WARD, L I; WAITE, R; BOONHAM, N; FISHER, T; PESCOD, K; THOMPSON, H; CHANTAWANNAKUL, P; BROWN, M (2007) First detection of Kashmir bee virus in the UK using real time PCR. *Apidologie* 38: 181-190.

WHITCOMB, R F (1983) Culture media for spiroplasmasa. *Methods in Mycoplasmology* 1: 147-158.

WILSON, W T; PETTIS, J S; HENDERSON, C E; MORSE, R A (1997) Tracheal mites. In *R A Morse; K Flottum (Eds). Honey bee pests, predators, and diseases (3rd Ed.).* A I Root Co.; Medina, OH, USA. pp. 253-278.

Journal of Apicultural Research 52(4): (2013)
DOI 10.3896/IBRA.1.52.4.21

REVIEW ARTICLE

Standard methods for *Tropilaelaps* mites research

I B R A
INTERNATIONAL BEE
RESEARCH ASSOCIATION

Denis L Anderson[1*] and John M K Roberts[1]

[1]CSIRO Ecosystem Sciences, PO Box 1700, Canberra ACT 2601, Australia.

Received 2 February 2013, accepted subject to revision 15 May 2013, accepted for publication 31 May 2013.

*Corresponding author: Email: denis@beesdownunder.com

Summary

Mites in the genus *Tropilaelaps* are native brood parasites of the non-domesticated giant Asian honey bees, *Apis dorsata*, *A. breviligula* and *A. laboriosa*. They spread onto the managed European honey bee (*A. mellifera*) some time after humans introduced that bee into Asia. Nowadays, *A. mellifera* is kept for beekeeping throughout Asia and *Tropilaelaps* mites are one of its most damaging pests. At present, these mites remain confined to Asia and bordering areas but are recognized as emerging threats to world apiculture. In spite of their important pest-status, *Tropilaelaps* mites remain poorly studied and much remains to be learned about them. The methods reviewed here are intended to assist future research efforts to better understand these parasites. The review begins with an introduction that clarifies why *Tropilaelaps* mites are worthy of immediate research. It is followed by an outline of the mites' taxonomy and descriptions of various methods (including their pros and cons) by which mites can be collected, identified and aspects of their life history and behaviour studied. The role that microbial pathogens play in the pathogenicity of *Tropilaelaps* mites is briefly discussed and a list of future research priorities is suggested.

Métodos estándar para la investigación de ácaros *Tropilaelaps*

Resumen

Los ácaros del género *Tropilaelaps* son parásitos originales de la cría de las abejas de la miel gigantes asiáticas no domesticadas *Apis dorsata*, *A. breviligula* y *A. laboriosa*. Se propagaron hacia la abeja de la miel europea (*A. mellifera*) en algún momento después de que el ser humano introdujera esta abeja en Asia. Hoy día, *A. mellifera* es utilizada para la apicultura en toda Asia y el ácaro *Tropilaelaps* es una de sus plagas más dañinas. Actualmente estos ácaros se mantienen confinados en Asia y sus zonas fronterizas pero están reconocidos como una amenaza emergente para el mundo de la apicultura. A pesar de su importante estatus como plaga, el ácaro *Tropilaelaps* permanece escasamente estudiado y aún hay mucho que aprender sobre él. Los métodos revisados aquí son un intento para ayudar en los futuros esfuerzos de la investigación para entender mejor este parásito. La revisión comienza con una introducción que aclara porqué el ácaro *Tropilaelaps* debe ser tenido en cuenta para una investigación inmediata. Continúa con un esquema sobre la taxonomía del ácaro y la descripción de varios métodos (incluyendo pros y contras) para su colecta, la identificación y aspectos sobre la vida y comportamiento ya estudiados. Se discute brevemente el papel que juegan los microbios patógenos en la patogenicidad de *Tropilaelaps* y se sugiere una lista de prioridades para futuras investigaciones.

Footnote: Please cite this paper as: ANDERSON, D L; ROBERTS, J M K (2013) Standard methods for *Tropilaelaps* mites research. In *V Dietemann; J D Ellis; P Neumann (Eds) The COLOSS* BEEBOOK, *Volume II: standard methods for* Apis mellifera *pest and pathogen research. Journal of Apicultural Research* 52(4): http://dx.doi.org/10.3896/IBRA.1.52.4.21

小蜂螨研究的标准方法

摘要

Tropilaelaps 属的螨是一类幼体寄生虫，原始寄主为野生大型亚洲蜜蜂 *Apis dorsata, A. breviligula* 和 *A. laboriosa*。在亚洲引入欧洲蜜蜂（*A. mellifera*）之后的某个时间，该蜂螨传播至饲养的欧洲蜜蜂。现如今，*A. mellifera*的饲养已遍及亚洲，小蜂螨是其危害性最大的害虫之一。目前，这些螨仍仅限于亚洲及其周边地区，但已被公认为世界养蜂业新出现的威胁因素。虽然*Tropilaelaps*是十分重要的虫害，但关于它的研究甚少，诸多方面有待进一步了解。本章综述了相关方法，旨在有助于将来对其更深入的探索。本综述首先在前言部分阐明为何小蜂螨值得当前研究，其次概述了该螨的分类，描述了蜂螨采集、鉴定、生活史和行为研究的不同方法（包括它们的优缺点）。简要探讨了微生物致病体在小蜂螨致病性中发挥的作用，并列举了一些未来需要重点研究的内容。

Keywords: COLOSS, *BEEBOOK*, honey bee, *Tropilaelaps mercedesae, Tropilaelaps clareae, Tropilaelaps koenigerum, Tropilaelaps thaii, Apis mellifera*, giant Asian honey bees, research method, protocols, identification, taxonomy

Table of Contents

1. Introduction

Mites in the genus *Tropilaelaps* are native brood parasites of the non-domesticated giant Asian honey bees, *Apis dorsata*, *A. breviligula* and *A. laboriosa*. They spread onto the managed European honey bee (*A. mellifera*) some time after humans introduced that bee into Asia. Nowadays, *A. mellifera* is kept for beekeeping throughout Asia and *Tropilaelaps* mites are one of its most damaging pests (Burgett *et al.*, 1983; Woyke, 1985a; Anderson and Morgan, 2007; Dainat *et al.*, 2009). At present, these mites remain confined to Asia and bordering areas but are recognized as emerging threats to world apiculture.

A question that is constantly asked of *Tropilaelaps* mites, the answer to which justifies or counters the need to direct scarce resources to study them, is: can they become a serious global pest of *A. mellifera*, like *Varroa destructor*? The answer is emphatically yes and it lies in how the mites feed and reproduce on their bee hosts, how they disperse, how they spread among bee colonies and how they might survive on *A. mellifera* in temperate zones outside of Asia.

The breeding cycle of *Tropilaelaps* mites on their honey bee hosts superficially resembles that of *Varroa* mites, in that a mature mated female enters a bee brood cell that contains a developing bee larva that is in the process of being capped by worker bees with a wax covering. Safely concealed within the sealed cell, the mother mite produces several offspring that all feed on blood (haemolymph) of the developing bee. Eventually the mites are released from the cell when the developing bee (which by now may or may not be physically damaged) chews its way out of the cell through the wax capping (Woyke, 1987; 1994). Unlike *Varroa* mites, the survival of *Tropilaelaps* mites depends solely on them having regular access to bee brood (larvae or pupae) on which to feed, as their mouthparts and body shape do not allow them to feed on adult bees, as do *Varroa* mites.

The mites can only survive for a few days in the absence of bee brood (Woyke, 1984; Koeniger and Musaffar, 1988; Rinderer *et al.*, 1994). This limited food source restricts their ability to disperse, as they can only disperse on adult bees on which they cannot feed.

The ever-increasing global trade of live honey bees, which provides a potential pathway for *Tropilaelaps* mites to disperse out of Asia, has not yet contributed to any increase in their geographical range. This is probably because live bee trade involves movements of adult bees (in the form of 'package bees') and live adult queen bees, on which *Tropilaelaps* mites cannot feed or survive for more than about 74 hours (Wilde, 2000). Because the mites cannot feed on adult bees, very few are found on them at any one time, even in heavily infested bee colonies (Woyke, 1984; 1985b). Those mites that do venture onto adult bees can spread to neighbouring bee colonies by various means, such as on swarms, on worker bees that rob resources from other colonies, on foraging bees that become disorientated and enter the wrong colony or simply by moving between forager bees from different colonies that visit the same flowers. In the Philippines, it has been suggested that *Tropilaelaps* mites spread between *A. breviligula* (their natural host) and *A. mellifera* colonies by interspecific robbing (Laigo and Morse, 1969).

As *Tropilaelaps* mites venture onto adult bees at some stage of their life, then live bee exports from Asia could potentially carry them, albeit at low numbers, and therefore aid their dispersion. However, for mites to survive this pathway, the exported bees would need to be moved quickly to their destination country and, on arrival, come in contact with local brood-right bee colonies into which mites could disperse and feed on brood before they starve to death. Another new potential pathway for the mites' spread out of Asia was recently uncovered in Australia, when *A. dorsata* worker bees were detected on air cargo that arrived at an international airport from Malaysia. These bees were probably night-foragers that had become disoriented by airport lights in Malaysia and rested and became stranded on cargo that was being loaded into an airplane bound for Australia.

Even given these and other potential pathways for *Tropilaelaps* mites to spread, they nevertheless still remain restricted to Asia and bordering areas. They currently occur as far west as Afghanistan-Pakistan and as far east as the large Melanesian island of New Guinea, where they were introduced in the 1980s in brood-right hived colonies of *A. mellifera* imported from Java (Delfinado and Aggarwal, 1987; Anderson, 1994; Baker *et al.*, 2005). They were also reported from Kenya during the early 1990s (Kumar *et al.*, 1993; Matheson, 1997), but this report has not been verified and it may have resulted from a false identification, as recent testing in Australia of mites that had been collected from honey bees in Kenya, and assumed to be *Tropilaelaps*, were found to be plant mites (Anderson, unpublished data).

So the question remains: does the current restricted distribution of *Tropilaelaps* mites reflect their inability to survive outside of Asia in temperate zones in the absence of their native bee hosts? The successful establishment of *Tropilaelaps* mites on *A. mellifera* in New Guinea suggests the answer is no, and that they can survive in temperate zones given an important proviso.

New Guinea is located to the north of Australia and it contains no native *Apis* species. Humans introduced colonies of *A. mellifera* to the island last century and their descendants have since thrived in the cool temperate-like highland regions, but not so well in the hotter humid tropical lowland regions (Clinch, 1979). Since their introduction to New Guinea in the 1980's, *Tropilaelaps* mites have become an endemic damaging pest of *A. mellifera* in the western half of the island (Irian Jaya). Their success is thought to be due to the unbroken year-round production of brood by the *A. mellifera* colonies, which provides a continuous food source for the mites and an ideal environment for their reproduction.

Hence, the New Guinea situation confirms that *Tropilaelaps* mites can survive and become an endemic pest of *A. mellifera* in temperate zones in the absence of their native hosts provided they have access to *A. mellifera* brood on a year-round basis. Such conditions are found in many temperate countries, such as parts of the USA, Australia and Europe. It has also been suggested that, in coming years, temperate areas in which *A. mellifera* can produce brood all year round may increase, as colder regions become warmer, due to the effects of climate change (Le Conte and Navajas, 2008).

This all adds up to a situation where it appears that good fortune has played a major role in restricting the distribution of *Tropilaelaps* mites and that it may be only a matter of time until they spread outside of Asia to cause hardship for temperate zone beekeepers. It is therefore no surprise that the mites are currently recognized as emerging threats to world apiculture (OIE, 2004) and hence deserve

immediate attention from the global research community. The methods presented here should assist those future research efforts.

2. Taxonomy and host-specificity
2.1. Taxonomy

The first species in the genus was described more than 50 years ago and the most recent in 2007. The taxonomy of the genus has recently been revised and it currently stands as follows (Lindquist *et al.*, 2009; Anderson and Morgan, 2007):

Kingdom:	Animalia
Phylum:	Arthropoda
Class:	Arachnida
Subclass:	Acari
Superorder:	Parasitiformes
Order:	Mesostigmata
Family:	Laelapidae
Genus:	*Tropilaelaps*
Species:	*T. clareae* Delfinado and Baker (1961)
	T. koenigerum Delfinado-Baker and Baker (1982)
	T. mercedesae Anderson and Morgan (2007)
	T. thaii Anderson and Morgan (2007)

2.2. Host specificity

Some behavioural and morphological features of *Tropilaelaps* mites, such as their fast movement and 'pincer-shaped' chelicerae, suggests they may have only recently adopted a parasitic life-style on honey bees and that they may not be very host-specific. Indeed, when they were first discovered in the Philippines in the 1960's they were found inside an *A. mellifera* colony and on field rats nesting nearby (Delfinado and Baker, 1961). It is likely that those rats picked up the mites after they entered the *A. mellifera* colonies, which rats often do in the tropics, because evidence gathered since that initial discovery indicates that *Tropilaelaps* mites are highly specialized parasites of honey bees.

Tropilaelaps mites are now recognized as common natural parasites of giant honey bees distributed throughout Asia. They have not colonized any other host organism, other than *A. mellifera*. Molecular studies have confirmed that *T. clareae* is native to *A. breviligula* in the Philippines (except on Palawan Island), *T. mercedesae* and *T. koenigerum* to *A. dorsata* and *A. laboriosa* in other parts of Asia (including Palawan Island) and *T. thaii* to *A. laboriosa*, in mountainous region of Mainland Asia (Laigo and Morse, 1969; Delfinado-Baker and Baker, 1982; Tangjingjai *et al.*, 2003; Anderson and Morgan, 2007).

Tropilaelaps mercedesae and *T. clareae* colonized *A. mellifera* after humans introduced that bee into Asia. However, in contrast with *Varroa* mites, of which only a few genotypes have switched-host to *A. mellifera*, many different genotypes of *T. mercedesae* and *T. clareae* now utilize *A. mellifera* as a host. In this respect, *Tropilaelaps* mites are less host-specific than *Varroa* mites but they are still relatively host-specific compared to some mites, such as the water mite *Protzia eximia* that parasitizes a wide variety of insect hosts (Walter and Proctor, 1999). Evidence suggests that *T. koenigerum* and *T. thaii* are restricted to their Asian bee hosts and are harmless to *A. mellifera* (Anderson and Morgan, 2007).

Very occasionally, *Tropilaelaps* mites are found in *A. cerana* and *A. florea* colonies in Asia, but in none of these instances have the mites been found to be producing offspring (Otis and Kralj, 2001). The exception is a report of a single female *T. mercedesae* found parasitizing and producing offspring on *A. cerana* brood in Thailand (Anderson and Morgan, 2007). The authors, when commenting on this find, stated that their observation was the exception rather the rule, as *T. mercedesae* mites are rarely found in *A. cerana* colonies and, when they are, they are not found inside brood cells or with offspring. Obviously, other factors, other than the ability to produce offspring on that bee, are responsible for preventing *T. mercedesae* from colonising *A. cerana*.

3. Sample collection

3.1. Mite appearance

Adult *Tropilaelaps* mites are small (< 1 mm long), light brown, and hold their first pair of legs upright, resembling antennae (Fig. 1). They can often be seen moving quickly over the surface of combs in infested colonies. Their body shape is quite different from that of *Varroa* mites, being much longer than it is wider (Fig. 2).

There are clear morphological differences between the sexes of the different species. Males of *T. thaii* have not yet been discovered. Males of *T. mercedesae*, *T. clareae* and *T. koenigerum* are slightly smaller than their female counterparts and their epigynial thoracic plates are also shorter and sharply pointed toward their posterior end (Fig. 3). In any collection of *Tropilaelaps* mites, males are usually much less common than females (Rath *et al.*, 1991; Anderson and Morgan, 2007). Males can be easily identified in the field using a magnifying glass to observe the chela spermatodactyl (sperm transfer organ), which in *T. mercedesae* and *T. clareae* is long with a spirally coiled apex and in *T. koenigerum* is short with a 'pig-tail' loop at its apex (Fig. 4).

The nymph stages of *Tropilaelaps* are brilliant white and are easily observed with the naked eye (Fig. 5).

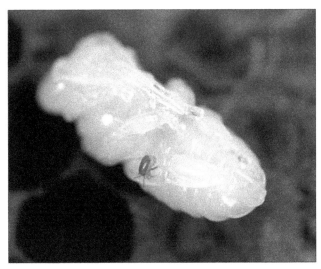

Fig. 1. A gravid *T. mercedesae* adult female feeding on an *A. mellifera* pupa. Note the first pair of legs of the mite is held upright, resembling antennae. Photo: Denis Anderson.

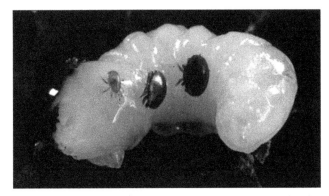

Fig. 2. Comparisons of a female *T. mercedesae* (left) with two female *V. jacobsoni* (right) on an *A. mellifera* larva. Photo: Denis Anderson.

3.2. Where to find mites

Adult females are most easily found inside of capped worker and drone bee brood cells of infested colonies, where they reproduce. They do not appear to markedly favour either cell type for their reproduction. This is also the only place where nymphal stages can be found. To find a mother mite and her nymph offspring, simply uncap a bee brood cell of an infested colony and remove the developing bee inside. Tilt the entire comb so that the ambient light will be directed into a cell. Any mites present will be easily seen with the naked eye in the bottom of the cell or on the cell wall.

The presence of adult mites with offspring inside bee brood cells is clear evidence that they have reproduced and have not simply entered the bee cell after being transported (say on robbing bees) from another colony of a sympatric bee species, which might confuse host-specificity attributed to them. Male mites are best found inside of capped cells in which the developing bee is about to emerge or else in random collections of adult mites found moving on the surfaces of combs.

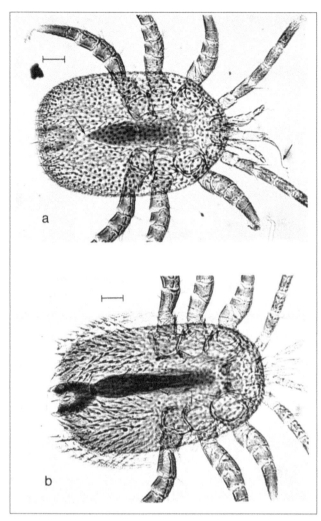

Fig. 3. Comparison of mounted specimens of a *T. mercedesae* male (top) and female (bottom). Anterior arrow on male points to the corkscrew like spermodactyl (sperm transfer organ) and posterior arrow points at the non-overlapping epigynial thoracic plate.
Bars = 0.1 mm.   Photo: Denis Anderson.

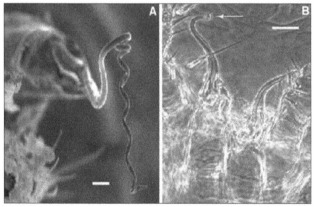

Fig. 4. Scanning electron micrographs of the sperm transfer organs of male *T. clareae* (left) and *T. koenigerum* (right). Bar = 10 μm.
                                                          Photo: Denis Anderson.

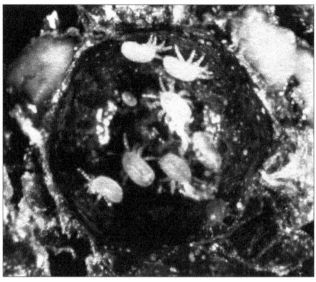

Fig. 5. A family of mites (*T. mercedesae*) with mother mite (light brown) and different stages of offspring (white) at the bottom of a cell from which the honey bee pupa was removed.   Photo: Denis Anderson.

Adult *Tropilaelaps* mites are much more mobile than adult *Varroa* mites and can be seen moving quickly across the surface of infested brood combs. In this situation they are hard to collect. When they are not moving, they also become well camouflaged against the background colour of the wax combs and are hard to see. When the brood comb is lifted out of the colony into the open light mites also quickly enter open cells and remain still on the cell walls, where again, they are hard to see.

Adults of both sexes can also be found on the bodies of adult bees, but only in extremely low numbers, even in heavily infested colonies. This is probably because mites avoid adult bees as much as possible because they cannot feed on them and hence can only survive for a few days (Woyke, 1984; Wilde, 2000). Nevertheless, at some stage they have to move on to adult bees in order to disperse from the colony.

3.3. Collecting mites from bee brood

Phoretic mites that are moving on the surfaces of bee combs are relatively hard to collect with forceps. In these situations it is best to collect them with a fine bristle bush slightly wetted with honey, water or alcohol (not human spittle as it will contaminate the sample with human DNA) or a small mouth aspirator.

When large numbers of mites are required, it is best to collect them from infested capped bee brood cells. This can be done in the field, or if large numbers of colonies needs to be sampled, combs can be collected and sampled later in the comfort of the laboratory. To do this:

1. Remove the wax cappings from a large number of bee brood cells on one side of a comb all at once.

2. Remove the developing bee brood from these cells.

3. Invert the comb over a sheet of white paper and tap it relatively hard on its upper surface to dislodge mites from the cells onto the paper.

4. Collect the mite from the paper into small vials containing 70% ethyl alcohol, using a fine paint brush wetted in honey, alcohol or water (not human spittle) or with a fine pair of tweezers, as shown for *Varroa* mites in Dietemann *et al.* (2013). *Tropilaelaps* mites (similar to *Varroa* mites) will immediately sink to the bottom of a container when immersed in ethyl alcohol. The white paper onto which mites fall can also be substituted with brown paper if mite collection is targeted at nymphal stages.

There are benefits in sometimes collecting live mites into hot water before transferring them into alcohol for storage (Section 4.1 below).

Pros: Mites collected into ethyl alcohol can be subsequently tested in the laboratory in a range of different tests, such as morphometric and DNA analyses.

Cons: Mites used in inoculation or behavioural studies cannot be placed into ethyl alcohol after collection. But even then, mites collected by this method are of little use for those types of studies, as the development stage of the collected mites cannot be ascertained with certain. Live mites to be used in those kinds of studies are best collected from very recently capped bee brood cells that contain late larval or prepupal stages. This ensures that the mites collected are mature adults that are at a specific stage of development – that is, at the pre-reproductive stage. These mites can be kept alive until needed by keeping them in a small Petri dish or a glass bottle with late stage bee larvae.

3.4. Collecting mites from adult bees

As mentioned, adult *Tropilaelaps* mites are usually only present in low numbers on adult bees in any given bee colony. Hence, it is usually a waste of time trying to manually find them on individual bees. A simple method for collecting them from adult bees is:

1. Collect a sample of 200 or so adult bees from an infested colony into a transparent container (such as a plastic bottle) that contains 70% ethyl alcohol.

2. Secure the lid on the container and shake it vigorously for about 1 minute.

3. Collect mites from the bottom of the container into containers containing fresh 70% ethyl alcohol, as described for *Varroa* mites in Dietemann *et al.* (2013).

Mites can also be removed from adult bees by washing the bees in soapy water or by treating them by the 'sugar-shake' method as also described for *Varroa* mites in Dietemann *et al.* (2013). Also, an alternative method to placing adult bees into ethanol, soapy water or sugar prior to shaking to dislodge mites, is to place the bees into plastic bags, label accordingly and freeze until the bees can be visually examined. However, this is much slower than the shaking methods.

Pros: A quick and simple method for finding mites on adult bees in any given infested bee colony.

Cons: Bee colonies need to be relatively highly infested with mites for this method to succeed. To increase the chances of collecting mites from adult bees, one can select a frame with emerging brood to collect mites emerging with the bees. However, it should be remembered that some of those mites will be new adults and may not be mated or fully mature.

3.5. Storage of collected mites

Methods for storing *Tropilaelaps* mites are identical for those described for *Varroa* mites in Dietemann *et al.* (2013).

3.6. Shipping collected mites

Methods for shipping *Tropilaelaps* mites are identical to those also described for *Varroa* mites in Dietemann *et al.* (2013).

4. Methods for identifying mites

4.1. Morphological methods

Morphological analyses are best carried out on mite specimens that have been cleared of their body tissue and mounted on glass microscope slides. Methods for clearing and mounting *Tropilaelaps* mites are the same as those described for *Varroa* mites (Dietemann *et al.*, 2013). Generally these specimens have been previously collected and stored in alcohol. However, sometimes it helps to collect specimens that are destined for morphological analyses into hot water before they are transferred in alcohol. This relaxes their internal body tissues and exposed hard-to-see organs, such as chelicerae, which are normally hard to see in mounted specimens that have simply been collected into alcohol.

Identification of specimens to the species level using morphology can be troublesome. Hence, the starting-point for identifying them should always be from information known about the location and host bee from which they were collected. Mites found on either giant Asian honey bees or European honey bees in the Philippines (except on Palawan Island) will in all probability be *T. clareae*, whereas those found on European honey bees in other parts of Asia (and Melanesia) will be *T. mercedesae*. The two other species, *T. koenigerum* and *T. thaii*, have been found only on their native Asian bee hosts, *A. dorsata, and A. laboriosa* respectively.

Table 1. Key to identification of *Tropilaelaps* mites. Males of *T. thaii* have not been discovered, but this species is restricted to *A. laboriosa*.

1 a. Collection sites: The Philippines (except Palawan Island) and Sulawesi Island (in Indonesia).	*Tropilaelaps clareae* Parasitizes *Apis breviligula* and *Apis mellifera* in the Philippines and *Apis dorsata binghami* and *Apis mellifera* in Sulawesi.
1 b. Collection sites: All other localities (including Palawan Island).	
2 a. Body length: ≤ 700 µm.	*Tropilaelaps koenigerum* Primarily found parasitizing *A dorsata dorsata* in mainland Asia.
2 b. Body length: ≥ 840 µm	
3 a. Female chelicerae: No apical tooth.	*Tropilaelaps thaii* Parasitizes *Apis laboriosa* in mountainous regions of mainland Asia.
3 b. Female chelicerae: One apical tooth present	*Tropilaelaps mercedesae* Primarily parasitizes *Apis dorsata dorsata* and *Apis mellifera* in

Many of the physical characters used to identify mites (such as their body size, sensilli and structures on body plates) are highly variable within the genus and even among member of the same species (Anderson and Morgan 2007). A key for identifying the different species of *Tropilaelaps* is given in Table 1.

The 4 main physical characters most useful for identifying *Tropilaelaps* mites to the species levels are:

1. Body length.

   T. koenigerum is the smallest member of the genus with a body length of < 700 µm for females and ~575 µm for males. Female *T. mercedesae*, *T. clareae* and *T. thaii* are much longer at ~ 950-990 µm, ~870-885 µm and ~ 890 µm respectively, while the body lengths of male *T. mercedesae* and *T. clareae* are slightly smaller than their respective females at 907-927 µm and 852-858 µm, respectively. Males of *T. thaii* have yet to be discovered.

2. Shape of the anal plate.

   Male and female *T. koenigerum* have a distinct pear-shaped anal plate, female *T. Thaii* have a bell-shaped anal plate, while the anal plates of male and female *T. mercedesae* and *T. clareae* are highly variable and of little use as a taxonomic reference.

3. Structure of the male sperm transfer organ (chela spermatodactyl or spermadactyl).

   Males of *T. koenigerum* have a 'pig-tail' loop at the apex of the chela spermatodactyl, while the apex of the chela spermatodactyls of male *T. mercedesae* and *T. clareae* are long cork-screw-like structures (Fig. 4).

4. Placement and shape of teeth on female chelicerae.

   Female *T. koenigerum* have a single sub apical tooth on the chelicerae with a characteristic groove near its anterior base, whereas *T. mercedesae* and *T. clareae* females also have the sub apical tooth, but without the groove. Female *T. thaii* lack the sub apical tooth (Fig. 6).

4.2. Molecular methods and systematics

Molecular technology was first used in *Tropilaelaps* research in the 1990s to examine genetic variation between so-called *T. clareae* and *T. koenigerum* (Tangjingjai *et al.*, 2003). It has since proved very useful to examine genetic variation within the genus and to help re-define known species and describe new species and new types within species (Anderson and Morgan, 2007).

DNA sequence obtained from a 538 base-pair fragment of *Tropilaelaps* mtDNA *cox1* gene, as well as from the entire nuclear ITS1-5.8s-ITS2 genes, provides a useful means for identifying mites to the species level. The *cox1* sequence is also useful for looking at genetic variation within a species.

Methods for extracting, amplifying and sequencing *Tropilaelaps* DNA are the same as those described for *Varroa* in Dietemann *et al.* (2013). The DNA primers used to amplify the *cox1* and ITS1-5.8s-ITS2 gene regions are shown in Table 2.

Table 2. Forward (F) and reverse (R) primer sequences (and their names) used in *Tropilaelaps* research to amplify fragments (base pairs) of specific genes.

Gene region	Fragment size (bp)	Primer sequences (5'-3')	Primer name
CoxI	538	(F) CTATCCTCAATTATTGAAATAGGAAC	TCF1
		(R) TAGCGGCTGTGAAATAGGCTCG	TCR2
ITS1-5.8S-ITS2	522-526	(F) GGAAGTAAAAGTCGTAACAAGG	ITS5
		(R) TCCTCCGCTTATTGATATGC	ITS4

Once a DNA sequence is obtained from a particular mite specimen, it is compared to other sequences of the same region that have been deposited in the GenBank database (Dietemann *et al.*, 2013). The ITS region shows no genetic variation within a particular species of *Tropilaelaps*, whereas *cox1* sequence shows from 1-4% variation within species and from 11-15% between species. Mites of particular

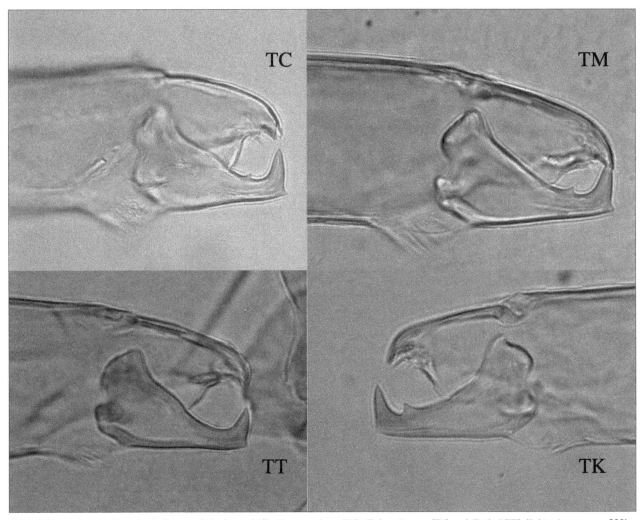

Fig. 6. Comparisons of female chelicerae of *T. clareae* (TC), *T. mercedesae* TM), *T. koenigerum* (TK) and *T. thaii* (TT) (light microscopy, x800).

Photo: Denis Anderson.

species that vary in the *cox1* gene sequence are referred to as 'haplotypes'. The concept of 'haplogroup', as described for *Varroa* mites in Dietemann *et al.* (2013), has not been adopted for *Tropilaelaps* mites, as there is much more variation in the *cox1* gene of *Tropilaelaps* than in *Varroa* (Anderson and Trueman, 2000).

Tropilaelaps mites can also be identified to the species level by digesting amplified fragments of their *cox1* gene with a combination of the *Fau*I *Bsr*I *Bst*NI and *Swa*I restriction enzymes, without the need for sequencing the fragments. Products produced from these digestions are visualized as bands in 2% agarose gels (Anderson and Morgan, 2007). The results of using these restriction enzymes to digest the cox1 gene fragment of the 4 known species are shown in Fig. 7. In summary, *Fau*I only digests *cox1* fragments obtained from *T. koenigerum*, *Bsr*I only digests fragments from *T. koenigerum* and *T. mercedesae* (2 bands are produced from each species, but the smaller band produced from *T. mercedesae* is larger than that of *T. koenigerum*), *Bst*NI only digests fragments from *T. clareae*, while *Swa*I only digests fragments from *T. thaii*.

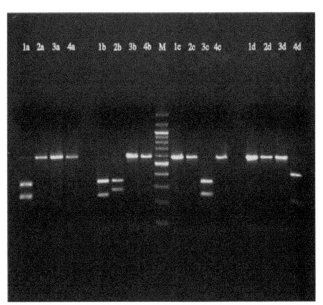

Fig. 7. Mites can be identified by digesting fragments of their cox1 gene with *Fau*I, *Bsr*I, *Bst*NI and *Swa*I restriction enzymes (labeled a–d respectively). The numbers 1-4 represent: *T. koenigerum*, *T. mercedesae*, *T. clareae* and *Thaii* respectively. M = 100 bp DNA Ladder.

Photo: Denis Anderson.

As *cox1* gene sequence can resolve *Tropilaelaps* mites to the species level it is useful in phylogenetic studies. Methods used to carry out phylogenic analyses on *Tropilaelaps cox1* gene sequence are the same as those used for *V. destructor* and other species, described in Dietemann *et al.* (2013). A phylogenetic tree of all the currently known and published *Tropilaelaps* haplotypes is shown in Fig. 8.

5. Life cycle and rearing

Much remains to be learned about the life cycle of *Tropilaelaps* mites. Reports published to date refer to the life cycle of *T. clareae* on *A. mellifera*, but with the recent taxonomic revision of the genus, which separated *T. clareae* into 2 species (Anderson and Morgan, 2007), it is now clear that those reports refer to *T. mercedesae* on *A. mellifera*. There are no reports of the life cycles of *T. clareae*, *T, koenigerum* or *T. thaii*, although they are thought to be very similar to that of *T. mercedesae* on *A. mellifera*. Obviously, this is an area ripe for research.

5.1. Life cycle of *T. mercedesae* on *A. mellifera*

As mentioned, to begin their reproduction, mated female mites enter *A. mellifera* worker or drone brood cells that are in the process of being capped (Burgett *et al.*, 1983; Ritter and Schneider-Ritter, 1986). There is no marked difference in the type of cell type that female mites choose to reproduce in. However, much variation has been reported in the timing of different events during the reproduction phase. After entering a cell a single female mite lays from 1-4 eggs (but typically 3-4) about 1 day apart. A comparison of the time in hours after cell capping that the first egg, larva, protonymph, deutonymph, and young adult mite appear in a cell, as reported by different authors, is shown in Table 3.

At the end of the reproduction phase, the mother mite and her offspring exit the cell when the developing bee chews its way out of the cell through the wax capping. They then enter a brief phoretic phase, in which they move about the comb, probably mate (as recently emerged male and female mites have been observed mating in glass test tubes (Woyke, 1994)) and spend time on adult bees, before they commence a new reproduction phase. The phoretic phase of *T. mercedesae* mites is much shorter than that of *Varroa* mites, and may be as short as 1-2 days. This means that *T. mercedesae* mites have quicker reproductive cycles than *Varroa* mites and hence their population buildup within a bee colony is much faster than that of *Varroa* mites, said by a UK Government source to be in the order of

about 25:1 in favour of *Tropilaelaps* (DEFRA, 2005). A diagram that summarizes data on the life cycle of *T. mercedesae* on *A. mellifera* in New Guinea, reported by Saleu (1994), is shown in Fig. 9.

5.2. Methods for studying mite development on *A. mellifera*

No artificial media has been described for rearing *Tropilaelaps* mites and hence aspects of their reproduction must be studied within a bee colony. To study their life cycle in an *A. mellifera* colony one first needs to select a colony that is already reasonably infested with mites. The general method is as follows:

1. Determine mite infestation levels in a number of *A. mellifera* colonies by removing the cell cappings from 300 capped cells/colony and determining how many contain mites.

2. Select a colony with the highest mite infestation and use this for further studies.

3. Remove 2 brood combs from the middle of the brood area of the selected colony and replace them with 2 brood combs from a non-infested colony that contains larvae that are 2-3 days away from being capped (Note: these brood combs can be obtained from other nearby colonies in which queen bees have been restrained to combs with a queen excluder for 24 hours. The queens will fill the cells of the combs with eggs, meaning that all brood that develops for the eggs with be within 24 hours of the same age).

4. Once the 2 brood combs have been capped, remove 1 at regular intervals to the laboratory (usually every 12 or 24 hours is best).

5. In the laboratory uncap a number of cells, remove the bee brood and determine the number of adult mites, eggs and mite nymphs within each cell.

6. Once all cells on this comb have been inspected, move on to the second comb.

7. Collate the recorded data. There are many methods by which this can be done but a simple method is simply to produce a diagram of the capping phase of the bee and plot the data recorded directly onto it.

Pros: This method is relatively simple and will show the progression of mite development, from egg to adult.

Cons: The method requires access to both infested and no-infested bee colonies.

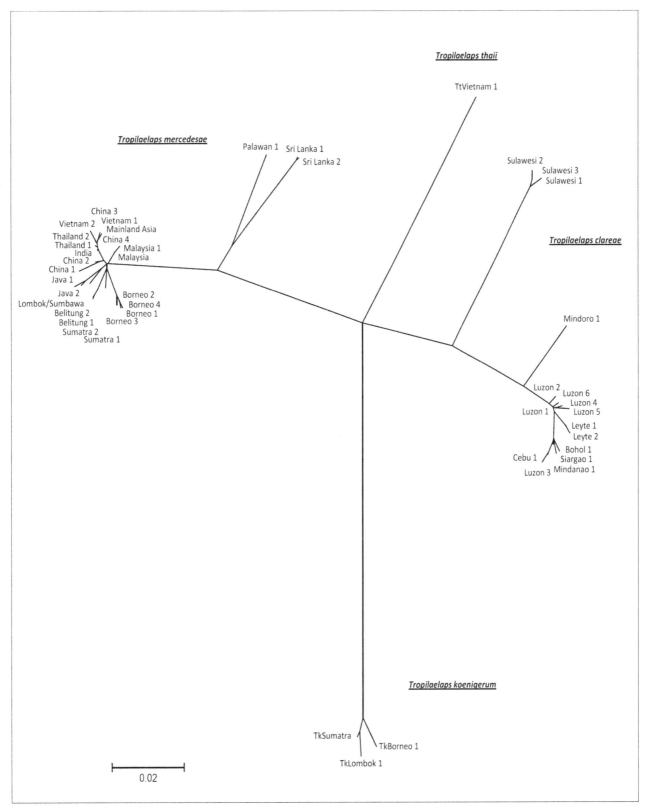

Fig. 8. A phylogenetic tree of all the currently known and published haplotypes of *Tropilaelaps*.   Photo Denis Anderson and John Roberts.

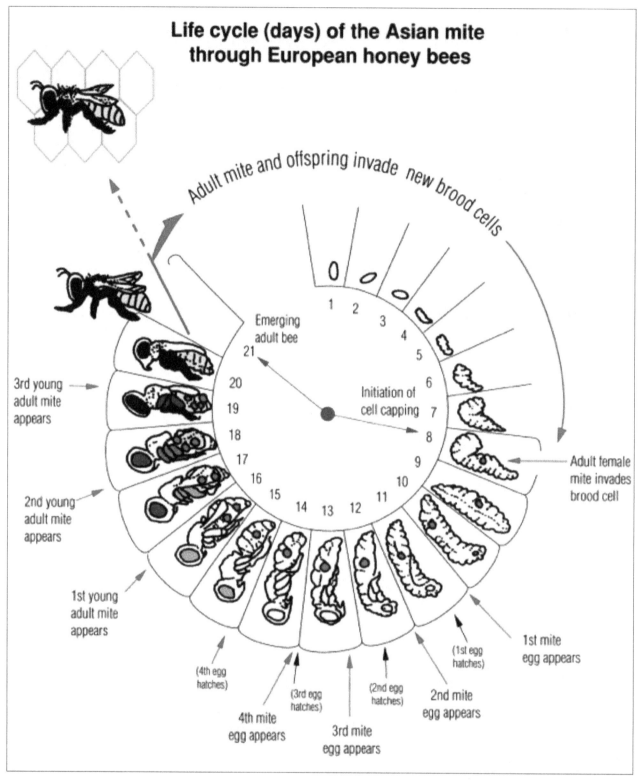

Life cycle (days) of the Asian mite through European honey bees

Adult mite and offspring invade new brood cells

Emerging adult bee

Initiation of cell capping

Adult female mite invades brood cell

3rd young adult mite appears

2nd young adult mite appears

1st young adult mite appears

1st mite egg appears

2nd mite egg appears

(1st egg hatches)

3rd mite egg appears

(2nd egg hatches)

4th mite egg appears

(3rd egg hatches)

(4th egg hatches)

Fig. 9. Life cycle of *T. mercedesae* on *A. mellifera*. Diagram was constructed from data reported by Saleu (1994).   Photo: Denis Anderson.

Table 3. Comparison of time (in hours) after bee brood cell capping of the first appearance of different developmental stages of *T. mercedesae*, as reported in 4 different studies: Saleu, 1994 (New Guinea); Kiprasert, 1984 (Thailand); Woyke 1984, 1985a (Afghanistan and Vietnam); Kumar et al., 1993 (India). The numbers in brackets represent honey bee brood development time in days.

Mite development Stage	Study 1	Study 2	Study 3	Study 4
Egg:	72	96	24-48	96
	(10-11)	(12)	(9-10)	(12)
Larva:	96	96	24-48	96
	(11-12)	(12)	(9-10)	(12)
Protonymph:	96	192	72-96	192
	(11-12)	(16)	(11-12)	(16)
Deutonymph:	168	216	120	192
	(14-15)	(17)	(13)	(16)
Young Adult:	216	312	192	298
	(16-17)	(22)	(16)	20)

5.3. Methods for studying mite development on giant Asian honey bees

There are virtually no reported studies on the life cycle of different species of *Tropilaelaps* on their native Asian giant bee hosts. This is a pity, as mites of the different species do not typically cause as much harm to their giant honey bee hosts as they do to *A. mellifera*. Hence studies on the life cycles of mites on the giant honey bees could provide clues as to how to control them on *A. mellifera*.

Studies on the life cycle of *Tropilaelaps* mites on their giant honey bee hosts in Asia can be done as follows:

1. First, and most importantly, locate a local 'honey-hunter' to gain access to a nest, as virtually all giant honey bee nests in the wild in Asia are 'owned' by a local hunter and accessing them without permission (or help) can lead to serious problems.
2. Once a nest is located, smoke the bees off the single comb (typically most of the bees will fly off the comb when it is smoked, but they will return later).
3. Remove a section of the comb that contains capped brood using a sharp knife (in this way the comb will not be significantly damaged and the bees will soon return to it and resume their normal activities).
4. Transport the comb to the comfort of the laboratory for examination.
5. In the laboratory open each capped cell and remove the developing bee noting the approximate age of the bee (prepupa, white –eyed pupa, pink-eyed pupa, and so on).
6. Record how many individuals of the various mite stage are present.
7. Collate the recorded data.

Pros: This method is relatively simple and will show the progression of mite development, from egg to adult.

Cons: It is often difficult to gain access to wild nests.

5.4. Methods for studying mite mating behaviour

Mating behaviour can be studied by placing a single adult male mite and female mite inside a clear plastic Petri dish (or glass bottle) that contains a bee larva on which the mites can feed. Mites are kept alive in the dish by continually replacing the bee larva.

Pros: This method allows for mating behaviour to be observed and videoed.

Cons: Behaviours of mites *in-vitro*, such as those shown in plastic or glass containers, may not necessarily be those shown naturally inside the bee colony.

6. Pathogenicity, control and association with pathogens

Tropilaelaps mercedesae (formerly known as *T. clareae*) may infest as much as 90% of the brood in *A. mellifera* colonies (Kiprasert, 1984), but smaller brood infestation levels of 3 to 6% have been consistently reported from *A. dorsata* colonies (Underwood, 1986) and adult *A. dorsata* and *A. cerana* workers show greater resistance to the mite than *A. mellifera* workers (Khongphinitbunjong *et al.*, 2012). High infestations of *A. mellifera* brood by this species often results in callow adult bees with deformed wings (De Jong *et al.*, 1982; Burgett *et al.*, 1983) and reduced body weights (Kiprasert, 1984) (see Fig. 10). Untreated infestations rapidly increase to high levels and invariably lead to the death of entire colonies (Atwal & Goyal, 1971; Ritter, 1988; Woyke, 1985a, 1985b).

Fig. 10. Damage caused by *T. mercedesae* to *A. mellifera* brood.

Photo: Denis Anderson.

The control of *Tropilaelaps* mites in *A. mellifera* colonies has been reviewed by De Jong *et al.* (1982) and Ritter and Sneider-Ritter (1988), but is in need of revision. Many of the various synthetic chemical acaricides used to control *Varroa* mites are also effective against *Tropilaelaps* mites (Pichai, *et al.*, 2008). Sulphur, formic acid and thymol have also proved satisfactory (Atwal and Goyal, 1971; Raffique *et al.*, 2012). Non-chemical means of controlling *T. mercedesae* in *A. mellifera* colonies have been achieved by interrupting the brood cycle of the bees. For example, Woyke (1984; 1985a & b) controlled *T. mercedesae* in *A. mellifera* colonies by removing all brood for 2 days, but removal for 5 days is recommended in order to kill all mites. Such methods would probably not be viable for commercial beekeepers that manage large numbers of colonies and they are very time-consuming.

There have been few studies on the pathogens associated with *Tropilaelaps*. However, like *Varroa* mites, *Tropilaelaps* mites have been associated with spread and infection of deformed wing virus in *A. mellifera* colonies (Dainat *et al.*, 2009; Forsgren *et al.*, 2009). Methods involved with studying bee viruses can be found in Dietemann *et al.* (2013).

7. Future research priorities

Much remains to be learnt about *Tropilaelaps* mites. The taxonomy of the genus is one area that has been relatively well studied, and it now seems to be well resolved. Given that each species is closely associated with a particular giant Asian bee species or sub-species, and that most of these bees have been examined for mites, it is unlikely that further species will be discovered. Nevertheless, more haplotypes of each species will undoubtedly be found, as has recently been demonstrated in China (Luo *et al.*, 2011). With this in mind, the follow list highlights areas that warrant immediate research effort.

1. Morphological descriptions of the different life stages of all species.
2. Life habits and reproductive behaviour of all species on both their native and adopted hosts.
3. Control of *T. mercedesae* and *T. clareae* on *A. mellifera*.
4. Associations of microbial pathogens with *T. mercedesae* and *T. clareae* infestations on *A. mellifera*.
5. Resistance (or tolerance) mechanisms of Asian bees to *Tropilaelaps* mites.
6. Genomic sequence information on *T. mercedesae* and *T. clareae*.

8. Concluding comments

Tropilaelaps mites, particularly *T. mercedesae* and *T. clareae*, present a serious threat to world beekeeping and hence they deserve the immediate attention of the global research community. Of the two species, *T. mercedesae* is the more likely to spread out of Asia, as it is found throughout mainland Asia and South East Asia, while *T. clareae* is confined just to the Philippines.

Evidence suggests that while these two species may be as, if not more, pathogenic to *A. mellifera* as *V. destructor*, they may be easier to control for small-scale beekeepers. However, large-scale commercial beekeepers that manage hundreds and thousands of bee colonies, and cannot afford the time to keep them broodless for even short periods, will find these mites as difficult to control as *V. destructor*.

Pre-emptive research carried out on *Tropilaelaps* mites before they spread globally is far more desirable than reactive research carried out once the mites have spread, as it reduces potential future losses and hardship for beekeepers.

9. Acknowledgements

The COLOSS (Prevention of honey bee COlony LOSSes) network aims to explain and prevent massive honey bee colony losses. It was funded through the COST Action FA0803. COST (European Cooperation in Science and Technology) is a unique means for European researchers to jointly develop their own ideas and new initiatives across all scientific disciplines through trans-European networking of nationally funded research activities. Based on a pan-European intergovernmental framework for cooperation in science and technology, COST has contributed since its creation more than 40 years ago to closing the gap between science, policy makers and society throughout Europe and beyond. COST is supported by the EU Seventh Framework Programme for research, technological development and demonstration activities (*Official Journal L 412, 30 December 2006*). The European Science Foundation as implementing agent of COST provides the COST Office through an EC Grant Agreement. The Council of the European Union provides the COST Secretariat. The COLOSS network is now supported by the Ricola Foundation - Nature & Culture.

10. References

ANDERSON, D L (1994) Non-reproduction of *Varroa jacobsoni* in
    Apis mellifera colonies in Papua New Guinea and Indonesia.
    Apidologie 25: 412-421.

ANDERSON, D L; MORGAN, M J (2007) Genetic and morphological
    variation of bee-parasitic *Tropilaelaps* mites (Acari: Laelapidae):
    new and re-defined species. *Experimental and Applied Acarology*
    43: 1-24.

ANDERSON, D L; TRUEMAN, J W H (2000) *Varroa jacobsoni* is more
    than one species. *Experimental and Applied Acarology* 24: 165-189.

ATWAL, A A; GOYAL, N P (1971) Infestation of honey bee colonies
    with *Tropilaelaps*, and its control. *Journal of Apicultural Research*
    10: 137-142.

BAKER, A B; HICK, A; CHMIELEWSKI, W (2005) Aspects of the history
    and biogeography of the bee mites *Tropilaelaps clareae* and *T.
    koenigerum*. *Journal of Apicultural Science* 49: 13-19.

BURGETT, M D; AKRATANAKUL, P; MORSE, R A (1983) *Tropilaelaps
    clareae*: a parasite of honey bees in South-East Asia. *Bee World*
    64: 25-28.

CLINCH, P (1979) *Nosema apis* and mites in honey bee colonies in
    Papua New Guinea. *Journal of Apicultural Research* 18: 298-301.

DAINAT, B; KEN, T; BERTHOUD, H; NEUMANN, P (2009) The
    ectoparasitic mite *Tropilaelaps mercedesae* (Acari, Laelapidae) as
    a vector of honey bee viruses. *Insectes Sociaux* 56: 40-43.
    http://dx.doi.org/10.1007/s00040-008-1030-s

DEFRA (2005) Tropilaelaps: *parasitic mites of honey bees*. UK Department
    for environment, Food and Rural Affairs; London, UK. 14 pp.

DEJONG, D; MORSE, E W; EICKWORT, G C (1982) Mite pests of
    honey bees. *Annual Review of Entomology* 27: 229-252.

DELFINADO, M D; BAKER, E W (1961) *Tropilaelaps*, a new species of
    mite from the Philippines (Laelapidae [s. lat]: Acarina). *Fieldiana
    Zoology* 44: 53-56.

DELFINADO-BAKER, M; AGGARWAL, K (1987) Infestations of
    Tropilaelaps clareae and *Varroa jacobsoni* in *Apis mellifera*
    colonies in Papua New Guinea. *American Bee Journal* 127: 443.

DELFINADO-BAKER, M D; BAKER, E W (1982) A new species of
    Tropilaelaps parasitic on honey bees. *American Bee Journal* 122:
    416-417.

DIETEMANN, V; NAZZI, F; MARTIN, S J; ANDERSON, D; LOCKE, B;
    DELAPLANE, K S; WAUQUIEZ, Q; TANNAHILL, C; FREY, E;
    ZIEGELMANN, B; ROSENKRANZ, P; ELLIS, J D (2013) Standard
    methods for varroa research. In *V Dietemann; J D Ellis;
    P Neumann (Eds) The COLOSS* BEEBOOK, *Volume II: standard
    methods for* Apis mellifera *pest and pathogen research. Journal of
    Apicultural Research* 52(1):
    http://dx.doi.org/10.3896/IBRA.1.52.1.09

FORSGREN, E; DE MIRANDA, J R; ISAKSSON, M; WEI, S; FRIES, I
    (2009) Deformed wing virus associated with *Tropilaelaps
    mercedesae* infesting European honey bees (*Apis mellifera*).
    Experimental and Applied Acarology 47: 87-97.
    http://dx.doi.org/10.1007/s10493-008-9204-4

KHONGPHINITBUNJONG, K; DE GUZMAN, L I; BURGETT, M D;
    RINDERER, T E; CHANTAWANNAKUL, P (2012) Behavioural
    responses underpinning resistance and susceptibility of honey
    bees to *Tropilaelaps mercedesae*. *Apidologie* 43: 590-599.
    http://dx.doi.org/10.1007/s13592-012-0129-x

KIPRASERT, C (1984) Biology and systematics of the parasitic bee
    mite *Tropilaelaps clareae* Delfinado and Baker (Acarina:
    Laelapidae). MSc thesis, Kasetsart University; Bangkok, Thailand.

KOENIGER, N; MUSAFFAR, N (1988) Lifespan of the parasitic honey
    bee mite *Tropilaelaps clareae* on *Apis cerana, dorsata* and
    mellifera. *Journal of Apicultural Research* 27: 207-212.

KUMAR, N R; KUMAR, R; MBAYA, J; MWANGI, R W (1993)
    Tropilaelaps clareae found on *Apis mellifera* in Africa. *Bee World*
    74: 101-102.

LAIGO, F M; MORSE, R A (1969) Control of the bee mites *Varroa
    jacobsoni* Oudemans and *Tropilaelaps clareae* Delfinado and
    Baker with chlorobenzilate. *Philippine Entomologist* 1: 144-148.

LE CONTE, Y; NAVAJAS, M (2008) Climate change: impact on bee
    populations and their illnesses. *Revue Scientifique Et Technique –
    Office International Des Epizooties* 27: 485-497.

LINDQUIST, E E; KRANTZ, G W; WALTER, D E (2009) Order
    Mesostigmata. In *G W Krantz; D E Walter (Eds). A manual of
    acarology.* Texas Technical University Press. pp 124-232.

LUO, Q H; ZHOU, T; WANG, Q; DAI, P L; WU, Y Y; SONG, H L (2011)
    Identification of *Tropilaelaps* mites (Acari, Laelapidae) infesting
    Apis mellifera in China. *Apidologie* 42: 485-498.
    http://dx.doi.org10.1007/s13592-011-0028-x

MATHESON, A (1997) Country records for honey bee diseases,
    parasites and pests. Appendix 11. In *R M Morse; K Flottum (Eds).
    Honey bee pests, predators and diseases (3ᵈ Ed.).* A I Root Co.,
    Medina OH, USA. pp 586-602.

OIE (2004) *Tropilaelaps* infestation of honey bees (*Tropilaelaps
    clareae, T. koenigerum*). In *Manual of diagnostic tests and
    vaccines for terrestrial animals (mammals, birds and bees) (5th
    Ed.).* Office International des Epizooties; Paris, France.

OTIS, G W; KRALJ, J (2001) Parasitic brood mites not present in North
    America. In *T C Webster; K S Delaplane (Eds). Mites of the honey
    bee.* Dadant and Sons; Illinois, USA. pp 251-272.

PICHAI, K; POLGAR, G; HEINE, J (2008) The efficacy of Bayvarol®
    and CheckMite+® in the control of *Tropilaelaps mercedesae* in
    the European honey bee (*Apis mellifera*) in Thailand. *Apiacta* 43:
    12-16.

RATH, W; DELFINADO-BAKER, M; DRESCHER, W (1991) Observations on the mating behaviour, sex ratio, phoresy and dispersal of *Tropilaelaps clareae* (Acari: Laelapidae). *International Journal of Acarology* 17: 201-208.

RAFFIQUE, M K; MAHMOOD, R; ASLAM, M; SARWAR, G (2012) Control of *Tropilaelaps clareae* mite by using formic acid and thymol in honey bee *Apis mellifera* colonies. *Pakistan Journal of Zoology* 44: 1129-1135.

RINDERER, T E; OLDROYD, B P; LEKPRAYOON, C; WONGSIRI, S; THAPA, R (1994) Extended survival of the parasitic mite *Tropilaelaps clareae* on adult workers of *Apis mellifera* and *Apis dorsata*. *Journal of Apicultural Research* 33: 171-174.

RITTER, W (1988) *Varroa jacobsoni* in Europe the tropics and subtropics. In *G R Needham; R E Page, Jr; M Delfinado-Baker; C E Bow (Eds). Africanized honey bees and bee mites*. Halsted Press; Chichester, UK. pp 349-359.

RITTER, W; SCHNEIDER-RITTER, U (1986) *Varroa jacobsoni* und *Tropilaelaps clareae* in bienenvolkern von *Apis mellifera* in Thailand. *Apidologie* 17: 384-386.

RITTER, W; SCHNEIDER-RITTER, U (1988) Differences in biology and means of controlling *Varroa jacobsoni* and *Tropilaelaps clareae*, two novel parasitic mites of *Apis mellifera*. In *G R Needham; R E Page, Jr; M Delfinado-Baker; C E Bow (Eds). Africanized honey bees and bee mites*. Halsted Press; Chichester, UK. pp 378-395.

SALEU, L (1994) Studies of the parasitic mite *Tropilaelaps clareae* in *Apis mellifera* colonies in Papua New Guinea. MSc thesis, The Australian National University; Canberra, Australia 96 pp.

TANGJINGJAI, W; VERAKALASA, P; SITTIPRANEED, S, LEKPRAYOON, C (2003) Genetic differences between *Tropilaelaps clareae* and *Tropilaelaps koenigerum* in Thailand based on ITS and RAPD analyses. *Apidologie* 34: 514-524. http://dx.doi.org/10.1051/apido:2003042

UNDERWOOD, B A (1986) The natural history of *Apis laboriosa* Smith in Nepal. MSc thesis, Cornell University; Ithaca, NY, USA.

WALTER, D E; PROCTOR, H C (1999) *Mites: ecology, evolution and behaviour*. UNSW Press. 322 pp.

WILDE, J (2000) Is it possible to introduce *Tropilaelaps clareae* together with imported honey bee queens to Europe? *Pszczeln. Zesz. Nauk.* 44: 155-162.

WOYKE, J (1984) Survival and prophylactic control of *Tropilaelaps clareae* infesting *Apis mellifera* colonies in Afghanistan. *Apidologie* 15: 421–434.

WOYKE, J (1985a) *Tropilaelaps clareae*, a serious pest of *Apis mellifera* in the tropics, but not dangerous for apiculture in temperate zones. *American Bee Journal* 125: 497–499.

WOYKE, J (1985b) Further investigation into control of the parasitic bee mite *Tropilaelaps clareae* without medication. *Journal of Apicultural Research* 24: 250–254.

WOYKE, J (1987) Length of successive stages in the development of the mite *Tropilaelaps clareae* in relation to honey bee brood age. *Journal of Apicultural Research* 26: 110–114.

WOYKE, J (1994) Mating behaviour of the parasitic honey bee mite *Tropilaelaps clareae*. *Experimental and Applied Acarology* 18: 723–733.

Journal of Apicultural Research 52(1): (2013)
DOI 10.3896/IBRA.1.52.1.09

REVIEW ARTICLE

Standard methods for varroa research

IBRA
INTERNATIONAL BEE
RESEARCH ASSOCIATION

Vincent Dietemann[1,2*], Francesco Nazzi[3], Stephen J Martin[4], Denis L Anderson[5], Barbara Locke[6], Keith S Delaplane[7], Quentin Wauquiez[1], Cindy Tannahill[8], Eva Frey[9], Bettina Ziegelmann[9], Peter Rosenkranz[9] and James D Ellis[8]

[1]Swiss Bee Research Centre, Agroscope Liebefeld-Posieux Research Station ALP, Bern, Switzerland.
[2]Social Insect Research Group, Department of Zoology and Entomology, University of Pretoria, Pretoria, South Africa.
[3]Dipartimento di Scienze Agrarie e Ambientali, Università di Udine, vi delle Scienze 206, 33100 Udine, Italy.
[4]School of Environment and Life Sciences, University of Salford, Manchester, UK, M5 4WT
[5]CSIRO Entomology, Canberra, ACT 2601, Australia.
[6]Department of Ecology, Swedish University of Agricultural Sciences, Uppsala, Sweden.
[7]Department of Entomology, University of Georgia, Athens, GA 30602, USA.
[8]Honey Bee Research and Extension Laboratory, Department of Entomology and Nematology, University of Florida, Gainesville, Florida, USA.
[9]University of Hohenheim, Apicultural State Institute, 70593 Stuttgart, Germany.

Received 16 May 2012, accepted subject to revision 25 October 2012, accepted for publication 14 November 2012.

*Corresponding author: Email: vincent.dietemann@alp.admin.ch

Summary

Very rapidly after *Varroa destructor* invaded apiaries of *Apis mellifera*, the devastating effect of this mite prompted an active research effort to understand and control this parasite. Over a few decades, varroa has spread to most countries exploiting *A. mellifera*. As a consequence, a large number of teams have worked with this organism, developing a diversity of research methods. Often different approaches have been followed to achieve the same goal. The diversity of methods made the results difficult to compare, thus hindering our understanding of this parasite. In this paper, we provide easy to use protocols for the collection, identification, diagnosis, rearing, breeding, marking and measurement of infestation rates and fertility of *V. destructor*. We also describe experimental protocols to study orientation and feeding of the mite, to infest colonies or cells and measure the mite's susceptibility to acaricides. Where relevant, we describe which mite should be used for bioassays since their behaviour is influenced by their physiological state. We also give a method to determine the damage threshold above which varroa damages colonies. This tool is fundamental to be able to implement integrated control concepts. We have described pros and cons for all methods for the user to know which method to use under which circumstances. These methods could be embraced as standards by the community when designing and performing research on *V. destructor*.

Métodos estándar de la investigación en varroa étodos

Resumen

Poco tiempo después de que el ácaro *Varroa destructor* invadiera las colmenas de *Apis mellifera*, su efecto devastador produjo un efectivo esfuerzo investigador para comprender y controlar este parásito. En unas pocas décadas, la varroasis se ha extendido a la mayoría de los países que explotan a *A. mellifera*. Como consecuencia, un gran número de equipos han trabajado con este organismo desarrollando diversos métodos de investigación. A menudo, se han utilizado diferentes enfoques para lograr el mismo objetivo. La diversidad de métodos hizo que los resultados fueran difíciles de comparar, lo que dificulta la comprensión de este parásito. En este artículo se proporcionan protocolos fáciles de usar para la recolección, identificación, diagnóstico, cría, cruzamiento, marcaje y medición de los índices de infestación y la fertilidad de *V. destructor*. También se describen los protocolos experimentales para el estudio de la orientación y la alimentación de los ácaros, la infestación de colonias o células y para medir la susceptibilidad del ácaro a los acaricidas. Cuando es pertinente, se describe qué ácaro se debe utilizar para los bioensayos puesto que su comportamiento está influido por su estado fisiológico. También proporcionamos un método para

Footnote: Please cite this paper as: DIETEMANN, V; NAZZI, F; MARTIN, S J; ANDERSON, D; LOCKE, B; DELAPLANE, K S; WAUQUIEZ, Q; TANNAHILL, C; FREY, E; ZIEGELMANN, B; ROSENKRANZ, P; ELLIS, J D (2013) Standard methods for varroa research. In *V Dietemann; J D Ellis; P Neumann (Eds) The COLOSS BEEBOOK, Volume II: standard methods for Apis mellifera pest and pathogen research. Journal of Apicultural Research* 52(1):
http://dx.doi.org/10.3896/IBRA.1.52.1.09

determinar el umbral de daño más allá del cual varroa causa daños a las colonias. Esta herramienta es fundamental para poder poner en práctica el concepto de control integrado. Hemos descrito los pros y los contras de todos los métodos para que el usuario sepa qué método utilizar según las circunstancias. Estos métodos podrían ser adoptados como estándares por la comunidad para el diseño y la realización de investigaciones sobre *V. destructor*.

大蜂螨研究的标准方法

自狄斯瓦螨侵袭西方蜜蜂蜂场以来，其带来的毁灭性危害促进了该领域的研究工作。在过去的几十年里，大蜂螨已分布到大多数饲养西方蜜蜂的国家。由此许多研究团队开展了蜂螨的研究工作，并形成了多种研究方法。但往往是运用不同的方法解决了同一问题，同时也造成了实验结果难以比较，妨碍了我们对大蜂螨的认知。本文我们提出了一些简单实用的实验方案，可用于开展大蜂螨感染率和生殖力方面的研究，包含了收集、鉴定、诊断、饲养、育种、标记和检测技术。还提供了研究蜂螨定位和饲养蜂螨，蜂螨侵染蜂群、侵染巢房以及蜂螨对杀螨剂的耐药性的相关实验方案。在相关内容中还描述了如何选择蜂螨开展生物学实验，因为蜂螨的行为在其不同的生理阶段是不同的。还给出了测定蜂螨对蜂群危害的临界值的方法，这是实施蜂螨综合治理的基本工具。对所有的方法我们都描述了其优、缺点，以帮助研究者选择合适的方法开展工作。这些方法也可作为标准方法介绍给广大从事大蜂螨研究或治理的工作者。

Keywords: COLOSS, *BEEBOOK*, *Varroa destructor*, *Apis mellifera*, research method, protocol, orientation, feeding, marking, taxonomy, bioassay, damage threshold, acaricide, artificial infestation, breeding, honey bee

Table of Contents

Table of Contents — continued

1. Introduction

Most honey bee researchers consider the ectoparasitic mite *Varroa destructor* to be the most damaging enemy of the honey bee. It has been recently identified as one of the major factor responsible for colony losses worldwide (e.g. Brodschneider *et al.*, 2010; Chauzat *et al.*, 2010; Dahle, 2010; Gensch *et al.*, 2010; Guzman-Novoa *et al.*, 2010; Schäfer *et al.*, 2010; Topolska *et al.*, 2010; vanEngelsdorp *et al.*, 2011; Martin *et al.*, 2012; Nazzi *et al.*, 2012). Both the development of new and innovative control methods against the mite and further studies on the complex interaction with the honey bee should be a priority in bee health research (Dietemann *et al.*, 2012). The use of standardised methods by those studying the mite will greatly increase the impact of such work. When reviewing the literature, researchers should take note that publications prior to 2000 mention *V. jacobsoni* instead of *V. destructor*. The species name was changed after Anderson and Trueman (2000) demonstrated with molecular tools that the invasive population was not the species from Indonesia described by Oudemans in 1904.

2. Taxonomy and systematics

2.1. Taxonomy

Varroa mites were first discovered more than 100 years ago on the Asian honey bee (*Apis cerana*) in Java, Indonesia and named *Varroa jacobsoni* (Oudemans, 1904). They were assigned to a new genus, *Varroa*, and eventually to a new family, Varroidae (Delfinado-Baker and Baker, 1974). At present the genus contains four species. Since the initial discovery, it has become clear that varroa mites are native brood parasites of a group of cavity nesting Asian honey bees that are closely related to *A. cerana*. These include, *A. cerana* itself (which is distributed throughout most of Asia), *A. koschevnikovi* (Borneo and surrounding regions), *A. nigrocincta* (Sulawesi) and *A. nuluensis* (Borneo). These bees are still undergoing taxonomic revision as seen by the recent proposal to elevate the plains honey bee of south India to a new species, *A. indica*, and separate it from *A. cerana* (Lo *et al.*, 2010). At present, varroa mites are only known to infest *A. cerana*, *A. koschevnikovi* and *A. nigrocincta*, although very few surveys for mites have been reported for *A. nigrocincta*, *A. nuluensis* or *A. indica* and those mites that have been found on *A. nigrocincta* in Sulawesi were most likely not native to that bee, but rather to sympatric *A. cerana* (Anderson and Trueman, 2000).

It is not exactly certain when the European honey bee (*A. mellifera*) first came in contact with varroa but it certainly occurred after that bee was introduced into Asia by man (De Jong *et al.*, 1982a). There are specimens of varroa in the Acarological Collection at Oregon State University, USA, that were collected from *A. mellifera* in China during the middle of the last century (Akratanakul and Burgett, 1975). The varroa mites that have since utilized *A. mellifera* as a host are all members of *V. destructor*, the most recently described species of the genus, and are native to *A. cerana* in northeast Asia (Anderson and Trueman, 2000). Hence, the current four recognized species of varroa came about through a long process of speciation on Asian honey bee hosts and, given the rather uncertain taxonomic status of those bees, it is possible that new varroa species await discovery. Prolonged co-evolution of *V. destructor* and *A. mellifera* may yet see these mites also becoming genetically diverse (Oldroyd, 1999), particularly as they gradually adapt to exist on isolated populations of *A. mellifera*. However, the movement of bee stocks around the world by man and the beekeeping practice of re-queening large numbers of *A. mellifera* colonies on a regular basis with queens from a common source will, to some extent, counter natural evolutionary processes that may eventually lead to varroa speciation on *A. mellifera*.

Various methods have been used over the years to determine variation within varroa, all of which have contributed to the current level of taxonomic understanding. The most common and simple methods of identifying species have been those that provide measurements of mite physical characteristics (morphology). These methods are discussed below. The initial discoveries of *V. jacobsoni* on *A. cerana*, *V. underwoodi* on *A. cerana* and *V. rindereri* on *A. koschevnikovi* all resulted from morphological studies.

More recently, molecular methods have helped clarify varroa taxonomy and have proven particularly useful for identifying genetic variation within species and even identifying cryptic species. These methods, also described below, played a crucial role in the discovery of a new species, *V. destructor*, and in showing that it was that species, not *V. jacobsoni* as previously thought, that had colonized *A. mellifera* after its introduction into Asia (Anderson and Trueman, 2000).

The current taxonomy of varroa on Asian honey bees can be summarized as follows (after Lindquist *et al.*, 2009):

Kingdom: Animalia

Phylum: Arthropoda

Class: Arachnida

Subclass: Acari

Superorder: Parasitiformes

Order: Mesostigmata

Family: Varroidae

Genus: *Varroa*

Species:

> *V. jacobsoni* (Oudemans, 1904)
>
> *V. underwoodi* (Delfinado-Baker and Aggarwal, 1987)
>
> *V. rindereri* (De Guzman and Delfinado-Baker, 1996)
>
> *V. destructor* (Anderson and Trueman, 2000).

The taxonomic status of three genetically distinct varroa types that infest *A. cerana* in the Philippines remains unresolved at this time (Anderson, 2000; Anderson and Trueman, 2000).

Mites of just two 'haplogroups' of *V. destructor* (see section 2.4.5. 'Haplogroup and haplotype identification') have colonized *A. mellifera* globally. Of the two, those belonging to a Korea haplogroup are the most common and widespread on *A. mellifera*, being present in Europe, the Middle East, Africa, Asia, the Americas and New Zealand. Mites of a Japan haplogroup are less common on *A. mellifera*, and are only found in Thailand, Japan and the Americas (Anderson and Trueman, 2000; Warrit *et al.*, 2006). At the present time Australia remains the only large landmass on earth on which the resident *A. mellifera* are free of varroa.

2.2. Collection of mites for identification

The best varroa specimens for laboratory analyses are those that have been collected live and preserved immediately. A benefit of sampling live mites is that they can be submerged in hot water prior to their preservation. This relaxes internal body tissues and exposes hard-to see organs, such as the chelicerae, which usually remain hidden from sight in mites collected directly into alcohol.

2.2.1. Mite appearance

Adult females are large (about 1.5 mm in width) and reddish-brown in colour, whereas males and female nymph stages are smaller and cream or white in colour. All stages are easily seen by the naked eye (Fig. 1). Each of the different life stages may be carefully removed from cells with the aid of a fine pair of forceps (such as #55 biologie forceps, Cat. No. 11255, from FST Fine Science Tools Inc.; Canada; Fig. 2) or soft paintbrush and dunked immediately into preserving fluid in a collection vial. Mites dunked into a vial of alcohol will immediately die and sink to the bottom, whereas those dunked into a vial of RNAlater will float on the surface and crawl around the inside of the vial before eventually dying some time later.

2.2.2. Where to find mites

Live adult mites, nymphs and eggs are most easily found in capped brood cells of bee colonies in which adult female mites are reproducing. In *A. cerana* colonies this is restricted to drone cells, but in *A. mellifera* colonies it may be either drone or worker cells. After removing the wax cappings and bee brood, the presence of white faecal deposits on cell walls (Fig. 3) is a sure indicator of the presence of reproducing females. Collecting mites from brood cells with offspring also provides evidence that these mites indeed reproduce on the bee species they have been collected from, as mites sometimes drift to and from colonies of foreign species on which they are unable to reproduce (Anderson and Trueman, 2000; Koeniger *et al.*, 2002), which might confuse the host-specificity attributed to them. Only live adult female varroa can be collected from broodless bee colonies. These are generally found on the bodies or in body cavities of worker bees.

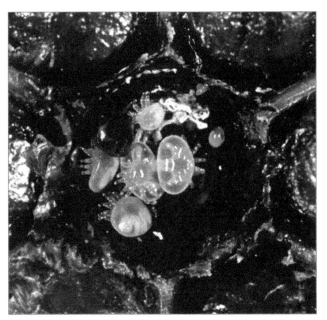

Fig. 1. A mite family with mother mite (reddish brown) and different stages of offspring at the bottom of a cell from which the honey bee pupa was removed.   Photo: Denis Anderson.

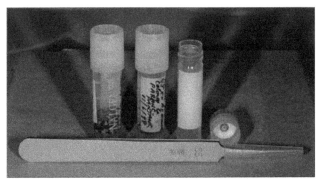

Fig. 2. Tool kit to collect *Varroa* spp. mites.   Photo: Denis Anderson.

Fig. 3. In this section of a cell (the bottom is on the right side), the pearly white faeces deposit is visible on the upper and back walls. Mature and immature varroa mites are also visible.

Photo: Swiss Bee Research Institute

2.2.3. Sampling techniques

Varroa spp. mites can be sampled from brood or adult workers. Sampling techniques are described below in section 3.1. 'Collecting mites'.

2.2.4. Storage of mite samples

2.2.4.1. Storage medium and conditions

Mites collected in the field should be preserved immediately in 70-95% ethyl alcohol or RNAlater. This ensures the specimens are not damaged and, even if they are kept this way at room temperature, are good for morphological analyses for at least a few months, but often much longer. However, if specimens are to be used in DNA analysis, they should be stored in a cool environment, such as a fridge at 4°C or freezer at -20°C, within a few days of collection to slow the degradation of DNA in body tissues. Specimens frozen at -20°C remain viable for several years, but to remain viable longer, they should be stored at -70°C (see the section on 'Storing dead adults' in the *BEEBOOK* paper on miscellaneous methods (Human *et al.*, 2013)).

2.2.4.2. Storage and collection container

Ideal containers for collecting mites are small and made from tough plastic, such as the small plastic 1.5 ml cryogenic vial supplied by Nalgene®, shown in Fig. 2. This vial may hold hundreds of mite specimens and has a large white-coloured area on its outside for a label. Importantly, its lid is secured on a thread that runs down the outside of the vial. This ensures that no preserving fluid is forced from the vial as it is being closed, which could result in smudging or complete removal of the label. The label should contain essential information, such as the date of collection, name of host bee, location and name of collector, using a fine point permanent marker pen. To overcome external labels becoming removed from the collection vial, a small piece of paper on which the collection data have been written with a pencil (alcohol resistant) may be inserted in the vial, with the sample.

2.2.5. Sample shipping

Specimens should be transported to their destination as soon as possible after collection. Some airlines prohibit the carriage of biological specimens preserved in alcohol on aircraft, whilst others are less stringent. It pays to check airline policy in this regard before attempting to send or carry specimens preserved in alcohol. A convenient way to avoid this problem is to pour the alcohol off the specimens shortly before transportation. In this way the specimens will still remain covered with a very small amount of alcohol and thus remain saturated in alcohol and preserved during transport. However, upon arrival the specimens should be again well-covered in fresh alcohol before storage. Some transportation courier services have arrangements in place with airlines to transport biological specimens preserved in alcohol on aircraft.

Some countries (e.g. Australia and the USA) require an official quarantine import permit to accompany imported varroa mite specimens. Other countries (e.g. Brazil) may prohibit the exportation of specimen due to specific laws on biopiracy. Therefore, before sending or transporting specimens to a particular country, that country's policy on importing biological specimens should be checked and followed.

2.3. Morphological methods for identifying varroa

The four recognized species of varroa are readily identified morphologically and are shown for comparison in Fig. 4.

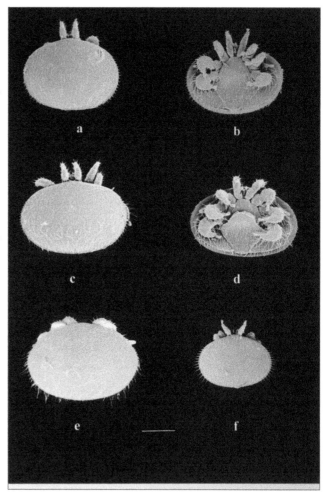

Fig. 4. The four species of *Varroa*: **a.** *V. jacobsoni* dorsal view; **b.** *V. jacobsoni* ventral view; **c.** *V. destructor* dorsal view; **d.** *V. destructor* ventral view; **e.** *V. rindereri*; **f.** *V. underwoodi*.

Photo: Denis Anderson.

2.3.1. Sample preparation

Morphological analyses are best carried out on mite specimens that have been mounted on glass microscope slides. For this, a specimen must first be cleared of its soft tissues before being mounted on a slide.

This is achieved as follows.

1. Remove specimen from preserving medium.
2. Immerse specimen in Nesbitt's Solution (see recipe below) in the depression of a concave slide.
3. Wait until the specimen becomes saturated with Nesbitt's Solution, and then push it under the surface of the solution to make it sink to the bottom, using a fine needle.
4. Place a cover slip over the depression of the slide.
5. Warm the slide in an oven for 1 hour at 45°C.
   The specimen should become free of body tissue and appear transparent, but older specimens may require further clearing in the oven for several hours or overnight.

This procedure can be speeded-up by warming the slide over a flame or hotplate for a few seconds, instead of placing it in an oven. However, extreme care should be taken to avoid boiling the Nesbitt's Solution, which will destroy the specimen. Laboratory gloves and coat should be worn when clearing specimens.

The cleared specimen is then mounted as follows:

6. Remove specimen from the Nesbitt's Solution and transfer it to a drop of Hoyer's Mounting Medium (see recipe below) on a glass microscope slide.
   Note: the drop should be just large enough to form a thin layer when a cover slip is placed on top, without overflowing around the edges of the cover slip.
7. Push the specimen down through the Hoyer's so that it rests on the slide, using a fine needle.
8. Gently lower a cover slip (thickness No. 1, diameter 16 mm) over the drop of Hoyer's, starting from the edge of the drop and letting it slowly settle over the drop under its own weight, spreading the Hoyer's as it goes.
9. Place the slide horizontally to cure in an oven at 45°C for at least 2 weeks.
10. Label and store slide.

Hoyer's medium does not completely harden and remains water-soluble, so that the slide can be reheated and specimen floated off the slide for dissection or re-mounting. For long-term storage or for transporting, the edges of the cover slip should be sealed with some water-resistant material, such as clear fingernail varnish. Laboratory gloves and coat should be worn when mounting specimens.

2.3.1.1. Recipe for Nesbitt's Solution:

- 60 g of chloral hydrate.
- 10 ml of concentrated (35.4%) hydrochloric acid.
- All dissolved in 100 ml of distilled water.

Note: care should be taken in preparing this solution, as it is highly corrosive to skin and microscope.

2.3.1.2. Recipe for Hoyer's medium:

- 30 g of gum Arabic.
- 200 g of chloral hydrate.
- 20 ml of glycerol.
- All dissolved in 50 ml of distilled water.

Note: the mixture needs to be stirred and warmed gently to allow the gum Arabic to dissolve, then filtered through muslin and stored in an airtight container, but not a container with a screw cap, as the cap will become permanently stuck.

2.3.2. Sample identification

Mounted mite specimens are best examined with dissecting or compound light microscopes that have been fitted with ocular micrometers. The following measurements should be considered.

- Body size (length and width).
- Structure and setation (i.e. stiff hair, bristle) of dorsal shield.
- Structure and chaetotaxy of the sternal, epigynal, anal and metapodal shields, peritreme, tritosternum and hypostome (see Fernandez and Coineau, 2007 for a description of varroa morphology).
- Number, arrangement and morphology of setae on the legs and palps.

The two species *V. destructor* and *V. jacobsoni* are morphologically similar, except in body size and shape. *V. jacobsoni* is much smaller and more circular in shape than *V. destructor* (Fig. 4). Nevertheless, some *V. jacobsoni* (e.g. those found on *A. cerana* in Laos, mainland Asia) are much larger than other *V. jacobsoni*. Hence it is always best to confirm a diagnosis of either of these species with additional molecular information.

In case varroa work is conducted in Asia where several species cohabit, we provide a determination key adapted from Oldroyd and Wongsiri (2006) and Warrit and Lekprayoon (2011) to differentiate those mites. Varroa mites have body as wide or wider as it is long. This characteristic distinguishes it from other Asian parasitic mite genera *Tropilaelaps* (with a body longer than it is wide) and *Euvarroa* (triangular shaped body).

2.4. Molecular methods and systematics

Molecular technology was first used in varroa research during the 1990s to look for variation within and among mite populations (Kraus and Hunt, 1995; De Guzman *et al.*, 1997, 1998, 1999; Anderson and Fuchs, 1998). Initially it was expensive and was only used by specialised laboratories. Currently, the landscape has changed and a number of quick and easy commercial kits can be purchased for extracting DNA from tissue and any number of laboratories will sequence DNA for a reasonable fee within hours of its extraction.

Sequence data from small DNA fragments (< 1,000 base pairs) has been particularly useful in providing 'snap-shots' of genetic variation across the entire varroa genome and for use in phylogenetic

Key to identification of varroa species * 'gutters' protruding from the spiracle on the ventral side, towards the edge of the body at the level of the third pair of legs (see Fernandez and Coineau, 2007).

1. a. Peritremes* are long and looping up from the ventral side, extending beyond the lateral margin of the dorsal shield and thus sometimes visible on a dorsal view.	*Varroa rindereri* (Fig. 4e) primarily found parasitizing *A. koschevnikovi*
b. Peritremes not extending beyond the lateral margin of the dorsal shield.	2.
2. a. Setae of the lateral margin long and slender	*Varroa underwoodi* (Fig. 4f) primarily found parasitizing *A. dorsata, A. laboriosa* and *A. breviligula*
b. Setae shorter and stout.	3.
3. a. Body size ratio (width to length) 1.2-1.3:1	*Varroa jacobsoni* (Fig. 4a, b) parasitize *A. cerana* on Sundaland, including *A. nigrocincta* on Sulawesi
b. Body size ratio ≥ 1.4	*Varroa destructor* (Fig. 4c, d) parasitize *A. cerana* on Mainland Asia, and *A. mellifera* worldwide

analyses or as molecular markers (Avise, 2004). Sequencing involves three basic steps described below in sections 2.4.1., 2.4.2. and 2.4.3.:

1. Extraction of total DNA from mite tissue.
2. Amplifying (making copies of) fragments of that DNA using Polymerase Chain Reaction (PCR).
3. Sequencing the amplified fragments.

2.4.1. DNA extraction

DNA is sourced from the tissue of *Varroa* spp. mites that have been collected and preserved in 70% ethyl alcohol or RNAlater (as described in section 2.2.4.1. 'Storage medium and conditions'). Any tissues can be used, but if the tissue is dissected from a single appendage (such as a leg), the rest of the mite can be used for other purposes. See the section on 'CTAB genomic DNA extraction from adult bees' of the *BEEBOOK* paper on molecular methods (Evans *et al.*, 2013)) for extraction methods.

2.4.2. DNA amplification

DNA amplification requires a PCR machine (such as an Eppendorf Mastercycler®) and a set of specific forward and reverse primers. The machine is initially programmed to carry out a number of cycles to amplify the DNA (see the section 'DNA methods' of the *BEEBOOK* paper on molecular methods (Evans *et al.*, 2013).

A commonly used method for amplifying *Varroa* spp. DNA consists of:

1. PCR thermo-cycles of 5 min pre-denaturation at 94°C.
2. 35 cycles of denaturation at 94°C for 1 min.

3. 1 min annealing at 52°C.
4. 2 min extension at 72°C.
5. Final extension at 72°C for 5 min.

2.4.3. DNA sequencing

Amplified DNA can be then sent to a laboratory for sequencing and the sequence can then be compared with sequences in GenBank using BLAST (see the section 'Obtaining and formatting sequences of interest for phylogenetics' of the *BEEBOOK* paper on molecular methods (Evans *et al.*, 2013)). The critical part of DNA sequencing is to decide which gene to sequence for a particular outcome, and only trial and error will determine this. Fortunately, studies have already shown that sequences obtained from specific regions of the mitochondrial DNA (mtDNA) of varroa are useful for examining inter and intra-species variation (Anderson and Fuchs, 1998; Anderson and Trueman, 2000; Navajas *et al.*, 2010).

2.4.4. Species identification

A 458 DNA base-pair fragment of varroa mtDNA *cox1* gene has proved useful in identifying mites to a particular species (Anderson and Trueman, 2000). To do this, a sequence of the fragment is obtained from a mite and compared to other sequences of the same region deposited in the GenBank database. If this sequence shows 2% or less difference from the one in the database, then it is considered to be a member of this particular species. Fragments from each of the 4 recognized species differ from each other by about 6%

(Anderson and Trueman, 2000). Sequences of the *cox1* gene fragment have been obtained from all *Varroa* spp. mites that have been identified to date by molecular methods. Hence, sequences of this fragment should be incorporated into all new molecular studies on varroa mites, as it places this new work in context with what has gone before.

2.4.5. Haplogroup and haplotype identification

The varroa *cox1* gene marker is also useful for identifying mites of large discreet populations within a species (such as island populations) (Anderson and Trueman, 2000). As smaller populations within these larger populations can be identified by concatenated (joined) sequence data obtained from the mtDNA *cox3*, *atp6* and *cytb* genes, the larger discreet populations have been referred to as 'haplogroups' and the smaller populations within them 'haplotypes' (Navajas *et al.*, 2010). The primer sequences for amplifying all these fragments, together with the size of fragments amplified, are shown in Table 1.

As the varroa *cox1* gene marker has proved useful for resolving mites from haplogroups to a particular species it has also been useful in phylogenetic studies. A phylogenetic tree of the all the currently known and published haplogroups with species is shown in Fig. 5 (see the section 'Phylogenetic analysis of sequence data' of the *BEEBOOK* paper on molecular methods (Evans *et al.*, 2013) for methods to perform phylogenetic analyses).

2.4.6. Kinship determination with microsatellites

Microsatellites are useful markers for measuring kinship or paternity relationships within varroa populations. These consist of repeating sequence of base-pairs DNA (such as CACACA) at a single locus (see the section 'Microsatellites' in the *BEEBOOK* paper on molecular methods (Evans *et al.*, 2013)). Loci with long repeats have more alleles than loci with short repeats and therefore often allow for a progenitor of a particular allele to be identified. Microsatellite loci in varroa have been published by Evans (2000) and Solignac *et al.* (2005).

2.5. Perspectives on the taxonomy of *Varroa* spp.

More research is needed to clarify the taxonomy of varroa mites from *A. cerana* in the Philippines. This will require examinations of nuclear DNA sequences obtained from these mites, as their mtDNA sequences do not provide the resolution needed to determine their identity (Fig. 5). From published research, there are three distinct mite types in the Philippines, two from the northern island of Luzon and another from the southern island of Mindanao (Anderson and Trueman, 2000). These could well be distinct species. New varroa species may also yet be found on other Asian honey bees, particularly on *A. nigrocincta* in Sulawesi, *A. nuluensis* in Borneo and *A. indica* from southern India.

An interesting feature of varroa mites on Asia honey bees is that most of them lack the ability to reproduce on *A. mellifera*. This is not from lack of trying though, for when *A. mellifera* colonies are introduced to different regions of Asia female mites that are indigenous to the local Asian honey bee readily invade the introduced colonies and enter brood cells that are about to be capped, in preparation for reproduction. However, they do not go on to lay eggs or produce offspring. Since at least the middle of the last century, only a few mite types have been able to reproduce on *A. mellifera*, the most successful of which is the Korea type of *V. destructor*. This suggests that female varroa mites must recognize specific signals on the host bee in order to successfully reproduce. Even though these signals may be fundamentally the same between different honey bee types and species, they may vary between honey bee populations. Identifying these signals and the genes that control them, could lead to the genes being targeted for particular purposes, such as control. This kind of research will require a good understanding of both the parasite and host genomes. Even though our understanding of the honey bee genome has improved in recent years, studies have only recently commenced on sequencing the varroa genome (Cornman *et al.*, 2010). As our understanding of the varroa genome improves, too will our understanding of varroa taxonomy and ways by which the mite can be controlled on European honey bees.

Table 1. Primer sequences (and their names) used in varroa research to amplify fragments of a particular size (base pairs) of mtDNA genes. From Anderson and Fuchs (1998); Navajas *et al.* (2010).

mtDNA Gene	Fragment Size (bp)	Primer Sequence	Primer Name
Cox1	458	GG(A/G) GG(A/T) GA(C/T) CC(A/T) AAT (C/T)T(A/T) TAT CAAC	COXF
		CCT GT(A/T) A(A/T)A ATA GCA AAT AC	COXRa
Cox1	929	CTT GTA ATC ATA AGG ATA TTG GAAC	10KbCOIF1
		AAT ACC AGT GGG AAC CGC	6,5KbCOIR
Atp6-cox3	818	GAC ATA TAT CAG TAA CAA TGAG	16KbATP6F
		GAC TCC AAG TAA TAG TAA AACC	16KbCOIIIR
Cytb	985	GCA GCT TTA GTG GAT TTA CCT AC	10KbCytbF-1
		CTA CAG GAC ACG ATC CCA AG	10KbCytbPRIM

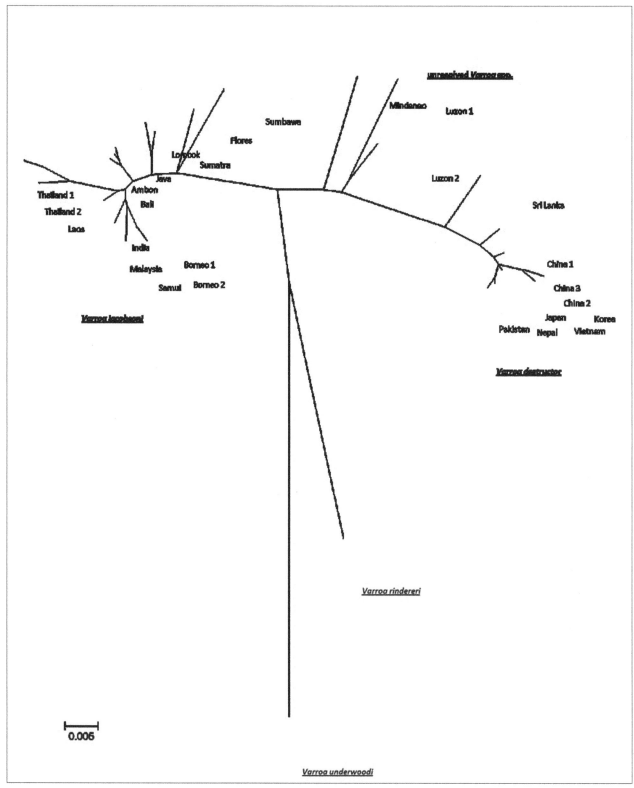

Fig. 5. A phylogenetic tree of all the currently known and published haplogroups.

3. Laboratory techniques

3.1. Collecting mites

There are several ways to collect varroa mites for experiments. Some methods described below provide mites of unknown age, which have reproduced an unknown number of times. Other methods provide mites in which oogenesis has been triggered. Which method is adopted depends on the physiological state of mites needed for the experiment (see section 3.6. 'Bioassays').

3.1.1. Manual collection

Phoretic mites can be picked up by hand from their host with a fine bristle brush or a small mouth aspirator.

1. Collect honey bees from a colony. Manual collection is easier when the colony is highly infested (see section 4.6. 'Breeding mites in colonies' for a method to obtain regular supply of highly infested colonies), but to collect 'healthy' mites it is recommended that the host colony does not have symptoms of extreme infestation such as crippled bees.

2. Catch honey bees one by one and examine them for the presence of mites.
   Mites may run freely over the bee's body or be hidden between two sternites. Finding and collecting them sometimes necessitates grasping the bee by the thorax and sting apparatus with forceps to stretch the abdomen, thus making the mites visible and reachable.

3. Honey bees can be treated with CO_2 or cooled down to facilitate the physical collection. CO_2 affects the bees' physiology (Czekońska, 2009), but recent results indicate that a short treatment with CO_2 does not affect fertility and fecundity of varroa female artificially introduced into brood cells (Rosenkranz et al., unpublished data). The effect of cooling on mites is not known and might affect mite survival. An alternative to CO_2 and cooling treatment is: (i) to let the bees crawl out of their container one by one so they can be caught easily or (ii) to cut off the head of the bees; mites tend to leave dead bees within a short time.

4. Place the mites collected in a mite-tight container with a source of humidity (a wet cotton plug or ball of paper) to prevent the mites desiccating.

Pros: allows for the collection of mites that have not been stressed by a treatment with water or powdered sugar (see sections 3.1.2. 'Icing sugar' and 3.1.3. 'Washing with water'). This is advantage if mites are used in long lasting experiments.

Cons: tedious, few mites can be sampled in a short time.

3.1.2. Icing sugar

Icing sugar can be used to detach mites from their host collected in a jar (Macedo et al., 2002) or still in the colony.

Material needed: a wide mouthed jar with a lid of which the centre part is replaced by a 2mm hardware cloth or mesh (Fig. 6a).

1. Place 300 bees in the jar and close the lid.
2. Pour 1 heaping table spoon (at least 7g) of powdered sugar through the mesh or cloth (Fig. 6a).
3. Roll the jar to cover all the bees with sugar (Fig. 6b).
4. Let stand for 1 min.
5. Turn jar upside down over a white surface (Fig. 6c).
6. Shake for 1 min.
7. Place the fallen mites and sugar (Fig. 6d) in a sieve and rinse with 1X phosphate-buffered saline (or other similar saline solution) to rid them of icing sugar particles (Fig. 6e).
8. Place mites on absorbent paper to help them dry up (Fig. 6f).
9. Place the mites collected in a mite-tight container to prevent them escaping.
   Place a source of humidity in the container to prevent the mites desiccating until they are used for experiments.

This can also be done using the entire colony fitted with a mesh floor:

1. Remove each frame containing adult bees.
2. Sprinkle with icing sugar so that the frames are all covered.
3. Place back into the colony.
4. Remove the excess icing sugar with the mites from the floor at 10-20 min intervals.
5. Pour over a sieve to remove the sugar and collect the mites.
6. Rinse with 1X phosphate-buffered saline (or other similar saline solution) to rid them of icing sugar particles.
7. Place mites on absorbent paper to help them dry up (Fig. 6f).
8. Place the mites collected in a mite-tight container to prevent them escaping.
   Place a source of humidity in the container to prevent the mites desiccating.

Pros: fast and allows for several hundreds of mites to be collected in short time. The treatment is bee-friendly since few individuals die during the process. Workers collected in the jars can be placed back in their colonies where they will be cleaned by their nestmates.

Cons: decreases lifespan of mites (Macedo et al., 2002). This can be a problem if they need to be used for long lasting experiments (> 3 days).

3.1.3. Washing with water

1. Collect bees from a colony in a bee tight container.
2. Fill the container with 1X phosphate-buffered saline (or other similar saline solution) to prevent the bees flying away and shake.

Fig. 6. c. the jar is turned upside down and shaken to dislodge the mites.   Photo by V. Dietemann

Fig. 6. Collecting mites with icing sugar: **a.** a heaped table spoon of powdered sugar is poured on 300 honey bees kept in a jar through the lid equipped with a mesh.   Photo by V. Dietemann

Fig. 6. d. mites (2 darker points) and sugar fallen through the mesh on the paper.   Photo by V. Dietemann

6. Place the mites collected in a mite-tight container with a humidity source to prevent the mites desiccating.

Pros: fast and allows for several hundreds of mites to be collected in a short time.

Cons: effect on lifespan of mites unknown; this can be a problem if they need to be used for long lasting experiments. The treatment it is not bee-friendly since many can die during the process.

3.1.4. Collecting mites from brood

3.1.4.1. Collecting mites from L5 larvae

Mites at a similar physiological stage can be collected from recently capped brood cells (after Chiesa *et al.*, 1989)

1. Remove a brood comb with L5 larvae ready to be capped in the evening of the day preceding the experiment.
2. Mark the capped cells with a convenient marker (e.g. correcting fluid, queen marker, felt pen).
3. Replace the comb in its colony of origin. Bees will continue capping mature cells.
4. The following morning, transfer the comb to the laboratory and unseal the unmarked cells that have been capped overnight.

Fig. 6. b. rolling the jar on its side ensures that bees are covered with the sugar.   Photo by V. Dietemann

3. Pour the content of the container over a first sieve (aperture: 2000 µm) to collect all the bees.
4. Place a second sieve (aperture < 0.5 mm) underneath to collect the mites.
5. Place mites on absorbent paper immediately after washing them off to help them dry up (Fig. 6f).

Pros: easy collection, all mites are at the same physiological stage.

Cons: there is no knowledge of the mite's age and of how many reproductive cycles she already performed.

3.1.4.2. Collecting mites from capped cells
3.1.4.2.1. Opening each cell

Brood mites can be picked up by hand from their host with a fine bristle brush or a small mouth aspirator after opening the cells they infest and removing the pupa. To obtain mites at a given time during the reproductive cycle, the collection can be made from brood of known age (see the section 'Obtaining brood and adults of known age' in the *BEEBOOK* paper on miscellaneous methods (Human *et al.*, 2013)). For this the queen is caged on an empty frame at the necessary date (see also section 4.6. 'Breeding mites in colonies').

1. Uncap the cell with fine forceps or scalpel.
2. Push away the cell walls to free the developing larva or pupa.
3. With soft forceps pull the larva or pupa out.
4. Carefully look on the larva or pupa and on the cell walls for mites.
5. Place the mites collected in a mite-tight container.
6. Place a source of humidity in the container to prevent the mites desiccating.

Pros: This is the less damaging collection method for the mite.
Cons: It is the most time consuming collection method.

3.1.4.2.2. Opening large number of cells and washing the brood

A quicker method for collecting large numbers of live mites from capped brood cells it to uncap large quantities of brood and force the mites out by knocking them out of the cells or by washing them off the brood. For this:

1. Uncap a large number of cells.
2. Remove all developing bee brood.
   These two steps can be done at once using an uncapping fork used for honey extraction.
3. Turn the comb upside-down over a sheet of white paper.
4. Tap on its upper surface to dislodge mites from the cells.
5. Collect the dislodged mites, sometimes in the hundreds, from the paper.

An alternative method to increase the number of mites sampled is to:

1. Uncap the brood cells.
2. Flush the comb with lukewarm water that will dislodge the brood and mites.
3. Collect dislodged brood and mites in a first sieve (5 mm mesh) that will retain the bees.

Fig. 6. e. the mites and sugar collected are placed in a sieve over which saline solution is poured to rid the mites from sugar particles.

Photo by V. Dietemann

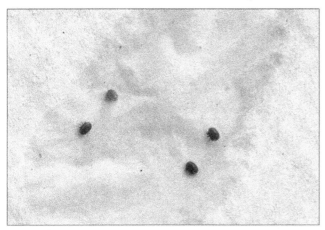

Fig. 6. f. the mites are placed on an absorbing paper to accelerate their drying.   Photo by V. Dietemann

5. Place the comb in an incubator at 34.5°C, 60-70% RH.
6. Infested and non-infested larvae deprived of the capping spontaneously emerge from brood cells in a short time.
7. Collect mites that have fallen from their cells with their host.
8. Place the mites collected in a mite-tight container to prevent them escaping.
9. Place a source of humidity in the container to prevent the mites desiccating.

4. Place a second sieve (0.5 mm mesh) underneath the first to retain mites.

5. Flush with more water.

6. Dry the mites by placing them on absorbent paper (Fig. 6f).

Pros: easy collection of high mite numbers (depending on colony infestation rate).

Cons: mites of unknown physiological stage; possible shortening of life expectancy after washing from brood; a large number of mites remain in the comb with the knocking method.

3.2. Rearing mites in the laboratory

It is relatively easy to maintain varroa mites during their phoretic phase (on honey bees or pupae) in the laboratory. Their maintenance in cages kept in incubator is necessary for screening of varroacides, or while marking before transfer to colonies for example. In contrast, few laboratory breeding methods for varroa are available. The conditions reigning in their breeding environment, i.e. in the cells in which honey bees develop, are so particular that it is very difficult to replicate them artificially. Obtaining a full life cycle in the laboratory therefore remains a challenge that few overcame (Donzé and Guérin, 1994; Nazzi and Milani, 1994). This section describes methods to keep or breed varroa in the laboratory. These methods do not yet allow their breeding in large quantities as would be desired for experimentation, but allow observation of mite behaviour and testing of products that may affect the mite's life cycle.

3.2.1. Maintaining mites in the laboratory
3.2.1.1. Maintaining mites on adult honey bees
Mites can be maintained on bees in hoarding cages (see section 'Cages to keep bees *in vitro* in the laboratory' in the *BEEBOOK* paper on maintaining adult bees *in vitro* (Williams *et al.*, 2013)). A temperature of around 33ºC and RH of 60-70% is also adapted for mites. Mortality can be high in case phoretic mites are used. Their age is unknown at the time of collection and the variability in their life expectancy is therefore high. However, mites will commonly survive for 1 week or longer in bee cages established in this way. It is recommended using mite-tight cages or keeping each cage in a dish to avoid any escaped mites from entering another cage of a different treatment group. This procedure can be used in acaricide toxicity *in vitro* assays, in assays where the investigator is attempting to trace the movement of a compound (such as dsRNA) from a bee food (sugar water, pollen patties, etc.), through bees, and into mites, or in other similar assays.

A common method for establishing *in vitro* studies with varroa-infested adult honey bees involves individually selecting worker bees carrying mites from combs/frames in colonies and placing them in cups or cages. This method is relatively time-consuming and can be particularly difficult for researchers with limited bee experience. An

alternative method includes the separate collection of bees and mites followed by the infestation of the bees with the collected mites. Mites can be transferred onto the caged bees using a brush or, if mites were collected in a container, they can be introduced in a cage with bees. The mites readily spread across the bees within minutes. A primary benefit of the latter method is that it is feasible for a single, inexperienced experimenter to accomplish quickly. Additionally, this method obviates the need for maintaining colonies with high varroa populations (which are preferred when infested bees are chosen individually from frames, see section 4.6. 'Breeding mites in colonies').

Collecting bees and mites:

1. Collect worker bees from frames and place in cages as described in section 'Cages to keep bees *in vitro* in the laboratory' of the *BEEBOOK* paper on maintaining adult bees *in vitro* (Williams *et al.*, 2013)).
   The number of bees placed in each cage can vary depending on experimental requirements.

2. Collect mites using the methods described in section 3.1. 'Collecting mites'.

In vitro infestation of bees with varroa:

3. Add 5-10 mites to a small (5 cm) Petri dish containing a filter paper circle wetted with 1:1 sugar syrup (by volume) or water.

4. Place the Petri dish with mites on the bottom of the cage containing worker bees.

5. Tap the cage lightly against the laboratory bench or table to cause the bees to drop from the top of the cage to the bottom and contact the mites.
   This artificial infestation procedure relies on the questing behaviour of mites, which readily attach to bees that contact them. The sugar syrup impregnated in the filter paper will keep the bees for an increased duration at the bottom of the cage to give more time for mites to find a host.

6. Remove the Petri dish after all of the mites have attached to bees.

7. Repeat steps 5-6 until the desired number of mites has been added to the cage. Partitioning the mite additions, rather than trying to add them all to the cup at one time, results in a more even distribution of mites on bees. Be aware that some bees will inevitably carry multiple mites and some bees will have no mites.
   More mites should be added to each cage than are needed for the experiment. For example, add 30 mites to a cage when the experiment calls for the live recovery of 25 mites later in the study. This is necessary because some mites will fail to attach to any bee. Regardless, the number of varroa mites per cage or per bee can vary based on the predetermined experimental criteria.

8. Add a sugar syrup feeder and water supply to the cage (see section 'Cages to keep bees *in vitro* in the laboratory' in the *BEEBOOK* paper on maintaining adult bees *in vitro* (Williams *et al.*, 2013)) after the mites have been added.

    Bees must stay hydrated and fed to ensure mite survival.

If keeping mites on brood is possible in Eppendorf tubes (see section 3.2.1.2. 'Maintaining mites on honey bee brood'), keeping mites on adult bees works better in hoarding cages. Ventilation within the Eppendorf tubes is poor and the health of the bees kept in them decreases rapidly.

3.2.1.2. Maintaining mites on honey bee brood

Adult mites can be kept for many days on bee larvae or pupae in the laboratory, under standardized thermo- and hygrometric conditions necessary for the brood (34.5°C and 60-70% RH). Brood must be replaced regularly with younger individuals before they reach the adult stage (Beetsma and Zonneveld, 1992). Small Petri dishes with a few immature honey bees can be used for this purpose. In order to improve mite feeding and survival, a close contact between the mite and the brood item is desirable. Some authors confine mites individually on a larva or pupa in smaller containers (1ml Eppendorf tube). To increase humidity in the Eppendorf tube, a wet piece of cotton wool is pushed into the bottom of the tube. Excess water accumulating on top of the cotton wool needs to be removed by shaking the tube to prevent mites from adhering to droplets. Piercing the lid with 1-2 holes (diameter < 1 mm) will improve ventilation and respiration. Up to 8 mites can be kept in one tube. Mites can be transferred into the Eppendorf tube using a brush. Once all mites are transferred, the pupa/larva can then be placed in the tube.

Pros: easy for keeping mites alive.

Cons: does not result in oviposition by the mites.

3.2.1.3. Artificial diet

To facilitate the keeping and rearing of the mite in the laboratory, attempts were made to develop rearing methods based on artificial diets that mites could suck through a synthetic membrane (Bruce *et al.*, 1988; Bruce *et al.*, 1991). Unfortunately, despite both membrane and diet seeming suitable for the purpose, satisfactory survival and reproduction were not achieved.

3.2.2. Breeding mites in the laboratory

In order to obtain the whole life cycle of varroa within the cells under laboratory conditions for observations and experimentation, rearing methods within artificial cells made of different materials were designed. Tested materials for cells include wax, glass, plastic, gelatin (Nazzi and Milani, 1994 and citations therein). In general, reproduction is very difficult to obtain, due to the seemingly high number of cues

necessary for the mite to reproduce successfully. However, in some cases reproduction rates close to the natural ones were obtained. In particular, Donzé and Guérin (1994, 1997) obtained complete reproductive cycles in artificial cells. The cells used where first kept in the hive until sealing, then brought to the laboratory and placed in an incubator. Despite the low acceptance and infestation rate of these cells, it allows using cells of natural size and naturally infested. Nazzi and Milani (1994) developed a method that allow normal mite reproduction under laboratory conditions in cells in which larvae were introduced and which were infested artificially to allow more control on the process.

3.2.2.1. Natural infestation

Material needed: cylindrical transparent polystyrol cells (internal dimensions: 5.1 mm diameter x 14 mm length for workers and 6.7 mm diameter x 16 mm long for drones)

1. Incorporate the cells at an inclination of 5-10 degrees in groups of 60-70 in wax combs.
2. Coat with honey to increase acceptance by the workers and stimulate cleaning behaviour.
3. In a heavily varroa infested colony, confine the queen on the artificial cells for 12 h.
4. Release the queen after this period.
5. Record the time of cell capping at 1-2 h interval some 8.5 days after oviposition by the queen.
6. Remove the cells from the colony and the comb after they have been capped.
7. Place in an incubator at 34.5°C and 60-70% RH.

Pros: transparent cells allowing observation of behaviour, natural infestation, natural cell size.

Cons: tedious, low acceptance and infestation rates.

3.2.2.2. Artificial infestation

1. Collect mites and L5 bee larvae from natural brood combs as described in section 3.1.4. 'Collecting mites from brood'.
2. Place the larva in a gelatin cell by holding its dorsum between thumb and first finger to get it stretched.
3. Insert the mite with a fine paint brush.

    Gelatin cells of different diameters were tested; as a general rule, the narrower the cell diameter the higher the reproduction, but the higher the chance of injuring the larva while inserting into the cell; the best compromise is achieved with gelatin cells of 6.5 mm diameter.
4. Place in an incubator at 34.5°C, 75% RH.
5. Place the cells so that pupae are laying on their back.

    Geotaxis is an important cue for varroa behaviour (Donzé and Guerin, 1994).
6. Fix the cell to a substrate to avoid rolling and manipulate only occasionally for observations.

Pros: high percentage of fertile mites and offspring number close to natural infestations in colonies can be obtained; transparent cells allow observation of behaviour; *in vitro* procedure allows complete control over the infestation state of the bee since workers do not have the opportunity to remove infested brood.

Cons: tedious; non-natural infestation; non-natural cell size.

Perspectives: With the aim of developing a complete rearing method for the mite, Nazzi and Milani (unpublished data) and Dietemann, Zheng and Su (unpublished data) carried out preliminary trials aiming at obtaining several reproduction cycles in the laboratory, i.e. artificially breeding mites that were born under laboratory conditions. Attempts were discouraging and continuing efforts are needed.

3.3. Assessing reproduction in the laboratory

3.3.1. Assessing fertility

Assessing fertility of mites reproducing in artificial (see section 3.2. 'Rearing mites in the laboratory') or natural cells follows the same principle than described for field methods (see section 4.3.1. 'Assessing reproductive success'). A difference is that if transparent cells are used (Donzé and Guérin, 1994; Nazzi and Milani, 1994), one does not need to open the cell to count offspring. Offspring production and survival can thus be monitored over time.

3.3.2. Assessing oogenesis

Assessing oogenesis of laboratory reared mites requires dissection of female mites and tissue dyeing (see the section on marking techniques 3.4.1. 'Oogenesis').

3.4. Marking techniques

3.4.1. Oogenesis

Activation of the oocyte (i.e. oogenesis) is followed by the incorporation of euplasmatic material and/or yolk proteins. Marking a whole mount of the mites' ovaries with toluidine blue is a rapid method to confirm such incorporation and therefore initiation of oogenesis (Garrido *et al.*, 2000).

1. Remove ventral shield of the mites with thin dissecting forceps under a binocular microscope.
2. Excise ovary together with spermatheca and lyrate organ.
3. Place in PBS buffer (phosphate buffered saline, pH 7.2–7.4).
4. Fix in formalin (4%) for 30 min.
5. Wash three times with PBS buffer.
6. Incubate in toluidine blue (0.005%) for 30 min.
   The duration of incubation might need optimization, which can be tested on the coloration of activated oocytes approximately 12 h after cell sealing.

7. Rinse with PBS buffer for 15 min.
8. Repeat rinsing twice more.
9. Verify the colouring of the oocytes under a microscope at 400x magnification.

Pros: easier and faster than the alternative histological method. Detects initiation of oogenesis with high resolution.

Cons: somewhat tedious; subjective grading of oocyte colour.

3.4.2. Feeding site

Varroa mothers pierce the cuticle of honey bee larvae and pupae on which they (Fig. 7a) and their offspring feed (Fig. 7b). In late pupae, the wound can be seen under the binocular thanks to the scarring process of the cuticle (Fig. 7c). It can also be located by observation of the feeding mites, events that are relatively rare and need an artificial in vitro system to be observed (see section 3.2.2. 'Breeding mites in the laboratory'). In most cases, no wound can be seen on larvae or pupae and a staining method is necessary to find it (Fig. 7d). The ability to locate the wound might be necessary for behavioural studies of feeding behaviour or reproduction or for secondary disease transmission studies (Kanbar and Engels, 2003). By extension, this method can also be used for all cases in which a perforation of the cuticle of immature honey bees has to be made visible (e.g. injection of pathogens or hormones).

Kanbar and Engels (2004) designed a vital staining method that allows the visualisation of feeding sites. They used Trypan blue, a dye that enters damaged cells (Roche, 1999), i.e. cells around the hole pierced by mother mites in the late 5th instar larvae, prepupae or pupae cuticle. Feeding sites could thus be stained durably on live individuals and observed over time (Herrmann *et al.*, 2005). Staining can be detected until the stage when the cuticule darkens to the point of hiding the dyed blue cells. At this point the dyeing is not any longer necessary.

Staining method:

1. Sample larva, pre-pupae or pupae to be stained from varroa infested cells (see section 3.1. 'Collecting mites').
2. Immerse them for 30 min in a volume of Ringer-based staining medium sufficient to cover the major part of the body surface.
   The larvae and pupae survive this treatment (Kanbar and Engels, 2003).
3. Rinse in pure Ringer solution for 3 min.
   Ringer solution: see Table 1 of the *BEEBOOK* paper on cell cultures (Genersch *et al.*, 2013) for a recipe.
   Vital staining medium: 100ml Ringer solution, 0.01 g Trypan blue adjusted to pH 6.8 with KOH (0.1M).

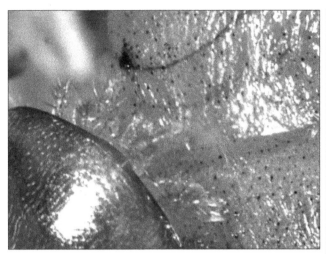

Fig. 7. a. Adult varroa mite sucking haemolymph of a pupa at the feeding site (black dot on top, left from centre).

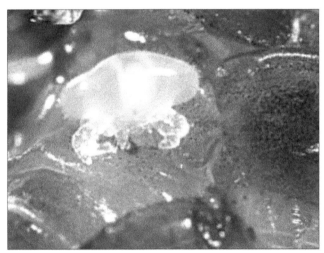

Fig. 7. b. varroa nymph sucking haemolymph of a pupa atthe feeding site (between the nymph's legs).

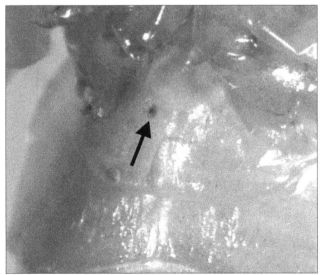

Fig. 7. c. feeding site with melanisation (arrow). It is visible without staining. Such instances are more frequent in older pupae.

Fig. 7. d. a feeding site (blue dot) on a white pupa after staining with Trypan blue.   Photos: Swiss Bee Research Institute.

3.4.3. Marking mites

Varroa mites can be marked with paint markers, enamel paint for models, correction fluid polyester glitter or fluorescent pigment (Schultz, 1984; Harris, 2001; Kirrane *et al.*, 2012). Methods with paint are faster to use and have been shown safer for the mites. Toxicity is mainly due to the solvents incorporated in the products.

For paint application, soft tools are preferable to toothpick or other hard tools to avoid injuring the mites. A droplet of paint can be placed on a microscope slide and little quantities collected with a very thin paintbrush or a fishing line (Kirrane *et al.*, 2012) for application on the mite. The hair of a paintbrush can be cut off leaving enough hair to obtain the desired size. The right size of the application tool is obtained when the paint dot is visible, but does not impair the behaviour of the mite.

If the mites are used for behavioural observations during the phoretic or reproductive stage, care should be taken to produce a flat paint mark that enables the mites to push herself between the bees' sternites and feed, or to allow the mite's free movement within the restricted space between pupae and cell walls.

Before use in large scale experiment, a toxicity test should be performed to ensure that the paint chosen is not toxic to the mite: marked and sham-treated mites should be kept in similar conditions and their longevity compared. Marked mites should live as long as unmarked mites. Refer to section 3.2. 'Rearing mites in the laboratory' for rearing methods. To minimise the risk of the paint dot coming off the cuticle and thus prevent recognition of the marked mite at the end of the experiment, preliminary tests with different brands should be made to select a long lasting paint.

3.5. Infecting varroa with secondary diseases

3.5.1. Microinjection

Microinjection techniques specific to varroa have not yet been developed. For basic microinjection techniques please refer to the *BEEBOOK* paper on miscellaneous methods (Human *et al.*, 2013). Campbell *et al.* (2010) mention various problems when testing such methods. The manipulation is laborious, requires specialised equipment and resulted in the death of most mites injected.

3.5.2. Dipping

Dipping mites in a solution containing dsRNA resulted in gene knockdown (Campbell *et al.*, 2010), demonstrating the potential of this procedure as a way to infect or contaminate mites with micro-organisms or genetic material. This study showed that when proper osmolality of the dipping solution is established, a high number of varroa mites (80%) can survive long immersion periods (14 h at 4°C) that might be necessary for infection/contamination. To achieve the right osmolality, 0.9% NaCl was used. Dipping methods have rarely been used for varroa, and optimisation work should be done to improve the survival of the mite after immersion.

3.6. Bioassays

The influence of both physical and chemical factors on *V. destructor* has been studied by means of different bioassays. So far, bioassays have been essential in two fields of varroa research: the study of the semiochemicals involved in the interactions between the mite and honey bee and the study of mite resistance towards acaricides.

3.6.1. Experimental conditions

The literature on the subject reveals some critical aspects that must be considered when conducting bioassays with varroa, including the environment, the chemicals tested and the origin of mites.

3.6.1.1. Environment

The mite spends its life in the hive, under strictly controlled environmental conditions; in order to get a realistic representation of the reactions of the mite towards a given chemical, this should be tested under the same conditions, that is a temperature around 32-35°C and a relative humidity around 70%.

3.6.1.2. Dosage of chemicals

In order to avoid any misjudgement about their real effect, semiochemicals should be tested at doses that are close to their biological range. For example, most chemicals, when tested at too high a dose, become repellent. Unfortunately, this aspect has often been overlooked in the study of varroa chemical ecology (for some examples see Milani, 2002). Dose-response studies, where the biological activity of a given compound is tested at different doses in a logarithmic scale, are mandatory in current research.

3.6.1.3. Mites to be used in the tests

In chemical ecology studies, the mites that are used in the bioassays should be those involved in the process under study (e.g. if the study is about cell invasion, the mites that invade the brood cells are used). In efficacy study of acaricides, the life stage of mites tested must correspond to that that will be exposed to the product under scrutiny.

In contrast to many other arthropods whose ecology is studied by means of bioassays, no artificial rearing method is currently available for the varroa mite (see section 3.2. 'Rearing mites in the laboratory'); thus standardisation is a difficult task when it comes to collecting mites that are homogenous for age, physiological condition, mating status etc. A solution to this problem is to use mites that are at the same stage of their life cycle (see section 3.1. 'Collecting mites').

3.6.2. Bioassays in varroa chemical ecology

Bioassays are a fundamental resource in the study of behaviour modifying chemicals, all the way from the demonstration of their existence, through all steps of isolation, until the final confirmation of their identity. Some of the bioassays used so far in varroa chemical ecology were simple adaptations of those already used for the study of arthropod semiochemicals (Baker and Cardé, 1984). In particular, the response of the varroa mite towards different odour sources and pure compounds was tested using several general purpose setups, including four-arms olfactometers (Le Conte *et al.*, 1989), servospheres (Rickli *et al.*, 1992), Y-mazes (Kraus, 1993), wind-tunnels (Kuenen and Calderone, 2002) and observation arenas (Rickli *et al.*, 1994), in other cases, bioassays were specifically designed for the varroa mite. In this section, we will concentrate on the latter. The chemical stimuli that influence the behaviour of the mite during the following stages of the mite's biological cycle have been studied by means of bioassays: cell invasion, mating, oviposition, phoretic phase.

3.6.2.1. Cell invasion

In this case, the attention is on the cues influencing the entrance of the mite into the brood cell containing an L5 bee larvae (see the section 'Obtaining brood and adults of known age' in the *BEEBOOK* paper on miscellaneous methods (Human *et al.*, 2013) for a description of larval stages), in the 20-60 hours preceding capping (Rosenkranz *et al.*, 2010). After the first studies with the star olfactometer (see Le Conte *et al.*, 1989), bioassays better adapted to the varroa mite were used; here a bioassay that appeared to be suitable to test both attractants linked to worker cell invasion (Rosenkranz, 1993; Nazzi *et al.*, 2001; Nazzi *et al.*, 2004; Aumeier *et al.*, 2002) and repellents involved in the avoidance of queen cells (Nazzi *et al.*, 2009) is presented.

3.6.2.1.1. Mites to be used

What determines the end of the phoretic phase of mites is not known. Using phoretic mites for this assay is therefore not ideal since the randomly sampled mites might not be 'motivated' to enter cells and respond to the stimulus provided. It is therefore possible to use mites which have recently expressed this behaviour and will repeat it again given the right circumstances. However, several experiments confirm that during the season most of the phoretic mites are willing to reproduce (e.g. Rosenkranz and Bartalszky, 1996; Martin and Cook, 1996; Garrido and Rosenkranz, 2003; Frey *et al.*, in preparation) and might also be used for these kinds of experiments (see section 3.1. 'Collecting mites' for a description of how to obtain them).

3.6.2.1.2. Experimental setup

This bioassay was described by Nazzi *et al.* (2001) and represents a modification of the device used by Rosenkranz (1993).

1. Use an arena consisting of a glass plate with four wells (7 mm diameter; 8 mm deep) equidistant (1 cm) from the centre.
2. Mount a glass lid on a circular metal ring (5.6 cm diameter) to confine the mites in the arena.
3. Apply the treatment to two opposite wells while the other two wells are used as controls.
4. Place one bee larva or dummy into each well.
5. Place one adult female mite in the centre of the arena between the four wells with a fine paint brush (Fig. 8).
6. Keep the arenas in a chamber at 34.5°C and 60-70% RH for the duration of the bioassay.
7. Note the position of the mites every 5 min for 30 min.

In order to obtain sufficient sample size, twenty arenas are used at a time and tests are replicated on different days for several times (typically a minimum of three) using a total of at least 60 mites.

3.6.2.1.3. Data analysis

A sampled randomization test is used because the distribution of the variables to be compared is unknown. Such test is preferred over a conventional parametric statistics since they often lead to an overestimate of the significance of differences.

For each arena, the number of times the mite is observed in the treated and control wells, respectively, over the 30 min period are used as scores for the statistical analysis. This is done regardless of whether the mite had changed wells between observations or just stayed in the same well. Then, a matrix is constructed with as many rows as the number of mites used in the bioassay, and two columns containing the scores for treated and control wells for each of the tested mites. The treated and control scores in a given set of data are compared by a sampled randomization test (Manly, 1997; Sokal and Rohlf, 1995). The randomization distribution should be resampled a sufficient number of times (e.g. 10^6 times).

3.6.2.2. Oogenesis

About 70 h after cell sealing, the mite begin egg-laying, which continues at 30 h intervals until 5-7 offspring are produced (Rosenkranz *et al.*, 2010). The cues influencing oviposition were

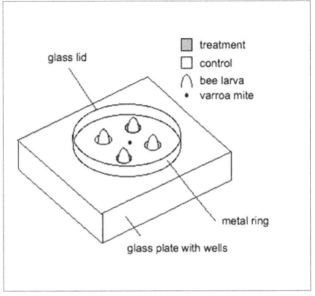

Fig. 8. Arena used for the bioassays on cell invasion behaviour.

studied using artificial brood cells made of different materials (e.g. wax, glass, gelatine) (Milani and Chiesa, 1990; Trouiller and Milani, 1999).

3.6.2.2.1. Mites used in the bioassay

To study the factors that affect oogenesis, mites that have not yet been stimulated to start oogenesis must be used. Indeed, if a mite has been in contact with a L5 larva for 0-6 h, oogenesis is initiated (Garrido *et al.*, 2000). Phoretic mites must therefore be used. However, not all phoretic mites will initiate reproduction and oogenesis since they are of various age and physiological status. To study factors triggering oviposition once mite oogenesis has been activated and completed, mites must be collected at least 24 h after capping, but no later than 132 h (after which they stop reproducing, Nazzi unpublished) and introduced in cells containing a pupa young enough to allow the normal reproductive cycle. See section 3.1.4.2.1. 'Opening each cell' for methods to obtain mites from their cells. For these studies, the different phases of the mite´s ontogenetic development and the time-dependent course of the gonocycle must be considered (Steiner *et al.*, 1994).

3.6.2.2.2. Experimental setup to test the activation of oogenesis

Bioassay to assess the effect of volatile or non-volatile compounds for their effect on oogenesis activation.

3.6.2.2.2.1. In the field

1. Treat brood cells in colonies with the stimulus under testing or treat mites. Beware of solvent toxicity for bees and mites.
2. Open cells a minimum of 70 h after mite introduction.

3. Investigate the presence of eggs in the cells.

   Opening the cells later (e.g. between 7 and 10 days) will facilitate detection of eggs and offspring, but increase the chances of removal by hygienic workers.

Pros: less time consuming compared to laboratory bioassay (see section 3.6.2.2.2.2.). Many brood cells can be treated under natural conditions.

Cons: stimuli cannot be tested independently from other factors (larvae, nurse bee activity).

3.6.2.2.2.2. In the laboratory

This protocol follows Garrido and Rosenkranz (2004) to test the effect of volatiles on activation of oogenesis:

1. Offer volatile compounds on a piece of filter paper placed in a 0.2 ml PCR tube.
2. Add a female mite in the tube.
3. Prevent the mite reaching the filter paper with a plastic gauze.
4. Remove the mites eight hours after the exposure to the putative triggering factor.
5. Dissect the reproductive tract of mites.
6. Dye the reproductive tract to determine the development stage of the terminal oocytes (see marking techniques section 3.4.1. 'Oogenesis').

Pros: test of single stimulus possible.

Cons: time consuming. Due to the lack of nutrition of the mite the test can only be performed for about 8 h.

3.6.2.2.3. Experimental setup to test oviposition

1. Treat the cell or the mite with the compound under testing. Beware of solvent toxicity for bees and mites.
2. Transfer the mites into cells containing a host at an early enough developmental stage, which allows reproduction by the mite to proceed normally.
3. Monitor reproduction after a given interval of time.

   When fecundity measurement is needed, the cell should be opened one day before emergence. This period can be shorter if only fertility (i.e. answering the reproduction yes or no) is of interest, but it must be passed 70 h after capping (Table 2).

3.6.2.3. Orientation inside the sealed cell

Careful observations of the behaviour of the mite inside the brood cell carried out by Donzé and Guérin (1994) revealed a well-structured spatial and time allocation of its activity. The chemicals involved in this crucial stage of the mite's life cycle were studied mostly using bioassays based on observation arenas (Donzé *et al.*, 1998; Calderone and Lin, 2001).

3.6.2.3.1. Mites to be used

Mites should be sampled from naturally infested cells at the stage at which the behaviour under scrutiny is expressed (see section 3.1. 'Collecting mites').

3.6.2.3.2. Experimental setup

Donzé *et al.* (1998) used a modification of the bioassay described by Rickli *et al.* (1994) to study the chemicals inducing arrestment of the mite.

1. Use a glass plate cleaned with acetone and pentane.
2. Draw three concentric circles with 12 and 24 and 36 mm diameter on the underside of the glass (Fig. 9).
3. Apply the compound tested on top of the glass plate on the ring delimited by the 12 and 24 mm circles.
4. Place a mite in the centre of the circles (mites not reaching the treated area within 300 s are not considered).
5. Observe the mite's walking activity.
6. Note the time spent in the treated area.

   This time is used as a measure of the arrestment activity of the stimulus under testing.
7. Stop the assay when the mite crosses the outer 36 mm circle, or after 300 s.

3.6.2.3.3. Data analysis

In order to check for significant differences between treatments and controls, the times spent inside the area treated with different stimuli are compared using the non-parametric tests of Mann–Whitney and Friedman (for simple and repeated experiments, respectively). Due to some unbalanced replications, a generalization of the Friedman test is used (Del Fabbro and Nazzi, 2008).

3.6.2.4. Phoretic phase

After emerging from the brood cell the varroa mite enters the phoretic phase, that is spent on adult bees and lasts until a new brood cell is invaded for reproducing. Bioassays have been used by several authors to work on the chemical cues affecting the mite on the adult bees.

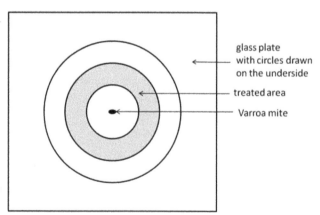

Fig. 9. Test arena for arrestment bioassays.

Table 2. Minimal time after capping for which various mite reproduction parameters can be accurately measured. Adapted from Martin, 1994 and 1995a. *see Fig. 16 for a physogastric mite. In comparison, the segments of a non-physogastric mite appear connected and not separated by white borders (intersegmental membrane).

Mite reproduction classification	Time from cell capping (hours)	Comments
abnormal	> 60	after first egg should have been laid
abnormal with only single male	> 140	after second egg should have hatched
non-reproducing	30-50	if mother mite non-physogastric*
	> 70	absence of eggs
mite dead trapped in cell wall	> 30	after cocoon spinning by larva is complete
mite dead in cell	> 0	at any time

Kraus (1994) used a simple two-choice bioassay to test several chemicals for their effect on the mite as a screening procedure to identify possible substances to be used in biological control methods. He and others used laboratory bioassays to investigate the stimuli affecting the host choice by the mite (Hoppe and Ritter, 1988; Kraus, 1990, 1994; Del Piccolo *et al.*, 2010). These bioassays are all based on the same kind of setup. Here the bioassay described by Del Piccolo *et al.* (2010), that was used to study the preference of the varroa mite for pollen and nurse bees, is presented.

3.6.2.4.1. Mites to be used

Mites are sampled with the host that carries them. Mites are separated from their host bee by means of a mouth aspirator or a paintbrush. Collection of mites with sugar powder method is not recommended given the possible effects of the sugar on mite vitality (see section 3.1. 'Collecting mites').

3.6.2.4.2. Experimental setup

1. Clean a small glass Petri dish (60 mm diameter) with acetone and hexane or pentane.
2. Place 2 dead adult bees at 2 diametrically opposite sides of the Petri dish, close to the walls (Fig. 10).
3. Treat one bee with the substance tested, treat the other (control) bee with the solvent used to transfer the tested substance on the first bee.
   Use a volume of solvent as small as possible to avoid perturbing the layer of cuticular hydrocarbons. In case of a removal / restoration bioassay the bees' cuticle need be washed with a solvent to remove the hydrocarbons before the tested profile is applied.
4. Place the Petri dishes in a thermostatic cabinet, in darkness, at 34.5°C and 60-70% RH.
5. Place one adult female mite in the centre of the Petri dish.
6. Note mite position every 10 min for 60 min.
   Three positions are considered: mite on the treated bee, mite on the control bee, mite not on bees.
7. Test 10 mites in different Petri dishes simultaneously and replicate 6 times.
   Alternate side of treated and control bees for each replicate to control for the influence of external factors on mite locomotion.

3.6.2.4.3. Data analysis

For each Petri dish, a score is calculated summing the number of mites that were found on the bees during the six observations. This figure can vary between 0 and 6 and is representative of the time the varroa mite spends on the bees. The score can thus be considered as a measure of the preference of the mite for the stimulus under testing. Data from all the replicates are organized in a matrix with as many rows as the number of mites used in the bioassay, and 2 columns containing the scores of the 2 stimuli to be compared. As the variables under study have an unknown distribution, the scores of different stimuli in a data set are compared by a sample randomization test (Sokal and Rohlf, 1995; Manly, 1997). The randomization distribution should be re-sampled a sufficient number of time (e.g. 10^6 times).

Active chemicals identified by means of laboratory bioassays can be tested in the field. For methodologies see section 4.5.4. 'Testing varroacides in the field'. This guideline describes the testing of acaricidal effects; however, it can also be used when using substances that do not kill mites, but disturb their orientation and reproduction.

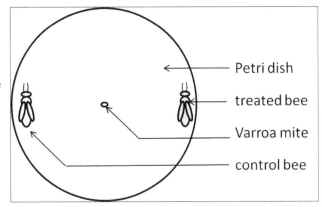

Fig. 10. Test arena for phoretic mite attraction cues.

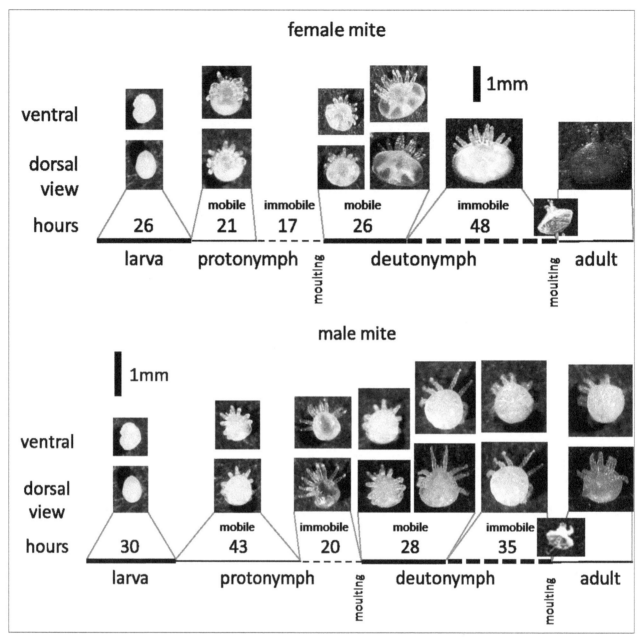

Fig. 11. Ventral and dorsal views of developmental stages of *Varroa destructor* females (above) and males (below) *on A. mellifera* brood. Approximate developmental time is given above the horizontal lines of different thicknesses which delimit the stages. Solid lines denote mobile phases, dashed lines immobile phases prior moulting (after Donzé *et al.*, 1994). Immobile and mobile phases can only be distinguished in live material, not in frozen samples.
Photos: R Nannelli and S J Martin

The effect expected is not mite death, but a reduction in mite population size in the colony which can also be detected with this method.

3.6.2.5. Mating bioassays

This bioassay allows the observation and analysis of the mating behaviour of mites under laboratory conditions and is also suitable for testing substances which might stimulate or disturb the mating behaviour.

3.6.2.5.1. Mites used in the bioassay

Adult males and all other relevant mite stages for the mating bioassay can be found in worker brood cells 8-9 days after cell capping. See section 3.1.4.2.1. 'Opening each cell' for the description of how to collect mites from cells. Females shortly after the adult moult should be used for the general observation of the mating behaviour and disturbance experiments. Deutonymphs (Fig. 11) are not attractive for males and can be used as "dummies" when stimulating cues are tested.

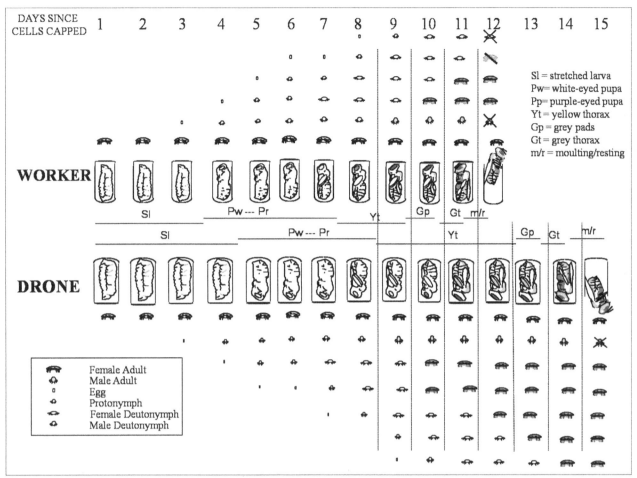

Fig. 12. Development chart of varroa mites and their honey bee host, *A. mellifera*.

The mites should be separated according to sex (see section 4.3.3. 'How to measure reproductive success' and Figs 11 and 12) and kept in groups of maximum 5 individuals at 28-30°C in order to avoid unwanted copulations and a decrease of fitness.

3.6.2.5.2. Experimental setup

This bioassay is described by Ziegelmann *et al.* (2012).

1. Queen cell cups (e.g. Nicot system®) can be used as test arena.
   It is recommended to embed the cell cups in a glass Petri dish with wax.
2. Ensure a temperature of 28-30°C in the cell cup by placing the setup on a hotplate.
3. Transfer the relevant mite stages into the cell cup.
   When extracts or single substances are tested, follow steps 4 and 5, if only behaviour is observed, go to 6.
4. Apply volatile test substances to a piece of filter paper (size: 1.5 mm x 15 mm).
5. Place volatile substances in the vicinity of the female; apply non-volatile substances directly to the female mite.
   For the application of test substance, chose a solvent which does not harm or repel the female.

6. Cover with a glass plate to prevent mites from escaping.
7. Record the male responses with e.g. the Observer software (Noldus Information Technology) for 5 or 10 minutes.
8. Categorise male responses as follows: 1. movement around female; 2. mounting the female's dorsum; and 3. copulation attempt on the female's venter.

3.6.3. Bioassays to quantify the susceptibility of the varroa mite to acaricides

Acaricide resistance represents a dramatic problem for apiculture and has been related to widespread losses of bee colonies. To reduce the impact of such losses, a prompt detection of resistant varroa population is vital and reliable methods for testing the susceptibility of the varroa mite to different acaricides are a fundamental resource, notwithstanding the possible use in basic research on the mode of action of pesticides (Milani, 1999). There is also a need to discover new varroacidal substances. For both purposes a simple and fast bioassay is necessary. A convenient bioassay was devised by Milani (1995) for the study of acaricides that are active by contact (i.e. the active ingredient contaminates the cuticle of the bees and is taken up by the mite by indirect contact). This is the case of most acaricides used currently (e.g. pyrethroids and some organic acids). The bioassay described has been used to test the activity of several acaricides including tau-fluvalinate, flumethrin (Milani, 1995), perizin and Cekafix (Milani and Della Vedova, 1996), oxalic and citric acids (Milani, 2001), as well as for the study of reversion of resistance (Milani and Della Vedova, 2002). Other acaricides that are widely used for the control of the varroa mite are air-borne and the bioassay above is not suitable. For these cases a new bioassay developed by the honey bee laboratory of the University of Udine is presented in section 3.6.3.3.

3.6.3.1. Mites used in susceptibility bioassays

For this assay, mites that would be exposed to the product in colonies in the apiaries should be used. Some products affect only phoretic mites, others also affect mites in brood cells. Adult mites at the reproductive stage might have different susceptibility compared to phoretic mites or to their offspring because their cuticle is not hardened or because of physiological differences. Milani (1995) and Milani and Della Vedova (1996) tested compounds only affecting phoretic mites, but showed that mortality was more homogeneous in mites collected from the brood. This might be due to a more homogeneous physiological status compared to phoretic mites. It is therefore recommended to test all life stages to obtain a complete picture of mite susceptibility.

When brood mites are tested, they are collected from combs (or pieces of comb) of infested colonies after opening and inspection of capped cells. Mites parasitizing brood of different developmental stage have different susceptibilities to acaricides (Milani and Della Vedova, 1996). For the tests, they are therefore grouped according to the age of their brood host and assayed separately. The age of larvae or pupae inhabiting these cells can be pre-determined by marking at the capping stage and opening it at a given time. Alternatively, the approximate age of the brood can be inferred on the basis of the morphology and pigmentation of the larva or the pupa (see section 'Obtaining brood and adults of known age' in the *BEEBOOK* paper on miscellaneous methods (Human *et al.*, 2013)). Varroa mites from different brood stages can be pooled if previous results indicate no differences among development stages (Milani, 1995). Differences between mite developmental stages might also influence their susceptibility to active ingredients, independently of host development, but this has not been shown yet.

Mites are kept on their host larva or pupa in glass Petri dishes until a sufficient number is collected. This ensures they can feed if hungry and the availability of their own host ensures that their physiological status is not changed. Since mites might stray from their host larva or pupa and climb onto another, only hosts at the same development stage should be kept in any given dish.

3.6.3.2. Bioassays for contact substances

1. A stainless steel ring (56 mm inner diameter, 2–3 mm height) and 2 glass circles (62 mm diameter; Na-Ca glass) are cleaned with acetone and hexane or pentane to form the testing arena.

2. Apply the product to be tested and the control solution on the arena pieces.

   Various concentrations of the products are tested. See the *BEEBOOK* paper on methods for toxicological studies (Medrzycki *et al.*, 2013) to define these concentrations. The application of the active ingredient on the arena pieces varies according to its physico-chemical properties.

2.1. For water soluble active ingredients (polar compound):

   2.1.1. Mix the active ingredient with a convenient solvent.

   2.1.2. Spray the glass disks and ring with a solution of the compound as evenly as possible.

   This can be done by means of a "Potter precision spray tower" (e.g. Burkard Manufacturing Co; UK) (Milani, 2001). To do so, the reservoir is loaded with 1 ml of solution; the distance of the sprayed surface from the bottom end of the tube is set at 11 mm and a nozzle 0.0275 inches is used. The pressure is adjusted (usually in the range 350–500 hPa) until the amount of solution deposited is 1 ± 0.05 mg/cm^2. Alternatively, if such piece of equipment is not available a glass Petri dish can be used as arena. The solution (active ingredient in a solvent of low boiling point) is poured in the dish so as to cover the whole bottom of the dish and left to evaporate under a fume hood. Depending on the surface tension of the active ingredient, this will result in an uniform layer of substance at the bottom of the dish. Varroa mites can thus be exposed to the substance in the Petri dish. This method cannot be used when the surface tension of the ingredient is too high and droplets are formed on the Petri dish.

2.2. For lipid soluble (apolar compound) active ingredients:

   2.2.1. Melt 10 g of paraffin wax (e.g. Merck 7151, melting point 46-48°C) in a glass container kept in a water bath at 60°C.

   2.2.2. Dissolve the required amount of the active ingredient in a convenient solvent (e.g. hexane or acetone).

   2.2.3. Add this solution to the melted wax.
   The solvent alone is added to the control wax.

   2.2.4. Weigh glass disks and iron rings before coating it with the wax.

   2.2.5. Stir the mixture for 30 min.

   2.2.6. Immerse the steel rings in the molten paraffin wax, one side of the glass disks is coated by lowering the disks onto the molten paraffin.

   2.2.7. Weigh glass disks and iron rings after coating.
   Discard the arenas (ring + glass circles) with a total amount of coating outside the range 1.6-2.0 g.

   2.2.8. Keep the arena pieces for at least 24 h at room temperature to allow for the solvent to evaporate.

   2.2.9. Store at 32.5°C until they are used.

3. Place the ring between the glass circles so as to build a cage. The cages are used within 60 h of preparation, for not more than three assays.

4. Introduce 10 to 15 varroa mites in this cage and bind the pieces together with droplets of melted wax.

   Mites collected from spinning larvae, stretched larvae, white eyed pupae and dark eyed with white and pale body are used. See the section 'Obtaining brood and adults of known age' estimating pupa age of the *BEEBOOK* paper on miscellaneous methods (Human *et al.*, 2013) for determining the age of pupae.

5. After 4 h transfer mites into a clean glass Petri dish (60 mm diameter) with two or three worker larvae taken from cells 0–24 h after capping (obtained as described in section 3.1.4. 'Collecting mites from the brood') or with two or three white eye pupae (4-5 days after capping).

6. Observe the mites under a dissecting microscope, 4 (i.e. at the time of transfer into the Petri dish), 24 and 48 h after the beginning of the treatment and classify as:

   6.1. Mobile: they walk around when on their legs, non-stimulated or after stimulation.

   6.2. Paralysed: they move one or more appendages, non-stimulated or after stimulation, but they cannot move around.

   6.3. Dead: immobile and do not react to 3 subsequent stimulations.

   A clean tooth pick or needle can be used to stimulate the mites by touching their legs. New tooth picks or cleaned needles should be used for stimulating control groups to avoid their contamination with residues of active ingredients from treated mites.

The assays are carried out at 32.5°C and 60-70% RH. If the mortality in the controls exceeds 30%, the replicate is excluded. Each experiment is replicated with a sufficient number of series of cages. To determine the sample size, refer to the *BEEBOOK* paper on methods for toxicological studies (Medrzycki *et al.*, 2013) and the *BEEBOOK* paper on statistics (Pirk *et al.*, 2013). If mites are scarce, more replications are carried out and more mites are assayed at doses around the median lethal density, to increase statistical resolution in this region.

3.6.3.3. Bioassays for volatile substances

1. Dissolve the active ingredient (e.g. thymol) in a suitable solvent (e.g. diethyl ether) at the concentration 0.5 g/ml.

2. Treat a circular area (diameter = 6 cm) of the inner side of the lid of a glass Petri dish (diameter = 14 cm) with 250 µl of the solution.

3. Let the solvent evaporate.

4. Place 10 to 15 varroa mites on the bottom of the Petri dish and keep different groups inside the closed container for 0, 15, 30, 45, 90, 135 min at room temperature.

   Mites of the same origin as for the bioassay for susceptibility to contact substances are used (see section 3.6.3.2. 'Bioassays for contact substances').

5. After each interval, transfer the group into small Petri dishes (diameter = 6 cm) with one bee larva for every five mites.

6. Place in an incubator at 34.5°C and 60-70% RH.

7. Monitor mite survival at 48 hours.

3.6.3.4. Data analysis

The data are analysed using the probit transformation. The natural mortality rate is taken into account using the iterative approach, according to Finney (1949). The concentrations which kill a given proportion of mites and their fiducial limits are computed according to Finney (1971). Refer to the *BEEBOOK* paper on toxicological methods (Medrzycki *et al.*, 2013) for these calculations.

4. Field methods

4.1. Diagnostic techniques

The OIE manual describes three methods to diagnose the presence of varroa mites in colonies (OIE *Terrestrial Manual* 2008). Debris, adult and brood examination are reported here.

4.1.1. Debris examination

Hives must be equipped with a bottom board on which debris are collected. The board must be protected by a mesh to prevent bees from discarding the dead mites. The mesh size should allow the mites to fall through. To increase probability to detect mites, colonies can be treated with a varroacide. After a few days, dead mites can be observed on the boards. In case a large quantity debris prevents easy detection of mites, debris can be cleaned from the varroa board and examined using a flotation procedure:

1. Dry debris for 24 h.

2. Flood with industrial grade alcohol.

3. Stir continuously for 1 minute or up to 10-20 min if debris contain wax or propolis particles.

4. Investigate the surface of the alcohol for the presence of mites.

4.1.2. Brood examination

Since varroa mites prefer drone brood, the probability to detect them on male pupae is higher than on worker brood. However, in absence of drone brood worker brood is used. When a large number of samples are examined, a rough determination of the degree of infection can be obtained.

1. Remove the cappings of the brood cells with a knife or fork.

2. Flush the pupae out of the combs with a stream of warm water over a sieve (mesh width 2–3 mm).

3. Collect the mites in a second sieve (mesh width 1 mm) placed below the first.

4. Examine the contents of the second sieve on a bright plate, where the mites can be easily identified and counted.

When a smaller number of samples are examined,

1. Open individual cells.

2. Remove larva, pre-pupa or pupa.

3. Examine cell walls using an appropriate source of light.

4. Identify infected cells by the presence of small white spots – the faeces of the mite (Fig. 3).

5. Confirm the presence of the mites themselves in the cell or on the brood.

4.1.3. Bee examination

1. Collect 200–250 bees from both sides of at least 3 unsealed brood combs.

2. Kill the bees in a container filled with alcohol.

3. Stir the container for 10 min.

4. Separate the bees from the mites by pouring the alcohol over a sieve with a mesh size of approximately 2–3 mm.

4.2. Measuring colony infestation rate

Three methods to estimate colony infestation have been designed (Ritter, 1981; De Jong *et al.*, 1982b). Acaricides can be used to kill all mites in a colony. Mites will fall to the bottom of the hive and can be counted (Branco *et al.*, 2006). Without the use of acaricides, the natural mortality can be quantified from the bottom of the hive to determine the population size of the live mites. Alternatively, the infestation rates of adults and brood can be estimated from adult and brood samples. When the first two methods are used, ants must be prevented access to bottom boards. Their scavenging habit will result in the disappearance of dead mites before they can be counted and will thus bias the results (Dainat *et al.*, 2011). Such a protection can be obtained by preventing access to the whole hive or to the bottom board. Hive protection can be achieved by using a stand with feet smeared with grease or resting in containers containing a liquid over which ants cannot walk (water or oil). Here it is important to regularly verify that dirt does not accumulate in the container or on the grease, allowing ants to reach the hive. Blades of grass can also form bridges and should be cut in the surrounding of the hives. Alternatively, the varroa board itself can be protected against ants. This is achieved by covering the board with sticky material (e.g. Vaseline, glue, absorbent paper impregnated with vegetable oil). Such 'sticky boards' can be purchased or homemade.

All three methods (using acaricides, monitoring natural mite fall and assessing infestation levels) were found to provide comparable

results (Branco *et al.*, 2006). For the adult infestation rate estimate, the sample size in relation to the level of precision required by the experimenter has been determined by Lee *et al.* (2010a). Their study provides methods with different workloads permitting to achieve several levels of precision. We present here the method with optimal time and effort investment ratio that is necessary to reach the precision necessary to researchers. Since researchers are mostly interested in the infestation rates of particular colonies rather than of whole apiaries, we do not describe the latter method here, but refer to Lee *et al.* (2010a) for the number of colonies to sample from in order to obtain a representative figure at apiary level.

The methods based on mite fall or on evaluating infestation rates from adult or brood samples are only reliable for colonies with medium to high infestation rate. The methods show imprecision when colonies have less than 3,000 brood cells, when the brood infestation rate is < 2% (unless very large samples are taken, see the *BEEBOOK* paper on statistics (Pirk *et al.*, 2013)) or when the colony is collapsing (due to decreased amount of brood) (Branco *et al.*, 2006; Lee *et al.*, 2010a). In these cases, the acaricide treatment can be used. Using synthetic acaricides to estimate parasite population size in the host is reliable provided a product with high efficiency (> 95%, taking possible resistance by the mite into account) is used. However, it is destructive and can only be used for a quantification /diagnostic purpose. The mites being killed by the treatment and the hive being contaminated with acaricide residues, the treated colony cannot be used as source or host of mites.

4.2.1. Acaricide treatment

Use an effective acaricide > 95% product as per manufacturer recommendation. Beware of resistance of mites to this product, see section '3.6.3. Bioassays to quantify the susceptibility of the varroa mite to acaricides ' for methods on how to test mite susceptibility to acaricides.

1. A protected bottom board should be used to prevent bees removing the fallen mites.

    The protection is typically a wire screen with 3-4 mm holes covering the whole surface of the board, leaving no access for bees to the fallen mites.

2. Ant protection should be put in place to prevent their access to the hives and predation on fallen mites and therefore biasing the number of mites counted.

3. Given the rapid action of efficient acaricides and to ease counting, mite fall should be assessed daily.

    See sections 4.2.4. 'Natural mite fall' and 4.2.5. 'Sub-sampling mites', to count mites on a bottom board.

If the active ingredient used is persistent enough (i.e. the treatment still in place or if residues persist in the hive) and do not penetrate in the cell through the capping (e.g. most synthetic acaricides), the mites that entered cells just before the treatment become exposed upon their emergence with their bee host and die within a few days. Mite fall should thus be counted for 3 weeks, this period covering the development times of pupae and the time necessary for mite fall to decrease to pre-treatment levels. The same counting period should be covered if a non-persistent acaricide is used that also kills mites in the cells (e.g. formic acid). Indeed, mites dead in the cells will only be released and fall on the bottom board to be counted upon emergence of their host bee. In case the product is not persistent and does not affect mites in cell (e.g. oxalic acid), colonies without capped brood must be treated. Absence of capped brood can be obtained by caging the queen 22 days before the planned treatment. All mites being in the phoretic phase, mites should fall for a shorter period (since none are trapped in cells). Mite fall count can therefore stop when it decreased to pre-treatment levels.

Pros: efficient, relatively low workload.

Cons: slow, dead mites, and in case of use of persistent acaricides, contaminated colonies cannot be used further; in case of queen caging, the development of varroa population before treatment can be slightly affected by the interruption of brood rearing.

4.2.2. Whole colony estimate

This method requires killing the whole colony. This is necessary when the real infestation rate of a colony is needed. Indeed, the use of acaricides is under these circumstances not appropriate since their efficiency is not 100%.

1. When all foragers are in the colony (early in the morning, late in the evening or at night) close the hive so that no bees can escape.
2. Place the whole colony in a freezer.

   Depending on nutritional status and size, colony survival in a freezer will vary. To determine when the colony died, workers from the centre of the cluster can be sampled and left to thaw. If they do not wake up, the whole colony can be considered dead and used for mite counts. In case the colony is of large size, gazing with CO_2 is required before freezing. This will prevent the bees thermoregulation and entering in the cells. Thermoregulation extends the duration needed to kill the colony and if bees get into the cells, they will be more difficult to collect for mite counting.
3. Refer to section 4.2.3. 'Measuring the infestation rate of brood and adult bees' for phoretic and brood infestation rate measurement.

If a measurement of total infestation rate is needed in summer, the colony can be made broodless by caging the queen for three weeks. When all the brood runs out (after 21 days if only worker brood was present or 24 days if drone brood was present), all mites have become phoretic. There is then no need to look for mites in the brood.

Pros: provides the exact total number of mites in a colony.

Cons: destructive, high workload, tedious.

4.2.3. Measuring the infestation rate of brood and adult bees

4.2.3.1. Infestation rates of adult bees

4.2.3.1.1. Sampling

Material: a rectangular graduated container in which 300 bees fit. Three hundred bees occupy a volume of 100 ml water. Fill this volume of water in a container and mark a line at the water surface (Lee *et al.*, 2010b, www.beelab.umn.edu). Given that bee sizes change with race, this volume should be verified and adapted for the particular bee under scrutiny.

1. Hold the frame at approximately 10 degrees from the vertical.
2. On the upwards facing side, slide the graduated container downwards on the back of the bees so that they tumble in it, making sure the queen is not one of them.
3. Rap the cup on a hard surface to be sure the bees are at the marked line; add or subtract bees as needed.
4. Collect 3 x 300 workers from any three frames in the first brood box.

Sampling such a large number of bees takes into account variations among frames to obtain an average infestation rate, and does not damage the colony if a non-destructive method is used to loosen the mites from the bees (see section 4.2.3.1.2.1. 'Powdered sugar'). Strong colonies (> 10,000 bees) are not dramatically affected by the removal of this amount of bees and will quickly recover. However, for analysing varroa population dynamics throughout the whole season with frequent and distructive sampling of bees (e.g. at 3-week intervals), lower numbers of individuals (300 bees per sampling date) should be used.

4.2.3.1.2. Dislodging mites from bees:

There are several ways to dislodge the mites from the bees. Some were already presented in section 3.1., but not all of them are adapted to estimating the infestation rate of the colony. Indeed, these methods must be standardised and deliver repeatable results.

4.2.3.1.2.1. Powdered sugar

After step five of section 3.1.2. describing how to dislodge mites from honey bees kept in a jar, perform the following steps:

6. Count the mites fallen out of the jar (e.g. 43).
7. Count the number of bees in the sample washed (e.g. 310)
8. Divide the number of mites counted by the number of bees in the sample (310) and multiply by 100 to determine the number of mites per 100 bees (e.g. (43/310)x100 = 14.3 mites per 100 bees).

Pros: practical, low cost, non-destructive (the bees can be reintroduced in the colony and will be cleaned by their nestmates), environmentally friendly.

4.2.3.1.2.2. Ether wash:

This method is modified from Ellis *et al.* (1988).

Material needed: a jar with a screen raised 2-3 cm above the bottom, automotive starter fluid

1. Spray the jar for two seconds with starter fluid to kill bees and mites.
   Dying bees regurgitate consumed nectar or honey that will make the wall of the jar sticky.
2. Shake the jar for 1 min to dislodge the mites from the bees.
3. Lay the jar sideways and roll three times completely along its vertical axis.
4. Count the mites stuck to the sides of the jar.
5. Count the number of bees in the sample washed.
6. Divide the number of mites counted by the number of bees in the sample to determine the proportion of infested individuals.
7. Multiply by 100 to obtain the number of mites per 100 bees.
   Caution: ether is highly flammable!

Pros: fast, low workload.

Cons: environmentally unfriendly, expensive, destructive, dangerous.

4.2.3.1.2.3. Warm/soapy water or ethanol (75%):

This method follows the protocol by Fries *et al.* (1991a). Since mites do not have to be collected alive as allows the method described in section 3.1. 'Collecting mites', soap can be added to water or ethanol can be used to improve the efficiency of mite dislodging.

1. Warm/soapy water or ethanol is added to jars to cover the 300 honey bees.
2. The jars are shaken for 20 s to dislodge the mites from the adult honey bees.
3. The content of the jar is poured over a first sieve (aperture: 3-4 mm) to collect all the bees.
4. Check the jar for mites sticking to the sides.
5. Place a second sieve (aperture < 0.5 mm) underneath the first to collect the mites.
6. Flush the bees and mites with large amounts of warm water. Strength of the water stream or volume of water used for rinsing as well as duration of rinsing should be standardised.
7. Count the mites remaining on the second sieve (e.g. 13).
8. Count the bees in the sample washed (e.g. 303).
9. Divide the number of mites counted by the number of bees in the sample to determine the proportion of infested individuals (13/303 = 0.043).
10. Multiply by 100 to obtain number of mites per 100 bees (4.3).

Pros: water based method: low environmental impact, low cost.

Cons: not practical on remote apiaries (large amount of water for rinsing and heat source needed); alcohol based: expensive, environmentally unfriendly.

None of these three methods is distinctly superior to the other and they can all be considered as reliable given that mite separation is done in a standardised manner (water always at the same temperature, or containing a standardized amount of soap etc.) and that the efficiency of the method is determined as described below.

4.2.3.1.2.4. Assessing the efficiency of dislodging method

It is important to dislodge mites in all samples in a standardized manner so as to being able to compare the infestation rates measured between samples. When comparisons of infestation rates between samples are aimed at, calculation of washing efficiency is not needed. In contrast, when the absolute number of mites is important, the efficiency of the washing method should be assessed to correct the figure obtained for errors. In addition, it is necessary to obtain absolute numbers in order to compare the figures obtained with other studies and therefore calculating efficiency in all cases is recommended.

1. Perform additional washes (with same or other solvent) until no more mites are found (optional).
2. Check bees manually for the presence of mites after the wash(es) or sugar treatment.
3. Add mites found after repeated washes and/or manually (e.g. 1) to those of the first wash/sugar treatment to obtain the total number of mites in the sample (e.g.10).
4. Divide number of mites of the first wash/sugar treatment by the total number of mites to obtain the method efficiency (10/11 = 0.91).
5. Repeat with 5-10 samples to obtain an average efficiency (e.g. 0.9).
6. Divide the number of mites obtained in samples of interest (X) by the average efficiency to obtain the corrected figure (Y) (Y = X/0.9).

4.2.3.2. Infestation rates of brood

1. Cut out 200 randomly selected capped cells from a brood frame. Sampling cells from several frames will account for the spatially irregular infestation by varroa.
2. Open each cell and examine it for mite infestation.
   Mite infestation can be diagnosed by observation of mites themselves or of their dejection (white rubbery material located most of the times on the two upper walls, towards the bottom of the cell, Fig. 3).
3. Count the total number of cells opened.

4. Count infested cells.

5. Divide the number of infested cells (e.g. 15) by the total number of opened cells (e.g. 212) to obtain the proportion of mite infested cells (0.071).

6. Multiply this figure by 100 to obtain the brood mite infestation rate in mites per 100 cells (7.1).

Depending on the question addressed, more detailed observation of cell infestation can be done (see section 4.3.3. 'How to measure reproductive success')

4.2.3.3. Evaluation of total mite population size in the colony

From the infestation rate of adult bees or from that of brood, it is possible to calculate the infestation rate of the whole colony (Martin, 1998). However, it is more accurate to assess both parameters based on samples and revert to the total for the colony after estimating the amount of adults and brood.

1. Measure colony strength (see the *BEEBOOK* paper on estimating colony strength by Delaplane *et al.* (2013)) on the same dates as sampling for determining mite infestation rates, so as to accurately calculate the total colony infestation rate (Fries *et al.*, 1991b).

2. Multiply the total number of bees in the colony (e.g. 9,356) by the proportion of infested workers in the sample investigated (e.g. 0.107) to obtain the size of the varroa population in the phoretic phase in the colony (1,001).

3. Multiply the total number of sealed brood cells in the colony (e.g. 12,035) by the proportion of infested cells in the sample investigated (0.071) to obtain the size of the varroa population in the reproductive phase in the colony (855).

4. Add adult and brood infestation numbers to obtain mite population size in the colony (1,001 + 855 = 1,856).

With this information, the mite distribution between the phoretic and reproductive phases can be determined as the proportion of either mites on adults, or mites in brood, in relation to the total mite population within the colony (in our example 54 and 46% respectively). Mite distribution patterns can be used to determine the effect of brood attractiveness on varroa for example.

4.2.4. Natural mite fall

This method is based on the quantification of naturally dead mites. Counting can be exhaustive (more accurate when done with a guide, Fig. 13) when mite fall is low to medium or can be sub-sampled (using a checked pattern board; see section 4.2.5. 'Subsampling mites to count on a bottom board') when mite fall is high. In both cases, a formula needs to be applied to calculate the total infestation rate of the colony. Various studies gave contradictory conclusions regarding the accuracy of the natural fall method to determine total infestation rate since natural mite fall is largely determined by the amount of

Fig. 13. A guide is placed above the varroa bottom sheet to help guide the eye when counting and thus avoid double counts.

Photo: Swiss Bee Research Centre

emerging infested brood (Lobb and Martin, 1997), but it is in general considered as a good indicator of colony infestation (see Branco *et al.*, 2006).

Material needed: screened floor board, guide or sheet with checked pattern see 4.2.5. 'Sub-sampling mites to count on a bottom board'. Note: make sure the hives are inaccessible to ants or use a sticky board.

1. Pull the bottom board from underneath the colony.

2. When using non-sticky boards, shelter from the wind to count.

3. Place a guide above the board to avoid counting the same mites (Fig. 13) or count mites from the selected squares on a checkered board (see section 4.2.5. 'Sub-sampling mites to count on a bottom board').

4. If dead bees are present on the board, check them as they act as magnets to fallen live mites.

5. Adapt counting frequency to mite fall rate since many mites on the board are difficult to count and since increasing amount of debris accumulating over time makes counting difficult.
   If frequency of visit cannot be increased and high mite numbers must be counted, use the checkered board method (see section 4.2.5. 'Subsampling mites to count on a bottom board').

6. Collect data for 2 weeks, average the figure to obtain mean weekly mite fall.
   This period covers natural variation in mite fall due to population dynamics cycles within the host.

7. Calculate the total colony infestation rate from the weekly mite fall by multiplying the daily mite drop by 250–500 or 20-40 when brood is absent or present, respectively (Martin, 1998). These correction factors are valid for central European conditions.

Pros: non-destructive/non-invasive, fast, no need to open the hive depending on design; reliable for non-collapsing colonies with brood; this is a method providing relative quantification that can be used for comparison between colonies within an experiment, not across studies.

Cons: sometimes unreliable, death rate may vary according to colony status, season, bee race, climate (regional variations possible), amount of brood; little is known on the influence of these factors on mite death; specific equipment necessary.

4.2.5. Sub-sampling mites to count on a bottom board

This method is useful when a high number of mites have to be counted on the board. However, it provides reliable counts for all mite densities on the board. If more than 1,000 mites are present, it is sufficient to count 22% of the cells to obtain a highly reliable figure (Ostiguy and Sammataro, 2000).

Material needed: a grid with 2 cm cells printed onto a floor board or sheet fitting the size of the board. Note: make sure the hives are inaccessible to ants or use a sticky board.

1. Cells are grouped into blocks of nine.
2. Three cells per block are randomly selected and greyed (10% shade) to indicate the cells in which mites are to be counted.
3. The number of mites in the shaded squares are counted.
4. The total number of mites is then divided by the number of cells counted and multiplied by the total number of cells on the board.
5. The total number of mites is divided by the number of days the board was under the hive to obtain the mite fall per day.

Pros: can be more accurate than exhaustive counts when mite fall is high.

Cons: need to design / build / buy the checkered board.

4.3. Estimating reproduction parameters

4.3.1. Assessing reproductive success

Mite reproductive success is defined as the ability of a mother mite to produce at least one viable mated female offspring before the developing bee pupa hatches as an adult. Successful mite reproduction requires the maturation of at least two eggs laid by a reproducing mother mite inside the brood cell: a male mite and a sister female mite, which must mate together before emergence of the host bee. The male mite offspring will die when the bee hatches from the cell, but any mated mature daughter mites will enter the colony's mite population along with their mother to find a new brood cell for reproduction. A mother mite that lays no eggs, lays only one egg, produces no male offspring, or begins egg-laying too late in relation to larval development, will not contribute any progeny to the mite population. The fecundity (number of eggs laid) is an additional parameter that can determine variation in the number of viable

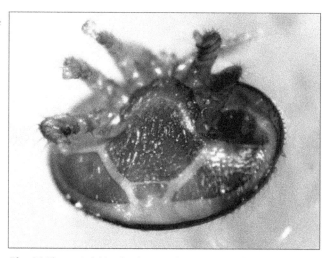

Fig. 14. The ventral side of a physogastric varroa mite. The intersegmental membranes are stretched due to the presence of oocytes in the mite's reproductive tract, which becomes partially visible.

Photo: Eva Frey

females each mother mite contributes to the population. Fecundity does not necessarily contribute to the mite's ability to reproduce successfully. Instead, it represents only the number of eggs laid without accounting for the age of the offspring or the likelihood of them reaching maturity. Therefore, fecundity may not be independent from the incidence of delayed egg-laying since any mother mite that begins laying eggs late may consequently lay fewer eggs.

Therefore, information on the following parameters are required to assess successful reproduction: the fertility (whether the mother mite laid eggs); the presence or absence of male offspring; the proportion of dead offspring; and the incidence of delayed egg-laying by mother mites (identified by relating the developmental stage of mite offspring to the developmental appearance and thus the determined age of the infested pupa; see the section 'Obtaining brood and adults of known age' in the *BEEBOOK* paper on miscellaneous methods (Human *et al.*, 2013)). These are important since relatively small differences in reproductive factors can have a large effect on the population dynamics at a population level (Martin and Medina, 2004).

4.3.2. When to measure reproductive success

The most accurate evaluation of mite reproductive success can be obtained just prior to worker emergence. However, the hours before emergence is not optimal since the last stage of bee development makes mite recognition difficult. It is therefore recommended to assess this parameter latest one day before emergence and to project in the future the development of the varroa offspring by comparing the developmental stage of the pupae to the developmental stage of the mite offspring. If direct comparisons between studies for several reproductive traits are required then the data must be collected over the same time period. That is, fecundity cannot be accurately measured until 220 hours after cell capping in workers and 240 hours in drone cells, as it is only after this period that all the potential eggs may be laid. It is only possible to determine various abnormal reproductive patterns 60 hours post capping. Table 2 gives the time after capping for which several traits related to reproductive success can be accurately measured.

To obtain a high amount of brood of relevant age in which reproductive success of the mite can be evaluated, cage the queen of a highly infested colony on an empty frame for several hours at the appropriate time before measurement.

Pros: mites are easily recognisable on the pupae.

Cons: a drawback is that the male or the daughters might die before emergence and been counted as part of reproductive success; determining of pupal age based on appearance is approximate.

4.3.3. How to measure reproductive success

The reproductive success of a mite is assessed by reconstructing the mite family in infested cells (Martin, 1994, 1995a). See section 4.3.2. 'When to measure reproductive success' for the optimal time when to measure this parameter.

It is important to examine cells infested by single mother mites as opposed to multiply infested cells since multiple infestations add an additional effect on the success of mite reproduction. Whether only one 'mother' mite reproduces can be tricky to ascertain in cells with multiple red mites, but using the mite development guide (Figs. 11 and 12) will indicate what is expected and give a good idea of the number of mother mites present. For statistical significance, at least 30 single mite infested cells per colony should be examined.

Different aspects of mite reproduction can be measured. Within each pupal cell, the following information can be collected: sex, developmental stage and vitality. Live or dead mites are easily identified if fresh material is used. If frozen material is used, mite appearance must be relied on. A shrivelled individual or abnormal appearance means it was dead prior to freezing. This characteristic can also be used to distinguish a dead individual from an individual in an immobile developmental stage in live samples. In addition, death rate can be estimated indirectly by comparing the developmental stages obtained with the reference: the development of any dead individual is prematurely arrested and can therefore be identified with the help of Figs. 11 and 12.

1. Using forceps, carefully open each capped cell by peeling back its top, papery seal and push away the walls of the cell.
2. Remove the pupa from the cell.
   This is best done by sliding the forceps each side of the neck between head and thorax and gently lifting up. When getting the pupa out of cell, place it on a microscope slide next to the cell to avoid mites dropping in the comb.

3. Record its developmental stage based on the appearance description given in Fig. 12.
   It is important to examine the pupae under a stereomicroscope once it is removed from the cell to make sure that mite progeny are not discarded with the pupae. Cell walls should also be inspected meticulously for mites and exuviae. The use of an optic fibre light source is particularly suitable to direct the beam to the bottom of the cell and inspecting it after the removal of the pupa.
4. Remove complete mite families together with exuviae from their cells and pupae using a fine brush.
5. Examine under a stereomicroscope.
6. Classify all female offspring into developmental stage groups using Figs. 11 and 12 as a guide.
   Protonymphs can be distinguished from similar looking young deutonymphs by the number of hair in the intercoxal region (between the 4 pairs of legs on the ventral side). Male and female protonymph have 3 and 4 pairs of hair, respectively, whereas deutonymphs have 5-6 pairs (see drawings pp 55-57 in Fernandez and Coineau, 2007). This information allows the mite family to be reconstructed in birth order or to check for multiple infestation and normality of development. Protonymphs are usually not sexed, but the number of hairs on the intercoxal region and the patchy hair pattern on the dorsum of males compared to the homogeneous and dense pattern of females provide recognition traits (Fernandez and Coineau, 2007). For the number of eggs laid by mites and to allow a more accurate comparison between studies, only mites that laid two or more eggs are compared. Mites that produce no eggs (non-reproductive) or one egg (single male) are considered as distinct reproductive categories (see b and c below). The mortality of the mite offspring in worker and drone cells is calculated by comparing the number of live and dead offspring at each position in birth order, i.e. first offspring, second offspring, etc. Then the average number of surviving females (unfertilised and fertilised) is calculated using only the levels of offspring mortality.
7. All infested cells are analysed by placing the mother mites into one of the following six categories:
   a) mother dead
   b) mother (alive), no offspring
   c) mother plus only male offspring
   d) mother plus (live) mature male and female offspring so mating is assumed
   e) mother plus (live) female offspring and dead male offspring
   These female offspring might remain unfertilised since the male might have died before they were mated (Harris and Harbo, 1999). Dissection of the spermatheca and microscopic

examination of its content can help determine their mating status (see Steiner *et al.*, 1994 for pictures of reproductive tract with spermatheca).

f) mother without (live) female offspring.

When working with frozen material, beware of repeated freezing and thawing cycles since this damages the samples. Only take out of the freezer the amount of material can be dealt with at a given time.

Pros: if this type of data is collected it can be compared with previous studies (e.g. Martin, 1994, 1995a, b; Medina and Martin, 1999; Martin and Kryger, 2002), where the same data for reproductive parameters of susceptible mites were obtained by using the exact same methodologies.

Cons: tedious, time consuming.

4.3.4. Assessing oogenesis

For estimates of oogenesis in mites, refer to section on marking methods 3.4.1. 'Oogenesis'.

4.4. Estimating damage thresholds

This section describes methods used to measure varroa damage at the colony level and to associate that damage with economic damage thresholds.

One of the goals of varroa Integrated Pest Management (IPM) is the reduction of beekeeper reliance on pesticides in the bee hive. Mite eradication is not a necessary goal, as IPM philosophy recognizes that eradication may require practices that are excessively toxic, invasive, or impractical. Varroa IPM places a premium on non-chemical management practices that eliminate mites from a colony (such as drone brood trapping) or slow the rate of mite population growth (such as genetic host resistance). The most promising varroa IPM practices have been reviewed by Rosenkranz *et al.* (2010). Unfortunately, it has been shown anecdotally as well as in computer simulations (Hoopingarner, 2001; Wilkinson *et al.*, 2001) that few if any of these practices can by themselves or indefinitely keep mites at non-damaging levels. Thus at this point it seems best to think of IPM as a means to delay, not necessarily eliminate, the application of acaricides (Delaplane, 2011). Many benefits accrue if the time between chemical treatments is delayed – namely, reduced toxin delivery to bees and the environment, reduced chemical residues in honey, and relaxed selection pressure for chemical resistance and the conservation of susceptible alleles in the mite population that prolong the commercial life of an acaricide.

Since the focus of IPM is not mite eradication, but rather mite management, the whole system hangs on the existence of criteria that can distinguish mite densities that are tolerable from those that are approaching damaging levels. Classical IPM tenets (Luckmann and Metcalf, 1982) identify the economic injury level (EIL) as that pest

Fig. 15. Adult bees with their phoretic mites from several colonies are mixed to homogenise the starting infestation level in the replicates.

Photo: Keith Delaplane

density at which point the grower is experiencing economic loss. The goal is to prevent this level from happening, in other words, to identify an earlier and lower pest density at which point a treatment could prevent reaching the EIL. This lower pest density is named variously the economic threshold, treatment threshold, action threshold, or damage threshold. Implicit in the use of damage thresholds is the recognition that some pest levels are tolerable and do not warrant the use of a pesticide. A good damage threshold will not only distinguish damaging from non-damaging pest densities, but also accommodate local variations due to geography and biology of host and parasite. Thresholds are simply a form of applied population modelling, and like any model, they are only as good as the data, detail, and specificity applied to their construction.

4.4.1. How to estimate damage thresholds

The following instructions are a synthesis from the field-derived damage thresholds of Delaplane and Hood (1997, 1999) and Delaplane *et al.* (2010) for the south-eastern USA, Strange and Sheppard (2001) for the north-western USA, and Currie and Gatien

(2006) for Manitoba, Canada. All of these studies used a design in which colonies of uniform strength and mite density are experimentally set up, applications of acaricide made at different times to create a spread of mite history and colony condition, and samples taken regularly to document mite levels and colony condition. With some, negative controls are present as colonies that are never treated, and positive controls present as colonies that are treated continuously. Thus, a range of mite densities exists across the experiment in space and time, and every colony has a history of known mite levels and colony strength at the time it was treated.

4.4.1.1. Colony establishment

1. Equalize field colonies in regard to bees, brood, food resources (see section 1.2.2.1 'Setting up experimental colonies of uniform strength' of the *BEEBOOK* paper on estimating colony strength (Delaplane *et al.*, 2013)) and mites within units of higher-order experimental replication, i.e. blocks or whole plots, usually based on geography.
   At least two more source colonies than the target number of experimental colonies should be used to account for bee loss through death or flight.

2. Collect adult bees and phoretic mites for experimental set-up by shaking workers from a diversity of source colonies into one large, common cage, allowing workers (and mites) to freely mix (Harbo, 1993) (Fig. 15).

3. Maintain the cage in cool conditions to prevent bee death from over-heating for at least 24 hours to allow thorough admixing of bees and mites, resulting in a uniformly heterogeneous mixture.

4. Distribute worker cohorts of equal size (preferably by weight, *ca.* 1 kg) into hive boxes pre-stocked with near-equal amounts of brood, honey, pollen, and empty cells (Fig. 16).

5. Provide each colony a sister queen reared from the same mother and open-mated in the same vicinity to minimize variation due to bee genetics.

6. Collect a sub-sample of *ca.* 300 workers from each incipient colony.

7. Weigh the samples.

8. Count the number of bees to derive a colony-specific measure of average fresh bee weight (mg).

9. Divide initial cohort size (kg) by average fresh weight of individuals (mg) to obtain initial bee population.

10. Collect a sample of *ca.* 300 worker bees (can be the same sample as above after it is weighed fresh) to obtain a measure of initial density of phoretic mites.

11. Separate the mites from the bees, see section 4.2.3.1. 'Infestation rates of adult bees' (Fig. 17).

12. Count the number of bees and number of mites to derive a colony-specific measure of mite / bee density.

Fig. 16. Boxes are pre-stocked with near-equal amounts of brood, honey, pollen, and empty cells to host the experimental colonies.

Photo: Keith Delaplane

Fig. 17. Number of bees and mites are counted to obtain a measure of initial density of phoretic mites. Photo: Keith Delaplane

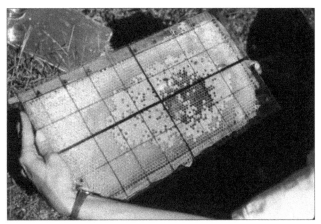

Fig. 18. Measuring the beginning quantity of sealed brood.

Photo: Keith Delaplane

13. Multiply this number by initial bee population to obtain phoretic mite population.

14. Store the samples in alcohol for future reference.

15. Collect brood for incipient experimental colonies from the same source colonies used to collect adults.

16. Assign a near-equal quantity of brood to experimental colonies without regard to source (see section 1.2.2.1.1. 'Classical objective mode' of the *BEEBOOK* paper on estimating colony strength (Delaplane *et al.*, 2013)).

17. Measure the beginning quantity of sealed brood to calculate beginning mite populations.

    This is done by overlaying on each side of every brood comb a grid pre-marked in cm^2 and visually summing the area of sealed brood (Fig. 18).

18. Convert the area (cm^2) of sealed brood to cells of sealed brood by multiplying cm^2 by the average cell density per cm^2 (see section 1.2.2.1.1. 'Classical objective mode' of the *BEEBOOK* paper on estimating colony strength (Delaplane *et al.*, 2013)).

    Cell density per cm^2 varies by geography; in the south-eastern USA conversion factors between 3.7 and 3.9 are used (Delaplane and Harbo, 1987; Harbo, 1993). This value should be verified with the honey bee lineage used for the experiment.

19. Measure incipient mite population in brood by opening at least 200 cells of sealed brood per experimental colony and examining under strong light and magnification to see and count sclerotized mites (see section 4.2.3.2. 'Infestation rates of brood').

20. Multiply the mite/cell density by cells of sealed brood to obtain the mite population in cells.

21. Sum phoretic mites + mites in cells to obtain the beginning total mite population.

22. Take corrective action should initial measures expose outliers in initial populations of bees, brood, or mites.

    In general, corrections aimed at minimizing experimental error are permissible until the point at which treatments are begun.

23. Once colonies are established and queens released and confirmed laying, it is recommended to collect one or more beginning relative mite measures. The most commonly recognized relative measures are mites recovered per 24 h on bottom board sheets (see section 4.2.4. 'Natural mite fall') and mites per adult bee recovered from samples (as described in section 4.2.3.1.2. 'Dislodging mites from bees'). It is to be hoped that these relative measures will correspond closely to real colony mite populations, and the investigator is encouraged to check the validity of the relative measures with regression analyses against total mite population.

24. Colony maintenance should include control of non-target diseases and disorders, swarm prevention, and feeding as necessary.

    The goal of these manipulations is to decrease experimental residual error.

4.4.1.2 Experimental treatments , sample size, and colony arrangements

1. Create a range of varroa colony densities by treating sets of colonies with acaricide at different points of a season, bracketing as widely as possible the months of bee activity particular to one's region. At the very least, one treatment should be early in the season when bees are emerging from winter senescence; one treatment should be at the peak of the active season, and another should occur in autumn which is typically the period of highest ratio of mites : bees. More intervals will improve the resolution of the resulting model. It is imperative that the design include a negative control – a set of colonies left untreated, and it is highly recommended that the design include a positive control – a set of colonies treated continuously (Table 3). Including both controls will provide the widest range of varroa densities possible within which the investigator can retrospectively search for mite densities that are damaging or non-damaging.

2. Use an average sample size (initial number of colonies per treatment) of 11.

   The literature indicates a range 7-20. Studies using sample sizes within this range never failed to detect treatment effects for at least some dependent variables (Table 3).

3. Stick to one mite control product.

   This avoids the risk of experimental confounding error due to variation in acaricide efficacy or unknown sublethal effects on host bees (investigators sometimes included different acaricides in their treatments, apparently with a view to fine-tuning control recommendations for their region).

4. Take into account resistance of mites to acaricides when choosing the treatment.

   Whether the population is resistant to a particular product can be tested following the method described in section 3.6.3. 'Bioassays to quantify the susceptibility of the varroa mite to acaricides'.

5. Control or at least monitor colony spatial arrangement.

   Varroa mites can spread horizontally through infested workers and drones drifting between colonies (Greatti *et al.*, 1992; Frey *et al.*, 2011) and exert a strong influence on results. Depending on one's objectives, one can set up apiaries to encourage drift (assign treatments within the same apiary) or discourage drift (assign treatments by apiary). The first scenario acknowledges that immigration may confound results, yet this condition is presumed more "real world" because modern beekeeping often encourages mite horizontal transmission with high-density apiaries. This option is however not relevant if treatments are made with persistent active ingredients that are distributed by contact between bees (e.g. fluvalinate). Drifters could contaminate colonies with different

Table 3. Experimental treatments, sample size, and colony spatial arrangements recommended and found in the literature on field-derived damage thresholds. [a] All studies cited here were performed in the Northern Hemisphere, so early season is Feb-May and autumn Sep-Oct; [b] initial colonies per treatment; [c] Delaplane and Hood (1997); [d] Delaplane and Hood (1999); [e] fluvalinate year 1 and thymol year 2; [f] Delaplane *et al.* (2010); [g] Strange and Sheppard (2001); [h] 10 each for May, 7 for Sep; [i] Currie and Gatien (2006).

Experimental treatments[a]	n[b]	Colony spatial arrangement	Reference
1. acaricide X early in the season 2. acaricide X at peak of season 3. acaricide X end of season 4. untreated colonies (negative control) 5. acaricide X continuous treatment (positive control)	12	according to objective (see step 5, section 4.4.1.2.)	recommended
1. fluvalinate Jun. 2. fluvalinate Aug. 3. fluvalinate Oct. 4. no treatment	18	divided equally between 2 states (ca. 120 km apart), treatments applied by apiary within state (minimize drift effect)	c
1. fluvalinate Feb. 2. fluvalinate Aug. 3. fluvalinate Feb.+Aug. 4. continuous fluvalinate 5. no treatment	12	divided equally between 2 states, treatments applied within apiary within state (maximize drift effect)	d
1. fluvalinate Feb. 2. fluvalinate Aug. 3. fluvalinate Feb.+Aug. 4. continuous fluvalinate 5. no treatment	8	divided equally between 2 states, treatments applied by apiary within state (minimize drift effect)	d
1. continuous treatment[e] beginning Jun. 2. treatment Aug. 3. treatment Oct.	20	divided equally between 2 states, treatments applied within apiary within state (maximize drift effect)	f
1. fluvalinate Apr. 2. fluvalinate Aug. 3. fluvalinate Oct. 4. fluvalinate Apr.+Oct. 5. continuous fluvalinate 6. no treatment 7. coumaphos Apr.	8	divided equally among groups (circles) of 8, each circle 15 m apart (minimize drift effect)	g
1. fluvalinate May 2. fluvalinate Sep. 3. 4 formic 4 d apart May 4. 4 formic 4 d apart Sep. 5. 4 formic 10 d apart May 6. 4 formic 10 d apart Sep. 7. coumaphos May 8. coumaphos Sep. 9. no treatment	10 or 7[h]	divided equally between 2 apiaries 8 km apart, within apiary colonies further subdivided into "low" initial varroa density or "moderate" initial density	i
1. fluvalinate May 2. 5 formic 1 week apart May 3. 1 formic slow release May 4. no treatment	7	treatments applied within 1 apiary of "high" initial varroa density	i

treatment regimes (Allsopp, 2006). The second scenario, in contrast, gives a more uncluttered description of the effects of delayed mite treatment. Including both conditions permits a test of the assumption that thresholds occur earlier under conditions of high mite immigration. In either case, it is recommended that the objective be explicit and the spatial arrangement designed accordingly: 1. if immigration is to be minimized then assign all colonies within one apiary the same treatment and space apiaries as widely apart as possible; or 2. if immigration is to be maximized then assign all treatments within the same apiary. Other drift-minimizing practices, such as painting symbols at hive entrances or arranging colonies in circles, will not substitute for wide spatial distances between colonies of different treatments.

4.4.1.3 Dependent variables and sampling protocols

A variety of parameters is measured to relate to damage thresholds of the colonies hosting different parasite population sizes (Table 4). For some of the most invasive measures, i.e., measures of bees or brood (Table 4), it is common during the mid-experiment sampling intervals to sample only a subset of colonies in each treatment. These numbers have ranged from 3 (Delaplane and Hood, 1999) to 4 (Delaplane and Hood, 1997) to 2 (Strange and Sheppard, 2001). It is important to remember that these mid-experiment measures are made not in pursuit of statistic differences, per se, but to provide a retrospective snapshot of colony condition at time of treatment or when threshold was achieved. They will also slightly increase within group variability of the parameters measured. All colonies are dismantled and fully measured at the end of the study when statistical rigor is desirable to identify the treatment regimens that optimized colony condition.

1. Place bottom board sheets under a wire mesh and positioned under the colony (see section 4.2.4. 'Natural mite fall').
2. Remove the sheet after 24-48 h.
3. Count number of mites and adjust number to mites recovered per 24 h.
4. Collect samples of 300-900 adult bees.
5. Count the number of mites and number of bees (see section 4.2.3.1. 'Infestation rates of adult bees').
6. Report the data as mites / 100 bees.
7. Derive fresh bee weight (mg) by sampling live bees off the comb into pre-weighed, or tared, containers, weighing the container, subtracting container weight to get net bee weight, counting bees, and dividing by net bee weight to get mg per bee.
    This sampling can be combined with sampling for mites per adult bee as described above.
8. Obtain ending colony bee population according to the methods described under section 1.2.2.2. 'Measuring colony strength at end of experiment: objective mode' of the

Fig. 19. The whole hive is weighed in the field to obtain final colony bee population.   Photo: Keith Delaplane.

BEEBOOK paper on estimating colony strength (Delaplane *et al.*, 2013), or from a variation from this method using net colony bee weight (kg) and average fresh bee weight (mg):

8.1. Obtain net colony bee weight by first closing the colony entrance with a ventilated screen in the evening or early morning before sampling to trap all bees inside.

8.2. Weigh the whole hive in the field (Fig. 19).

8.3. Open the hive.

8.4. Brush all bees off every comb and surface (usually into a temporary holding hive).

8.5. Re-weigh the hive without bees.

8.6. Calculate the difference in weight, which is the net weight of bees.

8.7. Divide this number by fresh bee weight to derive colony bee population.

9. Count the number of sealed brood cells as described in section 4.4.1.1., steps 17 and 18.
10. Colony mite population is derived from the methods described in section 4.4.1.1.

Table 4. Dependent variables and sampling protocols recommended and employed in the literature on field-derived damage thresholds. Months have to be considered according to the season at the location of the study. [a] Delaplane and Hood (1997); [b] Delaplane and Hood (1999); [c] fluvalinate year 1 and thymol year 2; [d] derived from regression predictions based on sticky board mite counts; [e] Delaplane *et al.* (2010); [f] Strange and Sheppard (2001); [g] Currie and Gatien (2006).

Experimental treatments	Dependent variables measured in each treatment group	Sampling intervals	Reference
1. acaricide X early in the season 2. acaricide X at peak of season 3. acaricide X end of season 4. untreated colonies (negative control) 5. acaricide X continuous treatment (positive control)	1. mites / 24 h on bottom sheets 2. mites / 300 bees 3. fresh bee weight (mg) 4. colony bee populations 5. number sealed brood cells 6. colony mite populations 7. visible brood disorders and other diseases 8. colony honey yield (kg) 9. subjective "survivability" score at beginning of the season of following year	at regular intervals in the year at mid-season	recommended
1. fluvalinate Jun. 2. fluvalinate Aug. 3. fluvalinate Oct. 4. no treatment	1. mites / 18 h on sticky sheets 2. mites / 300 bees 3. fresh bee weight (mg) 4. colony bee populations 5. number sealed brood cells 6. colony mite populations 7. visible brood disorders (Dec only)	Jun., Aug., Oct., Dec.	a
1. fluvalinate Feb. 2. fluvalinate Aug. 3. fluvalinate Feb.+Aug. 4. continuous fluvalinate 5. no treatment	1. mites / 20 h on sticky sheets 2. mites / 300 bees 3. fresh bee weight (mg) 4. colony bee populations 5. number sealed brood cells 6. colony mite populations 7. visible brood disorders (Sep.-Oct. only) 8. subjective "survivability" score (following Jan only)	Feb., May., Aug., Sep.-Oct.	b
1. continuous treatment[c] beginning Jun. 2. treatment Aug. 3. treatment Oct.	1. mites / 100 bees 2. fresh bee weight (mg) (Dec only) 3. colony bee populations 4. colony weight (kg) (Dec only) 5. cm^2 brood (all stages) 6. colony mite populations[d] 7. % bees infected with *Acarapis woodi* (Dec. only)	Aug., Oct., Dec.	e
1. fluvalinate Apr. 2. fluvalinate Aug. 3. fluvalinate Oct. 4. fluvalinate Apr.+Oct. 5. continuous fluvalinate 6. no treatment 7. coumaphos Apr.	1. mites / 20 hr on sticky sheets 2. mites / 300 bees 3. fresh bee weight (mg) 4. colony bee populations 5. number sealed brood cells 6. colony mite populations	Jun., Aug., Oct., Nov., Apr.	f
1. fluvalinate May 2. fluvalinate Sep. 3. 4 formic 4 d. apart May 4. 4 formic 4 d. apart Sep. 5. 4 formic 10 d. apart May 6. 4 formic 10 d. apart Sep. 7. coumaphos May 8. coumaphos Sep. 9. no treatment	1. mites per bee 2. colony honey yield (kg) (Aug. only)	At pre-treatment (May and Sep.), then post-treatment weekly for 3 weeks (spring) then biweekly thereafter	g

Once the investigator knows ending bee population, phoretic mites per bee, number of sealed brood cells, and mites per sealed brood cell, then one can extrapolate to (phoretic mite population + mite population in brood) = total colony mite population.

11. Quantify visible brood disorders.

Brood disorders are sometimes associated with varroa parasitism (Shimanuki *et al.*, 1994) and can contribute to colony damage.

11.1. Select two relatively contiguous patches of brood in the late larval – capped stages (stages more likely to express visible symptoms).

11.2. Overlay on each patch a 10-cm horizontal transect and 10 cm vertical transect intersecting at the centre (Fig. 20).

11.3. Examine along each transect every cell of brood under strong light and magnification for visible disorders, i.e., symptoms typical of American foulbrood (see the *BEEBOOK* paper by de Graaf *et al.*, 2013), European foulbrood (see the *BEEBOOK* paper by Forsgren *et al.*, 2013), sacbrood (see the *BEEBOOK* paper on honey bee viruses by de Miranda *et al.*, 2013), or chalkbrood (see the *BEEBOOK* paper on fungal diseases by Jensen *et al.*, 2013).

11.4. Report the parameter as percentage of brood expressing visible disorders.

12. When honey yield needs be considered as a parameter of the economic threshold, calculate colony yield (kg) by weighing honey supers before and after they are placed on colonies during a nectar flow.

13. Measure the infestation rate by *Acarapis woodi* in the colonies. This rate could be affected by the presence of varroa and its measure might be needed as economically relevant. For this sample collect 50 workers per colony and place in alcohol. See the *BEEBOOK* paper on tracheal mites (Sammataro *et al.*, 2013) for the method to detect the presence of these mites.

4.4.1.4 Analyses, interpretation, and pitfalls

The designs featured in this section lend themselves to a straightforward analysis of variance testing the effect of date of treatment on colony strength parameters at the end of the study (Table 4). Depending on the presence of higher-order replications such as blocks, the investigator should be alert to interactions between treatments and the blocking factor which, if present, prescribe that the investigator test main treatments separately by block. The number of surviving colonies by the end of the study (n) may differ across treatments, so it may be necessary to accommodate unequal sample sizes through use of harmonic means transformation or lsmeans. Treatment means are separated ($\alpha \leq 0.05$) by a conventional test such as Tukey's or Student-Newman-Keuls. If the investigator procured whole colony

Fig. 20. Selecting and inspecting brood cells for diseases.

Photo: Keith Delaplane

mite populations (see step 10 of section 4.4.1.3.) along with more user-friendly relative measures such as varroa board counts (see section 4.2.4. 'Natural mite fall') and mites per 100 bees (see section 4.2.3.1.2. 'Dislodging mites from bees'), then it is desirable to test the rigor of the relative measures at predicting real mite populations through the use of regression analyses testing linear, quadratic, and cubic terms. Ultimately, the investigator would like to deliver to beekeeper clients a user-friendly relative measure that accurately predicts real colony mite populations.

This analysis will permit the investigator to compare end-of-season colony condition across the various treatment regimens (times of acaricide application). The damage threshold is determined retrospectively as the highest average colony mite density at time of treatment associated with colony condition significantly non-different from positive controls at season's end. In one real example, the threshold was defined as conditions that prevailed when colonies were treated in August because August-treated colonies fared as well statistically at season's end as colonies treated continuously (Delaplane and Hood, 1999). The mid-season samplings permit the investigator to describe mite populations, user-friendly relative mite measures, and colony strength parameters that prevailed at the time thresholds were achieved. The highest, rather than lowest, retrospective mite density is used because of the conservative emphasis of IPM on prolonging the interval between treatments as long as possible. A low or zero pest tolerance is rare, unnecessary in the varroa / *Apis mellifera* IPM system, and more commonly associated with cropping systems for which pest-induced cosmetic damage is a problem with consumers.

This analysis will likewise identify mite densities that are irrecoverably damaging, in other words, mite densities at which point in time treatment does not prevent comparative colony deterioration by the end of the study. In another real example, it was shown that mite densities that prevailed in October exceeded a recoverable level because at season's end the October-treated colonies were in significantly worse condition than continuously-treated colonies (Delaplane and Hood, 1997).

Alternatively, Strange and Sheppard (2001) defined damage threshold as: 1. the mite levels corresponding to colony treatment groups at season's end with weight of bees less than initial starting levels (0.92 kg in this case); and 2. colonies with < 1150 cm^2 sealed brood – a number derived from regression analyses predicting the amount of brood that should be present in colonies with 0.92 kg bees. With these boundary conditions the authors were able to retrospectively identify legacy mite levels that were either tolerable or irrecoverably high.

One pitfall in the field studies described here is a confounding effect, inherent to the design, between season (time of treatment) and colony mite densities. Mite population growth is regulated by length of brood-rearing season, ratio of worker brood to drone brood, and number of brood cells (Fries *et al.*, 1994) and tends to increase over the course of the active season. Delaplane and Hood (1997) pointed out this confounding issue when they said, "Thus, our treatment threshold is reliable for August colonies meeting the conditions described in [the table showing retrospective colony descriptions], but may not be reliable for August colonies with significantly different amounts of bees or brood." A more highly-resolved field model would replicate each of these terms independently within month of treatment.

Another pitfall comes from the emerging realization that honey bee morbidity is not always the product of a simple linear process or one factor, but rather a web of interacting factors (vanEngelsdorp *et al.*, 2009). More sophisticated damage thresholds are needed that can integrate more than one morbidity factor and account for their possible interactions.

Pros: allows the definition of damage thresholds as basis of IPM implementation.

Cons: high workload, tedious.

4.4.2. Regional variations in reported damage thresholds.
Table 5 gives some published field-derived damage thresholds and their geography of origin.

4.5. Standardising field trials
4.5.1. Starting conditions
4.5.1.1. Obtaining mite free colonies
Mite free colonies can be obtained from varroa free areas. These colonies will also be residue free since acaricides are not used. Obtaining such colonies is usually not possible so the varroa population of the experimental colonies needs to be removed using a highly effective control method adapted to the region in which the study takes place. The occurrence of resistant populations needs to be taken into account

when choosing an acaricide for this purpose (see section 3.6.3. 'Bioassays to quantify the susceptibility of the varroa mite to acaricides' for methods to test for resistance). The efficacy of the treatment should be checked, as well as putative re-infestations from neighbouring apiaries (Greatti *et al.*, 1992).

Depending on the experiment planned, residues left behind by such treatment could bias the results by provoking delayed mortalities of mites. In such cases, residue free oxalic acid treatment can be used on swarms. No capped brood or frames with L5 larvae (see the section 'Obtaining adults and brood of known age' of the *BEEBOOK* paper on miscellaneous methods (Human *et al.*, 2013)) should be carried to the experimental hives in order not to bring in mites. Formic acid treatment can also be used on entire colonies with brood since the acid affects mites under the cell capping (Adelt and Kimmich, 1986; Koeniger *et al.*, 1987; Fries, 1991; Calis *et al.*, 1998). However, these two methods are only 95% efficient on average, which can influence the planed experiment. Oxalic and formic acid based products are available on the market and should be used as per manufacturer recommendations.

In such experiment, a control group of colonies treated continuously might be necessary. Such colonies need be separated from the experimental group since drifting and robbing bees could contaminate the test apiary (especially and mostly with synthetic acaricides; Allsopp, 2006). For the same reason, the control and experimental apiary need be separated by the same distance (~2km) from neighbouring uncontrolled apiaries. However, a compromise distance between control and experimental apiaries needs be found so that both are still subjected to equivalent environmental conditions.

4.5.1.2. Obtaining residue free hives
The presence of long lasting acaricide residues (Bogdanov *et al.*, 1998) in wax combs or honey can influence the results of experiments in which the survival of mites is a parameter of importance. Several methods to replace contaminated wax by residue free wax exist for beekeepers to switch to biological apiculture (Imdorf *et al.*, 2004). The method described here allows the decrease of residues below detectable levels within 1 year. This does not mean that all residues have disappeared, but that they have been diluted enough not to represent a problem for the quality of hive products. Whether the minimal quantities still present in the wax affect mite survival is however uncertain. This method based on comb removal can nevertheless be used to decrease the amount of residues in the wax for research purpose.

Complete removal of combs is best done at beginning of the bee season when comb is rapidly built by workers.

1. Split the colony.
2. Out of a split, create a broodless swarm with the old queen.

Table 5. Field-derived varroa damage thresholds and month of occurrence at various locations. [a]Delaplane and Hood (1999); [b]Delaplane *et al.* (2010); [c]Strange and Sheppard (2001); [d]Currie and Gatien (2006); [e]Gatien and Currie (2003).

Varroa damage thresholds and month of occurrence		Location	Reference
Feb.	Aug.		
1. colony mite population: 7-97	1. colony mite population: 3172-4261	Southeastern USA	a
2. 0.4 mites per 100 bees	2. 13 mites per 100 bees		
3. mites on overnight bottom board sheets: 0.6-10	3. mites on overnight bottom board sheets: 59-187		
	1. colony mite population: 1111 2. 20 mites per 100 bees	Southeastern USA	b
Apr.	Aug.		
1. 1 mite per 100 bees 2. mites on 48 hr bottom board sheets: 24	1. 5 mites per 100 bees 2. mites on 48 hr bottom board sheets: 46	Northwest USA	c
May	Late Aug. – mid Sep.		
1. 2 mites per 100 bees (to prevent honey loss)	1. 4 mites per 100 bees (to prevent winter loss)	Manitoba, Canada	d
	1. 5-8 mites per 100 bees	Manitoba, Canada	e

3. Scrape all propolis and wax, wash with soda and surface burn with a flame used hive parts to remove residues before introducing the colony; alternatively, use new hive parts.

4. Place swarm on residue free foundation (originating from location where no persistent miticides are used and originating from wax correctly sterilised by melting >121°C for >30 min).

5. Feed this first split.

6. In the other split, let the brood run out and a new queen be produced.

7. When the majority of the brood has emerged and the queen has started laying, remove all old combs.

8. Replace with residue free wax foundation.

9. Feed the second split.

   It is also possible to let bees build new combs from their own wax production rather than giving wax foundations.

Alternatively, the following can be done at the end of the bee season on whole colonies.

1. Trap the queen in a large cage made out of queen excluder allowing for the passage of workers.

2. Let the brood run out.

3. Scrape all propolis and wax, wash with soda and surface burn with a flame used hive parts to remove residues before introducing the colony; alternatively, use new hive parts.

4. Remove old combs.

5. Replace with residue free wax foundations.

6. Feed the colony.

Pros: decreases acaricide residues below detectable levels.

Cons: whether putative, but minimal residues remaining have a biological effect is unknown.

4.5.2. Artificial mite infestations

4.5.2.1. How many mites to introduce

The number of mites to introduce in colonies depends on the experiment performed. There are several factors to take into account:

- The statistical relevancy: a minimum number of successful infestations must be obtained (see the *BEEBOOK* statistics paper (Pirk *et al.*, 2013)).

- A higher number of mites introduced decreases the importance of resident residual/local mites.

- The infestation level depends on how long the colony should survive: the more mites are introduced, the quicker susceptible colonies might collapse.

- The method of introduction: introducing mites on the top of frames might result in high losses, but is easy. Alternatively, placing them on bees decreases this loss and allows a reduction in the number of mites used. Introducing mites in cells is a highly controlled method that requires few mites, but it is tedious.

- The rejection rate of mites by workers by grooming or hygienic behaviour.
- The sterility of some mites.
- The old age of mites of uncontrolled origin.
- The availability of mites.

In general, the number of mites to be introduced in experimental colonies should be overestimated to guarantee a sufficient sample size.

4.5.2.2. How to introduce varroa mites in colonies

There are two ways to obtain infested colonies: mites obtained from other colonies can be introduced or the existing mite population can be measured and the colony manipulated to obtained the desired infestation level. Bees can be taken out of a colony and the mite directly placed on its host. This can be done by pouring the collected mites on top of the workers in a cage or by picking mites up one by one with a paintbrush and placing directly on a worker. Time should be allowed for the mite to take refuge under the bees' abdominal plates before placing the latter back in its colony. This method is more efficient than dropping the mites onto the top of the frames since more mites can get attached on their host. Alternatively and if the level of infestation desired is not too different from the initial level of the colony, the latter can be split to obtain the desired level. If the level of infestation is above the desired level, brood combs (in which mites are trapped) can be removed.

4.5.2.3. How to introduce varroa mites in cells

An advantage of introducing mite directly in cells is to be able to monitor the events occurring in this particular cell. Cells can be manually infested or can be left to natural infestation if the infested cells can later be recognised. We here describe such artificial infestation methods.

4.5.2.3.1. Manual infestation

1. Using recently capped brood i.e. within 6 hours (see section 'Obtaining brood and adults of known age' of the *BEEBOOK* paper on miscellaneous methods (Human *et al.*, 2013)) make a small hole in the side of the capping.
2. Introduce the mite using a fine wetted paint brush.
3. Close and reseal the hole by pushing the capping down. Workers will seal the hole when the frame is reintroduced in the colony. Using melted wax to prevent the mite escaping is not recommended since it could damage the fragile larva.
4. Mark the location of the cell on a transparent sheet placed above the comb.

This method needs practice. From an initial 20% acceptance of artificially infested brood, one can rapidly reach 80%. This rate is however variable according to colony and experimenter. The success rate can be checked by removing frame after few hours and verifying

the status of the cell. Important: bees covering the combs used for artificial transfers must be carefully removed with a brush and not by shaking, which could damage pupae and mites. An opened and empty marked cell means that the workers removed the larva and the mite. Workers might also discard the old capping and reseal the cell without removing the larva. This can be recognised by a fresh capping deprived of cocoon layer. In this case the mite might have escaped or have been removed before resealing.

4.5.2.3.2. Natural infestation

Boot *et al.* (1992) designed a method that allows locating naturally infested cells. It is based on a one sided comb of which the cell walls where cut away from the bottom. The walls were then melted on a transparent sheet. These combs are consolidated by workers when replaced in the colonies and were accepted for oviposition by the queens. It might be necessary to cover the exposed side of the transparent sheet to prevent the bees building on it. Beetsma *et al.* (1994) also describe single rows of cells with two transparent walls that help locating and observing natural infestations.

4.5.3. Field bioassays of semiochemicals

Semiochemicals for which an effect on mite behaviour or physiology is proven in laboratory assays need to be tested under natural conditions in the hive. For example, semiochemicals involved in cell invasion and reproduction were tested with such method (Milani *et al.*, 2004).

4.5.3.1. Cell invasion

In the case of the compounds affecting cell invasion (either attractants or repellents), field testing involves treating brood cells with the chemical under study and evaluating the number of mites that entered the cell after it has been sealed.

1. Dissolve the compound to be tested in 1 µl of de-ionised water or other appropriate solvent.
   The dose used for the field bioassay is normally the most active in the laboratory bioassay. Beware that the solvent might dissolve the wax of the cell walls.
2. Select a highly infested colony.
3. Identify cells containing L5 larvae (see the section on obtaining brood and adults of known age in the *BEEBOOK* paper on miscellaneous methods (Human *et al.*, 2013)).
4. Apply the solution to these cells' walls with a 10 µl Hamilton syringe.
5. Treat an equal number of cells with 1 µl of solvent as a control.
6. Mark the position of the cells on a transparent sheet for subsequent tracking.
7. Open the sealed cells 18 h after treatment.
8. Inspect the cells for the presence of mites and count mites.

4.5.3.1.1. Data analysis

The proportion of treated and control cells that were infested are compared using the Mantel-Haenszel method after testing the homogeneity in the odds ratios of the replicated 2 × 2 tables. Any test that is suitable for comparing proportions could be used instead. However, if there are more replicates, using a certain number of cell each time, it is recommended to use a test that allows the analysis of strata. The number of mites in treated and control brood cells, in the hive bioassay, can be compared by a stratified sampled randomization test.

4.5.3.2. Mite reproduction

In the case of the compounds affecting mite reproduction, field testing involves treating brood cells with the chemical under study and evaluating both the fertility and fecundity of the mites reproducing in the cell.

1. Chose combs containing brood close to being capped.
2. Mark all the capped cells on a transparent sheet placed over the comb.
3. Replace the comb in the colony for two hours for workers to carry on capping cells.
4. Bring the combs to the laboratory after the two hours.
5. Dissolve the compound in an appropriate solvent.
   The dose used for the field bioassay is normally the most active in the laboratory bioassay.
6. Treat groups of freshly capped (unmarked) worker cells by injecting 1 µl of the solution with a 10 µl Hamilton syringe under the capping.
   Do not insert the syringe too deep into the cell to avoid hurting the larva. Beware that the solvent could dissolve the wax of the cell walls.
7. Treat an equal number of cells with 1 µl of the solvent as a control.
8. Choose groups of control and treated cells on both side of the comb, separated by at least one cell, which is left untouched to avoid contaminations.
9. Mark the position of the control and experimental cells on a transparent sheet placed over the comb.
10. Return the combs to the hive within 3 h.
11. Bring the comb to the laboratory 11 days later, when the bees are about to emerge.
12. Identify treated and control cells using the transparent sheet.
13. Count, uncap and inspect intact cells.
14. Note the condition of the infested pupae.
15. Collect mother mites and their offspring.
16. Mount on microscope slides and identify developmental stages as described in section 4.3.3. 'How to measure reproductive success'

In particular, the number of offspring and the number of mated daughters (i.e., the number of adult daughters in cells containing an adult male), are considered.

The effects of the solvent on the reproduction of *V. destructor* are studied by comparing the reproduction of mites in cells injected with 1 µl solvent and in sham-treated infested cells (syringe was introduced, but no solvent was injected). Proportions of reproducing mites out of the total mites found in cells are compared using G-tests (with the Williams' correction). The number of offspring and that of mated daughters per mother mite in treated and control groups can be compared using a two-sample randomisation test. The randomization distribution should be resampled an adapted number of times (e.g. 10^6 times).

4.5.4. Testing varroacides in the field

The European medicines agency has issued recommendations for the development of anti varroa treatment. These guidelines have been built on the knowledge accumulated by the Concerted Action 3686 (Commission of the European Communities European), which developed the commonly named 'alternative varroa control methods' based on the use of organic acids and essential oils. The aim of the guideline is to test and demonstrate the efficacy and safety of new miticides with the purpose of facilitating homologation. The original document (EMA/CVMP/EWP/459883/2008) should be consulted for legal issues and test for applicability of the treatment in various climatic regions. We here summarize and adapt the experimental design for research purposes at the local scale. Acaricides are considered efficient if the proportion of mites killed is at least 95% for synthetic substances and at least 90% for non-synthetic substances.

4.5.4.1. Preliminary tests

To facilitate and optimize efficacy test, it is recommended to perform dose finding and tolerance test on caged bees under controlled conditions in the laboratory. See section 3.6.3. 'Bioassays to quantify the susceptibility of the varroa mite to acaricides' and the *BEEBOOK* paper on toxicology methods (Medrzykci *et al.*, 2013). The highest concentration/quantity tolerated by the honey bees can be used as an indication for concentrations or quantities that can be used in subsequent dose-titration as well as dose-confirmation or field studies. Dose-titration studies should aim at identifying the minimum effective and maximum tolerated levels of active substance reaching bees and parasites and thus help establishing the dosage and dosing interval of the product. Implementation of dose-finding studies, carried out under controlled laboratory conditions is preferred, e.g. using 10 workers per cage, 3 cages per concentration, 3 untreated controls and one replicate, i.e. the studies should be carried out twice. See the *BEEBOOK* paper on toxicological methods (Medrzycki *et al.*, 2013).

Small scale outdoor pilot studies to confirm dose, efficacy and tolerance should be considered before large scale field studies are performed. It is thus possible to validate the results obtained in the laboratory in a situation closer to that of the field, but with a high reproducibility since variables can be better controlled in these small units compared to full size colonies. It also allows for troubleshooting before the investment in the full scale test is done. A minimum of five untreated control and five treated test colonies should be used. To ensure reproducibility in the pilot studies, colonies should be comparable with respect to environment, type and size of hive, level of varroa infestation, treatment history, age of queen, relatedness of queens (sister queens can be used to decrease variability between replicates, in contrast, unrelated queens can be used to consider a wider range of genotypes), presence of brood, and age distribution of worker bees.

4.5.4.2. Efficacy tests

4.5.4.2.1. Statistical analysis

Primary and secondary outputs, hypotheses, and statistical methods should be specified and justified in a protocol before beginning the experiments. Sample sizes, in terms of hives per area for climatologically different regions (when relevant), should be large enough to provide sufficient statistical power. Whenever possible, results of the analyses should be accompanied by confidence intervals. Refer to the *BEEBOOK* paper on statistics (Pirk *et al.*, 2013) and to the CVMP guideline on Statistical Principles for Veterinary Clinical Trials (EMEA/CVMP/816/00).

4.5.4.2.2. Hives

1. Type of hives should be homogenized.
2. Ant protected varroa boards should be installed under the hives for mite counting (see section 4.2.4. 'Natural mite fall').
3. Temperature and relative humidity inside the hive(s) as well as exposure to solar radiation can be recorded, if they can influence the performance of the product.

4.5.4.2.3. Colonies

The following parameters should be taken into account:

1. Do not include weak colonies or colonies affected by diseases other than varroa in the study.
2. Equalize or randomize bee breed depending on the aim of the test regarding genetic diversity.
3. Select sister queens or unrelated queens of same age.
4. Measure and equalize colony strength (see the *BEEBOOK* paper on estimating colony strength (Delaplane *et al.*, 2013)).
5. Measure and equalize the amount of brood (see the *BEEBOOK* paper on estimating colony strength (Delaplane *et al.*, 2013)). Presence and type of brood is determined by the mode of action of the product. The tests should thus be performed in the absence of sealed brood, unless the product is intended to be effective on mites in capped cells.

6. Initial varroa infestation level should be high enough (> 300 mites per colony) to be able to measure mite drop. It should however be below damage thresholds (e.g. for central Europe: < 3,000 mites per colony, see also Table 5 and section 4.4. 'Estimating damage thresholds') and comparable between hives included in the study.
7. Treatment history should be similar for all colonies to equalize the effect of past treatments on the results (e.g. type and amount of acaricide residues present in the wax); when possible use residue free wax.

4.5.4.2.4. Location

1. Apiaries tested should be sufficiently distant from neighbouring apiaries to avoid disturbance and reduce risk for re-infestation.
   Type and availability of food sources should be recorded.
2. As a general principle, if studies are carried out at different apiaries, habitats should be comparable (access to similar forage and exposure to similar climatic conditions). If these conditions are not met the sample size should be adapted (number of apiaries increased) to take these differences into account.
3. Depending on the mode of dispersion of the product, control and test apiaries should be distant enough to prevent the contamination of control groups by the tested product by drifting foragers and drones or robbers.
   Synthetic acaricides have been shown to contaminate control hives placed in the same apiary (Allsopp, 2006)

4.5.4.2.5. Treatment

The following parameters should be defined, as determined in the preliminary tests:

1. Treatment period.
   Treatment should preferably be carried out at outdoor temperatures > 5°C.
2. Number of treatments, if more than one treatment is carried out.
3. Treatment intervals, if more than one treatment is carried out.
4. Include an untreated control group in the study to establish the effect of handling and of natural variations on the level of infestation and thus to confirm that a decrease in mite population size observed is indeed due the product under investigation.

4.5.4.2.6. Observations and parameters

Studies should encompass a pre-treatment, a treatment and a post-treatment period. Monitoring begins with the pre-treatment 14 days before the first treatment is carried out. The post-treatment period should extend > 14 days after the last treatment. These periods encompass the pupal development time and allow taking into account the mites that are enclosed in the cells. The post-treatment period might need being prolonged, depending on the mode of action and persistence of the product tested.

4.5.4.2.6.1. Assessment of efficacy

1. Count dead mites on the bottom boards at regular intervals before, during and after treatment.

   The primary variable is mite mortality. During the treatment period dead mite counts should be carried out every 1-2 days given the high mortality expected. Pre- and post-treatment counts should be made 1-2 times per week depending on amount of mites falling, see section 4.2.4. 'Natural mite fall' and 4.2.5. 'Sub-sampling mite fall'. This allows the verification of efficacy since mite drop should peak during the treatment period.

2. Determine the amount of mites surviving the treatment with the product under investigation using a follow-up treatment with a chemically unrelated substance with > 95% documented efficacy.

   Follow-up treatment should be carried out in tested and control groups at the same time. This follow-up treatment should take place shortly after treatment with the test product, in order to keep the re-infestation level (and therefore the biasing of results) low when test apiaries and groups are not isolated by enough distance from neighbouring apiaries or hives. However, it is necessary to wait until mite drop returned to pre-treatment level in order to measure the full effect of the treatment and dissipation of the delayed mite mortality. This period is at least 14 days if the product kills mites in the cells or not. It is only after adult bee emergence that these mites will be released and fall on the bottom board or that they will get into contact with the product if the latter did not penetrate into the cell.

3. Count dead mites every 1-2 days in the week after follow-up treatment and 1-2 per week until mite drop returns to pre-treatment values.

4. Calculate treatment efficacy as follows:

   % mite reduction = (number of mites in test group killed by treatment x 100) / (number of mites in test group killed by treatment + number of mites killed in test group by follow-up treatment)

   Do not use data from colonies with abnormally high bee mortality in the efficacy evaluation.

5. Compare mite fall after treatment with untreated control to verify that the fall measured was not a natural phenomenon.

4.5.4.2.6.2. Assessment of safety of product for honey bees

1. Record bee mortality inside and adjacent to the hive daily or at least three times a week throughout the three stages of the experiment.

The use of dead-bee traps is recommended (see section 2. 'Estimating the number of dead bees expelled from a colony' of the *BEEBOOK* paper on miscellaneous methods (Human *et al.*, 2013)).

2. Monitor the morbidity, mortality, as well as the size and development of surviving colonies at the time of the first flight in spring and thereafter (see the *BEEBOOK* paper on estimating colony strength (Delaplane *et al.*, 2013)) if applicable (envisaged therapeutic use in autumn or winter).

3. Measure flight activity of bees during the trial (see the *BEEBOOK* paper on behavioural methods.

   This verifies whether the product influences foraging activity of the treated colonies.

4. Measure honey production.

   This verifies whether the product influences the productivity of the treated colonies

5. Quantify brood area of test colonies during the three phases and compare to the control group (see the *BEEBOOK* paper on estimating colony strength (Delaplane *et al.*, 2013)).

In cases in which the product is intended for use in colonies with brood, the demonstration of safety for all stages of brood should be carried out (see the *BEEBOOK* article on toxicology methods by Medrzycki *et al.* (2013)). An additional method to determine effect of the product tested on brood is to determine which of feeding incompetence of worker bees and direct adverse effects on eggs and larvae occurred. For this:

1. Leave frames with eggs and larvae to develop in the hive until a chosen stage of the larval stage after applying therapeutic doses of the test product.

2. Monitor feeding behaviour of these larvae by measuring the amount of food found in their cells.

   By comparing development of brood and amount of larval food and taking into account the ratio between quantity of brood and number of worker bees between control and test groups, it is possible to differentiate between effects of feeding incompetence of worker bees and direct adverse effects on eggs and larvae after application of the product.

3. Verify the presence of the live queen at the end of the experiment.

   A significant difference in queen survival between test and control groups indicates an effect of the treatment.

4.5.4.3. Resistance pattern

The possibility of resistance emerging after several treatments should be monitored. The product applications should cover several reproductive cycles of the parasite to show the development of resistance and the rate of such development. Such studies can be performed under laboratory (see section 3.6.3. 'Bioassays to quantify the susceptibility of the varroa mite to acaricides') and/or field conditions. Not only mites, but also bees might develop resistance against miticides after regular use for several bee generations. This translates in a change in dose-lethality relationship of the product or active substance(s) and therefore affects the safety of the product for bees which increases. See the *BEEBOOK* paper on toxicology methods (Medrzycki *et al.*, 2013) to evaluate acaricide toxicity for honey bees.

4.6. Breeding mites in colonies

A common problem for varroa research is obtaining mites in sufficient quantities for experiments. It is desirable to obtain mites already early in the season when their numbers in the colonies is still low and in large quantities for as long as possible thereafter. The method described here allows, within a short time, the regular harvesting of a high number of phoretic mites early in and throughout the season. The method is based on the trapping comb originally designed to control the mite (Fries and Hansen, 1993; Maul *et al.*, 1998). It consists in caging the queen from an infested colony, and letting all the brood emerge. Once the colony is broodless and all mites are in the phoretic stage, a comb of open brood is introduced. Just before capping, most of the phoretic mites looking for reproduction opportunities will enter the cells provided. Once the brood is capped, the comb is removed and placed in an incubator until the bees emerge. The newly emerged non-flying and non-stinging bees will be highly infested with varroa, making mite harvesting easy and fast. The infested comb can also be retrieved at any time to obtain mites at a particular developmental stage. This method is further developed to optimise logistical aspects according to the following protocol:

1. Prepare several hives as breeder colonies during the season preceding the experiments.
2. Adjust varroa treatment during the season preceding the experiments to ensure the survival of the colonies, but to also allow the survival of a relatively high number of mites over the winter.
   This makes it possible to keep a starter mite population for a fast growth in parasite numbers following the winter.
3. In the next year, when the colonies are well developed, the weekly natural mite fall is counted over a brood cycle (3 weeks).
4. Rank the colonies according to their mite load and strength. The most infested hives should be used first since they are susceptible to collapse before less infested colonies. The parasite population can still be left to grow in the less infested colonies until they are used for mite collection. Among several hives with the same range of infestation, those closer to swarming stage can be used first. This makes it possible to prevent swarming and the loss of mites.
   The breeding cycle can start:
5. Day 1: cage the queen from the colonies selected for mite rearing.

At day 22, all the brood present at day 1 will have emerged.
6. Day 12: prepare the trapping comb:
   6.1. Select a strong colony (brood provider) with an actively laying queen.
   6.2. Cage the queen on an empty dark comb (that queens prefer for egg-laying) and placed in the brood nest of her colony.
7. Day 13: after 24 h, remove the queen from the cage, but leave the comb in the cage to prevent further egg-laying by the queen.
   This comb contains brood of similar age in which varroa mites will later be trapped. To increase chances of obtaining enough brood for trapping, queens of several colonies can be caged and the comb with the most brood is used.
8. Day 19:
   8.1. Transfer the trapping comb that now contains 7 days old brood to the varroa rearing colony, of which the brood has emerged.
   8.2. Release the queen of the rearing colony so that she can resume her egg-laying activity.
9. Day 22: the brood cells of the trapping comb have been capped.
   9.1. Remove the comb from the rearing colony.
   9.2. Transport to the laboratory.
   9.3. Place in a well-ventilated bee tight box.
   9.4. Keep in an incubator at 34.5°C with 60-70% relative humidity until adult worker emergence.
   The comb should contain sufficient pollen and honey supplies so that the emerging bees can feed. Is it not the case, food should be supplied.
   Work on the rearing apiary should end with the collection of the trapping comb so that it does not remain for too long outside the colony before being placed in the incubator. To avoid damage to the brood transport should be done in a thermoregulated and moist container.
10. Day 33: Start collecting mites from the infested workers emerging this and the following day.
    For mite collection, bees can be held with forceps and the mites caught with a size 00 paintbrush or a mouth aspirator.

During the rearing cycle, the colony experiences 2 to 3 weeks without brood. After two subsequent brood cycles, the colony has usually regained strength and the varroa population will have increased again. Given that the varroa natural fall indicates a sufficiently large varroa population, the same colony can be used again to harvest mites. Furthermore, depending on the amount of mites needed, several colonies can be used at a time to increase the harvest. Breeding cycles on new colonies can be started every week. Thus, after 5 weeks, batches of mites can be harvested weekly. The

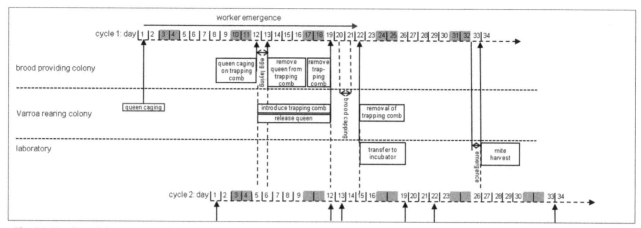

Fig. 21. Timeline of the rearing cycle. An additional cycle is depicted below the main timeline to illustrate how the various tasks (symbolized by arrows) can be combined between different rearing cycles to optimize the process.

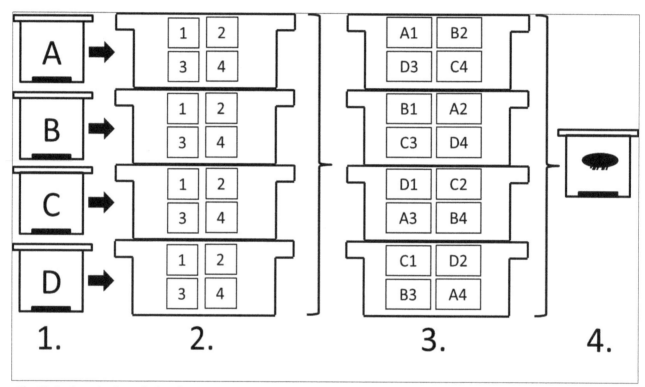

Fig. 22. A schematic of a method used to appraise varroa attraction to brood from different queen lines (modified from Ellis and Delaplane, 2001).

additional time axis shown at the bottom of Fig. 21 illustrates how most of the working days can be combined for colonies or groups of colonies at different stages in the cycle. By starting on a Thursday for example, no work on a Wednesday, Friday or weekend day is necessary and mite collection always occurs on a Tuesday.

Pros: the method allows the collection of mites indoors rather than on the apiary, prevents the danger of robbing by neighbouring colonies since the colonies do not remain open for mite sampling, necessitates few visits to the breeding apiary, allows the collection of mites on a particular day, facilitates sampling as the density of mite per emerging bee is high.

Cons: logistic intensive; if mites are collected from emerged workers older mother and young daughter mites are not separated.

4.7. Brood attractiveness

Varroa mites prefer drone over worker brood (Fuchs, 1990; Rosenkranz, 1993; Boot *et al.*, 1995). Honey bee lineages also vary in the attractiveness of their brood for the mites (Büchler, 1990; De Guzman *et al.*, 1995; Guzman-Novoa *et al.*, 1996). We here present a bioassay destined to compare brood attractiveness in comparable condition, i.e. in the same colony. This method is adapted from those of Ellis and Delaplane (2001) and Aumeier *et al.* (2002).

4.7.1. Procedure to test brood attractiveness

1. Queens from four different lines (Fig. 22, Step 1) tested are placed individually on a drawn, empty comb contained in a queen excluder cage.

2. Allow queens to oviposit for 24 h.

3. Twenty four hours later, the queens are removed from the combs, but the combs are left in the cages.
   This limits further queen oviposition in the target combs.

4. Leave each comb with eggs is in its respective line colony for 6 or 7 days (for worker / drone brood respectively).

5. After this, cut out of the comb squares or circles of comb containing L4 larvae (see the section on obtaining brood and adults of known age in the *BEEBOOK* paper on miscellaneous methods (Human *et al.*, 2013)) using a sharp knife or a metal spatula (Fig. 22, Step 2).

6. Take one section of larvae from each of the lines to be tested and randomly combine the sections of different lines in the centre of a fully drawn comb in which squares or circles of fitting sizes have been removed.
   Each frame now contains one brood section from each of the four genetic line colonies (Fig. 22, Step 3).

7. Prepare a varroa receiver colony (Fig. 22, Step 4) by removing the open brood in the colony to decrease target brood (the frame just created having the various queen lines represented on one comb) competition with the colony's own brood for the varroa present in the colony.
   Alternatively, the queen in the receiver colony could be caged 22 or 25 days earlier to let all worker and drone brood, respectively, emerge.

8. Introduce each reconstituted comb holding eggs from each genetic line into a receiver varroa infested colony (Fig. 22, Step 4).
   The colony can receive up to four frames generated from the procedure if the varroa populations are high enough. Mites invade worker cells from 15–50 h preceding cell capping (depending on the sex of the larvae).

9. Fifty hours post cell capping, remove the sections of capped pupae from the colonies.

10. If necessary chill in a freezer.

11. Assess varroa infestation rates (see section 4.2.3. 'Measuring the infestation rate of brood and adult bees').

Pros: brood attractiveness is assessed in comparable condition, i.e. in the same mite provider colony.

5. Acknowledgements

We are grateful to Diana Sammataro and Marla Spivak for their comments on an earlier version of this paper. We also acknowledge Anton Imdorf for letting us benefit from his experience on methods for varroa research and R. Nannelli for providing pictures of the immature stages of the mite.

6. References

ADELT, B; KIMMICH, K H (1986) Die Wirkung der Ameisensäure in die verdeckelte Brut. *Allgemeine Deutsche Imkerzeitung* 20: 382-385.

AKRATANAKUL, P; BURGETT, M (1975) *Varroa jacobsoni*: a prospective pest of honey bees in many parts of the world. *Bee World* 56: 119-121.

ALLSOPP, M (2006) Analysis of *Varroa destructor* infestation of Southern African honey bee populations. Dissertation, University of Pretoria. 285 pp.

ANDERSON, D L (2000) Variation in the parasitic bee mite *Varroa jacobsoni* Oud. *Apidologie* 31: 281-292.
http://dx.doi.org/10.1051/apido:2000122

ANDERSON, D L; FUCHS, S (1998) Two genetically distinct populations of *Varroa jacobsoni* with contrasting reproductive abilities on *Apis mellifera*. *Journal of Apicultural Research* 37: 69-78.

ANDERSON, D L; TRUEMAN, J W H (2000) *Varroa jacobsoni* (Acari: Varroidae) is more than one species. *Experimental and Applied Acarology* 24: 165-189. http://dx.doi.org/10.1023/A:1006456720416

AUMEIER, P; ROSENKRANZ, P; FRANCKE, W (2002) Cuticular volatiles, attractivity of worker larvae and invasion of brood cells by varroa mites: a comparison of Africanized and European honey bees. *Chemoecology* 12: 65–75.
http://dx.doi.org/10.1007/s00049-002-8328-y

AVISE, J C (2004) *Molecular markers, natural history and evolution (Second Edition)*. Sanauer; Sunderland, MA, USA. 684 pp.

BAKER, T C; CARDÉ, R T (1984) Techniques for behavioural bioassays. In *H Hummel; T A Miller (Eds). Techniques in Pheromone Research*. Springer-Verlag; New York, USA. pp. 45-73.

BEETSMA, J; ZONNEVELD, K (1992) Observation on the initiation and stimulation of oviposition by the varroa mite. *Experimental and Applied Acarology* 16: 303- 312.

BEETSMA, J; BOOT, W J; CALIS, J N M (1994) Special techniques to make detailed observations of varroa mites staying on bees and invading honey bee brood cells. In *L J Connor et al. (Eds). Asian apiculture*. Wicwas Press; Cheshire, USA. pp. 510-520.

BOGDANOV, S; KILCHENMAN, V; IMDORF, A (1998) Acaricide residues in some bee products. *Journal of Apicultural Research* 37: 57-67.

BOOT, W J; CALIS, J N M; BEETSMA, J (1992) Differential periods of varroa mite invasion into worker and drone cells of honey bees. *Experimental and Applied Acarology* 16: 295-301.

BOOT, W J; SCHOENMAKER, J; CALIS, J N M; BEETSMA, J (1995) Invasion of *Varroa jacobsoni* into drone brood cells of the honey bee. *Apidologie* 26: 109–118. http://dx.doi.org/10.1051/apido:19950204

BRANCO, M R; KIDD, N A C; PICKARD, R S (2006) A comparative evaluation of sampling methods for *Varroa destructor* (Acari: Varroidae) population estimation. *Apidologie* 37: 452-461. http://dx.doi.org/10.1051/apido:2006010

BRODSCHNEIDER, R; MOOSBECKOFER, R; CRAILSHEIM, K (2010) Surveys as a tool to record winter losses of honey bee colonies: a two year case study in Austria and South Tyrol. *Journal of Apicultural Research* 49(1): 23-30. http://dx.doi.org/10.3896/IBRA.1.49.1.04

BRUCE, W A; CHIESA, F; MARCHETTI, S; GRIFFITHS, D A (1988) Laboratory feeding of *Varroa jacobsoni* Oudemans on natural and artificial diets (Acari: Varroidae). *Apidologie* 19: 209-218.

BRUCE, W A; HENEGAR, R B; HACKETT, K J (1991) An artificial membrane for *in vitro* feeding of *Varroa jacobsoni* and *Acarapis woodi*, mite parasites of honey bees. *Apidologie* 22: 503-507.

BÜCHLER, R (1990) Possibilities for selecting increased varroa tolerance in central European honey bees of different origins. *Apidologie* 21:365–367.

CALDERONE, N W; LIN, S (2001) Arrestment activity of extracts of honey bee worker and drone larvae, cocoons and brood food on female *Varroa destructor. Physiological Entomology* 26: 341–350.

CALIS, J N M; BOOT, W J; BEETSMA, J; VAN DEN EIJNDE, J H P M; DE RUIJTER, A; VAN DER STEEN, J J M (1998) Control of varroa by combining trapping in honey bee worker brood with formic acid treatment of the capped brood outside the colony: putting knowledge on brood cell invasion into practice. *Journal of Apicultural Research* 37: 205-215.

CAMPBELL, E M; BUDGE, G E; BOWMAN, A S (2010) Gene-knockdown in the honey bee mite *Varroa destructor* by a non-invasive approach: studies on a glutathione S-transferase. *Parasites and Vectors* 3: 73. http://dx.doi.org/10.1186/1756-3305-3-73

CHAUZAT, M-P; CARPENTIER, P; MADEC, F; BOUGEARD, S; COUGOULE, N; DRAJUNEL, P; CLEMENT, M-C; AUBERT, M; FAUCON, J-P (2010) The role of infectious agents and parasites in the health of honey bee colonies in France. *Journal of Apicultural Research* 49(1): 31-39. http://dx.doi.org/10.3896/IBRA.1.49.1.05

CHIESA, F; MILANI, N; D'AGARO, M (1989) Observations of the reproductive behaviour of *Varroa jacobsoni* Oud.: techniques and preliminary results. In *R Cavalloro (Ed.). Proceedings of the Meeting of the EC-Experts' Group, Udine 1988.* CEC; Luxembourg. pp. 213–222.

COMMISSION OF THE EUROPEAN COMMUNITIES [2002] Concerted Action 3686: "Coordination in Europe of research on integrated control of varroa mites in honey bee colonies". Technical guidelines for the evaluation of treatments for control of varroa mites in honey bee colonies, Recommendations from the CA3686. Document prepared during discussions within the CA3686 working group: "Evaluation of treatment for control of varroa mites in honey bee colonies".

CORNMAN, R S; SCHATZ, M C; JOHNSTON, J S; CHEN, Y-P; PETTIS, J; HUNT, G; BOURGEOIS, L; ELSIK, C; ANDERSON, D; CROZINGER, C M; EVANS, J D (2010) Genomic survey of the ectoparasitic mite *Varroa destructor*, a major pest of the honey bee Apis mellifera. *BMC Genomics* 11: 602. http://dx.doi.org/10.1186/1471-2164-11-602

CURRIE, R W; GATIEN, P (2006) Timing acaricide treatments to prevent *Varroa destructor* (Acari: Varroidae) from causing economic damage to honey bee colonies. *Canadian Entomologist* 138: 238-252. http://dx.doi.org/10.4039/n05-024

CVMP [2001] *CVMP guidelines on 'Statistical principles for veterinary clinical trials'* (EMEA/CVMP/816/00).

CZEKOŃSKA, K (2009) The effect of different concentrations of carbon dioxide (CO_2) in a mixture with air or nitrogen upon the survival of the honey bee (*Apis mellifera*). *Journal of Apicultural Research* 48: 67-71. http://dx.doi.org/10.3896/IBRA.1.48.1.13

DAINAT, B; KUHN, R; CHERIX, D; NEUMANN, P (2011) A scientific note on the ant pitfall for quantitative diagnosis of *Varroa destructor. Apidologie* 42: 740-742. http://dx.doi.org/10.1007/s13592-011-0071-3

DAHLE, B (2010) The role of *Varroa destructor* for honey bee colony losses in Norway. *Journal of Apicultural Research* 49(1): 124-125. http://dx.doi.org/10.3896/IBRA.1.49.1.26

DE GRAAF, D C; ALIPPI, A M; ANTÚNEZ, K; ARONSTEIN, K A; BUDGE, G; DE KOKER, D; DE SMET, L; DINGMAN, D W; EVANS, J D; FOSTER, L J; FÜNFHAUS, A; GARCIA-GONZALEZ, E; GREGORC, A; HUMAN, H; MURRAY, K D; NGUYEN, B K; POPPINGA, L; SPIVAK, M; VANENGELSDORP, D; WILKINS, S; GENERSCH, E (2013) Standard methods for American foulbrood research. In *V Dietemann; J D Ellis; P Neumann (Eds) The COLOSS BEEBOOK, Volume II: standard methods for* Apis mellifera *pest and pathogen research. Journal of Apicultural Research* 52(1): http://dx.doi.org/10.3896/IBRA.1.52.1.11

DE GUZMAN, L I; RINDERER, T E; LANCASTER, V A (1995) A short test evaluating larval attractiveness of honey bees to *Varroa jacobsoni*. *Journal of Apicultural Research* 34: 89–92.

DE GUZMAN, L I; DELFINADO-BAKER, M (1996) A new species of *Varroa* (Acari: Varroidae) associated with *Apis koschevnikovi* (Apidae: Hymenoptera) in Borneo. *International Journal of Acarology* 22: 23-27. http://dx.doi.org/10.1080/01647959608684077

DE GUZMAN, L I; RINDERER, T E; STELZER, J A (1997) DNA evidence of the origin of *Varroa jacobsoni* Oudemans in the Americas. *Biochemical Genetics* 35: 327-335. http://dx.doi.org/10.1023/A:1021821821728

DE GUZMAN, L I; RINDERER, T E; STELZER, J A (1999) Occurrence of two genotypes of *Varroa jacobsoni* Oud. in North America. *Apidologie* 30: 31-36. http://dx.doi.org/10.1051/apido:19990104

DE GUZMAN, L; RINDERER, T E; STELZER, J A; ANDERSON, D (1998) Congruence of RAPD and mitochondrial DNA markers is assessing *Varroa jacobsoni* genotypes. *Journal of Apicultural Research* 37: 49-51.

DE JONG, D; DE ANDREA, R D; GONCALVES, L S (1982) A comparative analysis of shaking solutions for the detection of *Varroa jacobsoni* on adult honey bees. *Apidologie* 13: 297-304.

DE JONG, D D; MORSE, R A; EICKWORT, G C (1982) Mite pests of honey bees. *Annual Review of Entomology* 27: 229-252.

DE MIRANDA, J R; BAILEY, L; BALL, B V; BLANCHARD, P; BUDGE, G; CHEJANOVSKY, N; CHEN, Y-P; GAUTHIER, L; GENERSCH, E; DE GRAAF, D; RIBIÈRE, M; RYABOV, E; DE SMET, L; VAN DER STEEN, J J M (2013) Standard methods for virus research in *Apis mellifera*. In *V Dietemann; J D Ellis; P Neumann (Eds) The COLOSS* BEEBOOK, *Volume II: standard methods for* Apis mellifera *pest and pathogen research. Journal of Apicultural Research* 52(4): http://dx.doi.org/10.3896/IBRA.1.52.4.22

DEL FABBRO, S; NAZZI, F (2008) Repellent effect of sweet basil compounds on *Ixodes ricinus* ticks. *Experimental and Applied Acarology* 45: 219-228.

DEL PICCOLO, F; NAZZI, F; DELLA VEDOVA, G; MILANI, N (2010) Selection of *Apis mellifera* workers by the parasitic mite *Varroa destructor* using host cuticular hydrocarbons. *Parasitology* 137: 967-973.

DELAPLANE, K S (2011) Integrated pest management in varroa. In *N L Carreck (Ed.)*. *Varroa - still a problem in the 21st Century?* International Bee Research Association; Cardiff, UK, pp. 43-51.

DELAPLANE, K S; HARBO, J R (1987) Effect of queenlessness on worker survival, honey gain and defence behaviour in honey bees. *Journal of Apicultural Research* 26: 37-42.

DELAPLANE, K S; HOOD, W M (1997) Effects of delayed acaricide treatment in honey bee colonies parasitized by *Varroa jacobsoni* and a late season treatment threshold for the south-eastern United States. *Journal of Apicultural Research* 36: 125-132.

DELAPLANE, K S; HOOD, W M (1999) Economic threshold for *Varroa jacobsoni* Oud. in the south-eastern USA. *Apidologie* 30: 383-395. http://dx.doi.org/10.1051/apido:19990504

DELAPLANE, K S; ELLIS, J D; HOOD, W M (2010) A test for interactions between *Varroa destructor* (Acari: Varroidae) and *Aethina tumida* (Coleoptera: Nitidulidae) in colonies of honey bees (Hymenoptera: Apidae). *Annals of the Entomological Society of America* 103: 711-715. http://dx.doi.org/10.1603/AN09169

DELAPLANE, K S; VAN DER STEEN, J; GUZMAN, E (2013) Standard methods for estimating strength parameters of *Apis mellifera* colonies. In *V Dietemann; J D Ellis; P Neumann (Eds) The COLOSS* BEEBOOK, *Volume I: standard methods for* Apis mellifera *research. Journal of Apicultural Research* 52(1): http://dx.doi.org/10.3896/IBRA.1.52.1.03

DELFINADO-BAKER, M; AGGARWAL, K (1987) A new *Varroa* (Acari: Varroidae) from the nest of *Apis cerana* (Apidae). *International Journal of Acarology* 13: 233-237.

DELFINADO-BAKER, M D; BAKER, E W (1974) Varroidae, a new family of mites on honey bees (Mesostigmata: Acarina). *Journal of the Washington Academy of Sciences* 64: 4-10.

DIETEMANN, V; PFLUGFELDER, J; ANDERSON, D; CHARRIÈRE, JD; CHEJANOVSKY, N; DAINAT, B; DE MIRANDA, J; DELAPLANE, K; DILLIER, F-X; FUCHS, S; GALLMANN, P; GAUTHIER, L; IMDORF, A; KOENIGER, N; KRALJ, J; MEIKLE, W; PETTIS, J; ROSENKRANZ, P; SAMMATARO, D; SMITH, D; YAÑEZ, O; NEUMANN, P (2012) *Varroa destructor*: research avenues towards sustainable control. *Journal of Apicultural Research* 51: 125-132. http://dx.doi.org/10.3896/IBRA.1.51.1.15

DONZÉ, G; GUÉRIN, P M (1994) Behavioural attributes and parental care of varroa mites parasitizing honey bee brood. *Behavioural Ecology and Sociobiology* 34: 305-319.

DONZÉ, G; GUÉRIN, P M (1997) Time-activity budgets and space structuring by the different life stages of *Varroa jacobsoni* in capped brood of the honey bee, *Apis mellifera*. *Journal of Insect Behaviour* 10: 371-393.

DONZÉ, G; SCHNEYDER-CANDRIAN, S; BOGDANOV, S; DIEHL, P A; GUÉRIN, P M; KILCHENMANN, V; MONACHON, F (1998) Aliphatic alcohols and aldehydes of the honey bee cocoon induce arrestment behaviour in *Varroa jacobsoni* (Acari: Mesostigmata), an ectoparasite of *Apis mellifera*. *Archives of Insect Biochemistry and Physiology* 37: 129–145.

ELLIS, J D; DELAPLANE, K S (2001) A scientific note on *Apis mellifera* brood attractiveness to *Varroa destructor* as affected by the chemotherapeutic history of the brood. *Apidologie* 32: 603–604. http://dx.doi.org/10.1051/apido:2001105

ELLIS, M D; NELSON, R; SIMMONDS, C (1988) A comparison of the fluvalinate and ether roll methods of sampling for varroa mites in honey bee colonies. *American Bee Journal* April: 262-263.

EVANS, J D (2000) Microsatellite loci in the honey bee parasitic mite *Varroa jacobsoni. Molecular Ecology* 9: 1436-1438. http://dx.doi.org/10.1046/j.1365-294x.2000.00998-3.x

EVANS, J D; CHEN, Y P; CORNMAN, R S; DE LA RUA, P; FORET, S; FOSTER, L; GENERSCH, E; GISDER, S; JAROSCH, A; KUCHARSKI, R; LOPEZ, D; LUN, C M; MORITZ, R F A; MALESZKA, R; MUÑOZ, I; PINTO, M A; SCHWARZ, R S (2013) Standard methods for molecular research in *Apis mellifera*. In *V Dietemann; J D Ellis; P Neumann (Eds) The COLOSS BEEBOOK, Volume I: standard methods for* Apis mellifera *research. Journal of Apicultural Research* 52(4): http://dx.doi.org/10.3896/IBRA.1.52.4.11

FERNANDEZ, N: COINEAU, Y (2007) *Varroa, the serial bee killer mite.* Atlantica; Biarritz, France. 264 pp.

FINNEY, D J (1949) The estimation of the parameters of tolerance distributions. *Biometrika* 36: 139-256.

FINNEY, D J (1971) *Probit Analysis (3rd Ed.)*. Cambridge University Press; Cambridge, USA. 333 pp.

FORSGREN, E; BUDGE, G E; CHARRIÈRE, J-D; HORNITZKY, M A Z (2013) Standard methods for European foulbrood research. In *V Dietemann; J D Ellis, P Neumann (Eds) The COLOSS BEEBOOK: Volume II: Standard methods for* Apis mellifera *pest and pathogen research. Journal of Apicultural Research* 52(1): http://dx.doi.org/10.3896/IBRA.1.52.1.12

FREY, E; SCHNELL, H; ROSENKRANZ, P (2011) Invasion of *Varroa destructor* mites into mite-free honey bee colonies under the controlled conditions of a military training area. *Journal of Apicultural Research* 50(2): 138-144. http://dx.doi.org/10.3896/IBRA.1.50.2.05

FRIES, I (1991) Treatment of sealed honey bee brood with formic acid for control of *Varroa jacobsoni. American Bee Journal* 131: 313-314.

FRIES, I; HANSEN, H (1993) Biotechnical control of varroa mites in cold climates. *American Bee Journal* 133: 435–438.

FRIES, I; CAMAZINE, S; SNEYD, J (1994) Population dynamics of *Varroa jacobsoni*: a model and a review. *Bee World* 75: 5-28.

FRIES, I; AARHUS, A; HANSEN, H; KORPELA, S (1991a) Comparisons of diagnostic methods for detection of *Varroa jacobsoni* in honey bee (*Apis mellifera*) colonies at low infestation levels. *Experimental and Applied Acarology* 10: 279-287.

FRIES, I; AARHUS, A; HANSEN, H; KORPELA, S (1991b) Development of early infestations of *Varroa jacobsoni* in honey bee (*Apis mellifera*) colonies in cold climates. *Experimental and Applied Acarology* 11: 205-214.

FUCHS, S (1990) Preference for drone brood cells by *Varroa jacobsoni* Oud. in colonies of *Apis mellifera carnica. Apidologie* 21: 93–99.

GARRIDO, C; ROSENKRANZ, P (2003) The reproductive program of female *Varroa destructor* mites is triggered by its host, *Apis mellifera. Experimental and Applied Acarology* 31: 269–273 http://dx.doi.org/10.1023/B:APPA.0000010386.10686.9f

GARRIDO, C; ROSENKRANZ, P (2004) Volatiles of the honey bee larva initiate oogenesis in the parasitic mite *Varroa destructor. Chemoecology* 14: 193–197. http://dx.doi.org/10.1007/s00049-004-0278-0

GARRIDO, C; ROSENKRANZ, P; STÜRMER, M; RÜBSAM, R; BÜNING, J (2000) Toluidine blue staining as a rapid measure for initiation of oocyte growth and fertility in *Varroa jacobsoni* Oud. *Apidologie* 31: 559–566. http://dx.doi.org/10.1051/apido:2000146

GATIEN, P; CURRIE, R W (2003) Timing of acaricide treatments for control of low-level populations of *Varroa destructor* (Acari: Varroidae) and implications for colony performance of honey bees. *The Canadian Entomologist* 135: 749-763. http://dx.doi.org/10.4039/n02-086

GENERSCH, E; GISDER, S; HEDTKE, K; HUNTER, W B; MÖCKEL, N; MÜLLER, U (2013) Standard methods for cell cultures in *Apis mellifera* research. In *V Dietemann; J D Ellis; P Neumann (Eds) The COLOSS BEEBOOK, Volume I: standard methods for* Apis mellifera *research. Journal of Apicultural Research* 52(1): http://dx.doi.org/10.3896/IBRA.1.52.1.02

GENERSCH, E; VON DER OHE, V; KAATZ, H-H; SCHROEDER, A; OTTEN, C; BÜCHLER, R; BERG, S; RITTER, W; MÜHLEN, W; GISDER, S; MEIXNER, M; LIEBIG, G; ROSENKRANZ, P (2010) The German bee monitoring project: a long term study to understand periodically high winter losses of honey bee colonies. *Apidologie* 41: 332-352. http://dx.doi.org/10.1051/apido/2010014

GREATTI, M; MILANI, N; NAZZI, F (1992) Reinfestation of an acaricide-treated apiary by *Varroa jacobsoni* Oud. *Experimental and Applied Acarology* 16: 279-286.

GUZMÁN-NOVOA, E; SANCHEZ, A; PAGE, J R R; GARCÌA, T (1996) Susceptibility of European and Africanized honey bees (*Apis mellifera* L) and their hybrids to *Varroa jacobsoni* Oud. *Apidologie* 27: 93–103. http://dx.doi.org/10.1051/apido:19960204

GUZMÁN-NOVOA, E; ECCLES, L; CALVETE, Y; MCGOWAN, J; KELLY P G; CORREA-BENÌTEZ, A (2010) *Varroa destructor* is the main culprit for the death and reduced populations of overwintered honey bee (*Apis mellifera*) colonies in Ontario, Canada. *Apidologie* 41: 443-450. http://dx.doi.org/10.1051/apido/2009076

HARBO, J R (1993) Worker-bee crowding affects brood production, honey production, and longevity of honey bees (Hymenoptera: Apidae). *Journal of Economic Entomology* 86: 1672-1678.

HARRIS, J W (2001) A technique for marking individual varroa mites. *Journal of Apicultural Research* 40: 35-37.

HARRIS, J W; HARBO, J R (1999) Low sperm counts and reduced fecundity of mites in colonies of honey bees (Hymenoptera: Apidae) resistant to *Varroa jacobsoni* (Mesostigmata: Varroidae). *Journal of Economic Entomology* 92: 83-90.

HERRMANN, M; KANBAR, G; ENGELS, W (2005) Survival of honey bee (*Apis mellifera*) after with trypan blue staining of wounds caused by *Varroa destructor* mites or artificial perforation. *Apidologie* 36: 107-111. http://dx.doi.org/10.1051/apido:2004074

HOOPINGARNER, R (2001) Biotechnical control of varroa mites. In *T C Webster; K S Delaplane (Eds). Mites of the honey bee.* Dadant; Hamilton, USA. pp. 197–204.

HOPPE, H; RITTER, W (1988) The influence of Nasonov pheromone on the recognition of house bees and foragers by *Varroa jacobsoni. Apidologie* 19: 165–172.

HUMAN, H; BRODSCHNEIDER, R; DIETEMANN, V; DIVELY, G; ELLIS, J; FORSGREN, E; FRIES, I; HATJINA, F; HU, F-L; JAFFÉ, R; KÖHLER, A; PIRK, C W W; ROSE, R; STRAUSS, U; TANNER, G; TARPY, D R; VAN DER STEEN, J J M; VEJSNÆS, F; WILLIAMS, G R; ZHENG, H-Q (2013) Miscellaneous standard methods for *Apis mellifera* research. In *V Dietemann; J D Ellis; P Neumann (Eds) The COLOSS* BEEBOOK, *Volume I: standard methods for* Apis mellifera *research. Journal of Apicultural Research* 52(4): http://dx.doi.org/10.3896/IBRA.1.52.4.10

IMDORF, A; BOGDANOV, S; KILCHENMANN, V (2004) Wachsumstellung im Rahmen der Bioimkerei. *Schweizerische Bienen-Zeitung* 127: 15-18.

JENSEN, A B; ARONSTEIN, K; FLORES, J M; VOJVODIC, S; PALACIO, M A; SPIVAK, M (2013) Standard methods for fungal brood disease research. In *V Dietemann; J D Ellis, P Neumann (Eds) The COLOSS* BEEBOOK: *Volume II: Standard methods for* Apis mellifera *pest and pathogen research. Journal of Apicultural Research* 52(1): http://dx.doi.org/10.3896/IBRA.1.52.1.13

KANBAR, G; ENGELS, W (2003) Ultrastructure and bacterial infection of wounds in honey bee (*Apis mellifera*) pupae punctured by varroa mites. *Parasitology Research* 90: 349–354. http://dx.doi.org/10.1007/s00436-003-0827-4

KANBAR, G; ENGELS, W (2004) Visualisation by vital staining with trypan blue of wounds punctured by *Varroa destructor* mites in pupae of the honey bee (*Apis mellifera*). *Apidologie* 35: 25-29. http://dx.doi.org/10.1051/apido:2003057

KIRRANE, M J; DE GUZMAN, L I; RINDERER, T E; FRAKE, A M; WAGNITZ, J; WHELAN, P M (2012) A method for rapidly marking adult varroa mites for use in brood inoculation experiments. *Journal of Apicultural Research* 51(2): 212-213. http://dx.doi.org/10.3896/IBRA.1.51.2.10

KOENIGER, N; FUCHS, S; RAFIROIU, R (1987) Ameisensäure zur Behandlung von verdeckelter Brut. *Biene* 123: 286-289.

KOENIGER, G; KOENIGER; N; ANDERSON, D L; LEKPRAYOON, C; TINGEK, S (2002) Mites from debris and sealed brood cells of *Apis dorsata* colonies in Sabah (Borneo) Malaysia, including a new haplotype of *Varroa jacobsoni. Apidologie* 33 (2002) 15–24. http://dx.doi.org/10.1051/apido:2001005

KRAUS, B (1990) Effect of honey-bee alarm pheromone compounds on the behaviour of *Varroa jacobsoni. Apidologie* 21: 127-134.

KRAUS, B (1993) Preferences of *Varroa jacobsoni* for honey bees *Apis mellifera* L. of different ages. *Journal of Apicultural Research* 32: 57-64.

KRAUS, B (1994) Screening of substances for their effect on *Varroa jacobsoni*: attractiveness, repellency, toxicity and masking effects of ethereal oils. *Journal of Apicultural Research* 33: 34-43.

KRAUS, B; HUNT, G (1995) Differentiation of *Varroa jacobsoni* Oud. populations by random amplification of polymorphic DNA (RAPD). *Apidologie* 26: 283-290. http://dx.doi.org/10.1051/apido:19950402

KUENEN, L P S; CALDERÓNE, N W (2002) Positive anemotaxis by varroa mites: responses to bee odour plumes and single clean-air puffs. *Physiological Entomology* 23: 255–264.

LE CONTE, Y; ARNOLD, G; TROUILLER, J; MASSON, C; CHAPPE, B; OURISSON, G (1989) Attraction of the parasitic mite varroa to the drone larvae of honey bees by simple aliphatic esters. *Science* 245: 638–639.

LEE, K; MOON, R D; BURKNES, E C; HUTCHINSON, W D; SPIVAK, M (2010a) Practical sampling plans for *Varroa destructor* (Acari: Varroidae) in *Apis mellifera* (Hymenoptera: Apidae) colonies and apiaries. *Journal of Economic Entomology* 103: 1039-1050. http://dx.doi.org/10.1603/EC10037

LEE, K; REUTER, G S; SPIVAK, M (2010b) Sampling colonies for *Varroa destructor*. Poster #168 www.extension.umn.edu/honey bees

LINDQUIST, E E; KRANTZ, G W; WALTER, D E (2009) Order Mesostigmata. In *G W Krantz; D E Walter (Eds). A manual of acarology.* Lubbock; Texas, USA.

LO, N; GLOAG, R S; ANDERSON, D L; OLDROYD, B P (2010) A molecular phylogeny of the genus *Apis* suggests that the giant honey bee of the Philippines, *A. breviligula* Maa, and the plains honey bee of southern India, *A. indica* Fabricius, are valid species. *Systematic Entomology* 35: 226-233. http://dx.doi.org/10.1111/j.1365-3113.2009.00504.x

LOBB, N; MARTIN, S J (1997) Mortality of the mite *Varroa jacobsoni* during or soon after the emergence of worker or drone honey bees. *Apidologie* 28: 367-374. http://dx.doi.org/10.1051/apido:19970604

LUCKMANN, W H; METCALF, R L (1982) The pest-management concept. In *R L Metcalf; W H Luckmann (Eds). Introduction to insect pest management (2nd Ed.).* John Wiley and Sons; New York, USA. pp. 1-31.

MACEDO, P A; WU, J; ELLIS, M D (2002) Using inert dusts to detect and assess varroa infestations in honey bee colonies. *Journal of Apicultural Research* 41: 3-7.

MANLY, B F J (1997) *Randomization, bootstrap and Monte Carlo methods in biology.* Chapman and Hall; London, UK. 480 pp.

MARTIN, S J (1994) Ontogenesis of the mite *Varroa jacobsoni* Oud. in worker brood of the honey bee *Apis mellifera* L. under natural conditions. Experimental and Applied Acarology 18: 87-100. http://dx.doi.org/10.1007/BF000550033

MARTIN, S J (1995a) Ontogenesis of the mite *Varroa jacobsoni* Oud. in the drone brood of the honey bee *Apis mellifera* L. under natural conditions. *Experimental and Applied Acarology* 19(4): 199-210. http://dx.doi.org/10.1007/BF00130823

MARTIN, S J (1995b) Reproduction of *Varroa jacobsoni* in cells of *Apis mellifera* containing one or more mother mites and the distribution of these cells. *Journal of Apicultural Research* 34: 187-196.

MARTIN, S J (1998) A population model for the ectoparasitic mite *Varroa jacobsoni* in honey bee (*Apis mellifera*) colonies. *Ecological Modelling* 109: 267–281. http://dx.doi.org/10.1016/S0304-3800(98)00059-3

MARTIN, S J; COOK, C (1996) Effect of host brood type on the number of offspring laid by the honey bee parasite *Varroa jacobsoni*. *Experimental and Applied Acarology* 20: 387-390. http://dx.doi.org/10.1007/BF00130551

MARTIN, S J; KRYGER, P (2002) Reproduction of *Varroa destructor* in South African honey bees: does cell space influence varroa male survivorship? *Apidologie* 33: 51-61. http://dx.doi.org/10.1051/apido: 2001007

MARTIN, S J; MEDINA, L M M (2004) Africanized honey bees possess unique tolerance to varroa mites. *Trends in Parasitology*. 20: 112-114. http://dx.doi.org/10.1016/j.pt.2004.01.001

MARTIN, S J; HIGHFIELD, A C; BRETTELL, L; VILLALOBOS, E M; BUDGE, G C; POWELL, M; NIKAIDO, S; SCHROEDER, D C (2012) Global honey bee viral landscape altered by a parasitic mite. *Science* 336: 1304-1306. http://dx.doi.org/10.1126/science.1220941

MAUL, V; KLEPSCH, A; ASSMANN-WERTHMÜLLER, U (1988) Das Bannwabenverfahren als Element Imkerlicher Betriebsweise bei starkem Befall mit *Varroa jacobsoni* Oud. *Apidologie* 19: 139–154.

MEDINA, L M M; MARTIN, S J (1999) A comparative study of *Varroa jacobsoni* reproduction in worker cells of honey bees (*Apis mellifera*) in England and Africanized bees in Yucatan, Mexico. *Experimental and Applied Acarology* 23: 659-667. http://dx.doi.org/10.1023/A:1006275525463

MEDRZYCKI, P; GIFFARD, H; AUPINEL, P; BELZUNCES, L P; CHAUZAT, M-P; CLAßEN, C; COLIN, M E; DUPONT, T; GIROLAMI, V; JOHNSON, R; LECONTE, Y; LÜCKMANN, J; MARZARO, M; PISTORIUS, J; PORRINI, C; SCHUR, A; SGOLASTRA, F; SIMON DELSO, N; STEEN VAN DER, J; WALLNER, K; ALAUX, C; BIRON, D G; BLOT, N; BOGO, G; BRUNET, J-L; DELBAC, F; DIOGON, M; EL ALAOUI, H; TOSI, S; VIDAU, C (2013) Standard methods for toxicology research in *Apis mellifera*. In *V Dietemann; J D Ellis; P Neumann (Eds) The COLOSS BEEBOOK, Volume I: standard methods for* Apis mellifera *research. Journal of Apicultural Research* 52(4): http://dx.doi.org/10.3896/IBRA.1.52.4.14

MILANI, N (1995) The resistance of *Varroa jacobsoni* Oud. to pyrethroids - a laboratory assay. *Apidologie* 26: 415–429. http://dx.doi.org/10.1051/apido:19950507

MILANI, N (1999) The resistance of *Varroa jacobsoni* Oud. to acaricides. *Apidologie* 30: 229-234. http://dx.doi.org/10.1051/apido:19990211

MILANI, N (2001) Activity of oxalic and citric acids on the mite *Varroa destructor* in laboratory assays. *Apidologie* 32: 127–138. http://dx.doi.org/10.1051/apido:2001118

MILANI, N (2002) Chemical communication in the honey bee - varroa relationship. In *R Jones (Ed.). Proceedings of the sixth European Bee Conference "Bees without frontiers, Cardiff, UK, 1-5 July 2002.* pp 74–85.

MILANI, N; CHIESA, F (1990) Some stimuli inducing oviposition in *Varroa jacobsoni* Oud. In *Proceedings of the International Symposium on recent research on bee pathology, Gent, Belgium.* pp. 34-36.

MILANI, N; DELLA VEDOVA, G (1996) Determination of the LC50 in the mite *Varroa jacobsoni* of the active substances in Perizin and Cekafix. *Apidologie* 27: 175–184. http://dx.doi.org/10.1051/apido:19960306

MILANI, N; DELLA VEDOVA, G (2002) Decline in the proportion of mites resistant to fluvalinate in a population of *Varroa destructor* not treated with pyrethroids. *Apidologie* 33: 417–422. http://dx.doi.org/10.1051/apido:2002028

MILANI, N; DELLA VEDOVA, G; NAZZI, F (2004) (Z)-8-Heptadecene reduces the reproduction of *Varroa destructor* in brood cells. *Apidologie* 35: 265-273.

NAVAJAS, M; ANDERSON, D L; DE GUZMAN, L I; HUANG, Z Y; CLEMENT, J; ZHOU, T; LE CONTE, Y (2010) New Asian types of *Varroa destructor*: a potential new threat for world apiculture. *Apidologie* 41: 181-193. http://dx.doi.org/10.1051/apido:2009068

NAZZI, F; MILANI, N (1994) A technique for reproduction of *Varroa jacobsoni* Oud. under laboratory conditions. *Apidologie* 25: 579-584

NAZZI, F; MILANI, N; DELLA VEDOVA, G; NIMIS, M (2001) Semiochemicals from larval food affect the locomotory behaviour of *Varroa destructor*. *Apidologie* 32: 149–155. http://dx.doi.org/10.1051/apido:2001120

NAZZI, F; MILANI, N; DELLA VEDOVA, G (2004) A semiochemical from larval food influences the entrance of *Varroa destructor* into brood cells. *Apidologie* 35: 403–410. http://dx.doi.org/ 10.1051/apido:2004023

NAZZI, F; BORTOLOMEAZZI, R; DELLA VEDOVA, G; DEL PICCOLO, F; ANNOSCIA, D; MILANI, N (2009) Octanoic acid confers to royal jelly varroa-repellent properties. *Naturwissenschaften* 96: 309-314. http://dx.doi.org/10.1007/s00114-008-0470-0

NAZZI, F; BROWN, S P; ANNOSCIA, D; DEL PICCOLO, F; DI PRISCO, G; VARRICCHIO, P; DELLA VEDOVA, G; CATTONARO, F; CAPRIO, E; PENNACCHIO, F (2012) Synergistic parasite-pathogen interactions mediated by host immunity can drive the collapse of honey bee colonies. *PLoS Pathogens* 8(6): e1002735. http://dx.doi.org/10.1371/journal.ppat.1002735

OIE (2008) Chapter 2.2.7. Varoosis of honey bees. In *OIE. Manual of Diagnostic Tests and Vaccines for Terrestrial Animals (mammals, birds and bees), vol. 1, OIE, (Sixth Edition)*. Paris, France. pp 424-430.

OLDROYD, B P (1999) Co-evolution while you wait: *Varroa jacobsoni*, a new parasite of western honey bees. *Trends in Ecology and Evolution* 14: 312-315.
http://dx.doi.org/10.1016/S0169-5347(99)01613-4

OLDROYD, B P; WONGSIRI, S (2006) *Asian honey bees: biology, conservation and human interactions*. Harvard University Press.

OSTIGUY, N; SAMMATARO, D (2000) A simplified technique for counting *Varroa jacobsoni* Oud. on sticky boards. *Apidologie* 31: 707-716. http://dx.doi.org/10.1051/apido:2000155

OUDEMANS, A C (1904) Note VIII. On a new genus and species of parasitic Acari. *Notes Leyden Museum* 24: 216-222.

PIRK, C W W; DE MIRANDA, J R; FRIES, I; KRAMER, M; PAXTON, R; MURRAY, T; NAZZI, F; SHUTLER, D; VAN DER STEEN, J J M; VAN DOOREMALEN, C (2013) Statistical guidelines for *Apis mellifera* research. In *V Dietemann; J D Ellis; P Neumann (Eds) The COLOSS* BEEBOOK, *Volume I: standard methods for* Apis mellifera *research. Journal of Apicultural Research* 52(4):
http://dx.doi.org/10.3896/IBRA.1.52.4.13

RICKLI, M; DIEHL, P A; GUÉRIN, P M (1994) Cuticle alkanes of honey bee larvae mediate arrestment of bee parasite *Varroa jacobsoni*. *Journal of Chemical Ecology* 20: 2437–2453.

RICKLI, M; GUÉRIN, P M; DIEHL, P A (1992) Palmitic acid released from honey bee worker larvae attracts the parasitic mite *Varroa jacobsoni* on a servosphere. *Naturwissenschaften* 79: 320–322.

RITTER, W (1981) Varroa disease of the honey bee *Apis mellifera*. *Bee World* 62: 141-153.

ROCHE LEXICON MEDIZIN (1999) *Trypanblau, engl. trypan blue, zur Vitalprüfung (4th Ed.)*. Urban & Fischer Verlag; München, Germany.

ROSENKRANZ, P (1993) A bioassay for the test of the host finding behaviour of *Varroa jacobsoni*. *Apidologie* 24: 486–488.

ROSENKRANZ, P; AUMEIER, P; ZIEGELMANN, B (2010) Biology and control of *Varroa destructor*. *Journal of Invertebrate Pathology* 103: S96-S119. http://dx.doi.org/10.1016/j.jip.2009.07.016

ROSENKRANZ, P; BARTALSKY, H (1996) Reproduction of varroa females after long broodless periods of the honey bee colony during summer. *Apidologie* 27: 288-289.

SAMMATARO, D; DE GUZMAN, L; GEORGE, S; OCHOA, R (2013) Standard methods for tracheal mites research. In *V Dietemann; J D Ellis, P Neumann (Eds) The COLOSS* BEEBOOK*: Volume II: Standard methods for* Apis mellifera *pest and pathogen research. Journal of Apicultural Research* 52(4):
http://dx.doi.org/10.3896/IBRA.1.52.4.20

SCHULZ, A E (1984) Reproduction and population dynamics of the parasitic mite *Varroa jacobsoni* Oud. and its dependence on the brood cycle of its host *Apis mellifera* L. [in German]. *Apidologie* 15: 401-420. http://dx.doi.org/10.1051/apido:19840404.

SCHÄFER, M O; RITTER, W; PETTIS, J S; NEUMANN, P (2010) Winter losses of honey bee colonies (*Apis mellifera*): The role of infestations with *Aethina tumida* and *Varroa destructor*. *Journal of Economic Entomology* 103: 10-15.
http://dx.doi.org/10.1603/EC09233

SHIMANUKI, H; CALDERONE, N W; KNOX, D A (1994) Parasitic mite syndrome: the symptoms. *American Bee Journal* 134: 827-828.

SOKAL, R R; ROHLF, F J (1995) *Biometry: the principles and practice of statistics in biological research (3rd Ed.)*. Freeman and Co; New York, USA. 887 pp.

SOLIGNAC, M; CORNUET, J-M; VAUTRIN, D; LE CONTE, Y; ANDERSON, D; EVANS, J; CROS-ARTEIL, S; NAVAJAS, M (2005) The invasive Korea and Japan types of *Varroa destructor*, ectoparasitic mites of the western honey bee (*Apis mellifera*), are two partly isolated clones. Proceeding of the Royal Society B 272: 411-419. http://dx.doi.org/10.1098/rspb.2004.2853

STEINER, J; DITTMANN, F; ROSENKRANZ, P; ENGELS, W (1994) The first gonocycle of the parasitic mite (*Varroa jacobsoni*) in relation to preimaginal development of its host, the honey bee (*Apis mellifera carnica*). *Invertebrate Reproduction and Development*. 25: 175-183.

STRANGE, J P; SHEPPARD, W S (2001) Optimum timing of miticide applications for control of *Varroa destructor* (Acari: Varroidae) in *Apis mellifera* (Hymenoptera: Apidae) in Washington state, USA. *Journal of Economic Entomology* 94: 1324- 1331.

TOPOLSKA, G; GAJDA, A; POHORECKA, K; BOBER, A; KASPRZAK, S; SKUBIDA, M; SEMKIW, P (2010) Winter colony losses in Poland. *Journal of Apicultural Research* 49: 126-128.
http://dx.doi.org/10.3896/IBRA.1.49.1.27

TROUILLER, J; MILANI, N (1999) Stimulation of *Varroa jacobsoni* Oud. oviposition with semiochemicals from honey bee brood. *Apidologie* 30: 3–12.

VANENGELSDORP, D; HAYES JR, J; UNDERWOOD, R M; CARON, D; PETTIS, J (2011) A survey of managed honey bee colony losses in the USA, fall 2009 to winter 2010. *Journal of Apicultural Research* 50: 1-10. http://dx.doi.org/10.3896/IBRA.1.50.1.01

VANENGELSDORP, D; EVANS, J D; SAEGERMAN, C; MULLIN, C; HAUBRUGE, E; NGUYEN, B K; FRAZIER, M; FRAZIER, J; COX-FOSTER, D; CHEN, Y; UNDERWOOD, R; TARPY, D R; PETTIS, J S (2009) Colony Collapse Disorder: A descriptive study. *PLoS ONE* 4 (8): e6481. http://dx.doi.org/10.1371/journal.pone.0006481

WARRIT, N; LEKPRAYOON, C (2011) Asian honey bee mites. In *R Hepburn; S Radloff (Eds). Honey bees of Asia*. Springer; Heidelberg, Germany. pp 347-368

WARRIT, N; SMITH, D R; LEKPRAYOON, C (2006) Genetic subpopulations of varroa mites and their *Apis cerana* hosts in Thailand. *Apidologie* 37: 19-30.
http://dx.doi.org/10.1051/apido:2005051.

WILKINSON, D; THOMPSON, H M; SMITH, G C (2001) Modelling
biological approaches to controlling varroa populations. *American
Bee Journal* 141: 511–516.

WILLIAMS, G R; ALAUX, C; COSTA, C; CSÁKI, T; DOUBLET, V;
EISENHARDT, D; FRIES, I; KUHN, R; MCMAHON, D P;
MEDRZYCKI, P; MURRAY, T E; NATSOPOULOU, M E; NEUMANN,
P; OLIVER, R; PAXTON, R J; PERNAL, S F; SHUTLER, D; TANNER,
G; VAN DER STEEN, J J M; BRODSCHNEIDER, R (2013) Standard
methods for maintaining adult *Apis mellifera* in cages under *in
vitro* laboratory conditions. In *V Dietemann; J D Ellis; P Neumann
(Eds) The COLOSS BEEBOOK, Volume I: standard methods for
Apis mellifera research. Journal of Apicultural Research* 52(1):
http://dx.doi.org/10.3896/IBRA.1.52.1.04

ZIEGELMANN, B; LINDENMAYER, A; STEIDLE, J; ROSENKRANZ, P
(2012). The mating behaviour of *Varroa destructor* is triggered by
a female sex pheromone. Part 1: Preference behaviour of male
mites in a laboratory bioassay. *Apidologie.*
http://dx.doi.org/10.1007/s13592-012-0182-5

Journal of Apicultural Research 52(1): (2013)
DOI 10.3896/IBRA.1.52.1.10

REVIEW ARTICLE

Standard methods for wax moth research

James D Ellis[1]*, **Jason R Graham**[1] and **Ashley Mortensen**[1]

[1]Honey Bee Research and Extension Laboratory, Department of Entomology and Nematology, University of Florida, Steinmetz Hall, Natural Area Dr., P.O. Box 110620, Gainesville, FL, 32611, USA.

Received 7 July 2012, accepted subject to revision 16 July 2012, accepted for publication 14 November 2012.

*Corresponding author: Email: jdellis@ufl.edu

Summary

Greater (Lepidoptera: Pyralidae, *Galleria mellonella*) and Lesser (Lepidoptera: Pyralidae, *Achroia grisella*) wax moths are ubiquitous pests of honey bee colonies globally. The economic importance of wax moths has led to a number of investigations on wax moth life history, biology, behaviour, ecology, molecular biology, physiology, and control. Despite the importance of wax moths to the apicultural industry, they are investigated considerably more as a model organism for studies in insect physiology, genomics, proteomics, etc. Those studying wax moths from an apicultural perspective typically use only a small number of the total available research methods outlined in the literature. Herein, we describe methods associated with wax moth research that we feel are important from an apicultural research perspective. Ultimately, we hope that this paper will revitalize research on wax moths, since they remain both an important honey bee colony pest and an interesting colony symbiont.

Métodos estándar para la investigación de la polilla de la cera

Resumen

Las polillas de la cera grande (Lepidoptera: Pyralidae, *Galleria mellonella*) y pequeña (Lepidoptera: Pyralidae, *Achroia grisella*) son una plaga ubicua de las colonias de abejas al nivel mundial. La importancia económica de las polillas de la cera ha dado lugar a una serie de investigaciones sobre la historia de la vida de la polilla de la cera, la biología, el comportamiento, la ecología, la biología molecular, la fisiología y su control. A pesar de la importancia de la polilla de la cera en la industria apícola, se ha investigado mucho más como un organismo modelo para estudios de fisiología de insectos, genómica, proteómica, etc. Aquellos que estudian las polillas de la cera desde una perspectiva apícola suelen utilizar sólo un reducido número de métodos de investigación del total descrito en la literatura. En este documento, se describen los métodos asociados a la investigación de la polilla de la cera que creemos que son importantes desde una perspectiva de investigación apícola. En última instancia, esperamos que este documento revitalice la investigación sobre las polillas de la cera, ya que siguen siendo una plaga importante de las colonias de la abeja de la miel y un interesante simbionte de las colonias.

蜡螟研究的标准方法

大蜡螟（Lepidoptera: Pyralidae, Galleria mellonella）和小蜡螟（Lepidoptera: Pyralidae, Achroia grisella）是全球范围内蜂群中普遍存在的害虫。其重要的经济价值引起了在蜡螟生活史、生物学、行为学、生态学、分子生物学、生理学以及控制方法方面的研究。尽管蜡螟对养蜂业有重要影响，但它们还是更多的被用作模式昆虫研究昆虫生理学、基因组学、蛋白质组学等。从养蜂业的角度研究蜡螟仅涉及了大量文献中的少数研究方法。因此，我们挑选了我们认为对养蜂业而言十分重要的研究蜡螟的方法，希望能够促进对蜡螟的研究。至今，蜡螟仍是重要的蜂群害虫和蜂群共生生物。

Keywords: wax moth, *Galleria mellonella*, *Achroia grisella*, rearing, identification, control, *BEEBOOK*, COLOSS, honey bee

Footnote: Please cite this paper as: ELLIS, J D; GRAHAM, J R; MORTENSEN, A (2013) Standard methods for wax moth research. In *V Dietemann; J D Ellis; P Neumann (Eds) The COLOSS BEEBOOK, Volume II: standard methods for* Apis mellifera *pest and pathogen research. Journal of Apicultural Research* 52(1): http://dx.doi.org/10.3896/IBRA.1.52.1.10

1.   Introduction

Greater (Lepidoptera: Pyralidae, *Galleria mellonella*) and Lesser (Lepidoptera: Pyralidae, *Achroia grisella*) wax moths are ubiquitous pests of honey bee (*Apis mellifera*) colonies globally. The larvae of both moths are pests of honey bee colony wax combs, especially in stressed colonies, and can cause significant damage to stored beekeeping equipment. The economic importance of wax moths has led to a number of investigations on wax moth life history, biology, behaviour, ecology, molecular biology, physiology, and control.

Despite the importance of wax moths to the apicultural industry, they are investigated considerably more as a model organism for studies in insect physiology, genomics, proteomics, etc. This is especially true for greater wax moths. Consequently, there are thousands of literature references on wax moths and, correspondingly, possibly hundreds of research techniques associated with the insect. Those studying wax moths from an apicultural perspective typically use only a small number of the total available research methods outlined in the literature.

Herein, we describe research methods commonly used by people investigating wax moths from an apicultural perspective. It is important to note that developing a compendium of all methods related to wax moth research is beyond the scope and purpose of this paper. There simply are too many methods and manuscripts to include in such a reference. Indeed, research methods related to wax moths could be outlined in an entire book dedicated to the subject. Instead, we describe methods we feel are important from an apicultural research perspective. We hope that this paper will revitalize research on wax moths, since they remain both an important honey bee colony pest and an interesting colony symbiont.

2. Identification of greater and lesser wax moths

"Wax moth" is the common name for a variety of moths that invade, occupy and damage bee hives, though two species are known to impact honey bee colonies specifically. The wax moth has also been called the bee moth, the wax (or bee) miller, the waxworm or webworm. The greater wax moth is the more destructive and common comb pest whilst the lesser wax moth is both less prevalent and less destructive. Both wax moth species undergo complete metamorphosis. They have four stages of development: egg; larva; pupa; and adult. With proper training, one can recognize the differences between greater and lesser wax moths of all life stages. Most of our discussion of wax moth in this document concerns the greater wax moth, since it is the more investigated of the two species. Nevertheless, we do include information on lesser wax moths where known and appropriate, especially in Table 1 where diagnostic characteristics between greater and lesser wax moths are listed.

2.1. Wax moth eggs

Greater wax moth eggs are pearly white to light pink in colour and have a rough texture due to wavy lines running diagonally at regular intervals (Figs. 1 and 2). The surface texture of greater wax moth eggs differs from that of lesser wax moth eggs (Fig. 1; Table 1) and can be used as a diagnostic between the two. Other comparisons between eggs of the two species are made in Table 1. In most cases, greater wax moth females oviposit in clumps of 50-150 eggs (Williams, 1997). Throughout development, the egg changes from white to a yellowish colour. At approximately 4 days prior to hatching, the greater wax moth larva is visible as a dark ring within the egg. Twelve hours prior to hatching, the fully formed larva is visible through the egg chorion (Paddock, 1918). According to Williams (1997), greater wax moth eggs develop quickly at warm temperatures (29°C-35°C) and more slowly by about 30 days at cold temperatures (18°C). Eggs will not survive in extreme cold (at or below 0°C for 4.5 hours) or extreme heat (at or above 46°C for 70 minutes). SEM images comparing the eggs of the lesser and greater wax moths are available in Arbogast *et al.* (1980) and in Fig. 1.

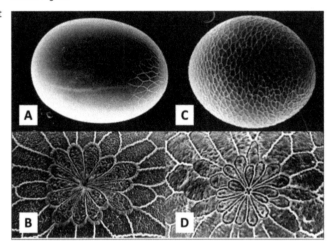

Fig. 1. The eggs of the greater and lesser wax moths. Lesser wax moth egg lateral view: ***A.*** magnification = 110x; and ***B.*** close up of micropylar area, magnification = 560x. Greater wax moth egg lateral view: ***C.*** magnification = 110x); and ***D.*** close up of micropylar area, magnification = 560x. From Arbogast *et al.*, 1980: original images provided by T Arbogast.

2.2. Wax moth larvae

Upon hatching, the greater wax moth larva is an off-white colour and 1-3 mm in length (Table 1; Fig. 2). The newly hatched larva immediately begins to eat and spin webbing (Fig. 3). The head capsule is yellowish and smaller than the more pronounced prothoracic segment (Paddock, 1918). The presence of stemmata on the head (Fig. 4) and the appearance of the spiracles (Fig. 5) can be used to differentiate between greater and lesser wax moth larvae. The thoracic legs are well developed when the larva first emerges but the abdominal legs are not visible until the larva is about 3 days old. A greater wax moth larva moults 7 times throughout its development.

Table 1. General characteristics of greater and lesser wax moth life stages.

	Lesser Wax Moth Eggs	Greater Wax Moth Eggs
size	0.41 ± 0.02 × 0.31 ± 0.01 mm (l × w)[1]	0.44 ± 0.04 × 0.36 ± 0.02 mm (l × w)[1]
description	nearly spherical creamy-white in color[2]	spheroid to ellipsoid, ovoid or obovoid, pink-cream white in clusters of 50-150 eggs[2]
length in life stage	7-22 days, depending on environmental conditions; 7.1 ± 1.0 days[3]	3 - 30 days depending on environmental conditions[2]
diagnostic characters	"Reticulation limited to anterior end, carinae surrounding primary cells conspicuously broader around outer margins of cells"[1] (Fig. 1)	"Reticulation at least faintly visible over entire surface, carinae surrounding primary cells of uniform width"[1] (Fig. 1)
	Lesser Wax Moth Larvae	**Greater Wax Moth Larvae**
Size	1-20 mm long; fully grown = 18.8 ± 0.4 mm (length)[3]	first instar = 1-3 mm (length) fully grown = 12-20 mm (length), 5-7 mm diameter[2]
description	narrow white bodies with brown heads and pronotal shields[2]	creamy-white with gray to dark gray markings, a small slightly pointed, reddish head[2] (Figs. 6, 7, and 9)
length in life stage	6-7 weeks at 29° to 32°C; 30.10 ± 2.5 days[3]	6-7 weeks at 29° to 32°C[2]
diagnostic characters	"Stemmata absent (Fig. 4); spiracle with black peritreme thicker on caudal margin"[4] (Fig. 5)	"Head with 4 stemmata on each side (Fig. 4); spiracle with yellowish peritreme of uniform thickness"[4] (Fig. 5)
	Lesser Wax Moth Pupae	**Greater Wax Moth Pupae**
Size	11.3 ± 0.4mm in length & 2.80 ± 1.89 mm in width[3]	12-20 mm in length & 5-7 mm in width[2]
description	yellow-tan pupa in a white cocoon often covered in frass and other debris[2]	dark reddish brown pupa in an off-white, parchment-thick cocoon[2] (Fig. 9)
length in life stage	37.3 ± 1.2 days[3]	6-55 days depending on environmental conditions[2]
	Lesser Wax Moth Adults	**Greater Wax Moth Adults**
size	male = 10 mm long female = 13 mm long	15 mm (length) with a 31 mm average wingspan
description	small, silver-bodied with a conspicuously yellow head, oval shaped forewings and heavily fringed hind wings[2]	heavy-bodied, reddish brown with mottled forewings and pale cream-colored lightly fringed hind wings[2]
lifespan	female = 6.90 ± 1.135 days male = 12.90 ± 1.30 days[3]	female = ~ 12 days male = ~ 21 days[2]
diagnostic characters	"Forewing breadth less than 5 mm; termen of forewing convex (hindwing of male with concave termen); Cu of hindwing apparently 3-branched; labial palps conspicuous though short (length not exceeding diameter of eye); labial palps of male transversely incurved, pincerlike"[4] (Figs. 11 and 12)	"Forewing breadth 5 to 7 mm; termen of forewing concave; Cu of hindwing apparently 4-branched; labial palp long (about as long as longest leg spur) and protruding" (Figs. 11 and 12)[4]

[1]Arbogast et al. 1980
[2]Williams, 1997
[3]Sharma et al. 2011
[4]Ferguson, 1987

Fig. 2. Greater wax moth eggs (cream-coloured, globular structures, left arrow) and 1st instar larva (right arrow).
Photograph: Lyle Buss, University of Florida.

Most of the growth and size increase happens during the final 2 instars. Larval development lasts 6-7 weeks at 29°-32°C and high humidity. A mature greater wax moth larva (Figs. 6 and 7) is approximately 20 mm in length (Paddock, 1918). Its body is grey in colour with a brown prothoracic shield having a broad band across it. The head is slightly pointed, small, and reddish with a v-shaped line opening towards the front of the head (Paddock, 1918). A greater wax moth larva goes through 8-9 stages (moults) over the course of its development at 33.8°C (Chase, 1921; Charriere and Imdorf, 1999).

Mature greater wax moth larvae are capable of boring into wood and often make boat-shaped indentations in the woodenware of the hive body or frames (Fig. 8). After finding a place in the hive to pupate, the larva begins spinning silk threads that will become the cocoon (Fig. 9), which they attach to the excavated indentations (Paddock, 1918). One often finds many of the cocoons congregated in areas

Fig. 3. Greater wax moth damage to wax comb. Note the larval frass and webbing.   Photograph: Lyle Buss, University of Florida.

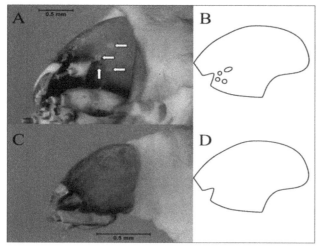

Fig. 4. Diagnostic characteristics on the head of greater and lesser wax moth larvae: **A.** The greater wax moth larvae head has four stemmata on both sides (small, pale ovals are arrowed); **B.** is redrawn from Ferguson 1987 and shows the location of the four stemmata. The lesser wax moth head: **C.** does not have the four stemmata (also shown in **D.** redrawn from Ferguson, 1987).

Photographs (A and C): Lyle Buss, University of Florida.

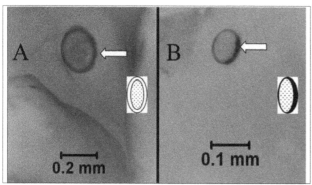

Fig. 5. Diagnostic characteristics on the spiracle of greater and lesser wax moth larvae: **A.** The greater wax moth larvae spiracle has a yellowish peritreme (arrowed, pale) of uniform thickness (also shown in the inset image redrawn from Ferguson 1987); **B.** The lesser wax moth spiracle has a black peritreme that is thicker on the caudal margin (arrowed, also shown in the inset image redrawn from Ferguson 1987).

Photographs: Lyle Buss, University of Florida.

Fig. 6. Greater wax moth larva in a wax cell from the brood nest.
Photograph: Lyle Buss, University of Florida.

Fig. 7. Greater wax moth larvae eating wax comb down to the plastic foundation. Notice the characteristic webbing and frass associated with the feeding behavior. Photograph: Lyle Buss, University of Florida.

Fig. 8. Wax moth damage to woodenware. The larvae excavate furrows in the wood and they attach their cocoons to these furrows. Notice the boat-shaped indentations in the wall of the hive.

Photograph: Ashley Mortensen, University of Florida.

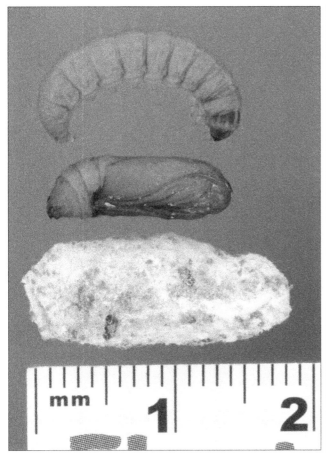

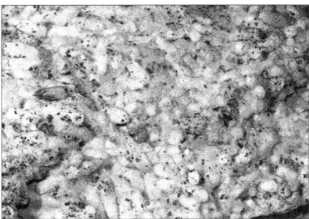

Fig. 10. Greater wax moth pupal cocoons. They are clumped together on the side wall of a brood super.

Photograph: Lyle Buss, University of Florida.

Fig. 9. Greater wax moth larvae (top), pupa (middle), and cocoon (bottom).   Photograph: Lyle Buss, University of Florida.

Fig. 11. Greater (left) and lesser (right) wax moth adults. (upper left) greater wax moth male, (lower left) greater wax moth female, (upper right) lesser wax moth male, (lower right) lesser wax moth female. Photograph is to scale.   Photographs: Lyle Buss, University of Florida.

around the perimeter of the bee nest in high infestations (Fig. 10). After hardening, the outer layer of the cocoon is somewhat tough while the inside remains soft and padded. Cocoon construction times can be variable due to temperature and humidity though the average cocoon construction takes 2.25 days to complete (Paddock, 1918). The larva becomes less active as the cocoon is constructed. The larva creates an incision point in the cocoon near the head through which to escape as a fully formed adult (Paddock, 1918). Greater wax moth larvae tend to congregate in the hive whereas the lesser wax moth larvae are more likely to be found individually in tunnels within the comb (Williams, 1997).

2.3. Wax moth pupae

The developmental time of greater wax moths from larvae to pupae within the cocoon ranges from 3.75 days to 6.4 days depending on temperature. Inside the cocoon, the newly formed pupa is white and becomes yellow after ~ 24 hours (Paddock, 1918). After 4 days have passed, the pupa becomes a light brown that gradually darkens, becoming dark brown by the end of pupation (Fig. 9). Pupae of the greater wax moth range in size from 5 mm to 7 mm in diameter and 12 mm to 20 mm in length (Paddock, 1918). A row of spines develops from the back of the head to the fifth abdominal segment and the bodyline curves downward (Paddock, 1918). The pupal development stage of

greater wax moths varies with season and temperature from 6 to 55 days (Williams, 1997).

2.4. Wax moth adult

The adult greater wax moth is approximately 15 mm long with a 31 mm average wingspan. The wings are grey in colour, though the hind third of the wing, normally hidden, is bronze coloured (Fig. 11). The wing venation patterns can be used as a diagnostic between greater and lesser wax moths (Ferguson, 1987; Fig. 12). Male greater wax moths are slightly smaller than females, lighter in colour, and have an indented, scalloped front wing margin in contrast to the females that have a straight front wing margin (Paddock, 1918). The female antennae are 10-20% longer than those of the male (Paddock, 1918). Greater wax moths emerge as adults in early evening and find a protected place to expand and dry their wings. Greater wax moths do not feed as adults and the females live ~12 days while the males live ~21 days (Paddock, 1918).

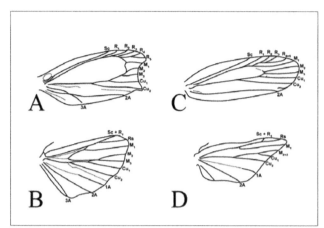

Fig. 12. The fore- and hind wings of the greater: **A.** forewing; **B.** hind wing; and lesser: **C.** forewing; **D.** hind wing wax moths. The forewing breadth is 5-7 mm for greater wax moths. The termen of the greater wax moth forewing is concave while the Cu of the hind wing is 4-branched. The forewing breadth is less than 5 mm for the lesser wax moth. The lesser wax moth forewing termen is convex and the Cu of the hind wing is 3-branched. Figure text and redrawn images are from Ferguson 1987.

2.5. Wax moth mating behaviour

Mating occurs shortly after adult emergence. Both the lesser and greater wax moth males attract the females by producing short ultrasonic signals. The male calls promote wing fanning by the females. This wing fanning causes pheromone release by males, leading to approach by females prior to copulation (Spangler, 1984, 1985, 1987; Jones *et al.*, 2002).

2.6. Wax moth oviposition

Female greater wax moths search for a crevice in which to lay their eggs. When a suitable location is found, the female extends her body in order to reach her ovipositor as deep into a crevice as possible. In laboratory studies, the females continued oviposition from 3-13 days (Paddock, 1918). The female greater wax moth can oviposit over 2,000 eggs in her lifetime, though the average is ~700 eggs (Warren and Huddleston, 1962). The female lesser wax moth will oviposit 250-300 eggs during her 7 day adult lifespan (Williams, 1997).

3. Rearing wax moths

Wax moth rearing methods are used in a variety of fields from molecular genetics and physiology to the simple production of wax moth larvae for reptile, bird food, and fish bait. Consequently, there are countless rearing methods available in the scientific literature as well as on hobbyist web sites, making it difficult to recognize a "standard" rearing method. Nevertheless, most rearing methods are very similar and share common components. We do our best to summarize a "standard" method for rearing greater wax moths. To begin a rearing programme,

the initial moths can be obtained from infested honey bee colonies or purchased commercially. Outlined here is the general rearing method of wax moths with modifications for method improvement indicated where appropriate.

3.1. Natural rearing method

1. Create a bee-free hive with frames of pulled, dark comb (dark comb is comb in which brood has been reared) containing honey and pollen.
2. Introduce three, late instar larval wax moths per frame to ensure wax moth presence (Hood *et al.*, 2004).
3. The hive and combs should be covered and under some type of shelter to protect it from rain. Darkness, warmth, and lack of ventilation promote colonization.
4. Unattended (bee-free) hives will be highly attractive to adult wax moths if they are present in the area (Hood *et al.*, 2004)
5. Provide additional used honeycomb containing honey and pollen as diet for rearing program as the food supply in the box is exhausted.
6. Moth eggs, larvae, pupae and adults can be collected from the hive with an aspirator, forceps, or a small, soft paintbrush. The latter should be used for the immature wax moth stages since they can be damaged easily.

3.2. *in vitro* rearing of wax moths

Most *in vitro* lab rearing techniques follow a simple series of events:

1. Place wax moth eggs on new diet.
2. Allow resulting larvae to feed on diet.
3. Harvest late instar larvae or pupa and place into a second container.
4. Allow late instar larva to pupate or pupa to emerge as adults.
5. Allow adults to mate and allow females to lay eggs.
6. Place eggs on new diet.

Methods to accomplish these steps are described in subsequent sections.

3.2.1. Diet

Both the greater and lesser wax moths feed only in the larval life stage. In nature, the larvae develop in bee colonies and feed on pollen, honey, cast larval skins and other debris incorporated into the wax comb. One method for feeding wax moth larvae is simply to provide them with sections of wax comb. This is useful because it provides the moths with what they ordinarily use. However, the production and use of wax comb can be expensive and unsustainable if a large number of wax moths are desired.

Correspondingly, many variations on a generalized artificial diet have been developed. We include three here. The first two are reported frequently in the literature while the third was provided by a reviewer with experience using the diet.

- Diet 1:
  1. Blend a mixture of:
     1.1. white honey (150 ml),
     1.2. glycerine (150 ml)
     1.3. tap water (30 ml).
  2. Add 420 g pablum (bran).
  3. Add 20 g ground brood comb.

The resulting diet has the consistency of damp sawdust (Bronskill, 1961). Coskun *et al.*, (2006) provide an analysis of this diet with several modifications resulting in larval weight gain or loss based on the modifications.

- Diet 2: (Jones *et al.*, 2002)
  1. Mix 300 ml liquid honey with
  2. 400 ml glycerol,
  3. Mix with 200 ml milk powder,
  4. 200 g whole-meal coarse flour
  5. 100 g dried brewer's yeast,
  6. 100 g wheat germ,
  7. 400 g bran.
- Diet 3:
  1. Mix seven parts (by volume) dry dog kibble,
  2. One part water,
  3. Two parts honey.
  4. You can adjust the vitamin A content to produce whitish larvae.

3.2.2. Environment

Wax moths, as adults, are nocturnal insects that fly at night and hide in dark places during the day. Wax moths thrive in dark, warm, poorly ventilated areas that are not well defended by honey bees. As such, ~30°C, ~70% RH and constant darkness are recommended in most manuscripts where rearing is discussed. Warren and Huddleston (1962) discuss the effect of humidity and temperature on various life stages of greater wax moths.

3.2.3. Containers

Several types of containers are recommended for use in rearing wax moths.

- Larval chamber - containing the eggs, developing larvae, and diet
- Mating chamber – where adults emerge from their pupal skins and cocoons and mate
- Oviposition chamber - where female moths will lay eggs

The size of the containers and method used will largely depend on the scope of the rearing program and the number of wax moths needed. Marston *et al.*, (1975) proposed a large mass-rearing program that spanned multiple rooms with diet prepared in a cement mixer and eggs collected by sieve. Waterhouse (1959) used plastic bags sealed with a paper clip. Metal, glass or plastic containers can be used, but wood, cardboard, and paperboard should be avoided as the larvae can chew through them.

3.2.4. Container sterilization

The containers should be sterilized before and between uses by boiling or autoclaving. Proper cleaning and sterilization of the cages will help to reduce the incidence of disease. Rearing wax moths in several containers will allow for infested batches to be discarded without shutting down overall production. It is best to discard containers with serious problems rather than attempt to salvage them. Cheap containers, such as those used commonly in kitchens to store food, can be discarded after first use.

3.2.5. Eggs

Multiple male and female moths should be placed in containers having diet mixtures. Females will begin laying eggs within hours of mating. Consider the temperature when designing an oviposition chamber to speed or slow egg development. Eggs develop quickly at warmer temperatures (29°C-35°C) and slowly (up to 30 days or more) at colder temperatures (18°C, Williams, 1997). The female will lay eggs on any surface but prefer surfaces that seem to protect the eggs and will preferentially lay in cracks and crevices. Several rearing programs recommended using crimped wax paper held together with a paper clip, as the eggs can be easily removed from the surface of the wax paper once unfolded (as in Burges and Bailey, 1968). About 1,000 eggs placed with about 1-1/2 pounds of diet should yield about 500 mature larvae (Marston *et al.*, 1975).

3.2.6. Larva

Crumpled paper towels, wax paper or corrugated cardboard can be added to the larval container after the first mature larvae begin to spin cocoons. The mature larvae will migrate to these materials to spin their cocoons. Eischen and Dietz (1990) observed prepupa spinning their cocoon inside cut soda straws, which facilitated subsequent handling, storing, and collection of the pupa. Pupae can be safely stored for 2-3 months at 15.5°C and 60% humidity (Jones *et al.*, 2002).

3.2.7. Pupa

If virgin females are needed, it is best to separate the females from males during the pupal stage as mating can occur shortly after adult emergence. The antennal and wing characteristics used to separate males and female adults (Table 1 and Section 2.4.) can also be seen in the pupal skins upon close examination. Smith (1965) provided two pupal characteristics which separate greater wax moth males and females:

1. The mesowing demarcation has a notch in the apical margin of male pupal cases and is straight in female pupal cases.
2. The sclerite of the 8th abdominal sternum is cloven in the female but not in the male pupal cases.

3.2.8. Adult

The adult will emerge from the cocoon. There are several helpful characteristics that can be used to distinguish between male and female greater wax moths (Table 1 and Section 2.4.). Adult moths will mate within hours of emergence and the mated females will begin egg-laying after mating (Jones *et al.*, 2002).

4. Quantification / qualification of wax moth damage and population

4.1. Qualification of wax moth damage in honey bee colonies and stored equipment

4.1.1. Damage to combs

Wax moth larvae feed on wax combs, cast larval skins, pollen, and some honey (Shimanuki *et al.*, 1992). Dark comb (comb in which brood has been reared) is preferred by the moth and subsequently suffers the most damage. The feeding habits of the larvae can reduce the wax combs to a pile of debris, wax moth frass, and webbing (Figs. 3 and 7).

4.1.2. Galleriasis

Greater wax moth larvae can tunnel and feed to the midrib of the wax comb. The midrib is the base of the comb on which the cells are constructed. The feeding larvae produce silken threads that can trap developing honey bee brood in the cells. Trapped bees will uncap their brood cell when ready to emerge as adults but will be unable to emerge. The result is a comb containing uncapped bees that struggle to emerge, a condition called galleriasis. Williams (1997) states that "entire combs of worker bees that have developed from brood of nearly the same age may be observed trapped in this way".

4.1.3. Bald brood

Lesser wax moths (and to a lesser extent greater wax moths) can cause "bald brood" in infected colonies. Lesser wax moth larvae will tunnel just below the surface of brood cells. The cells are uncapped and the developing bee pupae inside exposed (Fig. 13). This condition can be confused with general hygienic behaviour where adult bees detected disease / pest-compromised brood and uncap the cells. However, bald brood usually occurs over multiple cells in a linear pattern: uncapped brood cells that are adjacent to one another. The line of damage may turn in any direction based on the tunnelling habits of the larva. There may also be wax moth larva faecal pellets on the heads of the developing bee brood.

4.1.4. Damage to woodenware

Greater wax moth larvae can cause extensive damage to colony woodenware, including the frames and supers. After the moth larvae finish feeding, they look for a place attach their cocoons. Some moth

Fig. 13. Bald brood. Wax moth larvae tunnel under cell cappings, causing worker bees to remove the damaged cappings. Larval tunnels follow a definable pattern along rows of brood cells.

Photograph: Ashley Mortensen, University of Florida

larvae will chew away wood to create an area for cocoon attachment (Williams, 1997). This chewed area can be minor excavations or large holes (Fig. 8). Such damage is characteristic of wax moths and can weaken the structural integrity of the woodenware.

4.2. Quantification of wax moth damage in honey bee colonies and stored equipment

1. Cut a piece of plexi glass or wire mesh with desired mesh size (1 cm for example) to the size of frames or combs being used in the experiments (Hood *et al.*, 2003).
2. Scribe the plexi glass with a 1 cm grid.
3. Hold the plexi glass grid over both sides of all exposed frames
4. Quantify the total cm^2 of damage (see Section 4.1.).

4.3. Quantification of wax moth population drawn frames of comb

1. Carefully dissect comb to recover all larvae, pupae, and adults (James, 2011).
2. Quantify number of each life stage present and whether they are dead or alive.

Note: If mortality counts are not of interest, frames may be frozen and stored for later dissection.

4.4. Quantification of wax moth populations in whole colonies

We could not find detailed instructions for quantifying the population of wax moths in living colonies. The default method would be to freeze the entire colony for at least one week, to ensure wax moth death, and then dissect the combs (section 4.3.) for careful inspection for and collection of the various wax moth life stages. This procedure likely could be modified by removing the bees from the colony (via shaking or brushing the combs) prior to freezing the combs. However, removing bees from the combs carries with it the added risk of shaking moth eggs, young larvae, or adults from the comb, thus making it impossible to quantify the moth populations accurately.

5. Techniques associated with wax moth control

Beekeepers attempt to control wax moth populations in many ways. This section is not intended to outline all the methods related to controlling wax moths since these vary by region/country. Rather, this section focuses on techniques that are useful for purposes of studying wax moth control, i.e. these methods can be used to investigate potential methods of controlling wax moths.

5.1. Physiological parameters measured

Wax moths typically are considered a secondary pest of honey bee colonies. Consequently, there are comparatively fewer investigations on wax moth control than on more significant honey bee pests such as *Varroa destructor* (see the *BEEBOOK* paper on varroa, Dietemann *et al.*, 2013), *Aethina tumida* (see the *BEEBOOK* paper on small hive beetles, Neumann *et al.*, 2013), *Acarapis woodi* (see the *BEEBOOK* paper on tracheal mites, Sammataro *et al.*, 2013), etc. Most investigations on wax moth control determine the efficacy of the control based on its effects on the following measurable, physiological changes in the organism:

- Mortality: Death of the wax moth at any life stage. Sufficient time (a few hours to a few days depending on the target control method) must be allowed in an appropriate rearing environment to determine mortality in eggs and pupae.
- Diet consumption: The amount of diet consumed by developing larvae. It is ideal for test larvae to be housed individually if diet consumption is to be measured.
- Changes in development: This includes weight gain (i.e. daily, weekly, per instar), developmental time (oviposition to egg hatch, instar to instar, pupation to adult emergence, and/or total time from egg to adult), successful adult emergence, etc.
- Sterility: Daily and total fecundity of mated females.
- Post injection paralysis: The inability of a larva to return to a dorsal-ventral position when placed on its dorsum 30 min after injection.

5.2. Injection of test substances into the hemocoel

Potential wax moth control agents can be injected directly into the larval hemocoel (West and Briggs, 1968). Possible treatment compounds include bacterial toxins (such as *Bacillus thuringiensis*), fungal toxins (i.e. Vilcinskas *et al.*, 1997), insecticides, plant resins, etc. This procedure also can be used to initiate immune responses in wax moths and for other purposes beyond simple pest control.

1. Raise larvae per Section 3 to 100-200 mg/individual.
2. Prepare solutions (treatment and control) per the needs and conditions of the experiment.
3. Using a calibrated microinjection apparatus with a 27 gauge needle, insert the needle into the lateral integument about halfway down the body (be careful not to damage internal organs).

Note: Alternatively, microliter cemented needle syringes fitted with a 26 gauge needle may be used for microinjections.

4. Inject a consistent, desired volume into each larva.
5. Repeated injections are discouraged because of the size of the insect and possible associated damage (Stephens, 1959).
6. Observe specimens for desired change (see Section 5.1. for parameters).

Considerations: In microinjection experiments, care should be taken to maintain a clean workspace and equipment to limit physiological change due to contamination rather than the experimental treatment. One should also include controls for the study which include moths injected with Ringers solution. Solutions should be prepared so they are physiologically compatible with the larval hemocoel. It is possible for large injection volumes to cause non-treatment associated effects. West and Briggs (1968) had successful results injecting 20 ml bolus volumes though a range of injection volumes are reported in the literature.

5.3. Incorporation of test compound into the wax moth artificial diet

1. Prepare the treatment diet by adding the compound of choice to the wax moth artificial diet (Burges and Bailey, 1968; Eischen and Dietz, 1987). The diet should be prepared per Section 3.2.1. and the compound of interest added as experimental conditions necessitate.
2. Obtain moth eggs are by creating an egg laying surface for mature females (per Section 3.2.5.) out of a piece of wax paper. The wax paper is folded back and forth, making tight folds (accordion style), and held closed on the end by a paper clip. The female moths will oviposit in the folds.
3. Once eggs are laid, remove the folded wax paper.
4. Tap the eggs into a vial with no food.
5. As larvae hatch, move them carefully using a fine brush to a new vial provisioned with either treatment of control artificial diet.
6. Monitor larvae can be for physiological change at set times throughout their development.

Considerations: First instar larvae are very small and quite active. It is important that lids to containers housing individual larvae and eggs be secured tightly to prevent escape. The egg container should be monitored regularly for newly emerged larvae. The first instar larvae will starve quickly without food, and larvae that emerge or die overnight should be removed from the container each morning.

5.4. Comb treatment

The compound(s) of interest can be directly incorporated into melted wax prior to mill rolling of foundation sheets or applied to previously milled foundation (Burges and Bailey, 1968; Burges, 1976; Vandenberg and Shimanuki, 1990; Hood *et al.*, 2003, Ellis and Hayes, 2009).

1. Application to wax comb foundation: Based on the available form of the compound of interest, it may be sprayed, dipped, aerosolized, or dusted onto previously milled foundation sheets per the needs of the experiment. In the past, fogging (or aerosols) has been shown to be a less effective method for effective application - Vandenbergi and Shimanuki (1990).
2. Once dried, use treated foundation in experiments as is or sandwiched between untreated sheets of foundation and remilled to prevent direct exposure of honey bees in the colony to the test compound in the treated foundation.
3. Insert frames containing treatment and control foundation into healthy colonies for comb construction. The colonies may need to be fed a sucrose solution to encourage bees to construct comb on the foundation.
4. Once drawn, remove the experimental frames from the nest and any honey extracted from the comb.
5. Place newly hatched wax moth larvae (reared per Section 3) singly on a small piece of treated or control comb (comb produced on untreated foundation) in a dish container,
6. Incubate at 34°C,
7. Monitor for physiological changes.

Alternatively, hive boxes containing frames of treated comb, but no bees, can be inoculated with wax moths and the level of damage assessed (per Section 4).

Considerations: Compound concentrations should be determined for drawn comb after removal from the colony as honey bees will distribute wax from foundation throughout the comb (Burges and Bailey, 1968). Test compound impacts on honey bee colony fitness and/or behaviour should be accessed. Recommended methods for measuring colony strength parameters can be found in the *BEEBOOK* paper on measuring colony strength parameters (Delaplane *et al.*, 2013). If incorporating the compound directly into melted wax prior to milling, one must know heat impacts on the compound. The average temperature used to melt wax for milling machines has been reported to range from 77-99°C (Burges and Bailey, 1968).

5.5. Fumigation control

5.5.1. Standard crystal fumigation

Per Goodman *et al.* (1990). Beekeepers often use a similar method to protect stored combs against wax moths.

1. Prepare a super (a honey bee colony hive box) without bees, containing frames of drawn, empty comb with wax moth larvae.

This can be done either by removing one wax comb from the box and putting a frame cage (Section 6.7.1.) containing older moth larvae, pupae and/or adults in its place or by placing eggs and young larvae in dish cages (Section 6.7.2.) on top of the combs.

2. Place an additional open Petri dish containing the fumigant of choice on top of the frames.
3. Insert the super into a sealable container. The container must be large enough to prevent the death of the wax moths due to a build up of CO_2. This can be a large plastic container or even a plastic trash bag. The container should be sealed after the super is inserted.
4. Using silicone rubber and tape, fit one corner of the container with a plastic tube that has a removable, air tight cap. This is done easier if using a plastic bag as the container.
5. Remove the removable cap daily for the insertion of a Drager tube to measure the concentration of the fumigating agent.
6. At the end of the trial, the moth life stages should be monitored for mortality and physiological change (see Section 5.1.).

Considerations: A spacer must be used to prevent the trash bag or container lid from lying directly on top of the specimen and fumigation dishes. Fumigation efficacy is affected by gas leakage; care should be taken to control for this through diligent sealing of the container (Goodman *et al.*, 1990). The investigator can determine compound, dose, temperature, and time effects on moth mortality.

5.5.2. Controlled release of liquid or crystalline compounds
Per Burgett and Tremblay (1979).

5.5.2.1. Construction of dispersal packs

1. The test compound can exist as a crystal or be a liquid impregnated onto a piece of fibreboard or similar material.
2. Seal the compound, either in crystal or impregnated fibreboard form, in small packets of porous materials.

5.5.2.2. Determination of the permeation rate

1. Hang dispersal packets in a controlled environmental chamber and monitor for weight loss.
2. Calculate permeation rate = packet weight loss/elapsed time.
3. Obtain and average multiple permeation rates for each packet. Burgett and Tremblay (1979) monitored three test packets for each compound and weighed each packet a minimum of 5 times.

5.5.2.3. Larval Bioassay (per dispersal packet to be tested)

1. Insert various stages of moth development (reared and collected per Section 3) into a standard nucleus or full size honey bee colony with a dispersal packet (see Section 5.5.2.1.) and placed into a controlled environmental chamber.

Alternatively, individual dish containers (see Section 6.7.2.) of wax moths can be placed within hive boxes or stacks of boxes to simulate anticipated use by beekeepers.

2. Monitor the wax moths for physiological change (see Section 5.1.).

Considerations: Permeation rate varies with temperature, so it must be determined for the same temperature at which the experiment will be conducted. Under changing temperature regimens, mathematically weighted averages (estimated permeation rates based on the proportional amount of time spent at each temperature) approximate the actual dispersal packet weight loss $\pm$ 5% (Burgett and Tremblay, 1979).

5.5.3. Ozone Treatment

Per Cantwell *et al*. (1972) and James (2011).

5.5.3.1. Equipment needed

1. Ozone generator – The size and type of generator used will vary based on what level of ozone is desired/needed for the experiment.
2. Fumigation Chamber.
3. Ozone Analyser (Low Concentration Analyser). It must be able to detect and quantify the amount of ozone created by the ozone generator.
4. Data logger to measure temperature and humidity.
5. Ozone Destructor - eliminates ozone from the test facility. Ozone is potentially fatal to humans so care must be taken during its use.

5.5.3.2. Equipment establishment

1. Ozone is produced externally by the ozone generator and pumped into the fumigation chamber.
2. Measure ozone concentration, temperature, and humidity in the chamber by the ozone analyser and data logger.
3. Continually exhaust gas from the chamber via the ozone destructor.

5.5.3.3. Sample Protocol

1. Expose multiple moth life stages, contained in dish containers (Section 6.7.2.), to a range of ozone concentrations (measured in mg O_3/m^3) for a range of timed durations.
2. Exposure temperature may also be assessed for effect on treatment efficacy by incubating at multiple temperatures during fumigation.
3. It is best to recreate the environment under which the treatment, when applied by beekeepers, would normally occur.

For example, all moth life stages will be in and among the wax combs to be treated. So, it is best to place the dish of moths among combs, or in hive bodies as would be experienced in normal circumstances.

Considerations: The method could be adapted to fit other forms of gaseous treatment, i.e. carbon dioxide. Ozone is acutely toxic to humans and only should be used in sealed fumigation chambers (James, 2011).

5.6. Gamma–ray irradiation and sterilization

The ideal moth developmental stage for irradiation is the pharate adult (see Section 3). During this stage, the somatic cells have fully differentiated and germ cells are most actively dividing (Jafari *et al*., 2010). Not only does irradiation at this time minimize the likelihood of adult abnormalities like deformed wings (which would keep them from being useful in sterile male release campaigns), but the specimen is also very easy to handle without risk of escape or damage (North, 1975). Males are more resistant to gamma ray sterilization than females (Carpenter *et al*., 2005) and the effective irradiation doses are 350 Gy and 200 Gy, respectively (Flint and Merkle, 1983; Jafari *et al*., 2010). Specific methodologies for irradiation facilities and techniques are somewhat standard and will not be described beyond the parameters presented above.

5.7. Entomopathogenic control of wax moths

Many species of entomopathogenic nematodes can be reared and cause mortality in wax moth larvae. The moth larvae, in turn, can be infected with nematodes using various methods. The techniques described below can be used to test exposure time (how long the wax moth larva is exposed to nematodes), nematode dose (often measured in nematode "IJs" or infective juveniles), exposure temperature, and many other factors on infection and mortality rates of wax moths. These same techniques are used by nematologists to investigate nematode biology, though the end result often is moth mortality, making the methods applicable to apicultural research. There can be some concern over non-target effects, including on bees, but these can be minimized with proper screening.

5.7.1. Infecting single greater wax moth larvae with entomopathogenic nematodes

Per Molyneux (1985) and Fan and Hominick (1991). The method below can be used to screen for entomopathogenic nematodes that show action against wax moths. Though nematodes possibly can be used in wax moth control programs, the methods outlined below are also useful for nematologists who need an effective method for rearing nematode species of interest.

1. Wash sand with distilled water.
2. Autoclave
3. Oven-dry.
4. Filter through a 1.18 mm sieve.
5. Moisten the filtered sand with 1 ml of distilled water for every 25 ml of sand (4% V/V).
6   Place 25 ml of moistened sand in a 30 ml plastic tube.
7. Pipette nematodes diluted in 1 ml of water (per producer's instructions or experimental needs) into the sand in the tube.

The nematode/water solution brings the V/V content to 8%. Any desired number of nematodes can be introduced to the soil in this way, though including more infective juveniles in the inoculum typically results in greater infestation with nematodes.

8. Invert (turn upside down) the tube multiple times to disperse the nematodes in the sand.

9. Place a single wax moth larva on the sand surface in the tube (late instar larvae are 250-350 mg).

10. Replace the tube lid and invert the tube.

11. Leave the tube inverted for set time periods and temperatures per the needs of the study.

12. Recover the wax moth larva and wash it three times with distilled water.

13. Process (dissect, etc.) the larvae immediately or maintain on moistened filter paper at 20°C for a period of time before use.

5.7.2. Recovery of entomopathogenic nematodes from soil using greater wax moth larvae

Per Fan and Hominick (1991), this method can be used to screen local soils for the occurrence of entomopathogenic nematodes that infest wax moths.

1. Collect soil of interest for use.

2. Place 200-250 cm^3 of soil in a plastic or glass dish (~300 cm^3 in volume).

3. Place five late instar *G. mellonella* larvae (late instar larvae are 250-350 mg) on the soil surface.

4. Seal the dish with a tight lid to limit larvae escape.

5. Incubate the dish at 20°C.

6. Replace the larvae (alive or dead) every 4-6 days. This should be done until larvae in the dish no longer die (i.e. all the living nematodes in the soil are "harvested").

7. Dissect all harvested larvae in saline

8. Quantify the number of nematode adults.

5.7.3. Recovery of entomopathogenic nematodes from inoculated sand using greater wax moth larvae

Per (Fan and Hominick, 1991) and similar to the method outlined in 5.7.2., wax moth larvae can be used to recover entomopathogenic nematodes from inoculated sand.

1. Prepare sand and plastic tubes according to the protocol outlined in 5.7.1.

2. Inoculate the soil with any nematode species and/or any number of IJs of interest.

3. Add single wax moth larvae (late instar larvae are 250-350 mg) to the soil.

4. Keep the tubes at 20°C.

5. Replace the wax moth larva in the tube with a new individual weekly. This should be done until added larvae no longer die, indicating that no nematodes remain in the soil.

6. Dissect all harvested larvae in saline or maintain on moistened filter paper at 20°C for a period of time before use.

5.8. Protecting stored combs from wax moths

Wax moths are major pests of stored wax combs. Stored combs can be protected and/or made moth free using a number of techniques.

5.8.1. Protecting stored combs via freezing

1. Freeze supers of combs or individual combs (≤ 0°C) for > 24 hours. Other times/freezing temperatures include 2 hours at -15°C, 3 hours at -12°C, and 4.5 hours at -7°C (Charriere and Imdorf, 1999).

2. Once thawed, place the combs in plastic bags for storage or on strong colonies for protection from bees.

Note: Combs that are thawing need to be inaccessible to wax moths. Combs must be dry before bagging. Otherwise they can mold. Combs containing honey and/or pollen should remain in the freezer until use or placed on colonies for further protection from bees (see section 5.8.3.). This method can be used to start colonies "free" of wax moths (see section 6.8.).

5.8.2. Protecting stored combs via climate manipulation

1. Stored combs that are free of honey and/or pollen in supers.

2. Stack the supers in an "open shed" (a covered pavilion with only 1-3 walls).

3. Stack the supers in a crisscross pattern. To do this, place a super on a solid surface (such as a hive lid) that is situated on the ground. Place another super of combs on the one on the ground, orienting it at a 90° angle from the bottom super. Repeat this pattern until the stack of supers is a desirable height (a maximum height of 2 m is recommended).

The open shed and super stacking pattern ensure that light and air will penetrate the supers. This minimizes wax moth attraction since the moths do not like light/airflow.

Modifications of this method include stacking the supers in a climate controlled room with cool (0-15°C) temperatures, an oscillating fan, and constant light. This method is best used to protect white combs (combs in which no brood has been reared). Dark combs (combs in which brood has been reared) is best protected in a freezer (see section 5.8.1.) or on strong bee colonies (5.8.3.). Heat treatment is also possible. The combs must be stored for 80 minutes at 46°C or 40 minutes at 49°C (Charrière and Imdorf, 1999).

5.8.3. Protecting stored combs using strong colonies

Place supers of combs (containing no honey or pollen residues) directly onto strong colonies. Strong colonies can protect combs from wax moth infestation/damage. It is best if the stored combs contain no honey and/or pollen. Otherwise, the combs may be vulnerable to damage caused by small hive beetles (see the *BEEBOOK* paper on small hive beetles, Neumann *et al.*, 2013).

6. Miscellaneous techniques

6.1. Field collection of various wax moth life stages

1. Establish supers of moth-free, drawn, dark wax comb per Section 6.8.
2. Once the wax moth population has been established, collect all moth life stages present as described in Section 4.3.

6.2. Collecting greater wax moth haemolymph

Numerous investigations in the literature call for the collection and manipulation of wax moth haemolymph. Though the methods outlined to do this may not be immediately useful to those studying wax moths from an apiculture perspective, we feel that it is helpful to include methods related to haemolymph collection in this manuscript since it is such a popular technique and it is a technique used to answer many fundamental questions about wax moths. A method for collecting honey bee haemolymph is described in the *BEEBOOK* paper on physiology methods (Hartfelder *et al.*, 2013).

6.2.1. Method for collecting haemolymph

From Stephens (1962):

1. Larvae can be field-collected (see Section 6.1.) or reared *in vitro* (Section 3).
2. Anaesthetize the larvae with CO_2 until visible movement ceases. This makes it easier to handle larvae since they are other wise quite active.
3. Surface sterilize the larvae per Section 6.6. or with a hypo chlorite solution (24 ml Millendo bleach in 1 l distilled water) for 5 minutes.
4. Rinse the larvae twice with distilled water.
5. Dry the larvae on sterile blotters at 30°C until normal movement resumes.
6. Wax moth larvae can be bled by cutting a proleg from the body or puncturing the proleg with a sterile needle and collecting the haemolymph that pools at the wound.
7. Collect haemolymph by capillary action into sterile capillary tubes. Larvae from which only a small amount of haemolymph is collected can survive, complete their development, and reproduce normally.

Modifications

- The haemolymph can be transferred to pre-cooled Eppendorf tubes containing a few crystals of phenylthiourea. This prevents melanization (Vilcinskas *et al.*, 1997, Wedde *et al.*, 1998).
- 1 ml aliquots can be centrifuged twice at 10,000 g for 5 min to remove haemocytes (Wedde *et al.,* 1998).

6.2.2. Avoiding prophenoloxidase (PPO) activation while collecting haemolymph

Per Kopáček *et al.* (1995):

1. Precool the larvae for 15 min at 4°C.
2. Collect the haemolymph per Section 6.2.1.
3. Flush the haemolymph from the capillary into an Eppendorf tube kept on ice.
4. Add ice cold CA-CAC buffer (20 mM $CaCl_2$ and 10 mM Na-cacodylate, pH 6.5).
5. Vigorously agitate the tube.
6. Freeze immediately in liquid nitrogen.
7. Store the frozen haemolymph at -20°C.

6.2.3. Removing haemocytes from haemolymph

Per Fröbius *et al.* (2001):

1. Collect haemolymph from wax moth larvae per Section 6.2.1.
2. Transfer the haemolymph to chilled tubes. The tubes should contain a few crystals of phenylthiourea to prevent melanization (Vilcinskas *et al.*, 1997; Wedde *et al.*, 1998).
3. Centrifuge the haemolymph twice at 100 g for 10 min to remove the haemocytes.
4. Store the supernatants at -20°C until needed.

6.3. Eliciting immune responses in wax moth larvae

Per Wedde *et al.* (1998) and Fröbius *et al.* (2001):

1. Suspend 20 mg zymosan A (Sigma) in 1 ml of sterile, physio-logical saline (172 mM KCl, 68 mM NaCl, 5 mM $NaHCO_3$, pH 6.1, adjusted with HCl).
2. Homogenize the mixture with a vortex.
3. Centrifuge at 10,000 g for 5 min.
4. Inject the supernatant and solubilized content at 10 μl supernatant/larva following Section 5.2.

6.4. Alternative method for eliciting immune responses in wax moth larvae

Per Schuhmann *et al.* (2003):

1. Inject per Section 5.2. last instar larvae (250-350 mg) with 10 μl of bacterial lipopolysaccharide suspension (2 mg/ml in water; Sigma, Deisenhofen, Germany).
2. Keep the larvae at 30°C for desired amount of time. Schuhmann *et al.* (2003) allowed them to sit for 4, 6, and 8 h – per desired experimental conditions.

6.5. Collecting greater wax moth larva cuticle

Per Samšiňáková *et al.* (1971). Samšiňáková *et al.* (1971) collected the cuticle from the greater wax moth on which they tested the action of enzymatic systems of *Beauveria bassiana*. The cuticle was collected two ways. In the first method, all of the accompanying biological material was removed from the cuticle (see section 6.5.1.). For the second method, the authors attempted to keep the cuticle as natural as possible, leaving the deteriorated cuticle with adjacent epidermis (6.5.2.).

6.5.1. Complete isolation of the larval cuticle

1. Euthanize fully grown greater wax moth larvae with ether.
2. Boil them in 5% KOH for 3 h.
3. Wash the larvae with water.
4. Place overnight in 2% pancreatin at pH 8.5 and 37°C.
5. Wash the remaining material with water.
6. Remove any remaining tissues.
7. Immerse the cuticle in boiling water for 20 min.
8. Centrifuge to remove excess water.
9. Dry the cuticle in a stream of hot air to constant weight.

6.5.2. Rough isolation of the larval cuticle

1. Euthanize the wax moth larva.
2. Dissect away the larva's head.
3. Press the larva with a glass rod from the posterior end to the anterior end. This squeezes out the larva's viscera.
4. Rinse the remaining integument with distilled water
5. Dry the integument carefully.

6.6. Surface sterilization of wax moth larvae

Per Reddy *et al.* (1979):

1. Surface sterilize wax moth larvae with a wash (whole body) or rub (target body part) of 70% ethanol.
2. Manipulate (including dissection) the sterilized individual in sterile insect Ringers solution.

6.7. Containment of various moth life stages

6.7.1. Frame caging

Per Burgett and Tremblay (1979):

1. Construct circular cages by replacing the metal sealing lid of 2 Mason® jar screw caps with 11.5 mesh/10 mm wire gauze and taping the two open sides together.
2. Place contains diet medium (section 3.2.1.) in each cage.
3. Place wax moths (life stage dependent on the project goals) in each cage.
4. Secure up to nine cages, in rows of three, with large rubber bands in a standard "deep" Langstroth hive frame (480 × 29 × 230 mm; l × w × h) with no comb or foundation. Up to six cages, in rows of tree, can be secured to a standard "medium" Langstroth honey frame (480 × 29 × 160; l × w × h) with no comb or foundation.
5. The wax moth frame may be inserted into a nucleus or full sized brood box for trials.

Considerations: Frame caging is not ideal for bioassays involving eggs and early larvae. Dish containers (see Section 6.7.2.) are ideal for egg and early instar larval assays.

6.7.2. Dish caging

Per Goodman *et al.* (1990).

1. Place eggs and early instar larvae (collected per Section 3) in a small specimen tube. Goodman *et al.* (1990) used one that was 25 x 75 mm, with diet medium (for newly hatched larvae – prepared per Section 3.2.1.). The vial opening should be covered with 24 mesh/10 mm (or similar) wire gauze. First instar moth larvae are small so care should be taken to limit their escape from the dish cages.
2. Larvae, pupae, adults: a 13 mm hole is bored in the lid of an 85 mm diameter (or similar sized) plastic Petri dish. The hole is covered with 11.5 mesh/10 mm (or similar) wire gauze. Specimens are placed in the Petri dish with diet medium (diet prepared per Section 3.2.1.).

6.8. Creating wax moth free combs

Per Hood *et al.* (2003):

1. Remove drawn comb for honey bee colonies.

The comb should be dark (i.e. comb that has had brood reared in it at some point).

2. Extract any honey present.
3. Expose comb to foraging bees to remove any remaining honey residues.
4. Place all comb in a standard freezer (≤ 0°C) for at least 24 h to kill all wax moth life stages present (for more freezing temperatures and times, see section 5.8.1.).
5. Examine all frames, and select frames with no signs of wax moth activity for experimental trials.

7. Conclusion

Although we include a number of methods associated with the study of wax moths in this paper, there remain methodological gaps for this important pest of honey bees. For example, we failed to find a method to artificially infest field colonies with wax moths. Such a method may seem intuitive, (just open the colony and insert moths), but it is not considering the natural tendency for adult bees to eject immature moths from colonies. We also discovered no methods related to marking/ recapturing the various moth life stages, or how to determine damage thresholds for the moths. These are but a sample of methods that would prove useful to researchers, especially those investigating wax moths from an apicultural perspective.

In sharp contrast to applied methods related to wax moth research, there are a plethora of research methodologies related to basic investigations on wax moths. This is especially true of investigations focused on wax moth physiology, genomics, and proteomics. We considered adding these methods to our paper, but soon realized that an entire book (similar to the *BEEBOOK*) could be written just about wax moth research methods. Including a comprehensive bibliography of the wax moth literature seemed to be a good compromise, but we discovered that this could include many thousands of references. Such an inclusion would be beyond the scope of this paper, but we hope such a bibliography will be published in the future.

In conclusion, wax moths remain a vexing problem for beekeepers and honey bee colonies around the globe. The number of investigations related to wax moth control has dropped significantly, largely due to the perception of wax moths as a secondary pest of bee colonies. Regardless, they remain an important test model for entomologists, physiologists, and investigators from other disciplines. Based on current trends in wax moth research, we expect that wax moth usefulness to investigators will continue into perpetuity.

References

ARBOGAST, R T; LEONARD LECATO, G; VAN BYRD, R (1980) External morphology of some eggs of stored-product moths (Lepidoptera: Pyralidae, Gelechiidae, Tineidae). *International Journal of Insect Morphology and Embryology* 9(3): 165-177. ISSN: 0020-7322 http://dx.doi.org/10.1016/0020-7322(80)90013-6

BRONSKILL, J F (1961) A cage to simplify the rearing of the greater wax moth, *Galleria mellonella* (Pyralidae). *Journal of the Lepidopterists' Society* 15(2): 102-104.

BURGES, H; BAILEY, L (1968) Persistence of *Bacillus thuringiensis* in foundation beeswax and beecomb in beehives for the control of *Galleria mellonella*. *Journal of Invertebrate Pathology* 28(2): 217-222. http://dx.doi.org/10.1016/0022-2011(76)90125-7

BURGES, H; BAILEY, L (1968) Control of the greater and lesser wax moths (*Galleria mellonella* and *Achroia grisella*) with *Bacillus thuringiensis*. *Journal of Invertebrate Pathology* 11(2): 184-195. http://dx.doi.org/10.1016/0022-2011(68)90148-1

BURGETT, D M; TREMBLAY, A (1979) Controlled release fumigation of the greater wax moth. *Journal of Economic Entomology* 72: 616-617.

CANTWELL, G E; JAY, E; PEARMAN Jr, G P; THOMPSON, J (1972) Control of the greater wax moth, *Galleria mellonella* (L.), in comb honey with carbon dioxide: Part I. *American Bee Journal* 112: 302-303.

CARPENTER, J; BLOEM, S; MAREC, F (2005) Inherited sterility in insects. In *V A Dyck; J Hendrichs; A S Robinson (Eds) Sterile Insect Technique* pp. 115-146. http://dx.doi.org/10.1007/1-4020-4051-2_5

CHARRIERE, J-D; IMDORF, A (1999) Protection of honey combs from wax moth damage. *American Bee Journal* 139(8): 627-630.

CHASE, R W (1921) The length of the life of the larva of the wax moth, *Galleria mellonella* L., in its different stadia. *Transactions of the Wisconsin Academy of Sciences, Arts and Letters* 20: 263-267.

COSKUN, M; KAYIS, T; SULANC, M; OZALP, P (2006) Effects of different honeycomb and sucrose levels on the development of greater wax moth *Galleria mellonella* larvae. *International Journal of Agriculture and Biology* 8(6): 855-858.

DELAPLANE, K S; VAN DER STEEN, J; GUZMAN, E (2013) Standard methods for estimating strength parameters of *Apis mellifera* colonies. In *V Dietemann; J D Ellis; P Neumann (Eds) The COLOSS* BEEBOOK, *Volume I: standard methods for* Apis mellifera *research. Journal of Apicultural Research* 52(1): http://dx.doi.org/10.3896/IBRA.1.52.1.03

DIETEMANN, V; NAZZI, F; MARTIN, S J; ANDERSON, D; LOCKE, B; DELAPLANE, K S; WAUQUIEZ, Q; TANNAHILL, C; FREY, E; ZIEGELMANN, B; ROSENKRANZ, P; ELLIS, J D (2013) Standard methods for varroa research. In *V Dietemann; J D Ellis; P Neumann (Eds) The COLOSS* BEEBOOK, *Volume II: standard methods for* Apis mellifera *pest and pathogen research. Journal of Apicultural Research* 52(1): http://dx.doi.org/10.3896/IBRA.1.52.1.09

EISCHEN, F A; DIETZ, A (1987) Growth and survival of *Galleria mellonella* (Lepidoptera: Pyralidae) larvae fed diets containing honey bee-collected plant resins. *Annals of the Entomological Society of America* 80: 74-77.

EISCHEN, F; DIETZ, A (1990) Improved culture techniques for mass rearing *Galleria mellonella* (Lepidoptera: Pyralidae). *Entomological News* 101: 123-128.

ELLIS, A M; HAYES, G W (2009) Assessing the efficacy of a product containing *Bacillus thuringiensis* applied to honey bee (Hymenoptera: Apidae) foundation as a control for *Galleria mellonella* (Lepidoptera: Pyralidae). *Journal of Entomological Science* 44(2): 158-163.

FERGUSON, D C (1987) Lepidoptera. In *J R Borham (Ed.). Insect and mite pests in food: an illustrated key. USDA Agriculture Handbook* 655: 231-244.

FLINT, H; MERKLE, J (1983) Mating behaviour, sex pheromone responses, and radiation sterilization of the greater wax moth (Lepidoptera: Pyralidae). *Journal of Economic Entomology* 76: 467-472.

FRÖBIUS, A C; KANOST, M R; GÖTZ, P; VILCINSKAS, A (2000) Isolation and characterization of novel inducible serine protease inhibitors from larval haemolymph of the greater wax moth *Galleria mellonella*. *European Journal of Biochemistry* 267(7): 2046-2053. http://dx.doi.org/10.1046/j.1432-1327.2000.01207.x

GOODMAN, R; WILLIAMS, P; OLDROYD, B; HOFFMAN, J (1990) Studies on the use of phosphine gas for the control of greater wax moth (*Galleria mellonella*) in stored honey bee comb. *American Bee Journal* 130: 473-477.

HARTFELDER, K; GENTILE BITONDI, M M; BRENT, C; GUIDUGLI-LAZZARINI, K R; SIMÕES, Z L P; STABENTHEINER, A; DONATO TANAKA, É; WANG, Y (2013) Standard methods for physiology and biochemistry research in *Apis mellifera*. In *V Dietemann; J D Ellis; P Neumann (Eds) The COLOSS BEEBOOK, Volume I: standard methods for* Apis mellifera *research. Journal of Apicultural Research* 52(1): http://dx.doi.org/10.3896/IBRA.1.52.1.06

HOOD, W M; HORTON, P M; MCCREADIE, J W (2003) Field evaluation of the red imported fire ant (Hymenoptera: Formicidae) for the control of wax moths (Lepidoptera: Pyralidae) in stored honey bee comb. *Journal of Agricultural and Urban Entomology* 20(2): 93-103.

JAFARI, R; GOLDASTEH, S; AFROGHEH, S (2010) Control of the wax moth *Galleria mellonella* L. (Lepidoptera: Pyralidae) by the male sterile technique (MST). *Archives of Biological Sciences* 62(2): 309-313. http://dx.doi.org/10.2298/ABS1002309J

JAMES, R R (2011) Potential of ozone as a fumigant to control pests in honey bee (Hymenoptera: Apidae) hives. *Journal of Economic Entomology* 104: 353-359. http://dx.doi.org/10.1603/EC10385

JONES, G; BARABAS, A; ELLIOTT, W; PARSONS, S (2002) Female greater wax moths reduce sexual display behaviour in relation to the potential risk of predation by echolocating bats. *Behavioural Ecology* 13(3): 375-380. http://dx.doi.org/10.1093/beheco/13.3.375

KOPÁCEK, P; WEISE, C; GÖTZ, P (1995) The prophenoloxidase from the wax moth *Galleria mellonella*: Purification and characterization of the proenzyme. *Insect Biochemistry and Molecular Biology* 25: 1081-1091. http://dx.doi.org/10.1016/0965-1748(95)00040-2

MARSTON, N L; CAMPBELL, B; BOLDT, P (1975) Mass producing eggs of the greater wax moth, *Galleria mellonella* (L.). *Agricultural Research Service, US Department of Agriculture Technical Bulletin* 1510: 15 pp.

MOLYNEUX, A S (1985) Survival of infective juveniles of *Heterorhabditis* spp., and *Steinernema* spp. (Nematoda: Rhabditida) at various temperatures and their subsequent infectivity for insects. *Revue De Nématologie* 8: 165-170.

NEUMANN, P; PIRK C W W; SCHÄFER, M O; ELLIS, J D (2013) Standard methods for small hive beetle research. In *V Dietemann; J D Ellis, P Neumann (Eds) The COLOSS BEEBOOK: Volume II: Standard methods for* Apis mellifera *pest and pathogen research. Journal of Apicultural Research* 52(4): http://dx.doi.org/10.3896/IBRA.1.52.4.19

NORTH, D T (1975) Inherited sterility in Lepidoptera. *Annual Review of Entomology* 20: 167-182.

PADDOCK, F B (1918) *The beemoth or waxworm*. Texas Agricultural Experiment Station; USA. 44 pp.

REDDY, G; HWANG-HSU, K; KUMARAN, A K (1979) Factors influencing juvenile hormone esterase activity in the wax moth, *Galleria mellonella. Journal of Insect Physiology* 25: 65-71. http://dx.doi.org/10.1016/0022-1910(79)90038-6

SAMSINAKOVA, A; MISIKOVA, S; LEOPOLD, J (1971) Action of enzymatic systems of *Beauveria bassiana* on the cuticle of the greater wax moth larvae (*Galleria mellonella*). *Journal of Invertebrate Pathology* 18: 322-330. http://dx.doi.org/10.1016/0022-2011(71)90033-4

SAMMATARO, D; DE GUZMAN, L; GEORGE, S; OCHOA, R (2013) Standard methods for tracheal mites research. In *V Dietemann; J D Ellis, P Neumann (Eds) The COLOSS BEEBOOK: Volume II: Standard methods for* Apis mellifera *pest and pathogen research. Journal of Apicultural Research* 52(4): http://dx.doi.org/10.3896/IBRA.1.52.4.20

SCHUHMANN, B; SEITZ, V; VILCINSKAS, A; PODSIADLOWSKI, L (2003) Cloning and expression of gallerimycin, an antifungal peptide expressed in immune response of greater wax moth larvae, *Galleria mellonella. Archives of Insect Biochemistry and Physiology* 53: 125-133. http://dx.doi.org/10.1002/arch.10091

SHARMA, V; MATTU, V K; THAKUR, M S (2011) Infestation of *Achoria grisella* F. (wax moth) in honey combs of *Apis mellifera* L. in Shiwalik Hills, Himachal Pradesh. *International Journal of Science and Nature* 2(2): 407-408.

SHIMANUKI, H; KNOX, D; FURGALA, B; CARON, D; WILLIAMS, J (1992) Diseases and pests of honey bees In *J M Graham (Ed.). The hive and the honey bee*. Dadant and Sons; Hamilton, IL, USA. pp. 1083-1151.

SMITH, T L (1965) External morphology of the larva, pupa and adult of the wax moth *Galleria mellonella* L. *Journal of the Kansas Entomological Society* 38: 287-310.

SPANGLER, H G (1984) Attraction of female lesser wax moths (Lepidoptera: Pyralidae) to male-produced and artificial sounds. *Journal of Economic Entomology* 77: 346-349.

SPANGLER, H G (1985) Sound production and communication by the greater wax moth (Lepidoptera: Pyralidae). *Annals of the Entomological Society of America* 78: 54-61.

SPANGLER, H G (1987) Acoustically mediated pheromone release in *Galleria mellonella* (Lepidoptera: Pyralidae). *Journal of Insect Physiology* 33: 465-468. http://dx.doi.org/10.1016/0022-1910(87)90109-0

STEPHENS, J M (1959) Immune responses of some insects to some bacterial antigens. *Canadian Journal of Microbiology* 5(2): 203-228. http://dx.doi.org/10.1139/m59-025

STEPHENS, J M (1962) Bactericidal activity of the blood of actively immunized wax moth larvae. *Canadian Journal of Microbiology* 8: 491-499. http://dx.doi.org/10.1139/m62-064

VANDENBERG, J; SHIMANUKI, H (1990) Application methods for *Bacillus thuringiensis* used to control larvae of the greater wax moth (Lepidoptera: Pyralidae) on stored beeswax combs. *Journal of Economic Entomology* 83: 766-771.

VILCINSKAS, A; MATHA, V; GÖTZ, P (1997) Effects of the entomopathogenic fungus *Metarhizium anisopliae* and its secondary metabolites on morphology and cytoskeleton of plasmatocytes isolated from the greater wax moth, *Galleria mellonella. Journal of Insect Physiology* 43: 1149-1159.
http://dx.doi.org/10.1016/S0022-1910(97)00066-8

WARREN, L; HUDDLESTON, P (1962) Life history of the greater wax moth, *Galleria mellonella* L., in Arkansas. *Journal of the Kansas Entomological Society* 35: 212-216.

WATERHOUSE, D (1959) Axenic culture of wax moths for digestion studies. *Annals of the New York Academy of Sciences* 77: 283-289.
http://dx.doi.org/10.1111/j.1749-6632.1959.tb36909.x

WEDDE, M; WEISE, C; KOPACEK, P; FRANKE, P; VILCINSKAS, A (1998) Purification and characterization of an inducible metalloprotease inhibitor from the haemolymph of greater wax moth larvae, *Galleria mellonella. European Journal of Biochemistry* 255: 535-543.
http://dx.doi.org/10.1046/j.1432-1327.1998.2550535.x

WEST, E J; BRIGGS, J D (1968) *In vitro* toxin production by the fungus *Beauveria bassiana* and bioassay in greater wax moth larvae. *Journal of Economic Entomology* 61: 684-687.

WILLIAMS, J L (1997) Insects: Lepidoptera (moths). In *R Morse; K Flottum (Eds). Honey bee pests, predators, and diseases.* The AI Root Company; Ohio, USA. pp. 121-141.

XUEJUAN, F; HOMINICK, W M (1991) Efficiency of the galleria (wax moth) baiting technique for recovering infective stages of entomopathogenic rhabditids (Steinernematidae and Heterorhabditidae) from sand and soil. *Revue Nématol* 14: 381-387.

Journal of Apicultural Research 52(1)

Journal of Apicultural Research 52(1): (2013)
DOI 10.3896/IBRA.1.52.1.11

REVIEW ARTICLE

Standard methods for American foulbrood research

Dirk C de Graaf[1*], Adriana M Alippi[2], Karina Antúnez[3], Katherine A Aronstein[4], Giles Budge[5], Dieter De Koker[1 §], Lina De Smet[1], Douglas W Dingman[6], Jay D Evans[7], Leonard J Foster[8], Anne Fünfhaus[9], Eva Garcia-Gonzalez[9], Aleš Gregorc[10], Hannelie Human[11], K Daniel Murray[12], Bach Kim Nguyen[13], Lena Poppinga[9], Marla Spivak[14], Dennis vanEngelsdorp[15], Selwyn Wilkins[5] and Elke Genersch[9]

[1]Ghent University, Laboratory of Zoophysiology, K.L. Ledeganckstraat 35, B-9000 Ghent, Belgium.
[2]Universidad Nacional de La Plata, Facultad de Ciencias Agrarias y Forestales, Centro de Investigaciones de Fitopatología, calles 60 y 118, c.c. 31, 1900 La Plata, Argentina.
[3]Instituto de Investigaciones Biológicas Clemente Estable, Department of Microbiology, Avda. Italia 3318, 11600 Montevideo, Uruguay.
[4]USDA-ARS, Honey Bee Research Unit, Kika de la Garza Subtropical Agricultural Center, 2413 E. Hwy 83, 78596 Weslaco TX, USA.
[5]National Bee Unit, Food and Environment Research Agency, Sand Hutton, YO41 1LZ York, UK.
[6]The Connecticut Agricultural Experiment Station, Department of Biochemistry and Genetics, New Haven CT 06504, USA.
[7]USDA-ARS, Bee Research Lab, BARC-E Bldg 476, Beltsville MD 20705, USA.
[8]University of British Columbia, Department of Biochemistry & Molecular Biology, 2125 East Mall, V6T 1Z4 Vancouver BC, Canada.
[9]Institute for Bee Research, Friedrich-Engels-Str. 32, 16540 Hohen Neuendorf, Germany.
[10]Agricultural Institute of Slovenia, Hacquetova 17, 1000 Ljubljana, Slovenia.
[11]University of Pretoria, Department of Zoology and Entomology, Pretoria, South Africa.
[12]South Texas College, Biology Department, 400 N. Border, 78596 Weslaco TX, USA.
[13]Université de Liège, Gembloux Agro-Bio Tech, Entomologie fonctionnelle & évolutive, Passage des Déportés 2, B-5030 Gembloux, Belgium.
[14]University of Minnesota, Department of Entomology, 219 Hodson Hall, 1980 Folwell Ave., 55108 St. Paul MN, USA.
[15]University of Maryland, Department of Entomology, 3136 Plant Sciences, College Park, MD 20742, USA

§ *in memoriam*: ° July 24, 1987 - † May 30, 2012

Received 10 April 2012, accepted subject to revision 18 June 2012, accepted for publication 5 November 2012.

*Corresponding author: Email: Dirk.deGraaf@UGent.be

Summary

American foulbrood is one of the most devastating diseases of the honey bee. It is caused by the spore-forming, Gram-positive rod-shaped bacterium *Paenibacillus larvae*. The recent updated genome assembly and annotation for this pathogen now permits in-depth molecular studies. In this paper, selected techniques and protocols for American foulbrood research are provided, mostly in a recipe-like format that permits easy implementation in the laboratory. Topics covered include: working with *Paenibacillus larvae*, basic microbiological techniques, experimental infection, and "'omics" and other sophisticated techniques. Further, this chapter covers other technical information including biosafety measures to guarantee the safe handling of this pathogen.

Métodos para la investigación de la loque americana

Resumen

La loque americana es una de las enfermedades más devastadoras de la abeja melífera, causada por el bacilo, formador de esporas Gram-positivo *Paenibacillus larvae*. El reciente ensamblaje y anotación del genoma de este patógeno permite actualmente la realización de profundos estudios moleculares. En este trabajo, se proporcionan técnicas y protocolos seleccionados para la investigación de la loque americana, principalmente bajo la forma de protocolos de trabajo con una estructura similar al de las recetas, para facilitar su implementación en el laboratorio. Los temas desarrollados incluyen: el trabajo con *Paenibacillus larvae*, técnicas básicas microbiológicas, la infección experimental, y "'ómicas" y otras técnicas sofisticadas. Además, este capítulo abarca otro tipo de información técnica, incluyendo medidas de bioseguridad para garantizar la seguridad en el manejo de este patógeno.

Footnote: Please cite this paper as: DE GRAAF, D C; ALIPPI, A M; ANTÚNEZ, K; ARONSTEIN, K A; BUDGE, G; DE KOKER, D; DE SMET, L; DINGMAN, D W; EVANS, J D; FOSTER, L J; FÜNFHAUS, A; GARCIA-GONZALEZ, E; GREGORC, A; HUMAN, H; MURRAY, K D; NGUYEN, B K; POPPINGA, L; SPIVAK, M; VANENGELSDORP, D; WILKINS, S; GENERSCH, E (2013) Standard methods for American foulbrood research. In *V Dietemann; J D Ellis; P Neumann (Eds) The COLOSS BEEBOOK, Volume II: standard methods for* Apis mellifera *pest and pathogen research. Journal of Apicultural Research* 52(1): http://dx.doi.org/10.3896/IBRA.1.52.1.11

美洲幼虫腐臭病研究的标准方法

美洲幼虫腐臭病是最具毁灭性的疾病之一，由革兰氏阳性杆状菌 *Paenibacillus larvae* 引起。近年来，随着基因组学的开展，该病原体的基因组组装和注释已成为开展，深入的分子研究成为可能。本文提供了经选择的美洲幼虫腐臭病研究技术和实验程序，大多数以"食谱"的格式给出，很容易在实验室开展操作。覆盖的主题包括：*Paenibacillus larvae* 的处理技术，基本微生物技术、实验感染技术、"组学"以及其他的一些复杂技术。此外，本章还包含了生物安全的评价方法，以确保安全的开展该病原体的研究

Key words: honey bee, American foulbrood, *Paenibacillus larvae*, brood, disease, pathogen, technique, *BEEBOOK*, COLOSS

Table of Contents

1. Introduction

American foulbrood (AFB) is a devastating brood disease of the honey bee caused by the spore-forming, Gram-positive rod-shaped bacterium *Paenibacillus larvae*. AFB is one of the bee diseases listed in the *OIE* (Office International des Epizooties – the World Organization for Animal Health) *Terrestrial Animal Health Code* (2011) and member countries and territories are obliged to report its occurrence. In 2006, a draft of the *P. larvae* genome was published at an estimated 5-6x coverage (Qin *et al.*, 2006). Last year, this coverage was further extended and the genome sequence was further annotated with a combination of bioinformatics and proteomics (Chan *et al.*, 2011). These efforts will certainly help to usher in the next level of research for this economically important pathogen, ultimately allowing us to better understand the intimate relationship between the pathogen and its host. More generally, the honey bee / AFB system provides a wealth of opportunities and tools for addressing basic questions regarding microbe-microbe interactions, host immunity, strain virulence, and horizontal transmission, among others.

In the present paper, selected techniques and protocols in American foulbrood research are provided, mostly in a recipe-like format that permits easy implementation in the laboratory. The different topics that are covered include: working with *Paenibacillus larvae*, basic microbiological techniques, experimental infection and "'omics" and other sophisticated techniques. Thus, the chapter covers a broad set of technical information going from biosafety measures to guarantee the safe handling of this pathogen to the expression of heterologous proteins in *P. larvae*. Techniques exclusively related to the diagnosis of AFB are not included as they have been reviewed elsewhere (de Graaf *et al.*, 2006a; OIE, 2008).

2. Working with *Paenibacillus larvae*

2.1. Biosafety measures

In some countries, microbial species are categorized in different classes based on biosafety risk. Each biosafety risk class has its own recommendations with respect to facility design, safety equipment, and working practices (de Graaf *et al.*, 2008). This classification mostly takes into account the risk for human health, potential for dispersal of the disease, and the potential economic impact of the disease. However, a generally accepted biosafety risk classification has not been prescribed for *P. larvae*. Consideration of the severity of clinical American foulbrood infections in honey bee colonies, the contagiousness, the longevity of the spores, the legal context of AFB (a notifiable disease), and the economic value of honey bee pollination services, justifies *P. larvae* classification as an organism with 'high biosafety risk for animals'. Table 1 summarizes basic biosafety practices - mainly in accordance with the Belgian model

(http://www.biosafety.be/) - that should be considered when manipulating *P. larvae* for research or diagnosis. It should be noted that *P. larvae* has been safely cultured in the laboratory for decades using only standard bacteriological procedures (i.e. aseptic handling techniques and careful decontamination of biological waste), and to date, no known AFB outbreaks due to intentional/accidental laboratory release of this organism have been reported.

2.2. Strains

Various *P. larvae* strains can be obtained from bacterial culture collections. However, many of these collection agencies store exactly the same strains but with their own strain designations. This fact is relevant when comparing different *P. larvae* strains for research purposes, or when using strains as positive controls in diagnosis or species identification. Table 2 shows many of the important strains that are available in these culture collections and shows some alternate designations. With regard to obtaining strains, it should be noted that certain countries have import/export and interstate transport shipping regulations regarding the movement and storage of this pathogen (http://www.biosafety.be/RA/Class/ListBact.html).

2.3. Sampling for AFB monitoring or diagnosis

Testing for the presence of *P. larvae* may be carried out for different reasons – for example either as part of national monitoring or prevention programmes, or as part of scientific research projects such as epidemiological studies. The proper collection procedure depends on whether the testing is to be carried out following the observation of suspected clinical signs of AFB, or the testing is part of general surveillance to identify a potential sub-clinical presence of *P. larvae* in colonies within a population. These considerations are addressed in the *OIE Manual of Diagnostic Tests and Vaccines for Terrestrial Animals* vol.1 (OIE, 2008), and we advise readers to consult the AFB chapter in that resource.

3. Basic microbiological techniques

3.1. Cultivation

As *P. larvae* is a spore-forming bacterium, its isolation from biological samples is typically preceded by a heat treatment step to kill all vegetative microorganisms. This step significantly reduces the risk that *P. larvae* colonies will be masked by competitors. Different genotypes of *P. larvae* show variation in germination ability, and their response to heat treatment is variable (Forsgren *et al.*, 2008). MYPGP agar (Dingman and Stahly, 1983) is routinely used to cultivate *P. larvae* for AFB diagnosis. This medium makes incubation under CO_2 unnecessary although the presence of 5% CO_2 significantly increases germination (Nordstrom and Fries, 1995). Contaminants of the genera *Bacillus* and *Brevibacillus*, as well as other *Paenibacillus* species, are

Table 1. Biosafety rules mainly in accordance to the Belgian model; http://www.biosafety.be/. As this model is very restrictive, we have indicated the rules that are minimally (**M**) required. The other biosafety rules should be considered optional depending on the demands of the responsible authority.

Topic	Measurement
Facility design	- The laboratories are physically separated from other work areas in the building. - The access to the laboratories are locked if the zone or corridor access is not reserved. The doors have automatic closing if they open directly into a public area. - (**M**) The furniture has been designed to allow easy cleaning and disinfection, and easy insect and rodent control. - (**M**) There is a sink for washing and decontamination of hands in the laboratory. - (**M**) There are coat hooks or a dressing room equipped with protective clothing. Normal clothing and protective clothing should remain separated. - (**M**) The tables are easy to clean, water impermeable and resistant to acids, alkalis, organic solvents, disinfectants and decontamination
Safety equipment	- If the laboratory is equipped with a microbiological safety cabinet (MSC) of class II, it must be localized as such that it does not disturb the air flow in the room. It should be kept at a sufficient distance from windows, doors, and places with frequent passage, vents for air intake or outlet. The MSC should be checked and certified upon purchase or relocation, as well as at least once a year afterward. - There is an autoclave available in the building if the biological waste and/or biological residues shall be inactivated by steam sterilization. - The centrifuge that will be used is available in the containment zone. If this is not the case, and centrifugation is done outside the
Working practices	- Access to laboratories is restricted to persons approved by the responsible authority, and these persons have been informed of the biological risk. - Laboratory doors should display: the biohazard sign, the containment level (if applicable), the coordinates of the controller. - (**M**) Protective clothing is worn. This protective clothing should not be worn outside the laboratory. - There are gloves available for staff. - The windows must remain closed during experiments. - (**M**) Viable (micro-) organisms must be physically contained in closed systems (tubes, boxes, etc.), when they are not being manipulated. - (**M**) Splashes or aerosols should be minimized, and their spread must be controlled by appropriate equipment and work practices. - In no case may a horizontal laminar flow cabinet used for manipulation of pathogenic organisms. - (**M**) Mechanical pipetting is required. Pipetting by mouth is prohibited. - (**M**) Drinking, eating, smoking, use of cosmetics, handling contact lenses and storage of food for human consumption is prohibited in the laboratories. - A register of all manipulated or stored pathogenic organisms should be kept. - The control measures and equipment should be inspected regularly and in an appropriate manner. - (**M**) Hands should be washed when leaving the laboratory, when another activity is started, or when deemed necessary. - (**M**) After completion of the work, or when biological material has been spilled, the work surfaces should be disinfected. - There is a note available for the staff describing the correct use of the disinfectants. This memo specifies for a given purpose, the disinfectant that must be used, the necessary concentration and contact time. - The staff is trained in relation to biosafety issues and is regularly monitored and retrained. - A biosafety manual has been written and adopted. The staff is informed of the potential risks and must read the biosafety regulations that are applicable. Instructions that must be followed in case of accident should be posted in the laboratory. - The biohazard sign is posted in incubators, freezers, and nitrogen tanks containing biological material 'high biosafety risk for animals'. - An effective insect and rodent control program is applied.
Waste management	- (**M**) Contaminated biological waste and/or biological residues and contaminated disposable equipment should be inactivated by an appropriate, validated method before it is discharged--e.g., by autoclaving or by incineration. The incineration is performed by an authorized company. The waste is collected in secure and hermetically sealable containers. These should be closed for transport. - (**M**) Contaminated material (glassware, etc.) is inactivated by an appropriate, validated method before cleaning, reuse and/or destruction.

inhibited by nalidixic acid (Hornitzky and Clark, 1991) and pipemidic acid (Alippi, 1991; 1995). Apart from brood samples, food stores (honey, pollen and royal jelly), adult workers, and wax debris can also be used to detect the presence of *P. larvae* spores.

The outline of the cultivation procedure starting from brood samples is as follows:

1. Prepare an aqueous solution containing *P. larvae* spores by taking twice samples with a sterile swab from a brood comb (each time multiple brood cells should be sampled), and subsequently suspending them in 5 ml of phosphate buffered saline (PBS).

2. Incubate different aliquots of the spore suspension at 80, 85, 90, 95 and 100°C for 10 min (Forsgren *et al.*, 2008).

P. larvae occurs in two forms: vegetative cells and spores. Only

Table 2. Strains of *Paenibacillus larvae*.

Species	Old subsp. classification	Strain no.	Other designation	Source
Paenibacillus larvae	*larvae*	LMG 9820[T]	ATCC 9545[T]	Foulbrood of honey bees
			DSM 7030[T]	
			NRRL B-2605[T]	E C Holst #846
			LMG 15969[T]	
Paenibacillus larvae	*larvae*	LMG 14425	ATCC 25747	Ohio, USA, diseased honey bee larvae
Paenibacillus larvae	*larvae*	LMG 15969[T]	See LMG 9820[T]	See LMG 9820[T]
Paenibacillus larvae	*larvae*	LMG 16245	NRRL B-3650	Diseased honey bee larvae L Bailey, Rothamsted Expt. Station, Harpenden, UK. strain Australia ("Victoria")
Paenibacillus larvae	*larvae*	LMG 18149	Hornitzky 89/2302/4	Victoria, Australia, honey bee
Paenibacillus larvae	*pulvifaciens*	LMG 6911[T]	ATCC 13537[T]	Dead larvae honey bee
			DSM 3615[T]	
			IFO 15408[T]	
			NCIMB 11201[T]	
			NRRL B-3688[T]	
			NRRL B-3685[T]	
			NRRL B-3670[T]	
			LMG 16248[T]	
			LMG 15974[T]	
Paenibacillus larvae	*pulvifaciens*	LMG 14427	ATCC 25367	Unknown
Paenibacillus larvae	*pulvifaciens*	LMG 14428	ATCC 25368	Unknown
Paenibacillus larvae	*pulvifaciens*	LMG 15974[T]	See LMG 6911[T]	See LMG 6911[T]
Paenibacillus larvae	*pulvifaciens*	LMG 16247	NRRL B-3687	(1949), honey bee larvae, H Katznelson #754
Paenibacillus larvae	*pulvifaciens*	LMG 16248[T]	See LMG 6911[T]	See LMG 6911[T]
Paenibacillus larvae	*pulvifaciens*	LMG 16250	NRRL B-14154	Unknown
Paenibacillus larvae	*pulvifaciens*	LMG 16251	CCM 38	Unknown
			CCUG 7427	
			NCFB 1121	
			NRRL NRS-1283	Powdery scale, H Katznelson #113
Paenibacillus larvae	*pulvifaciens*	LMG 16252	DSM 8443	Dead honey bee larvae
			NRRL NRS-1684	

spores are infectious to honey bees. While *P. larvae* sporulates and grows efficiently in the haemolymph of bee larvae, most strains grow poorly in artificial media. Different culture media have been developed for *P. larvae* cultivation. MYPGP agar (Dingman and Stahly, 1983) yielded the highest percentage of spore recovery, while J-agar (Hornitzky and Nicholls, 1993), brain heart infusion agar (BHI) (Gochnauer, 1973), Columbia sheep blood agar (CSA) (Hornitzky and Karlovskis, 1989) proved to be less efficient in this respect (Nordström and Fries, 1995). Other media used for the cultivation of *P. larvae* are

PLA agar (Schuch *et al.*, 2001) and T-HCl-YGP agar (Steinkraus and Morse, 1996). PLA medium shows superior plating efficacy and also the advantage of inhibiting the majority of micro-organisms normally present in the hive and in bee products. T-HCl-YGP agar is the medium of choice for cultivation *P. larvae* starting from honey (Steinkraus and Morse, 1996). When starting from diseased larvae, nalidixic acid is necessary to avoid growth of *P. alvei*. When starting from other sources (e.g. honey), pipemidic acid prevents contamination with other spore-forming bacteria.

MYPGP agar (per litre):

- 10 g Mueller-Hinton broth (Oxoid CM0405)
- 15 g yeast extract
- 3 g K_2HPO_4
- 1 g Na-pyruvate
- 20 g agar
- Autoclave at 121°C/15 min.
- Add 20 ml 10% glucose (autoclaved separately).

BHI agar:

- Suspend 47 g brain heart infusion agar (Oxoid CM1136) in 1 litre of distilled water.
- Autoclave at 121°C for 15 min.
- Add 1 mg thiamine hydrochloride per litre.

CSA-agar:

- Dissolve 39 g Columbia blood agar base (Oxoid CM0331) in 1 litre distilled water.
- Autoclave at 121°C/15 min.
- Supplement with 50 ml sterile defibrinated blood (at 50°C).

T-HCl-YGP (per litre):

- 15 g yeast extract
- 1 g pyruvic acid
- 200 ml 0.1 M Tris-HCl, pH 7.0
- 20 g agar
- Autoclave at 121°C/15 min.
- Add 40 ml 10% glucose (autoclaved separately).

J agar (per litre):

- 5 g tryptone
- 3 g K_2HPO_4
- 15 g yeast extract
- 20 g agar
- Adjust pH to 7.3 to 7.5.
- Autoclave at 121°C/15 min.
- Add 20 ml 10% glucose (autoclaved separately).

PLA medium consists of three different media supplemented with egg yolk. Equal quantities (100 ml) of sterile, molten *Bacillus cereus* selective agar base (Oxoid CM617), trypticase soy agar (Merck 5458) and supplemented nutrient agar (SNA) are combined and mixed. SNA is composed of (per litre):

- 23 g nutrient agar
- 6 g yeast extract
- 3 g meat extract
- 10 g NaCl
- 2 g Na_2HPO_4
- Adjust pH to 7.4 ± 0.2.

All solid media are sterilized at 121°C for 15 min. After the three molten media are combined 30 ml of 50% egg-yolk suspension is added to form the PLA medium.

Cool the media to 50°C and add the antibiotics to a final concentration of 20 µg/ml for nalidixic acid and 10 µg/ml for pipemidic acid.

- Nalidixic acid stock solution (1 mg/ml) is prepared by dissolving 0.1 g in 2 ml of 1 M NaOH and diluting to 100 ml with 0.01 M phosphate buffer (pH 7.2).
- Pipemidic acid stock solution (2 mg/ml) is prepared by dissolving 0.2 g in 2 ml of 1 M NaOH and then diluting to 100 ml with 0.01 M phosphate buffer (pH 7.2).
- Both antibiotic solutions are filter sterilized.

The medium is poured (20 ml) into sterile Petri dishes and plates are dried before use (15 min).

Plates inoculated with 150 µl heat-shocked spore suspension are incubated at 35°C up to 6 days in either aerobic conditions or under an atmosphere of 5-10% CO_2. Vegetative bacteria are grown overnight at 35°C without heat-shock treatment.

The outline of the procedure starting from honey samples is as follows:

1. Dilute 20 g of honey in 20 ml PBS.
2. Shake vigorously.
3. Centrifuge the suspension 40 min at 6,000 x g to harvest the spores.
4. Resuspend the pellet in 1 ml of PBS.
5. Heat treat and plate this spore containing aqueous solution as described above.

3.2. Identification

Often the first step in the identification of *P. larvae* growing on solid media is the verification of its growth rate and colony morphology. Visible colonies may appear on the second day of incubation. However, if no colonies emerge it is advisable to extend the incubation time for a few more days. Two serial subcultures should be grown to insure culture purity. Pure *P. larvae* colonies have a characteristic morphology but this appears to be highly dependent on the medium that was used (see OIE, 2008). Using *P. larvae* reference strains is highly advisable.

Some non-molecular identification protocols exist and provide a good alternative for diagnostic purposes when sophisticated equipment is lacking (see OIE, 2008). However, for research purposes we recommend a PCR-based identification of *P. larvae*. Several PCR methods have been described (reviewed by de Graaf *et al.*, 2006a), but one in particular based on the 16S rRNA gene (Dobbelaere *et al.*, 2001) has proven its robustness in the past decade. A detailed description is given here below. Primers are listed in Table 3.

Table 3. Primer sets for identification and genotyping of *P. larvae* by PCR.

Name	Sequence	PCR-product size	Reference
AFB-F	5'-CTTGTGTTTCTTTCGGGAGACGCCA-3'	1106 bp	Dobbelaere *et al.*, 2001
AFB-R	5'-TCTTAGAGTGCCCACCTCTGCG-3'		
Primer 1	5'-AAGTCGAGCGGACCTTGTGTTTC-3'	973 bp	Govan *et al.*, 1999
Primer 2	5-'TCTATCTCAAAACCGGTCAGAGG-3'		
ERIC1R	5´-ATGTAAGCTCCTGGGGATTCAC-3´	Several amplicons	Versalovic *et al.*, 1994
ERIC2	5´-AAGTAAGTGACTGGGGTGAGCG-3´		
BOXA1R	5´-CTACGGCAAGGCGACGCTGACG-3	Several amplicons	Versalovic *et al.*, 1994
MBO-REP1	5´-CCGCCGTTGCCGCCGTTGCCGCCG-3	Several amplicons	Versalovic *et al.*, 1994

3.2.1. Bacterial DNA extraction

Bacterial DNA extraction can be done using commercialized kits/matrices (InstaGene matrix, Bio-Rad, Genersch and Otten, 2003; Genome DNA Extraction kit, Sigma, Antúnez *et al.*, 2007). However, heating the bacterial suspension at 95°C for 15 min works also fine for simple species identification.

3.2.2. Polymerase chain reaction

PCR reactions (modified from Dobbelaere *et al.*, 2001) are set up as 50 µl mixtures containing:

- 1-5 µl template DNA
- 50 pmol forward (AFB-F) and reverse primer (AFB-R); (primers used by Govan *et al.*, 1999 also work well)
- 10 nmol of each dNTP
- 1-2.5 U of *Taq* polymerase in the appropriate PCR buffer containing 2 mM MgCl$_2$.

Use the following PCR conditions: a 95°C (1-15 min) step; 30 cycles of 93°C (1 min), 55°C (30 sec), and 72°C (1 min); and a final cycle of 72°C (5 min).

3.3. Genotyping

The availability of standardized techniques that allow the discrimination of different *P. larvae* strains is essential for studying the epidemiology of AFB. This will allow scientists to identify outbreaks of the disease, determine the source of infection, determine the relationship between outbreaks, recognize more virulent strains, and monitor prevention and treatment strategies.

To date different techniques have been used in order to evaluate the diversity of *P. larvae* isolates. Some of them are based on the analysis of phenotypic characteristics, such as study of cell and colony morphology, analysis of whole bacterial proteins by SDS-PAGE or biochemical profile, among others (Hornitzky and Djordevic, 1992; Neuendorf *et al.*, 2004; de Graaf *et al.*, 2006a; Genersch *et al.*, 2006; Antúnez *et al.*, 2007). During the last decade methods based on genetic analysis have gained more attention. Different strategies have been used to evaluate the genetic diversity of *P. larvae*, including

restriction endonuclease fragment patterns (Djordjevic *et al.*, 1994; Alippi *et al.*, 2002), pulsed-field gel electrophoresis (Wu *et al.*, 2005; Genersch *et al.*, 2006), amplified fragment length polymorphism (de Graaf *et al.*, 2006b), ribotyping and denaturing gradient gel electrophoresis (Antúnez *et al.*, 2007). Nevertheless, some of these techniques differentiating *P. larvae* genotypes, have not been adopted by the scientific community.

An appropriate genotyping method should be highly discriminatory, but also demonstrate interlaboratory and intralaboratory reproducibility. It should be easy to use and interpret (Genersch and Otten, 2003). For these reasons, the most utilized method is rep-PCR, or PCR amplification of repetitive elements (Versalovic *et al.*, 1994), although presently its reproducibility in other labs has not been proven. There are three sets of repetitive elements randomly dispersed in the genome of bacteria, enterobacterial repetitive intergenic consensus (ERIC) sequences, repetitive extragenic palindromic (REP) elements, and BOX elements (which includes boxA, boxB, and boxC). Primers to amplify those elements have been reported and proved to be useful for subtyping of Gram-positive and Gram-negative bacteria (Versalovic *et al.*, 1994; Olive and Bean, 1999).

rep-PCR has been widely used for the study of *P. larvae* (Alippi and Aguilar, 1998a, 1998b; Genersch and Otten, 2003; Alippi *et al.*, 2004; Antúnez *et al.*, 2007; Peters *et al.*, 2006; Loncaric *et al.*, 2009). The most useful pair of primers are ERIC1R-ERIC2, which allowed the differentiation of four different genotypes (ERIC I, II, III and IV) (Genersch *et al.*, 2006). Genotypes ERIC I and II corresponds to the former subspecies *P. l. larvae* while genotypes ERIC III and IV corresponds to the former subspecies *P. l. pulvifaciens* (Genersch, 2010).

P. larvae genotype ERIC I is the most frequent genotype and is present in Europe and in America, genotype ERIC II seems to be restricted to Europe and genotypes ERIC III and IV have not been identified in field for decades, but exist as few isolates in culture collections (Genersch, 2010). In order to enhance the discrimination of strains, the analysis using ERIC primers can be complemented with the use of other primers. The use of BOXA1R primer allowed the discrimination of four banding patterns in America, all of them belonging to genotype ERIC I (Alippi *et al.*, 2004; Antúnez *et al.*, 2007) and three in Europe (Genersch and Otten, 2003; Peters *et al.*, 2006; Loncaric *et al.*, 2009). Primers BOX B1 and BOX C1 did not amplify *P. larvae* DNA (Genersch and Otten, 2003). When REP primers were used, four banding patterns were found in America and Europe although results could not be compared since different pairs of primers (REP1R-I and REP2-I and MBO REP1 primers) were used (Alippi *et al.*, 2004; Kilwinski *et al.*, 2004; Loncaric *et al.*, 2009). Protocols for subtyping of *P. larvae* are provided below.

Restriction fragment length polymorphic (RFLP) analysis of bacterial genomes, as visualized via pulsed-field gel electrophoresis (PFGE), is also a very effective procedure for bacterial genotyping (PFGE-typing). PFGE-typing of 44 *P. larvae* isolates, obtained from honey bee larval smears and honey samples collected in Australia and from Argentinean honey, has demonstrated resolution of this bacterium into 12 distinct genotypes when using restriction endonuclease *Xba*I (Wu *et al.*, 2005). Outlined below is a PFGE-typing procedure for *P. larvae*. This procedure is presented as a three-part operation of genomic DNA preparation, restriction digestion of DNA, and then electrophoresis of the digested DNA. Performance of PFGE-typing is labour intensive. Also, many factors can contribute to an unsuccessful electrophoresis run. Therefore, troubleshooting PFGE and helpful hints for performing this technique can be found in the protocols section of the Bio-Rad website (http://www.bio-rad.com/evportal/en/US/LSR/Solutions/LUSORPDFX/Pulsed-Field-Gel-Electrophoresis).

3.3.1. PCR amplification of repetitive elements

Polymerase chain reaction (according to Genersch and Otten, 2003):

1. Carry out PCR reactions in a final volume of 25 µl consisting of 1 × reaction buffer and a final concentration of 2.5 mM MgCl$_2$, 250 µM of each dNTP, 10 µM of primer, and 0.3 µg of Hot start Taq polymerase. Five to ten µl of template DNA is added to the reaction.

2. The cycling conditions are: an initial activation step at 95°C for 15 min, 35 cycles at 94°C for 1 min, 53°C for 1 min, and at 72°C for 2.5 min, and a final elongation step at 72°C for 10 min.

3. Analyse five µl of the PCR reaction by electrophoresis on 0.8% agarose gel in TAE or TBE buffer.

4. Stain the amplified bands by incubation of the agarose gel in ethidium bromide (0.5 µg/ml in water) for 30 min.

5. Visualize under UV light and photograph using a digital camera.

6. Compare obtained fingerprints visually or using specific analysis programs.

Independent PCR reactions should be performed using primers ERIC1R/ERIC2, BOXA1R and MBO-REP1 (Table 3).

Modifications of the present protocol, such as those reported by Alippi *et al.* (2004) or Antúnez *et al.* (2007) also resulted useful, allowing the differentiation of *P. larvae* genotypes.

Interpretation of the results:

- ERIC1R-ERIC2 primers: four different genotypes (ERIC I, II, III and IV) can be distinguished (Genersch *et al.*, 2006).
- BOXA1R primers: three different patterns can be found in Europe (A, a and ·) (Genersch and Otten, 2003) while four patterns can be found in America (A, B, C and D) (Alippi *et al.*, 2004).
- MBO REP1 primers: four band patterns can be found in Europe (В, b, β, Б) (Peters *et al.*, 2006).

3.3.2. Pulsed-field gel electrophoresis

3.3.2.1. Preparation of genomic DNA agarose plugs

1. Culture an isolate of *P. larvae* on an MYPGP agar plate for 48 h at 37°C.

2. Suspend a loopful of bacteria (2-3 colonies from an isolated area on the plate) into 500 µl MYPGP broth and centrifuge for 1 min at RT.

3. Remove the broth, suspend the pellet with 1 ml washing buffer (see recipe below), and centrifuge for 1 min at RT.

4. Remove the washing buffer.

5. **Completely** suspend the bacterial cell pellet in 0.30 ml washing buffer + 0.05 ml Proteinase K (0.5 mg/ml).

6. Warm the suspension to 50°C.

7. Mix the suspension with an equal volume of melted 2% SeaKem Gold agarose (prepared in washing buffer; Cambrex Bio Science Rockland, Inc., ME) at 50°C.

8. Quickly pipette into two wells of a plug mold (Bio-Rad Laboratories, Inc.) warmed to 37°C.

9. Solidify the plugs at 4°C for 30 min.

10. Remove the two plugs from the mold.

11. Add plugs to 5 ml preheated Proteinase K Solution in a 50 ml plastic culture tube.

12. Incubate overnight at 50°C with gentle agitation (shaker water bath).

13. Preheat 10 ml H$_2$O and 40 ml TE80 buffer (see recipe below) to 50°C.

14. Wash plugs (use a sterile BioRad green screened caps; part #170-3711) with 10 ml 50°C H$_2$O (add, swirl, and drain).

15. Wash plugs with 10 ml 50°C TE80 buffer 4X (15 min for each wash with gentle shaking).

16. Add 5 ml RT TE80 buffer to the plugs in the 50 ml culture tube and store at 4°C. Plugs are good for approximately 2 months in TE80 buffer at 4°C.

Wash Buffer (100 ml):

- 200 mM NaCl (4.0 ml of 5 M stock)
- 10 mM Tris-HCl (pH 7.5) (1.0 ml of 1 M stock)
- 100 mM EDTA (20.0 ml of 0.5 M stock)

Proteinase K Solution (100 ml):

- 50 mM EDTA (10.0 ml of 0.5 M stock)
- 1.0 g N-lauroylsarcosine
- 50 mg Proteinase K (final 0.5 mg/ml)
- 50 mg Lysozyme (final 0.5 mg/ml)

TE80 Buffer (100ml):

- 10 mM Tris-HCl (pH 8.0) (1.0 ml of 1 M stock)
- 1 mM EDTA (0.2 ml of 500 mM stock)

3.3.2.2. Restriction enzyme digestion

1. Aseptically remove a plug from the TE80 buffer and place onto the inner surface of a sterile petri dish.
2. Using a plastic ruler under the petri dish, cut two 2 mm slices from the plug with a sterile razor blade.
3. Place the two plug slices into 1 ml TE80 buffer in a 1.5 ml microcentrifuge tube.
4. Wash the slices 2X with 1 ml TE80 buffer (30 min for each wash with gentle agitation).
5. Quickly rinse the slices with 500 µl restriction digestion buffer (without restriction enzyme).
6. Add 100 µl of *Xba*I Digestion Mixture (see recipe below) and incubate 3 h at 37°C with gentle agitation.
7. Remove digestion mixture solution.
8. Rinse slices 2X with 1 ml 0.5X TBE Buffer (see recipe below).
9. Suspend the plug slices in 1 ml 0.5X TBE Buffer.
10. Store overnight at 4°C.

*Xba*I Digestion Mixture (10 samples):

- 10X New England BioLabs Buffer 4 (100 µl)
- 100X BSA (10 µl)
- *Xba*I enzyme (20 µl) New England BioLabs (20,000 U/ml)
- H_2O (870 µl)

5X TBE Buffer (500 ml):

- Tris Base (54 g)
- Boric Acid (27.5 g)
- 0.5 M EDTA (pH 8.0) (20 ml)
- H_2O to 500 ml volume

3.3.2.3. Gel loading and electrophoresis

1. Make 2.3 litres of 0.5X TBE buffer (230 ml 5X TBE + 2070 ml H_2O).
2. Add 1.0 g Seakem Gold agarose to 100 ml 0.5X TBE Buffer.
3. Carefully melt agarose in microwave oven. Save 5 ml (at 55°C) to seal the wells.
4. Assemble the gel forming tray, assure that the tray is level with the well comb in place, and pour melted agarose into forming tray (cool melted agarose to 55-60°C before pouring).

5. Allow gel to solidify for a minimum of 30 min.
6. Add remaining 2.2 litres of 0.5X TBE buffer to the CHEF electrophoresis box.
7. Chill the buffer to 14°C by circulation through the chiller-pump unit.
8. Adjust flow rate of buffer through the chiller-pump unit to between 0.75 and 1.0 l/min.
9. Once gel has solidified and the comb is removed, remove a digested plug slice from the microcentrifuge tube.
10. Place the slice onto inner surface of sterile Petri dish.
11. Carefully slide the plug slice onto the side of a sterile razor blade using a sterile microspatula.
12. Holding the razor blade (with the plug slice sticking to the side) at the edge of a well, carefully use the microspatula to slide the plug slice into the well. Use care not to introduce air bubbles in the well.
13. Wash and flame sterilize the razor blade and microspatula for continued loading of wells with other prepared DNA plug slices.
14. Add low range molecular size standards (4.9-120 kb) (Bio-Rad) to outer wells.
15. Seal the wells with the saved SeaKem Gold agarose. Use care not to introduce any air bubbles during sealing and fill any wells that do not contain slices.
16. Place the gel into the circulating pre-cooled buffer within the CHEF electrophoresis box.
17. Allow the gel to cool for 15 min prior to beginning electrophoresis.
18. Perform electrophoresis in the Bio-Rad CHEF DR III system using the following electrophoresis run parameters:
    - Switch angle: 120°
    - Switch time: 1-6 sec
    - Voltage: 6 V/cm
    - Temperature: 14°C
    - Run time: 16 hours
19. Following electrophoresis, remove and stain the gel in the dark with Sybr Green (Molecular Probes, Inc., Eugene, OR) (30 µl of 10,000X concentrate diluted into 300 ml TE pH 7.5) at RT for 45 min.
20. Photodocument the gel via UV transillumination/epi-illumination (254 nm).

3.4. *In vitro* sporulation of *Paenibacillus larvae*

Paenibacillus larvae and *Paenibacillus popilliae*, "catalase-minus" paenibacilli as defined by the loop test (i.e. scraping growth from a slant or plate with an inoculating loop, placing into 3% H_2O_2, and examining for bubble formation), have the characteristic of sporulating efficiently in their insect hosts, but usually exhibit very poor sporulation when general *in vitro* growth conditions are used.

For *P. larvae*, suppression of *in vitro* sporulation has been postulated to result from oxygen toxicity (Dingman and Stahly, 1984). Growth of strain NRRL B-3650 under limiting O_2 improves sporulation in liquid culture. Also, nutrient availability at time of sporulation has been shown to influence spore production (Dingman and Stahly, 1983).

Procedures promoting effective sporulation of *P. larvae* on solid, and in liquid, growth media have been developed (Dingman, 1983). By limiting colony number, many strains have been observed to sporulate well on MYPGP agar plates (see section 3.1. for recipe). However, Mueller-Hinton broth – one of its ingredients – was inhibitory to sporulation of strain NRRL B-3650 in liquid culture, rendering use of this growth medium in liquid form ineffective for sporulation. Development of a liquid growth medium (TMYGP; 1.5% Difco yeast extract; 0.4% glucose; 0.1% sodium pyruvate; 0.03M Tris -maleate, pH 7.0; in distilled water) and conditions that aided sporulation of *P. larvae* NRRL B-3650 has been reported (Dingman and Stahly, 1983). Unfortunately, other strains of this bacterium sporulated poorly, if at all, in this medium. However, Genersch *et al.* (2005) reported using the liquid part of Columbia sheep-blood agar slants for production of endospores from different *P. larvae* strains. Following are protocols using solid and liquid media growth conditions for *in vitro* sporulation of *P. larvae*.

3.4.1. Sporulation on solid growth medium

1. Create a 2-fold dilution series of the *P. larvae* bacterial culture being studied using MYPGP broth and spread each dilution onto several MYPGP agar plates (see section 3.1. for recipe).
2. Incubate plates at 37°C for 6-7 days and select plates exhibiting 50 to 5,000 colonies per plate. Note: When high numbers of colonies are present on a plate, sporulation efficiency can decline. Also, maximum sporulation obtained in relation to the plate colony number will vary between bacterial strains.
3. During the 6-7 days of incubation, microscopically monitor cellular growth and sporulation via single colony analysis.
4. After incubation, remove spores from the surface of the agar medium by washing three times (5 ml sterile H_2O per wash).
5. Combine the three washes. Once H_2O is added to a plate, use gentle rubbing of the agar surface with the sterile glass pipette to loosen spores from the surface.
6. To produce a spore stock following removal of spores from a plate surface, concentrate the spore suspension via centrifugation (12,000 x g, 15 min, 4°C).
7. Discard the supernatant.
8. Suspend the resulting spore pellet in 30 ml cold sterile H_2O.
9. Perform alternate centrifugation and pellet suspension four times.
10. Suspend the spore pellet in a final volume of 5 ml cold sterile H_2O.

11. Store at 4°C. Note: Spores must be removed from the solid growth medium for preservation. When left on the agar plates for extended time, heat-resistant counts decline rapidly (Dingman, 1983). Also, long-term survival of washed spores that have been desiccated is not known.
12. Obtain spore concentration by heating a portion of the spore stock at 65°C for 15 min.
13. Perform serial-dilution plating onto MYPGP or T-HCl-YGP (Steinkraus and Morse, 1996) agar plates.
14. Incubate inoculated plates 6-7 days at 37°C to determine colony counts (i.e., spore counts). Alternatively, determine spore number by direct microscopic counting. Note: The latter will give an overestimation. Heat resistant spore counts are usually about 6% of direct microscopic spore counts (Dingman and Stahly, 1983).

3.4.2. Sporulation in liquid growth medium

1. Inoculate TMYGP broth (6 ml in a 20 x 150 mm loosely capped screw-cap glass culture tube, see section 3.1. for recipe) with *P. larvae* NRRL B-3650. Note: Other bacterial strains must be tested separately because they may sporulate poorly in this medium and under these growth conditions.
2. Incubate the culture at 37°C in a rotary incubator shaker adjusted to 195 rpm. The culture tube is held at a 45° angle in a wire test tube rack during incubation and aeration.
3. Incubate for 3 to 4 days while microscopically monitoring cellular growth and sporulation. Other strains of *P. larvae* may require a longer incubation time for sporulation to occur. Alternatively, see Genersch *et al.* (2005) for sporulation of *P. larvae* in the liquid part of Columbia sheep-blood agar slants.
4. Collect and concentrate spores via centrifugation.
5. Wash the spores four times with 30 ml cold sterile H_2O (as described in section 3.4.1.).
6. Suspend the washed spores in a final volume of 5 ml cold H_2O.
7. Store at 4°C.
8. Obtain spore counts (i.e. heat resistant counts) by serial-dilution plating of the spore suspension onto MYPGP plates following heating of the suspension at 65°C for 15 min or determine counts by direct microscopic counting.

Note: Heat resistant spore counts are usually about 6% of direct microscopic spore counts (Dingman and Stahly, 1983).

3.5. Long term conservation of vegetative cells

Experimental work using *P. larvae* requires a readily available source of this bacterium in culture. However, *P. larvae* quickly dies in culture and some means of preserving an isolate must be used. Although short-term preservation works (see section 3.5.1.), long-term conservation is required to maintain the genetic integrity of the original isolate.

Production of frozen endospore stock suspensions is a very good procedure for long-term preservation. However, some isolates may not sporulate well *in vitro* and the slow rate of spore germination can hamper the start of experiments. Methods employing ultra-low freezing and lyophilization of vegetative cells (see sections 3.5.2. and 3.5.3. below) are suitable for long-term storage of this microbe.

Use of ultra-low freezing of *P. larvae* in glycerol has been routinely used and cultures exceeding five years of storage at -80°C remain viable (Douglas W Dingman; personal observations). No known research regarding preservation of *P. larvae* by lyophilization has been published. However, Haynes *et al.* (1961) developed a method to preserve *P. popilliae* by lyophilization and Gordon *et al.* (1973) imply that this method can be used for *P. larvae*. The protocol long used to preserve *P. larvae* strains at the National Center for Agricultural Utilization Research (USDA-ARS Culture Collection; NRRL) is similar to that described by Haynes *et al.* (1961) for *P. popilliae*.

3.5.1. Short-term preservation
1. Weekly transfer of an isolate onto fresh growth medium.
2. Incubate for two days at 37°C.
3. Store at 4°C.

3.5.2. Preservation via ultra-low freezing
1. Inoculate MYPGP broth (see section 3.1. for recipe) using a fresh culture of *P. larvae*.
2. Grow overnight at 37°C with moderate aeration.
3. Growth in the morning should show light to moderate turbidity. Note: Turbidity should not reach the point where the culture has become opaque. A light turbidity will place the culture in early to mid-exponential growth. Also, growth can be washed from the surface of a solid medium using fresh liquid medium.
4. Examine the culture microscopically to gauge contamination while chilling the culture on ice (optional).
5. Add an appropriate volume of a sterile solution of 100% glycerol to the culture to produce a bacterial suspension containing 20% glycerol (i.e. 0.25 ml of glycerol per 1.0 ml of culture).
6. Aliquot 0.5 ml of the bacterial / glycerol suspension into cryovials. Snap-top microcentrifuge tubes also work, but may result in faster loss of viability during storage.
7. Label and date the vials.
8. (Optional) Quickly freeze the aliquots in an ethanol/dry ice bath.
9. Place the bacterial suspensions in a pre-chilled storage box and store at -80°C.

3.5.3. Preservation via lyophilization
(J. Swezey, USDA-ARS Culture Collection (NRRL), Peoria , IL; personal communication)

To prepare lyophilization vials:
1. Cut Pyrex glass tubing (6 mm diameter glass with 1 mm wall thickness) 15 cm long and seal one end by melting.
2. Plug the open ends of the vials with cotton.
3. Autoclave. Note: Ampules using rubber septa and aluminium sealing rings may be used as an alternative to the above three steps. However, use of these ampules increase costs and the vacuum may be lost over time.
4. After 2-3 days incubation of the bacterium from an agar slant/plate, wash with sterile bovine serum (e.g. Colorado Serum Co., Denver, CO).
5. Place 0.1 ml aliquots of the cell/serum suspension into the sterilized vials, label, date.
6. Attach the vials to a lyophilizer apparatus.
7. Lower the vials into a solution of 50% ethylene glycol and 50% water (chilled to -50°C with dry ice) to freeze the cell suspensions.
8. Turn on the lyophilizer vacuum pump.
9. Evacuate for 3 hours while letting the temperature of the glycol/water bath gradually warm to -4°C.
10. Lift the vials from the glycol/water bath.
11. Allow the outer surfaces to dry for 1 hour at RT.
12. Carefully cut the vials from the lyophilizer apparatus using a dual tip burner, running on natural gas and oxygen, to melt the upper portion of the glass vial while sealing the contents under vacuum.
13. Store sealed vials refrigerated in the dark.

3.6. Measuring susceptibility/resistance to antibiotics of *Paenibacillus larvae*
In some countries, the antibiotic oxytetracycline (OTC) has been used by beekeepers for decades to prevent and control AFB in honey bee colonies as an alternative to the burning of infected beehives in areas where disease incidence is high. However, the intensive use of tetracyclines in professional beekeeping resulted in tetracycline-resistant (Tc^R) and oxytetracycline-resistant (OTC^R) *P. larvae* isolates. There is now general concern about widespread resistance involving horizontal-transfer via non-genomic (i.e. plasmid or conjugal transposon) routes and also induced resistance by the presence of sub-inhibitory concentrations of tetracycline (Alippi *et al.*, 2007). *P. larvae* highly resistant phenotypes have been correlated with the presence of natural plasmids carrying different Tc resistance determinants, including *tetK* and *tetL* genes (Murray and Aronstein, 2006; Alippi *et al.*, 2007; Murray *et al.*, 2007).

Most *Paenibacillus* species, including *P. larvae*, are highly susceptible to tetracyclines; it has been reported that the growth of *P. larvae* strains is inhibited at concentrations as low as 0.012 µg of oxytetracycline per ml of culture medium. Alternatively, when a disc containing 5 µg of oxytetracycline is placed on an agar plate previously spread with a bacterial suspension, the clear zones formed by the sensitive strains usually average 50 mm in diameter including the disc (Shimanuki and Knox, 2000). Any reduction of the inhibition zone or an increase in the minimal inhibitory concentration (MIC) required to prevent the growth of *P. larvae* would be evidence of the development of resistant strains.

3.6.1. Determination of minimal inhibitory concentrations (MICs)

Microorganisms can be tested for their ability to produce visible growth on a series of agar plates (agar dilution), in tubes with broth (broth dilution), or in microplate wells of broth (broth microdilution) containing dilutions of an antimicrobial agent. Additionally, gradient MIC tests are also commercially available. MIC is defined as the lowest antibiotic concentration that prevents visible growth of bacteria. MIC methods are widely used in the comparative testing of new agents, or when a more accurate result is required for clinical management. As there are no CLSI (formerly NCCLS) (www.clsi.org) nor EUCAST (www.eucast.org) recommendations for the determination of MICs of *P. larvae*, MIC values of tetracycline and other antibiotics can be determined by the agar dilution method using MYPGP as basal medium (see section 3.1. for recipe) as described as follows:

1. Obtain antimicrobial powders directly from the manufacturer or from commercial sources. The agent must be supplied with a stated potency (mg or International Units per g powder, or as percentage potency).

2. Store powders in sealed containers in the dark at 4°C with a desiccant unless otherwise recommended by the manufacturer.

3. Prepare antibiotic stock solutions by using the following formula:

$$= \frac{\text{Weight of powder (mg)}}{\text{Volume of solvent (ml) X Concentration (µg/ml)}}{\text{Potency of powder (µg /mg)}}$$

4. It is recommended that concentrations of stock solutions should be 1,000 µg/ml or greater. In the case of tetracyclines, the tested concentrations can be achieved by using two stock solutions of 5,000 µg/ml and 1,000 µg tetracycline/ml in ethanol, stored at -20°C in darkness until used.

5. Prepare MYPGP agar flasks and maintain them at 45°C until the antibiotic solutions are incorporated.

6. Pour 25 ml of culture medium onto each Petri dish of 90 mm in diameter to give a level depth of 4 mm ± 0.5 mm. If using 150 mm diameter Petri dishes, 70 ml of culture medium should be dispensed.

7. Prepare plates with increasing concentrations of tetracycline i.e.: 0.03, 0.06, 0.125, 0.25, 0.5, 1, 2, 4, 8, 16, 32, 64, and 128 µg/ml. For the controls, MYPGP agar without antibiotic is used.

8. Allow the plates to set at RT before moving them.

9. Dry the plates in a sterile laminar flow cabinet so that no drops of moisture remain on the surface of the agar; do not over-dry plates.

10. Incubate each *P. larvae* strain to be tested on MYPGP agar for 48 h at 36°C to obtain mainly vegetative cells.

11. Adjust the bacterial suspension until the OD_{620} (density of a culture determined spectrophotometrically by measuring its optical density at 620 nm) is about 0.4.

12. Each bacterial suspension of each strain must be inoculated onto the surface of the culture medium by adding drops of 5 µl each by means of an automatic micropipette (usually 15-20 drops per plate).

It is possible to test different strains on the same plate. This procedure must be repeated at least twice for each strain and tetracycline concentration, and control plates without antibiotic must be used. It is strongly recommended to include control strains with known MICs in each batch.

13. Place the plates open into a sterile laminar flow cabinet until the drops are absorbed.

14. Incubate the plates in inverted position at 36°C ± 1 for 48 h.

15. After incubation, ensure that each tested strain has grown on the antibiotic-free plate control.

16. Read the MIC endpoint for each strain as the lowest concentration of antibiotic at which there is no visible growth. The growth of one or two colonies or a fine film of growth should be disregarded.

Interpretation: for tetracyclines, *P. larvae* isolates should be considered as "susceptible" when there MICs are <4 µg/ml, "intermediate" for MICs between 4-8 µg/ml and "resistant'" for MICs ≥16. Examples of acceptable MIC values for control strains are: *Pseudomonas aeruginosa* (ATCC 27853): between 16-32 µg/ml (resistant); *Escherichia coli* (ATCC 25922): between 0.5-2 µg/ml (susceptible); *Staphylococcus aureus* (ATCC 29213): between 0.12-1 µg/ml (susceptible) and *Enterococcus faecalis* (ATCC 29212): between 16-32 µg/ml (resistant).

When examining a population of bacteria, it is suggested to calculate their values of MIC_{50} and MIC_{90} (minimum concentration necessary to inhibit the growth of 50 and 90% of microorganisms tested respectively).

3.6.2. Determination of antibiotic susceptibility testing by the disc diffusion method

The disc diffusion method of antibiotic susceptibility testing is the most practical method for determining antibiotic susceptibility/resistance of microorganisms to different antimicrobial agents. The accuracy and reproducibility of this test are dependent on maintaining standard procedures. As there are no CLSI (formerly NCCLS) (www.clsi.org) nor EUCAST (www.eucast.org) recommendations for the determination of susceptibility/resistance of *P. larvae* by the disc diffusion method, a method developed for this species is described below.

For determining tetracycline resistance, the agar diffusion procedure can be employed, using MYPGP agar (see section 3.1. for recipe) and 5 μg tetracycline discs (Oxoid® or BBL®). The discs can also be prepared in the laboratory by using S & S® or similar sterile discs (6 mm in diameter) impregnated with 5 μg tetracycline per disc.

3.6.2.1. Preparation of discs

1. Prepare a stock ethanol solution containing 500 μg/ml of tetracycline as explained in section 3.6.1.
2. Pipette 10 μl of the stock solution onto each sterile disc.
3. Dry the discs in a sterile laminar flow cabinet.
4. Store in sterile containers at -20°C in darkness until used.

Note: Other tetracyclines (e.g., oxytetracycline) can be tested in the same way. Discs containing antibiotics other than tetracyclines should be prepared at the concentrations suggested by CLSI or EUCAST.

3.6.2.2. Preparation of plates

1. Dispense MYPGP-agar cooled below 50°C into sterile Petri dishes to give a level depth of 4 mm ± 0.5 mm (25 ml in 90 mm diameter Petri dish, 70 ml in 150 mm diameter Petri dish).
2. Allow the agar to set before moving the plates.
3. Prepared plates can be stored at 4-8°C in sealed plastic no more than 7 days before using.
   The surface of the agar should be dry before use.
4. No drops of water should be visible on the surface of the agar when the plates are used. Plates must not be over-dried.

3.6.2.3. Determination of resistance/susceptibility

1. Incubate vegetative cells of each *P. larvae* strain to be tested on MYPGP agar for 48 h at 36°C ± 1.
2. Suspend the cells directly from MYPGP in screw capped tubes containing sterile distilled water or sterile saline.
3. Adjust the bacterial suspension until the OD_{620} is in the range of 0.4. Alternatively, individual colonies of each strain can be incubated at 36°C ± 1 in 2 ml aliquots of MYPGP broth for 26 to 75 h until the OD is in the range of 0.5. Optimally, use the adjusted suspension within 15 min of preparation and always within 60 min.
4. Vortex for 3 minutes.
5. Dip a sterile cotton swab in the bacterial suspension and remove excess fluid on the swab by turning it against the inside of the tube.
6. Spread the inoculum evenly over the entire surface of the plate prepared in 3.6.2.2. by inoculating in three directions.
7. Place antibiotic discs (pre-warmed to RT) on the plate. Discs should be in firm, even contact with the surface of the medium. At least three replications for each bacterial strain is recommended. It is possible to apply more than one disc per plate; in this case, discs should be spaced so that zones of inhibition in susceptible isolates do not overlap. Overlapping will impede the measurement of zone diameters.
8. Apply the "15-15-15 rule": Use the inoculum within 15 min of preparation and never beyond 60 min. Apply discs within 15 min of inoculating plates. Start incubation within 15 min of application of discs.
9. Incubate the plates inverted at 36°C ± 1 during 72 h.
10. Measure the resulting inhibition zone (clear area without bacterial growth including the disc) by using a calliper or an automated zone reading. Read the plates from the back against a black background illuminated with reflected light.
11. A correct inoculum and satisfactorily spread plates will result in a confluent lawn of growth in the absence of antibiotic. It is important that there is an even lawn of growth to achieve uniformly circular inhibition zones. If individual colonies can be seen, the inoculum is too light and the test must be repeated. In case of distinct colonies within zones, subculture the colonies, check purity and repeat the test if necessary.

Interpretation: for tetracyclines, an inhibition zone of less than 20 mm in diameter (including the disc) is considered as the separation point between resistant and susceptible strains as follows: "resistant" ≤ 14 mm; "intermediate": between 15-19 mm and "susceptible": ≥20 mm (Alippi *et al.*, 2007).

It is strongly recommended to use reference standard strains according to the indications of NCCLS or EUCAST with the only difference that MYPGP agar (see section 3.1. for recipe) should be used as basal medium.

4. Experimental infection

4.1. Infection of *in vitro* reared larvae for the analysis of virulence and pathomechanisms of *Paenibacillus larvae*

Exposure bioassays (Genersch *et al.*, 2005) are a reliable approach to determine virulence of different *P. larvae* strains in the laboratory. Briefly, this experiment consists in:

1. Rear first-instar larvae from various colonies in 24-well plates, with 10 larvae per well.
2. Provide larvae with larval food mixed with a determined quantity of infectious spores during the first 24 hours. Afterwards, larvae will receive normal food every day for the rest of the experiment. A control group receives normal larval diet throughout.
3. Each day dead larvae are recorded and examined for AFB infection.

Table 4. Estimated LC$_{50}$ and LT$_{100}$ values (min-max) for ERIC I and ERIC II. (Genersch *et al.*, 2005).

Genotype	LC$_{50}$ (CFU ml^{-1} larval diet)	LT$_{100}$ (days p.i.)
ERIC I	<<100-800	7-10
ERIC II	<<100-620	10-138.9

Spore preparations for experimental infection can be performed as described under section 3.4. or as described below (Genersch *et al.*, 2005):

1. Resuspend around 100 *P. larvae* colonies in 300 µl of brain heart infusion broth (BHI).
2. Inoculate on Columbia sheep blood agar (CSA) slants.
3. Incubate slants containing bacterial suspension at 37°C for 10 days or until sporulation has occurred.
4. Collect medium containing spores.
5. Calculate spore concentration in the medium in colony forming units after plating different dilutions of the collected medium.
6. Count colonies after 6 days of incubation.

4.1.1. Protocol for exposure bioassays

1. For grafting of larvae, dispense 300 µl of larval feed per well in a 24-well plate. Leave 6 wells empty (A1, A3, A5, D2, D4 and D6) and fill them with 500 µl of double distilled water to avoid desiccation.
2. Incubate plates 30 min at 35°C to warm them up.
3. Graft the larvae by inserting a grafting tool under the back of the larvae floating in royal jelly without touching it, and carefully deposit it on the surface of the larval feed prepared in the well plate. Deposit ten larvae per well. Note: Graft only L1 instar larvae, under 12 hours of age from hatching. Collecting larvae from as many different populations as possible will ensure randomization, achieve homogenous treatment groups, and avoid population-specific variations.
4. For starting the experimental phase, aliquot larval feed and add the necessary volume of spore suspension, which needs to contain a defined concentration of spores to adjust a defined final spore concentration in the larval feed, e.g., to the desired lethal concentration (LC) for the infection group (Table 4).
5. Dispense 300 µl per well of the spore-contaminated larval diet in 3 different wells. Leave 3 wells for the control group receiving non-spore contaminated larval diet during the entire experiment.
6. Set ten larvae per well.
7. Incubate plates at 35°C for 24 hours.
8. Groups of 30 larvae (in 3 wells) are treated as one replicate and at least three independent replicates should be performed for statistical analysis (see the *BEEBOOK* paper on statistics (Pirk *et al.*, 2013)).

9. After 24 hours of infection, transfer larvae to a pre-warmed, fresh normal larval diet plate.

    Use a different grafting tool for each treatment group to avoid reinfection. Thereafter, every treatment group receives fresh larval diet every 24 hours.
10. Analyze the plates each day under a stereo microscope to determine the health status of the larvae.
11. Transfer remaining (living) larvae to a new plate containing pre-warmed fresh larval diet.
12. Proceed with experiment until day 14.

    Since larvae increase in size during the experiment, the number of larvae per well must be decreased accordingly.
13. After defecation (at day 7-8, when light yellow secretion can be observed surrounding the larvae), transfer larvae to pupation plates. Prepare pupation plates by lining every well with laboratory tissues, leaving 6 wells free for double distilled water.
14. Larvae are classified as dead when they stop breathing (movement of tracheal openings stops) and lose body elasticity. The number of dead larvae should be reported every day.
15. To determine whether *P. larvae* infection caused the death of a larva, dead larvae are plated out on Columbia sheep blood agar (CSA) plates. Plates are incubated overnight at 37°C to allow the growth of vegetative bacteria only (spores need about 3 days to germinate under these conditions). Positive AFB infection will be confirmed by growth of *P. larvae* (see section 3.2.).
16. Further confirmation is provided by performing *P. larvae*-specific PCR-analysis of colonies grown from larval remains.
    16.1. Pick one colony.
    16.2. Dissolve it in 25 µl of double distilled water.
    16.3. Boil at 95° for 10 minutes.
    16.4. Centrifuge for 10 minutes at 9,500 rcf.
    16.5. Supernatant can be used as template for PCR (see section 3.2.2.).

4.1.2. Analysis of generated data

In order to evaluate compiled experimental data, two different analyses to measure virulence can be performed. The first virulence indicator is the lethal concentration (LC) value (Thomas and Elkinton, 2004), which indicates the spore concentration at which 50% (LC$_{50}$) or 100% (LC$_{100}$) of the individuals are killed. To calculate this measure, the proportion of dead larvae from the number of exposed larvae is plotted against spore concentration. From such graphs, one can to estimate the spore concentration needed to kill a given proportion of the exposed population. These graphs also allow to deduce - for a certain analysed strain - the approximate spore concentration present when a specific percentage of the exposed population is dead.

Another measure of virulence is the lethal time (LT) (Thomas and Elkinton, 2004) which is the time it takes the pathogen to kill 50% (LT_{50}) or 100% (LT_{100}) of the infected animals. In order to obtain this measure, the time course of infection must be determined. Cumulative mortality per day is calculated as percentage of all individuals which died from *P. larvae* infection during the course of the experiment (total number of *P. larvae*-killed animals until the end of the experiment). Average values are calculated every day from at least three independent replicates, and plotted against every time point (day post-infection).

4.2. Experimental infection of a bee colony

Colony assays of AFB prevalence can be used to determine the efficacies of antibiotics and other treatments as well as the resistance traits of specific colonies.

4.2.1. Inoculation with known spore concentration solution

Below is one protocol for a field evaluation of AFB prevalence, derived from Evans and Pettis (2005):

1. Establish clean colonies of equal strength by placing 1.2 kg of worker bees and a marked queen (from resistant or susceptible stock, if desired) into a standard (e.g. Langstroth) hive body.

2. Following establishment and the initiation of brood rearing (*ca.* 1 month), inoculate each colony twice (2-4 weeks apart) with spores from a fresh field isolate of *P. larvae*.

3. Inoculate by spraying approximately 2,000 immature bees (eggs, embryos, and first-and second-instar larvae) with a sucrose-water (1:10 weight/volume) suspension containing ca. 200 million *P. larvae* spores.

    Sufficient inoculant for > 50 colonies is prepared by macerating and suspending ca. 100 scales (dried larval remains) collected from symptomatic colonies.

4. Spray inoculation is effective at initiating AFB infection in colonies, with ca. 50% of managed colonies exhibiting AFB disease one month following a single inoculation, and nearly all of susceptible colonies showing symptoms by three weeks after the second inoculation.

4.2.2. Inoculation with diseased brood

If it is not important to inoculate colonies with a known quantity of spores (e.g. if the goal of the experiment is simply to induce clinical symptoms of AFB), a comb section (15 cm x 15 cm) can be cut from frames containing infective AFB scales (dried spores). At least 50% of the cells should contain scale. Introduce the comb sections with AFB scale into the middle frame in each colony.

4.3. Measuring colony resistance to AFB

4.3.1. Surveying inoculated colonies

1. Inspect colonies every two weeks to assess disease levels and colony growth, and to ensure that all colonies were successfully inoculated.

2. Approximately 90 days after colony establishment, colony growth can be measured by removing individual frames and estimating the area of sealed immature brood (see the *BEEBOOK* paper on methods for estimating strength parameters (Delaplane *et al.*, (2013)). Simultaneously, the level of foulbrood disease can be determined by making a visual inspection of all brood (immature larval and pupal honey bees) for evidence of infection. Severity of AFB infection can be quantified using a modification of a standard scoring method (Hitchcock *et al.*, 1970). Each frame with brood is rated on a 0-3 scale as to AFB infection:

    *   0 = no visible signs of disease
    *   1 = less than 10 cells/frame with visible AFB
    *   2 = 11-100 cells with AFB
    *   3 = greater than 100 cells with AFB

    A composite disease score is then generated by summing across all frames for each colony.

3. An overall severity score for each colony on each inspection date can be obtained by calculating the mean (± s.d.) of the individual frame scores. An overall score of 1 corresponds to a colony with only slight clinical symptoms, possibly not noted by cursory inspection. An overall score of 2 would indicate noticeable symptoms, and a score of 3 corresponds to a highly symptomatic colony.

4.3.2. Colony resistance

It is important to test the viability of the AFB spores used to challenge the colony. A lack of clinical symptoms after challenge may be due to colony resistance (Spivak and Reuter, 2001) or the use of non-viable spores.

*   An indirect way to test for colony resistance to AFB is to assay the colony for hygienic behaviour using the 24 h freeze-killed brood test, as outlined in the *BEEBOOK* paper on queen rearing and selection (Büchler *et al.*, 2013). Colonies that remove > 95% of the freeze-killed brood within 24 hours, over two repeated tests, are likely to demonstrate resistance to AFB.

*   It has not been established whether colonies that remove > 95% of pin-killed brood (another assay for hygienic behaviour) are also likely to be resistant to AFB. This needs to be tested.

5. 'omics and other sophisticated techniques

5.1. *Paenibacillus larvae* gene expression

P. larvae gene expression under different experimental conditions can be investigated at the transcriptional level by making use of quantitative reverse transcriptase polymerase chain reactions (qRT-PCR). The state-of-the-art-analysis of qRT-PCR data relies on normalized and calibrated relative quantities (Vandesompele *et al.*, 2002; Hellemans *et al.*, 2007). In order to normalize the qRT-PCR data on target genes, normalization factors (NFs), based on the geometric mean of converted threshold cycle values (Ct-values), need to be calculated for each sample. Before calculating NFs, one has to decide how many reference genes should be included in this calculation. This decision is based on the expression stability of candidate reference genes under all examined experimental conditions. Therefore, the protocol below describes how to select reference genes for *P. larvae*.

5.1.1. Reference gene selection

5.1.1.1. Sample collection and storage

1. Grow at least ten independent *P. larvae* cultures for each condition to be analysed with qRT-PCR.
2. Centrifuge cultures for 5 min at 8,000 x g and 4°C.
3. Pour off the supernatant.
4. Resuspend the bacterial pellet in RNA*later* solution (Ambion).
5. Incubate for approximately half an hour on ice.
6. Divide the suspension in aliquots.
7. Centrifuge aliquots for 2 min at 8,000 x g and 4°C.
8. Store at -80°C.

5.1.1.2. RNA and cDNA preparation

1. Centrifuge thawed aliquots for 2 min at 6,500 x g and 4°C.
2. Resuspend bacterial pellets in TE buffer with Lysozyme (15 mg/ml) and Proteinase K (Qiagen).
3. Vortex for 10 sec.
4. Incubate 10 min at RT with constant shaking.
5. Isolate RNA with RNeasy Plus Mini Kit and protocol "Purification of Total RNA for Animal Cells" (with on-column DNase I treatment).
6. Elute RNA with 30 µl RNase-free water.
7. Store at -80°C.
8. Convert RNA to cDNA with RevertAid First Strand cDNA Synthesis Kit (Fermentas), using random hexamer primers.

5.1.1.3. Primer design and secondary structures

- Design primers (80 – 150 bp) with Primer3Plus (Untergasser *et al.*, 2007).
- Evaluate secondary structures of amplicon with MFold (Zuker, 2003) for 60°C, 50 mM Na$^+$ and 3 mM Mg^{2+}. Selected candidate reference genes for *P. larvae* are listed in Table 5.

5.1.1.4. qRT-PCR reactions

For a single reaction, assemble the following components (total volume: 15 µl/reaction):

- 7.5 µl 2x Platinum SYBR Green qPCR SuperMix-UDG (Invitrogen)
- 0.03 µl of 100 µM forward and reverse primer each
- 1.0 µl cDNA template
- 6.5 µl distilled water

5.1.1.5. qRT-PCR program

Set program as follows:

- 50°C, 2 min, 1 cycle
- 95°C, 2 min, 1 cycle
- [95°C, 20 sec; 60°C, 40 sec] 40 cycles

After PCR amplification, perform a melt curve analysis by measuring fluorescence after each temperature increase of 0.5°C for 5 sec over a range from 65°C to 95°C.

5.1.1.6. qRT-PCR analysis

Analyze the reference gene stability with geNorm[PLUS] within qBase[PLUS] (Vandesompele *et al.*, 2002; Hellemans *et al.*, 2007) using target-specific amplification efficiencies.

5.1.2. Differential gene expression

Sample collection and storage (except for the recommendation of ten cultures), RNA and cDNA synthesis, *in silico* primer design and secondary structure evaluation, and qPCR experimental procedures (reactions, program and analysis, except for geNormPLUS analysis) are essentially the same for the study of differential gene expression by qRT-PCR as described in section 5.1.1. for reference gene selection.

5.2. Comparative genome analysis within the species *Paenibacillus larvae* using suppression subtractive hybridization

Suppression subtractive hybridization (SSH) is a powerful tool for elucidating genomic sequence differences among closely related bacteria. SSH is a PCR-based DNA subtraction method which was originally developed for generating differentially regulated or tissue-specific cDNA probes and libraries (Diatchenko *et al.*, 1996). However, it has also been successfully adapted for bacteria, especially for the identification of genes that contribute to the virulence of bacterial organisms (Akopyants *et al.*, 1998). For example, SSH has been successfully employed to compare genomes of pathogenic and non-pathogenic (Janke *et al.*, 2001; Reckseidler *et al.*, 2001) or virulent and avirulent strains of bacterial pathogens (Zhang *et al.*, 2000). SSH has also led to the identification of pathogenicity islands (Hacker *et al.*, 1997) in infectious bacteria (Agron *et al.*, 2002). Recently, SSH analysis of all four genotypes of *P. larvae* led to the identification of putative virulence factors like potent antibiotics belonging to the class

Table 5. Reference genes for *P. larvae*.

Name	Sequence	Predicted gene product
AFB_rpoD_fw	5'-AACTTGCCAAACGGATTGAG-3'	RNA polymerase sigma factor RpoD
AFB_rpoD_rv	5'-AAGCCCCATGTTACCTTCCT-3'	
AFB_gyrA_fw	5'-ATGCGGTCATCCCTATTGAG-3'	DNA gyrase subunit A
AFB_gyrA_rv	5'-GGTCATCTTCCCGCAAATTA-3'	
AFB_cmk_fw	5'-GTACAGGGCGATTACCTGGA-3'	Cytidylate kinase
AFB_cmk_rv	5'-GCCATCAACGAATACCTGCT-3'	
AFB_sucB_fw	5'-ATTGCCAAGGGTGTTGTAGC-3'	Succinyl-CoA synthetase subunit β
AFB_sucB_rv	5'-TTCAGCCCGGATTCATTTAG-3'	
AFB_eftu_fw	5'-TAACATCGGTGCCCTTCTTC-3'	Elongation factor Tu
AFB_eftu_rv	5'-CCACCCTCTTCGCTAGTCAG-3'	
AFB_fum _fw	5'-CCAAAATATGCGGAGCTGAT-3'	Fumarate hydratase
AFB_fum _rv	5'-GTGAACCGCAATTTCCCTTA-3'	
AFB_purH_fw	5'-TTCTCTCGGGGCTTTTGATA-3'	Bifunctional phosphoribosyl aminoimidazole carboxamide formyltransferase / IMP cyclohydrolase
AFB_purH_rv	5'-CTACTGTTGGCTCACGGTCA-3'	
AFB_adk_fw	5'-TCAACAGGTGATGCTTTTCG-3'	Adenylate kinase
AFB_adk_rv	5'-TGTGATTTCGTCAGGAACCA-3'	
AFB_gapdh_fw	5'-TGTTGAAGCTGGTGAAGGTG-3'	Glyceraldehyde-3-phosphate dehydrogenase
AFB_gapdh_rv	5'-TCCGCTTTTTCTTTTGCAGT-3'	

of non-ribosomal peptides and polyketides as well as several toxins and cytolysins (Fünfhaus *et al.*, 2009).

The principle of any SSH analysis is that one genome putatively containing additional genes (the so-called 'driver') is subtracted from the other genome ('tester') with the result that the additional sequences specific to the tester remain and can be visualized via PCR. For SSH analysis, genomic DNA of *P. larvae* grown in liquid culture of brain heart infusion broth (BHI broth, see section 3.1. for recipe) is used. Since the quality of the genomic DNA is of crucial importance, special kits for DNA extraction of Gram-positive bacteria (e.g. MasterPure Gram Positive DNA Purification Kit, Epicentre Biotechnologies) to isolate *P. larvae* DNA are recommended. The following protocol for SSH is based on the protocols published by Akopyants *et al.* (1998) using the PCR-Select Bacterial Genome Subtraction Kit (BD Clontech). SSH consists of two consecutive phases, the hybridization and the amplification. Within the phase of hybridization, the genomic DNA extracted from the 'driver' strain is hybridized with DNA extracted from the 'tester' strain. Sequences that are present in the tester strain but missing in the driver strain are then isolated in the amplification phase in which target genomic DNA fragments are amplified, while amplification of non-target DNA is simultaneously suppressed using the suppression PCR effect (Siebert *et al.*, 1995).

The protocol for the comparison of *P. larvae* genotypes involves the following steps:

1. Following the manufacturer's protocol, digest the isolated DNA from both genomes using restriction enzyme *Rsa*I, which has a recognition sequence of only four bases. Since this sequence occurs often in the genome of *P. larvae*, a high fragmentation of the bacterial DNA (100 to 1000 bp) with blunt ends is achieved.

2. After purification of the fragmented DNA with the MinElute Cleanup Kit (QIAGEN) according to the manufacturer's protocol, divide the DNA of the tester strain into two pools which are ligated with either Adaptor 1 (5'–CTAATACGACTCACTATAGGGCTCGAGCGGCCGCCCGGGCAGGT-3') or Adaptor 2R (5'–CTAATACGACTCACTATAGGGCAGCGTGGTCGCGGCCGAGGT-3'), catalysed by T4 DNA ligase.

The ends of the adaptors lack a phosphate group, so only one strand of each can be ligated to the 5' end of the tester DNAs.

3. For the first round of hybridization, mix each pool of adaptor-ligated tester-fragments with a 50-100-fold excess of driver-fragments.

4. Incubate the mixed samples at 98°C for 90 sec. (denaturation).

5. Incubate at 65°C for 90 min (annealing).

6. In the second round of hybridization, mix both pools without denaturation.

7. Incubate overnight at 65°C to allow free tester-fragments from both pools to form heterohybrids (hybridization of complementary tester DNAs with different adaptors (Akopyants *et al.*, 1998)).

8. Add freshly denatured driver to the mixture.

9. Allow the samples to hybridize.

    During this step, hybrid molecules are formed, but only DNA fragments which are exclusively present in the sample of the tester strain are amplified in the subsequent amplification phase.

10. For amplification, perform a nested PCR with adaptor specific primers.

    For the first PCR, the primer 1 (5'-CTAATACGACTCACTATAGGGC-3') is used. For the second PCR, the nested primer 1 (5'-TCGAGCGGCCGCCCGGGCAGGT-3') and the nested primer 2R (5'-AGCGTGGTCGCGGCCGAGGT-3') are used.

11. To obtain a library of tester-specific sequences, secondary PCR products - i.e., clone tester specific DNA fragments - are then cloned into appropriate vectors (e.g. pCR2.1TOPO vector, Invitrogen, containing an ampR gene).

12. Transform into competent *E. coli* (e.g. TOP10 cells, Invitrogen) following the manufacturer's instructions.

13. Plate transformation mixes on agar plates supplemented with Ampicillin (100 µg/ml)

14. Incubate overnight at 37°C.

15. Pick and grow clones individually overnight in Luria broth in the presence of Ampicillin (100 µg ml/ml) at 37°C and 300 rpm.

16. Extract plasmid DNA from these overnight cultures (e.g. by using the Qiagen plasmid mini kit).

17. The presence of inserted *P. larvae* DNA can be verified by restriction digestion. In the case of pCR2.1TOPO, perform digestion with *Eco*RI.

18. Sequence detected inserts from positive clones using appropriate primers.

    For pCR2.1TOPO, use primers M13 uni (-21) 5'-TGTAAAACGACGGCCAGT-3' and M13 rev (-29) CAGGAAACAGCTATGACC.

19. To verify the specificity of the DNA fragments for the tester strain, perform a PCR with fragment specific PCR primers on DNA from the tester strain and related *P. larvae* strains (i.e., representatives of the same genotype).

In each subtraction, all controls recommended by the manufacturer must be performed and all must test positive. Analysis of the tester-specific sequences is performed with BLASTx (Altschul *et al.*, 1990, 1997) followed by functional annotation based on the COG (cluster of orthologous groups) classification (Tatusov *et al.*, 1997, 2003).

5.3. Conventional proteomics using two-dimensional gel electrophoresis

Genomic sequences are not sufficient for explaining biological functions because there is no strict linear correlation between the genome and the proteome of an organism. For example, protein modifications and relative concentration of proteins cannot be determined by genomic analysis (Pandey and Mann, 2000). Furthermore, the DNA sequences give no information about conditions and time for translation as well as effects of up or down regulation of gene expression (Humphery-Smith *et al.*, 1997). Therefore, predictions on the basis of genetic information should be completed by expression data at the level of the transcriptome and the proteome. By means of proteome analysis, a holistic approach of protein expression under specific conditions is possible. The classical method of proteomics is the two-dimensional (2D) gel electrophoresis (O'Farrell, 1975; Klose, 1975). Recently, it was successfully used for a comparison of the *P. larvae* genotypes ERIC I and II (Fünfhaus and Genersch, 2012).

Sample preparation for 2D electrophoresis uses the following protocol (Fünfhaus and Genersch, 2012):

1. Cultivate *P. larvae* strains to be analysed on Columbia sheep blood agar (CSA; see section 3.1. for recipe) plates for three days at 37°C.

2. To obtain a pre-culture, inoculate 3 ml brain heart infusion broth (BHI, see section 3.1 for recipe) with one bacterial colony and the cells are grown overnight at 37°C with shaking at 200 x g.

3. To obtain a 10 ml main culture, inoculate 9 ml BHI with a pre-culture to achieve a final OD_{600} of 0.01 after adjustment to a final volume of 10 ml with BHI.

4. Incubate at 37°C with shaking at 200 x g.

5. Monitor growth continuously by measuring OD_{600}.

6. Stop growth in the late exponential phase (OD_{600} 0.65) by harvesting the cells via centrifugation (20 min, 5,000 x g, 4°C).

7. Wash the bacterial cell pellets three times with ice-cold PBS.

8. Resuspended in 1 ml lysis buffer (7 M urea, 2M thiourea, 4% (w/v) CHAPS, complete protease inhibitor cocktail (Roche)).

9. Disrupt cells by using a sonicator, e.g. ranson Sonifier 250 (duty control: 10%; output control: 1). Repeated sonication cycles (ten times for 30 sec) are interrupted by cooling phases for 60 sec.

10. Incubate samples for 1 h at RT to facilitate dissolving the proteins.

11. Separate crude protein extracts from cellular debris by centrifugation at 16,100 x g for 25 min at 4°C.

    The resulting supernatant contains the cytosolic proteins but also salts and other small charged molecules which need to be removed from the solution by precipitation.

12. Precipitate the cytosolic proteins of *P. larvae* with one volume 20% TCA, 2% Triton X-100 overnight at 4°C.

13. Pellet the precipitated proteins by centrifugation (16,100 x g, 25 min, 4°C).

14. Wash the pellets with 80% acetone to remove residual TCA.

15. Resuspend the washed pellet in 200 µl sample buffer (7 M urea, 2M thiourea, 4% (w/v) CHAPS, 100 mM DTT, 1% Bio-Lyte Ampholyte (Bio-Rad), complete protease inhibitor cocktail (Roche)).

16. Vortex for 1 h at RT.

17. Separate the insoluble material from the soluble proteins by centrifugation (15,000 x g, 1 h, 15°C) to obtain the soluble cytoplasmic fraction.

18. Determine the protein concentration by performing the Pierce® 660 nm Protein Assay (Thermo Scientific) according to the manufacturer's protocol.

19. Store samples at -80°C until further analysis.

These protein samples are then subjected to 2D-gel electrophoresis with the isoelectric focussing (IEF) as the first dimension followed by SDS-PAGE analysis as second dimension.

20. Dilute the samples with rehydration buffer (7 M urea, 2 M thiourea, 1% (w/v) CHAPS, 10 mM DTT, 0.25% Bio-Lyte Ampholyte (Bio-Rad)).

21. Determine the amount of protein to load on IPG strip according to the sample, the pH range, the length of the IPG strip and the staining method.

22. Focussing of proteins is best performed in commercially available, immobilized pH gradient strips (IPG strips) selecting a suitable pH gradient. For the analysis of *P. larvae* proteins IPG strips with a pH gradient 5-8 and a length of 7 cm (Bio-Rad) proved to be useful.

23. Load 60 µg cytosolic *P. larvae* proteins in a total rehydration volume of 125 µl on the IPG strips.

The following protocol for IEF is adapted to the PROTEAN IEF Cell (Bio-Rad) at 20°C:

24. Load the samples on the IPG strips by an active in-gel rehydration for 18 h at 50 V followed by a voltage profile with increasing values:

   • linear increase from 50 – 200 V for 1 min.
   • 200 V for 200 Vh.
   • linear increase from 200 – 500 V for 1 min.
   • 500 V for 500 Vh.
   • linear increase from 500 – 1000 V for 1 min.
   • 1000 V for 1000 Vh.
   • linear increase from 1000 – 2000 V for 1 min.
   • 2000 V for 2000 Vh.
   • linear increase from 2000 – 4000 V for 1 min.
   • 4000 V for 4500 Vh.

25. Subsequently, saturate the proteins separated in the IEF gel with SDS by equilibrating the IPG strips in equilibration buffer I (6M urea, 30% (v/v) glycerol, 5% (w/v) SDS, 0.05 M Tris pH 8.8, 1% (w/v) DTT) for 10 min.

26. Block free SH-groups of the separated proteins by equilibrating the IPG strips in equilibration buffer II (6M urea, 30% (v/v) glycerol, 5% (w/v) SDS, 0.05 M Tris pH 8.8, 5% (w/v) iodoacetamide) for 10 min.

27. For the SDS-PAGE as second dimension a 12% polyacrylamide gel run at 35 mA in a PROTEAN II XL Cell (Bio-Rad) proved to be suitable.

28. Gels are stained with Coomassie (Page Blue Protein Solution, Fermentas) according to standard protocol (see the *BEEBOOK* paper on physiological and biochemical methods (Hartfelder *et al.*, 2013)).

Analysis of the 2D gels can be performed by using software PDQuest 8.0 (Bio-Rad). Protein identification can be achieved by mass spectrometric analysis followed by comparison of peptide masses and sequence information of the sample with different databases (see the *BEEBOOK* paper on physiological and biochemical methods (Hartfelder *et al.*, 2013) and the *BEEBOOK* paper on chemical ecology (Torto *et al.*, 2013)).

5.4. Differential proteomics of *Paenibacillus larvae*

Liquid chromatography coupled directly to mass spectrometry provides another means for monitoring a proteome or changes in a proteome. Such global monitoring of changes in levels of proteins in response to a stimulus (e.g. a pathogen such as *P. larvae*) can provide very direct insight into the molecular mechanisms employed to respond to that challenge. For example, bee larvae up-regulate expression and activation of phenoloxidase in response to a *P. larvae* challenge (Chan *et al.*, 2009). Mass spectrometry is currently the favoured detection method for monitoring the entire protein component of a system (i.e. the proteome), and various methods exist for comparing protein expression in one state to that in another. The most quantitative approach involves the use of stable isotopes to introduce a 'mass tag' into the proteins from two or more different conditions and then in subsequent mass spectrometric analyses, the intensities of the differently tagged forms reflect the relative quantities in the original sample. Several such labelling methods exist and are reviewed elsewhere (Ong *et al.*, 2003), so here we focus on the method that has been used most extensively in honey bee proteomics. It involves the reductive dimethylation of primary amines in peptides using formaldehyde isotopologues.

Assumptions: a suitable, controlled experiment should be designed to compare untreated bees/cells to equivalent samples treated with a stimulus or challenged with a pathogen. If the following steps cannot be carried out immediately, then samples can typically be stored as a cell pellet, tissue, or whole bee at -80°C for weeks or months without protein degradation.

5.4.1. Sample preparation

5.4.1.1. Extract proteins

Extract proteins into 50 mM NH_4HCO_3 (pH 8.0) with 1% sodium deoxycholate (ABC/DOC).

Option 1: For cultured or primary cells:

1. Wash cells first in PBS.
2. Pellet at 600 x g.
3. Remove supernatant.
4. Solubilize the pellet in ABC/DOC at a ratio of 100 µl per 2E+07 cells.

The final protein concentration, measured by the BCA method, should be approximately 1 µg/µl.

Option 2: For any bee tissues:

1. Place the material in a beadmill tube with enough 6 M urea, 2 M thiourea and 50 mM Tris (pH 8.0) to fully immerse the tissue.
2. Pulverize the tissue in a beadmill.
   Determine the specific conditions empirically for each tissue and beadmill type.
3. After milling, pellet insoluble material at 16,000 x g for 10 min at 4°C.
4. Move the supernatant to a clean tube.
5. Precipitate the proteins in the supernatant by adding four volumes of 100% ethanol, 20 µg of molecular biology-grade glycogen and 50 mM sodium acetate (pH 5, from a 2.5 M stock solution).
6. Allow the solution to stand at RT for 90 min.
7. Pellet the proteins by centrifuging for 10 min at 16,000 x g.
8. Resuspend the pellet from this final step in ABC/DOC to bring the protein concentration to 1 µg/µl.

5.4.1.2. Reduce, alkylate and digest proteins to peptides

1. For each sample, add 1 µg dithiothreitol (from a stock solution of 1 µg/µl) to 50 µg total protein.
2. Incubate the solution for 30 min at 37°C.
3. Add 5 µg iodoaccetamide (from a stock solution of 5 µg/µl).
4. Incubate for 20 min at 37°C.
5. Add 1 µg trypsin (mass spectrometry-grade, Promega).
6. Incubate overnight at 37°C.

5.4.1.3. Clean up peptides

1. Adjust pH of overnight digestion to ~2 using 10% acetic acid.
2. For each sample, prepare a STAGE tip (Rappsilber *et al.*, 2007) by pushing 20 µl methanol through, followed by 20 µl 0.5% acetic acid.
3. Push 10 µg total peptide mass (based on original 50 µg input) through the STAGE tip.
4. Wash each tip with 20 µl 0.5% acetic acid.
5. Elute the peptides from each tip into a clean microfuge tube with 10 µl 80% acetonitrile, 0.5% acetic acid.
6. Evaporate the solvent from each sample in a vacuum centrifuge.

5.4.1.4. Label peptides with stable isotopes

1. Add 20 µl of 100 mM triethylammonium bicarbonate to each tube of dried peptides.
2. Sonicate for 5 min to solubilize peptides.
3. Add 20 µl of formaldehyde isotopologues to respective samples.
   200 mM CH_2O (L)-light to the first sample.
   200 mM C^2H_2O (M)- medium to the second sample.
   200 mM $^{13}C^2H_2O$ – heavy (H) to the third sample).
4. Add 2 µl of 1 M light ALD solution (sodium cyanoborohydride) to the first (L) sample and second (M) sample.
5. Add 2 µl of 1 M heavy ALD solution (sodium cyanoborodeuteride) to the third (H) sample.
6. Sonicate for 5 min.
7. If required, centrifuge for 1 min at low speed to collect all of the solution at the bottom.
8. Leave at ambient temperature in the dark for 90 min.
9. Add 20 µl of 3.0 M NH_4Cl.
10. Leave at ambient temperature in the dark for 10 min.
11. Add acetic acid until pH < 2.5.

5.4.1.5. Clean up peptides

1. From the final step in paragraph 5.4.1.4. 'Label peptides with stable isotopes', combine the two (for duplex labelling) or three (for triplex labelling) into a single tube.
2. Repeat all steps from paragraph 5.4.1.3. 'Clean up peptides' above, eluting the peptides in step 5 into a 96-well microwell plate.

5.4.1.6. Mass spectrometry analysis

1. Analyze the combined sample with at least a two-hour gradient into a liquid chromatography-tandem mass spectrometry system (LC-MS/MS).
   Most labs will access such a system through a core facility rather than operating the system themselves. Interested readers are referred elsewhere for details on setting up such a system (Forner *et al.*, 2007).
2. Search the LC-MS/MS data against a protein database using MaxQuant (Cox and Mann, 2008), Proteome Discoverer (from ThermoFisher) or SpectrumMill (from Agilent).

Background on the principles underlying such database searching has been covered elsewhere (Forner *et al.*, 2007). The database should contain all protein sequences from all species that might be in the sample. For example, if the experiment has involved infecting bees with *P. larvae*, the database should contain all *A. mellifera* and all *P. larvae* sequences available from their respective genome project pages in GenBank. The search engines mentioned above should also report the isotope ratios representing the relative amounts of each protein detected among the different samples.

5.5. Expression of heterologous proteins in *Paenibacillus larvae*

The understanding of phenotypic differences within the species *P. larvae* (Genersch and Otten, 2003; Genersch *et al.*, 2005; Neuendorf *et al.*, 2004; Rauch *et al.*, 2009) and the role of genotype-specific putative virulence factors (Fünfhaus *et al.*, 2009; Fünfhaus and Genersch, 2012) have been hampered by the lack of molecular tools allowing genetic manipulation of this pathogen.

To functionally analyse putative virulence factors during bacterial pathogenesis, the corresponding genes need to be disrupted, manipulated, or labelled for specific visualisation. Likewise, the expression of homologous or heterologous proteins in *P. larvae* would be a valuable tool to further study the molecular pathogenesis of *P. larvae* infections. Recently, the first protocol for the expression of a foreign protein (green fluorescent protein, GFP) in *P. larvae* has been described (Poppinga and Genersch, 2012). Since GFP expression is one of the most successful molecular tools to specifically label proteins and to visualize them under native conditions (Tsien, 1998), this can be considered a breakthrough in AFB research. Constitutive expression of GFP in *P. larvae* will help to visualize and quantify bacterial cells in larval experiments. Additionally, GFP-fused virulence factors that are expressed during infection can be detected and visualised in and/or outside the bacterial cell.

5.5.1. Transformation of Paenibacillus larvae

For this purpose plasmid pAD43-25 (BGSC, Bacillus Genetic Stock Center), carrying the gene sequence for a GFP variant (*gfp*mut3a), was transformed into wild type *P. larvae* strains ATCC9545 and 04-309 representing both relevant genotypes ERIC I and ERIC II, respectively (Genersch *et al.*, 2005, 2006). Plasmid pAD43-25 functions as an *E. coli*/Gram-positive shuttle vector and enables bacteria to constitutively express the GFP variant *gfp*mut3a (Dunn and Handelsman, 1999). High level constitutive expression of mutant GFP in vegetative cells is facilitated by the *Bacillus cereus* UW85 P*upp* promoter upstream of *gfp*mut3a. Mutant *gfp*mut3a has an optimal excitation wavelength of 498 nm and the plasmid contains a chloramphenicol resistance cassette.

The first critical step of the molecular manipulation of *P. larvae* is the uptake of foreign DNA, e. g. plasmid DNA. Because *P. larvae* is a gram-positive bacterium, transformation needs a high voltage electric pulse for a successful uptake of foreign plasmid DNA. For this purpose electrocompetent bacterial cells need to be prepared as described by Murray and Aronstein (2008):

1. Grow *P. larvae* cultures to early exponential phase (OD$_{600}$ 0.3).
2. Harvest by centrifugation.
3. Wash bacterial pellets three times using 0.625 M sucrose.

4. Resuspend the final bacterial pellet in 1/500 of the initial culture volume.
5. Store at -80°C in 40 µl aliquots.

Electrocompetent *P. larvae* cells can then be transformed with a plasmid containing a GFP-gene like pAD43-25 using the following protocol:

1. For pAD43-25 transformation, thaw competent cells at 4°C.
2. Add 500 ng of pAD43-25 in a maximum volume of 10 µl.
3. During an incubation time of 15-20 min gently mix each transformation tube every 5 min.
4. Pulse probes in ice cold 1 mm electroporation cuvettes (Eppendorf).
5. The best transformation results can be achieved with 9 kV/cm for representatives of the genotype ERIC I and 10 kV/cm for representatives of the genotype ERIC II. An average of 1.8E+04 and 1.1E+06 transformants per 0.5 µg DNA (3.6E+04 and 2.2E+06 transformants per µg DNA) for ERIC I and ERIC II, is obtained using these conditions.
6. Immediately transfer transformation tubes containing shocked cells to 960 µl pre-warmed MYPGP broth.
7. Incubated at 37°C for 16 hours with shaking (350 x g).
8. Dilute regenerated transformation mixes.
9. Plate on MYPGP agar supplemented with 5 µg/ml chloramphenicol.
10. Incubate agar plates at 37°C.
11. Determine the number of colony forming units (cfu) after 3 days.

5.5.2. Detection of GFP-expression

For GFP detection in the vegetative recombinant *P. larvae* isolates ATCC9545 (ERIC I) + pAD43-25 and *P. larvae* 04-309 (ERIC II) + pAD43-25, clones can be cultivated and analysed during different stages of growth in liquid MYPGP media (Dingman and Stahly, 1983), supplemented with 5 µg/ml chloramphenicol. In each bacterial stage of growth, 10 µl aliquots of the suspension are analysed bright field microscopically using differential interference contrast (DIC). Fluorescence activity is detected by a FITC filter block (e.g. Nikon Ti-E Inverted Microscope). All stages of growth are analysed and it can be observed that the rate of *gfp*mut3a expressing *P. larvae* clones remains stable throughout the logarithmic as well as the stationary growth phase, indicating that plasmid pAD43-25 is correctly replicated throughout bacterial growth.

Expression of GFP remained stable even after sporulation and germination. This is an essential prerequisite for using recombinant bacteria in infection assays because only spores can be used for infection, and the vegetative bacteria inside the larvae should still carry and express the introduced gene.

5.6. Fluorescent *in situ* hybridization for the detection of *Paenibacillus larvae*

Recently, Yue and collaborators (Yue *et al.*, 2008) described a *P. larvae* specific method based on a previously described general technique of fluorescence *in situ* hybridization (FISH) (Moter *et al.*, 1998). FISH methods are based on specific binding of fluorescent-labelled oligonucleotide probes to complementary sequences in fixed and permeabilized sections (Itzkovitz and van Oudenaarden, 2011). This approach provides a highly specific method to visualize *P. larvae* in infected larval histological sections, thereby allowing disease monitoring and observation of the life cycle of the bacterium inside the host. *P. larvae* FISH methodology uses a specific complementary *P. larvae* 16S rRNA targeted oligonucleotide probe coupled with fluorescein isothiocyanate (FITC), that specifically binds to *P. larvae* rRNA. Since bacterial cells are filled with ribosomes, this technique allows the "staining" of the vegetative bacteria. In order to simultaneously visualize the cytoplasm of larval cells, cyanine Cy3-labeled oligonucleotides universally detecting conserved 18S rRNA eukaryotic sequences are used. Finally, 4',6-diamidino-2-phenylindole (DAPI), a fluorescent dye that binds A-T rich regions of DNA, is employed to visualize cell nuclei.

The following protocol provides a method to fix and embed larval tissues, in order to obtain histological sections, as well as for further processing these sections by fluorescence *in situ* hybridization.

5.6.1. Preparation and embedding of larval tissues

1. Perform infection assays as described in section 4.1.
2. Wash each larva for 5 seconds with 100% ethanol.
3. Fix larvae by overnight incubation with 4% Roti Histofix (Roth, Karlsruhe) at 4°C at 20 rpm.
4. Wash samples overnight with 1X PBS solution containing 6.8% sucrose under the same conditions.
5. Wash samples with 100% acetone.
   During the first five minutes, change acetone until the solution becomes clear.
6. Incubate with the final change of acetone for 1 hour under the same conditions.
7. Mix equal volumes of 100% acetone and Technovit 8100 solution (Heraeus Kulzer, Wehrheim).
8. Incubate samples for 2 hours with this solution under previous conditions.
9. Perform infiltration and embedding using Technovit 8100 according to manufacturer´s instructions.
10. Cut 4 µm-sections using a microtome, straighten them with warm sterile water on slides.
11. Store slides at 4°C until further use.

5.6.2. Performing fluorescence in situ hybridization

1. Sections should be rehydrated and prepared for hybridization as follows in order to improve permeability:
   1.1. Wash slides with xylene 3 x 10 min.
   1.2. Wash slides with 100% ethanol 3 x 3 min.
   1.3. Wash slides with 90% ethanol 2 x 3 min.
   1.4. Wash slides with 80% ethanol 2 x 3 min.
   1.5. Wash slides with 70% ethanol 1 x 3 min.
   1.6. Wash slides with 50% ethanol 2 x 3 min.
   1.7. Wash slides with double distilled water (DEPC) 2 x 3 min.
   1.8. Incubate slides with 1 mg/ml Proteinase K in 0.2M Tris-HCl (pH 7.9) for 5 min at 37°C in a humid chamber.
   1.9. Incubate slides with 1 mg/ml lysozyme in DEPC double-destilled water for 15 min at 37°C in a humid chamber.
   1.10. Wash slides three times with 1X PBS.
2. Incubate slides with 100 ng of each probe (specific bacterial 16S rRNA-probe and universal eukaryotic 18S rRNA-probe) diluted in 20 µl of FISH-hybridization buffer (20% (v/v) deionized formamide, 0.9 M NaCl, 20 mM Tris-HCl pH 7.9, 0.01% (m/v) SDS).
3. Cover with slip and transfer to a Corning chamber.
4. Dispense double-distilled water into the cavities of the chamber and close it.
5. Incubate in a humid chamber at 46°C from 4 h to overnight.
6. Carefully remove cover slip in 1X PBS.
7. Wash slides three times with 1X PBS.
8. Add 50 µl of DAPI solution (1 µg/ml in methanol) to each slide.
9. Cover again.
10. Incubate at RT for 10 min.
11. Remove cover slip in 1X PBS and wash slides three times with 1X PBS.
12. Mount slides with antifade reagent.

6. Final remarks

In the past decade, different basic issues related to American foulbrood have been addressed, including the reclassification of its aetiological agent (Genersch *et al.*, 2006; Ashiralieva and Genersch, 2006), the development of methods for genotyping (Genersch and Otten, 2003; Alippi *et al.*, 2004; Genersch *et al.*, 2006; Antúnez *et al.*, 2007) and the annotation of its genome sequence (Qin *et al.*, 2006; Chan *et al.*, 2011). The latter permitted the introduction of techniques for studying the bacterial transcriptome and proteome. Candidate reference genes and the methods to select them have been described in this paper. The availability of a protocol for heterologous expression of foreign proteins in *P. larvae* can be considered a great breakthrough in AFB research (Poppinga and Genersch, 2012). Nevertheless, we are only at the start of understanding the intimate relationship between *P. larvae* and its host. We hope that methods that are presented in this paper can help scientists to further explore the secrets of American foulbrood disease in honey bees.

7. Acknowledgements

We mourn the passing of Drs. Dieter De Koker who contributed to this chapter. Dieter was a talented PhD student at Ghent University. Our thoughts are with his family and friends. DCdG and LDS gratefully acknowledge the Research Foundation of Flanders (FWO-Vlaanderen; G.0163.11) for financial support.

8. References

AGRON, P G; MACHT, M; RADNEDGE, L; SKOWRONSKI, E W; MILLER, W; ANDERSEN, G L (2002) Use of subtractive hybridization for comprehensive surveys of prokaryotic genome differences. *FEMS Microbiology Letters* 211: 175–182. http://dx.doi.org/10.1111/j.1574-6968.2002.tb11221.x

AKOPYANTS, N S; FRADKOV, A; DIATCHENKO, L; HILL, J E; SIEBERT, P D; LUKYANOV, S; SVERDLOV, E; BERG, D E (1998) PCR-based subtractive hybridization and differences in gene content among strains of *Helicobacter pylori*. *Proceedings of the National Academy of Sciences USA* 95 (22): 13108-13113. http://dx.doi.org/10.1073/pnas.95.22.13108

ALIPPI, A M (1991) A comparison of laboratory techniques for the detection of significant bacteria of the honey bee *Apis mellifera* L. in Argentina. *Journal of Apicultural Research* 30(2): 75–80.

ALIPPI, A M (1995) Detection of *Bacillus larvae* spores in Argentinian honeys by using a semi-selective medium. *Microbiologia SEM* 11: 343–350.

ALIPPI, A M; AGUILAR, O M (1998a) Characterization of isolates of *Paenibacillus larvae* subsp. *larvae* from diverse geographical origin by the polymerase chain reaction and BOX primers. *Journal of Invertebrate Pathology* 72: 21-27. http://dx.doi.org/10.1006/jipa.1998.4748

ALIPPI, A M; AGUILAR, O M (1998b). Unique fingerprints of *Paenibacillus larvae* subsp. *larvae* strains. *Journal of Apicultural Research* 37: 273-280.

ALIPPI, A M ; LOPEZ, A C; AGUILAR, O M (2002) Differentiation of *Paenibacillus larvae* subsp. *larvae*, the cause of American foulbrood of honey bees, by using PCR and restriction fragment analysis of genes encoding 16S rRNA. *Applied and Environmental Microbiology* 68(7): 3655-3660. http://dx.doi.org/10.1128/AEM.68.7.3655-3660.2002

ALIPPI, A M; LÓPEZ, A C; REYNALDI, F J; GRASSO, D H; AGUILAR, O M (2007) Evidence for plasmid-mediated tetracycline resistance in *Paenibacillus larvae*, the causal agent of a honey bee larval disease. *Veterinary Microbiology* 125: 290-303. http://dx.doi.org/10.1016/j.vetmic.2007.05.018

ALIPPI, A M; REYNALDI, F J; LÓPEZ, A C; DE GIUSTI, M R; AGUILAR, O M (2004) Molecular epidemiology of *Paenibacillus larvae larvae* and incidence of American foulbrood in Argentinean honeys from Buenos Aires Province. *Journal of Apicultural Research* 43: 135-143.

ALTSCHUL, S F; GISH, W; MILLER, W; MYERS, E W; LIPMAN, D J (1990) Basic local alignment search tool. *Journal of Molecular Biology* 215(3): 403-410. http://dx.doi.org/10.1006/jmbi.1990.9999

ALTSCHUL, S F; MADDEN, T L; SCHÄFFER, A A; ZHANG, J; ZHANG, Z; MILLER, W; LIPMAN, D J (1997) Gapped BLAST and PSI-BLAST: a new generation of protein database search programs. *Nucleic Acids Research* 25(17): 3389-3402. http://dx.doi.org/10.1093/nar/25.17.3389

ANTUNEZ, K; PICCINI, C; CASTRO-SOWINSKI, S; ROSADO, A S; SELDIN, L; ZUNINO, P (2007) Phenotypic and genotypic characterization of *Paenibacillus larvae* isolates. *Veterinary Microbiology* 124: 178-183. http://dx.doi.org/10.1016/j.vetmic.2007.04.012

ASHIRALIEVA, A; GENERSCH, E (2006) Reclassification, genotypes and virulence of *Paenibacillus larvae*, the etiological agent of American foulbrood in honey bees – a review. *Apidologie* 37(4): 411-420. http://dx.doi.org/10.1051/apido:2006028

BÜCHLER, R; ANDONOV, S; BIENEFELD, K; COSTA, C; HATJINA, F; KEZIC, N; KRYGER, P; SPIVAK, M; UZUNOV, A; WILDE, J (2013) Standard methods for rearing and selection of *Apis mellifera* queens. In *V Dietemann; J D Ellis; P Neumann (Eds) The COLOSS BEEBOOK, Volume I: standard methods for* Apis mellifera *research. Journal of Apicultural Research* 52(1): http://dx.doi.org/10.3896/IBRA.1.52.1.07

CHAN, Q W T; MELATHOPOULOS, A P; PERNAL, S F; FOSTER, L J (2009) The innate immune and systemic response in honey bees to a bacterial pathogen, *Paenibacillus larvae*. *BMC Genomics* 10: 387. http://dx.doi.org/10.1186/1471-2164-10-387

CHAN, Q W T; CORNMAN, R S; BIROL, I; LIAO, N Y; CHAN, S K; DOCKING, T R; JACKMAN, S D; TAYLOR, G A; JONES, S J M; DE GRAAF, D C; EVANS, J D; FOSTER, L J (2011) Updated genome assembly and annotation of *Paenibacillus larvae*, the agent of American foulbrood disease of honey bees. *BMC Genomics* 12: 450. http://dx.doi.org/10.1186/1471-2164-12-450

DE GRAAF, D C; ALIPPI, A M; BROWN, M; EVANS, J D; FELDLAUFER, M; GREGORC, A; HORNITZKY, M A Z; PERNAL, S F; SCHUCH, D M T; TITERA, D; TOMKIES, V; RITTER, W (2006a) Diagnosis of American foulbrood in honey bees: a synthesis and proposed analytical protocols. *Letters in Applied Microbiology* 43: 583-590. http://dx.doi.org/10.1111/j.1472-765X.2006.02057.x

DE GRAAF, D C; DE VOS, P; HEYNDRICKX, M; VAN TRAPPEN, S; PEIREN, N; JACOBS, F J (2006b) Identification of *Paenibacillus larvae* to the subspecies level: an obstacle for AFB diagnosis. *Journal of Invertebrate Pathology* 91(2): 115-123. http://dx.doi.org/10.1016/j.jip.2005.10.010

DE GRAAF, D C; BRUNAIN, M; JACOBS, F J (2008) Implementation of quality control and biosafety measurements in the diagnosis of honey bee diseases. *Journal of Apicultural Research* 47(2): 151-153.

DELAPLANE, K S; VAN DER STEEN, J; GUZMAN, E (2013) Standard methods for estimating strength parameters of *Apis mellifera* colonies. In *V Dietemann; J D Ellis; P Neumann (Eds) The COLOSS BEEBOOK, Volume I: standard methods for* Apis mellifera *research. Journal of Apicultural Research* 52(1): http://dx.doi.org/10.3896/IBRA.1.52.1.03

DIATCHENKO, L; LAU, Y F; CAMPBELL, A P; CHENCHIK, A; MOQADAM, F; HUANG, B; LUKYANOV, S; LUKYANOV, K; SVERDLOV, E D; SIEBERT, P D (1996) Suppression subtractive hybridization: a method for generating differentially regulated or tissue-specific cDNA probes and libraries. *Proceedings of the National Academy of Science USA* 93(12): 6025-6030. http://dx.doi.org/10.1073/pnas.93.12.6025

DINGMAN, D W (1983) *Bacillus larvae*: parameters involved with sporulation and characteristics of two bacteriophages. PhD thesis, University of Iowa, Iowa City, Iowa, USA.

DINGMAN, D W; STAHLY, D P (1983) Medium promoting sporulation of *Bacillus larvae* and metabolism of medium components. *Applied and Environmental Microbiology* 46: 860–869.

DINGMAN, D W; STAHLY, D P (1984) Protection of *Bacillus larvae* from oxygen toxicity with emphasis on the role of catalase. *Applied and Environmental Microbiology* 47: 1228-1237.

DOBBELAERE, W; DE GRAAF, D C ; PEETERS, J E; JACOBS, F J (2001) Development of a fast and reliable diagnostic method for American foulbrood disease (*Paenibacillus larvae* subsp. *larvae*) using a 16S rRNA gene based PCR. *Apidologie* 32: 363-370.

DJORDJEVIC, S; HO-SHON, M; HORNITZKY, M A Z (1994) DNA restriction endonuclease profiles and typing of geographically diverse isolates of *Bacillus larvae*. *Journal of Apicultural Research* 33(2): 95-103.

DUNN, A K; HANDELSMAN; J (1999) A vector for promoter trapping in *Bacillus cereus*. Gene 226: 297-305. http://dx.doi.org/10.1016/S0378-1119(98)00544-7

EVANS, J D; PETTIS, J S (2005) Colony-level impacts of immune responsiveness in honey bees, *Apis mellifera*. *Evolution* 59: 2270-2274. http://dx.doi.org/10.1554/05-060.1

FORNER, F; FOSTER, L J; TOPPO, S (2007) Mass spectrometry data analysis in the proteomics era. *Current Bioinformatics* 2(14): 63-93. http://dx.doi.org/10.2174/157489307779314285

FORSGREN, E; STEVANOVIC, J; FRIES, I (2008) Variability in germination and in temperature and storage resistance among *Paenibacillus larvae* genotypes. *Veterinary Microbiology* 129: 342-349. http://dx.doi.org/10.1016/j.vetmic.2007.12.001

FÜNFHAUS, A; ASHIRALIEVA, A; BORRISS, R; GENERSCH, E (2009) Use of suppression subtractive hybridization to identify genetic differences between differentially virulent genotypes of *Paenibacillus larvae*, the etiological agent of American foulbrood of honey bees. *Environmental Microbiology Reports* 1(4): 240-250. http://dx.doi.org/10.1111/j.1758-2229.2009.00039.x

FÜNFHAUS, A; GENERSCH, E (2012) Proteome analysis of *Paenibacillus larvae* reveals the existence of a putative S-layer protein. *Environmental Microbiology Reports* 4: 194-202. http://dx.doi.org/10.1111/j.1758-2229.2011.00320.x

GENERSCH, E; OTTEN, C (2003) The use of repetitive element PCR fingerprinting (rep-PCR) for genetic subtyping of German field isolates of *Paenibacillus larvae* subsp. *larvae*. *Apidologie* 34: 195-206. http://dx.doi.org/10.1051/apido:2003025

GENERSCH, E; ASHIRALIEVA, A; FRIES, I (2005) Strain- and genotype-specific differences in virulence of *Paenibacillus larvae subsp. larvae*, a bacterial pathogen causing American foulbrood disease in honey bees. *Applied and Environmental Microbiology* 71: 7551-7555. http://dx.doi.org/10.1128/AEM.71.11.7551-7555.2005

GENERSCH, E; FORSGREN, E; PENTIKÄINEN, J; ASHIRALIEVA, A; RAUCH, S; KILWINSKI, J; FRIES, I (2006) Reclassification of *Paenibacillus larvae* subsp. *pulvifaciens* and *Paenibacillus larvae* subsp. *larvae* as *Paenibacillus larvae* without subspecies differentiation. *International Journal of Systematic and Evolutionary Microbiology* 56(3): 501-511. http://dx.doi.org/10.1099/ijs.0.63928-0

GENERSCH, E (2010) American foulbrood in honey bees and its causative agent, *Paenibacillus larvae*. *Journal of Invertebrate Pathology* 103 Suppl. 1: 10-19. http://dx.doi.org/10.1016/j.jip.2009.06.015

GOCHNAUER, T A (1973) Growth, protease formation, and sporulation of *Bacillus larvae* in aerated broth culture. *Journal of Invertebrate Pathology* 22: 251-257.

GORDON, R E; HAYNES, W C; PANG, C H (1973) *The genus* Bacillus. Agriculture handbook no. 427, Agricultural Research Service, US Dept. of Agriculture, Washington, DC, USA.

GOVAN, V A; ALLSOPP, M H; DAVISON, S (1999) A PCR detection method for rapid identification of *Paenibacillus larvae*. *Applied and Environmental Microbiology* 65: 2243-2245.

HACKER, J; BLUM-OEHLER, G; MUHLDORFER, I; TSCHAPE, H (1997) Pathogenicity islands of virulent bacteria: structure, function and impact on microbial evolution. *Molecular Microbiology* 23: 1089-1097. http://dx.doi.org/10.1046/j.1365-2958.1997.3101672.x

HARTFELDER, K; GENTILE BITONDI, M M; BRENT, C; GUIDUGLI-
LAZZARINI, K R; SIMÕES, Z L P; STABENTHEINER, A; DONATO
TANAKA, É; WANG, Y (2013) Standard methods for physiology
and biochemistry research in *Apis mellifera*. In *V Dietemann; J D
Ellis; P Neumann (Eds) The COLOSS* BEEBOOK, *Volume I:
standard methods for* Apis mellifera *research. Journal of Apicultural
Research* 52(1): http://dx.doi.org/10.3896/IBRA.1.52.1.06

HAYNES, W C; ST JULIAN, G; SHEKLETON, M C; HALL, H H;
TASHIRO, H (1961) Preservation of infectious milky disease
bacteria by lyophilization. *Journal of Insect Pathology* 3: 55-61.

HELLEMANS, J; MORTIER, G; DE PAEPE, A; SPELEMAN, F;
VANDESOMPELE, J (2007) qBase relative quantification framework
and software for management and automated analysis of real-time
quantitative PCR data. *Genome Biology* 8: R19.
http://dx.doi.org/10.1186/gb-2007-8-2-r19

HITCHCOCK, J D; MOFFETT, J O; LACKETT, J J; ELLIOTT, J R (1970)
Tylosin for control of American foulbrood disease in honey bees.
Journal of Economic Entomology 63: 204–207.

HORNITZKY, M A Z; CLARK, S (1991) Culture of *Bacillus larvae* from
bulk honey samples for the detection of American foulbrood.
Journal of Apicultural Research 30: 13–16.

HORNITZKY, M A Z; DJORDEVIC, S (1992). Sodium dodecyl sulphate
polyacrylamide profiles and Western blots of *Bacillus larvae*.
Journal of Apicultural Research 31(1): 47-49.

HORNITZKY, M A Z; KARLOVSKIS, S (1989) A culture technique for
the detection of *Bacillus larvae* in honey bees. *Journal of
Apicultural Research* 28(2): 118–120.

HORNITZKY, M A Z; NICHOLLS, P J (1993) J-medium is superior to
sheep blood agar and brain heart infusion agar for the isolation of
Bacillus larvae from honey samples. *Journal of Apicultural Research*
32: 51–52.

HUMPHERY-SMITH, I; CORDWELL, S J; BLACKSTOCK, W P (1997)
Proteome research: Complementarity and limitations with respect
to the RNA and DNA worlds. *Electrophoresis* 18(8): 1217-1242.
http://dx.doi.org/10.1002/elps.1150180804

ITZKOVITZ, S; VAN OUDENAARDEN, A (2011) Validating transcripts
with probes and imaging technology. *Nature Methods* 8: S12-S19.
http://dx.doi.org/10.1038/NMETH.1573

JANKE, B; DOBRINDT, U; HACKER, J; BLUM-OEHLER, G (2001) A
subtractive hybridization analysis of genomic differences between
the uropathogenic *E. coli* strain 536 and the *E. coli* K-12 strain
MG1655. *FEMS Microbiology Letters* 199: 61–66.
http://dx.doi.org/10.1111/j.1574-6968.2001.tb10651.x

KILWINSKI, J; PETERS, M; ASHIRALIEVA, A; GENERSCH, E (2004)
Proposal to reclassify *Paenibacillus larvae* subsp. *pulvifaciens* DSM
3615 (ATCC 49843) as *Paenibacillus larvae* subsp. *larvae*. Results
of a comparative biochemical and genetic study. *Veterinary
Microbiology*. 104: 31-42.
http://dx.doi.org/10.1016/j.vetmic.2004.08.001

KLOSE, J (1975) Protein mapping by combined isoelectric focusing
and electrophoresis of mouse tissues. A novel approach to testing
for induced point mutations in mammals. *Humangenetik* 26(3):
231-243.

LONCARIC, I; DERAKHSHIFAR, I; OBERLERCHNER, J T;
KÄGLBERGER, H; MOOSBECKHOFER, R (2009) Genetic diversity
among isolates of *Paenibacillus larvae* from Austria. *Journal of
Invertebrate Pathology* 100(1): 44-46.
http://dx.doi.org/10.1016/j.jip.2008.09.003

MOTER, A; LEIST, G; RUDOLPH, R; SCHRANK, K; CHOI, B K;
WAGNER, M; GÖBEL, U B (1998) Fluorescence *in situ*
hybridization shows spatial distribution of as yet uncultured
treponemes in biopsies from digital dermatitis lesions.
Microbiology 144(9): 2459-2467.

MURRAY, K D; ARONSTEIN, K A (2006) Oxytetracycline-resistance in
the honey bee pathogen *Paenibacillus larvae* is encoded on novel
plasmid pMA67. *Journal of Apicultural Research*. 45: 207-214.

MURRAY, K D; ARONSTEIN, K A; DE LEÓN, J H (2007) Analysis of
pMA67, a predicted rolling-circle replicating, mobilizable,
tetracycline-resistance plasmid from the honey bee pathogen,
Paenibacillus larvae. Plasmid 58: 89-100.
http://dx.doi.org/10.1016/j.plasmid.2007.02.001

MURRAY, K D; ARONSTEIN, K A (2008) Transformation of the Gram-
positive honey bee pathogen, *Paenibacillus larvae*, by
electroporation. *Journal of Microbiological Methods* 75: 325-328.
http://dx.doi.org/10.1016/j.mimet.2008.07.007

NEUENDORF, S; HEDTKE, K; TANGEN, G; GENERSCH, E (2004)
Biochemical characterization of different genotypes of *Paenibacillus
larvae* subsp. *larvae*, a honey bee bacterial pathogen. *Microbiology*
150(7): 2381-2390. http://dx.doi.org/10.1099/mic.0.27125-0

NORDSTRÖM, S; FRIES, I (1995) A comparison of media and cultural
conditions for identification of *Bacillus larvae* in honey. *Journal of
Apicultural Research* 34: 97-103.

O'FARRELL, P H (1975) High resolution two-dimensional
electrophoresis of proteins. *Journal of Biological Chemistry* 250
(10): 4007-4021.

OIE (2008) Chapter 2.2.2. American foulbrood. In *OIE Manual of
Diagnostic Tests and Vaccines for Terrestrial Animals (mammals,
birds and bees), vol. 1 (Sixth Edition)*. OIE; Paris, France. pp 395-
404.

OIE (2011) Chapter 9.2. American foulbrood. In *OIE Terrestrial
Animal Health Code, vol. 2 (Twentieth Edition)*. OIE; Paris, France.
pp 504-506.

OLIVE, D M; BEAN, P (1999) Principles and applications of methods
for DNA-based typing of microbial organisms. *Journal of Clinical
Microbiology* 37(6): 1661-1669.

ONG, S E; FOSTER, L J; MANN, M (2003) Mass spectrometric-based
approaches in quantitative proteomics. *Methods*. 29(2): 124-30.
http://dx.doi.org/10.1016/S1046-2023(02)00303-1

PANDEY, A; MANN, M (2000) Proteomics to study genes and genomes. *Nature* 405(6788): 837-846.

PETERS, M; KILWINSKI, J; BERINGHOFF, A; RECKLING, D; GENERSCH, E (2006) American foulbrood of the honey bee: Occurrence and distribution of different genotypes of *Paenibacillus larvae* in the administrative district of Arnsberg (North Rhine-Westphalia). *Journal of Veterinary Medicine* 53(2): 100-104. http://dx.doi.org/10.1111/j.1439-0450.2006.00920.x

PIRK, C W W; DE MIRANDA, J R; FRIES, I; KRAMER, M; PAXTON, R; MURRAY, T; NAZZI, F; SHUTLER, D; VAN DER STEEN, J J M; VAN DOOREMALEN, C (2013) Statistical guidelines for *Apis mellifera* research. In *V Dietemann; J D Ellis; P Neumann (Eds) The COLOSS BEEBOOK, Volume I: standard methods for* Apis mellifera *research. Journal of Apicultural Research* 52(4): http://dx.doi.org/10.3896/IBRA.1.52.4.13

POPPINGA, L; GENERSCH, E (2012) Heterologous expression of green fluorescent protein in *Paenibacillus larvae*, the causative agent of American foulbrood of honey bees. *Journal of Applied Microbiology* 112: 430-435. http://dx.doi.org/10.1111/j.1365-2672.2011.05214.x

QIN, X; EVANS, J D; ARONSTEIN, K A; MURRAY, K D; WEINSTOCK, G M (2006) Genome sequences of the honey bee pathogens *Paenibacillus larvae* and *Ascosphaera apis*. *Insect Molecular Biology* 15(5): 715-718. http://dx.doi.org/10.1111/j.1365-2583.2006.00694.x

RAPPSILBER, J; MANN, M; ISHIHAMA, Y (2007) Protocol for micro-purification, enrichment, pre-fractionation and storage of peptides for proteomics using StageTips. *Nature Protocols* 2(8): 1896-906. http://dx.doi.org/10.1038/nprot.2007.261

RAUCH, S; ASHIRALIEVA, A; HEDTKE, K; GENERSCH, E (2009) Negative correlation between individual-insect-level virulence and colony-level virulence of *Paenibacillus larvae*, the etiological agent of American foulbrood of honey bees. *Applied and Environmental Microbiology* 75: 3344-3347. http://dx.doi.org/10.1128/AEM.02839-08

RECKSEIDLER, S L; DESHAZER, D; SOKOL, P A; WOODS, D E (2001) Detection of bacterial virulence genes by subtractive hybridization: identification of capsular polysaccharide of *Burkholderia pseudomallei* as a major virulence determinant. *Infection and Immunity* 69: 34–44. http://dx.doi.org/10.1128/IAI.69.1.34-44.2001

SCHUCH, D M T; MADDEN, R H; SATTLER, A (2001) An improved method for the detection and presumptive identification of *Paenibacillus larvae* subsp *larvae* spores in honey. *Journal of Apicultural Research* 40: 59–64.

SHIMANUKI, H; KNOX, D A (2000) *Diagnosis of honey bee diseases.* USDA Agricultural Handbook No. H-690, Washington DC, USA. 61 pp.

SIEBERT, P D; CHENCHIK, A; KELLOGG, D E; LUKYANOV, K A; LUKYANOV, S A (1995) An improved PCR method for walking in uncloned genomic DNA. *Nucleic Acids Research* 23: 1087-1088. http://dx.doi.org/10.1093/nar/23.6.1087

SPIVAK, M; REUTER, G S (2001) Resistance to American foulbrood disease by honey bee colonies, *Apis mellifera*, bred for hygienic behaviour. *Apidologie* 32: 555-565.

STEINKRAUS, K H; MORSE, R A (1996) Media for the detection of *Bacillus larvae* spores in honey. *Acta Biotechnologica* 16: 57-64.

TATUSOV, R L; KOONIN, E V; LIPMAN, D J (1997) A genomic perspective on protein families. *Science* 278(5338): 631-637. http://dx.doi.org/10.1126/science.278.5338.631

TATUSOV, R L; FEDOROVA, N D; JACKSON, J D; JACOBS, A R; KIRYUTIN, B; KOONIN, E V; KRYLOV, D M; MAZUMDER, R; MEKHEDOV, S L; NIKOLSKAYA, A N; RAO, B S; SMIRNOV, S; SVERDLOV, A V; VASUDEVAN, S; WOLF, Y I; YIN, J J; NATALE, D A (2003) The COG database: an updated version includes eukaryotes. *BMC Bioinformatics* 4: 41. http://dx.doi.org/10.1186/1471-2105-4-41

THOMAS, S R; ELKINTON, J S (2004) Pathogenicity and virulence. *Journal of Invertebrate Pathology* 85: 146-151. http://dx.doi.org/10.1016/j.jip.2004.01.006

TORTO, B; CARROLL, M J; DUEHL, A; FOMBONG, A T; NAZZI, F, GOZANSKY, K T; SOROKER, V, TEAL, P E A (2013) Standard methods for chemical ecology research in *Apis mellifera*. In *V Dietemann; J D Ellis; P Neumann (Eds) The COLOSS BEEBOOK, Volume I: standard methods for* Apis mellifera *research. Journal of Apicultural Research* 52(4): http://dx.doi.org/10.3896/IBRA.1.52.4.06

TSIEN, R Y (1998) The green fluorescent protein. *Annual Review of Biochemistry* 67: 509-44. http://dx.doi.org/10.1146/annurev.biochem.67.1.509

UNTERGASSER, A; NIJVEEN, H; RAO, X; BISSELING, T; GEURTS, R; LEUNISSEN, J A M (2007) Primer3plus, an enhanced web interface to primer3. *Nucleic Acids Research* 35: W71-W74. http://dx.doi.org/10.1093/nar/gkm306

VANDESOMPELE, J; DE PRETER, K; PATTYN, F; POPPE, B; VAN ROY, N; DE PAEPE, A; SPELEMAN, F (2002) Accurate normalization of real-time quantitative RT-PCR data by geometric averaging of multiple internal control genes. *Genome Biology* 3(7): 0034.1. http://dx.doi.org/10.1186/gb-2002-3-7-research0034

VERSALOVIC, J; SCHNEIDER, M; DE BRUIJN, F J; LUPSKI, J R (1994). Genomic fingerprinting of bacteria using repetitive sequence-based polymerase chain reaction. *Methods in Molecular and Cellular Biology* 5: 25-40.

WU, X Y; CHIN, J; GHALAYINI, A; HORNITZKY, M A Z (2005) Pulsed-field gel electrophoresis typing and oxytetracycline sensitivity of *Paenibacillus larvae* subsp. *larvae* isolates of Australian origin and those recovered from honey imported from Argentina. *Journal of Apicultural Research* 44: 87-92.

YUE, D; NORDHOFF, M; WIELER, L H; GENERSCH, E (2008) Fluorescence *in situ* hybridization (FISH) analysis of the interactions between honey bee larvae and *Paenibacillus larvae*, the causative agent of American foulbrood of honey bees (*Apis mellifera*). *Environmental Microbiology* 10: 1612-1620. http://dx.doi.org/10.1111/j.1462-2920.2008.01579.x

ZHANG, Y L; ONG, C T; LEUNG, K Y (2000) Molecular analysis of genetic differences between virulent and avirulent strains of *Aeromonas hydrophila* isolated from diseased fish. *Microbiology* 146: 999–1009.

ZUKER, M (2003) Mfold web server for nucleic acid folding and hybridization prediction. *Nucleic Acids Research* 31: 3406-3415. http://dx.doi.org/10.1093/nar/gkg595

Journal of Apicultural Research 52(1)

Journal of Apicultural Research 52(1): (2013)
DOI 10.3896/IBRA.1.52.1.12

REVIEW ARTICLE

Standard methods for European foulbrood research

Eva Forsgren[1]*, Giles E Budge[2], Jean-Daniel Charrière[3] and Michael A Z Hornitzky[4]

[1]Department of Ecology, Swedish University of Agricultural Sciences, PO Box 7044, SE-75007 Uppsala, Sweden.
[2]National Bee Unit, Food and Environment Research Agency, Sand Hutton, York, YO41 1LZ, UK.
[3]Swiss Bee Research Center, Agroscope Liebefeld-Posieux (ALP), Schwarzenburgstr. 161, 3003 Bern, Switzerland.
[4]New South Wales, Department of Primary Industries, Elizabeth Macarthur Agricultural Institute, PB 4008, Narellan NSW 2567, Australia.

Received 30 March 2012, accepted subject to revision 26 April 2012, accepted for publication 12 September 2012.

*Corresponding author: Email: eva.forsgren@slu.se

Summary

European foulbrood (EFB) is a severe bacterial honey bee brood disease caused by the Gram-positive bacterium *Melissocccus plutonius*. The disease is widely distributed worldwide, and is an increasing problem in some areas. Although the causative agent of EFB was described almost a century ago, many basic aspects of its pathogenesis are still unknown. Earlier studies were hampered by insensitive and unspecific methods such as culture based techniques. Recent advances in molecular technology are making it increasingly easy to detect and characterize microbes, and nucleic acid detection technologies are quickly displacing the traditional phenotypic assays in microbiology. This paper presents selected methodologies which focus on EFB and its causative agent *M. plutonius*.

Métodos estándar para la investigación sobre la loque europea

Resumen

La loque europea (LE) es una grave enfermedad bacteriana de la cría de la abeja de la miel causada por la bacteria Gram-positiva *Melissococcus plutonius*. La enfermedad se encuentra ampliamente distribuida en todo el mundo y es un problema creciente en algunas áreas. Aunque el agente causante de la LE fue descrito hace casi un siglo, muchos aspectos básicos de su patogénesis son aún desconocidos. Estudios anteriores se vieron obstaculizados por métodos poco sensibles e inespecíficos, tales como las técnicas basadas en cultivos. Los recientes avances en la tecnología molecular están haciendo cada vez más fácil la detección y caracterización de los microbios, y las tecnologías de detección de ácidos nucleícos están desplazando rápidamente a los ensayos fenotípicos tradicionales de microbiología. Este artículo presenta algunas metodologías seleccionadas que se centran en LE y en su agente causal *M. plutonius*.

欧洲幼虫腐臭病研究的标准方法

欧洲幼虫腐臭病是蜜蜂幼虫病中较为严重的细菌病，由革兰氏阳性细菌Melissocccus plutonium 引起。欧洲幼虫腐臭病在世界范围内广泛分布，成为一些地区日益突出的主要病害。虽然一个世纪前就已有相关病原体的描述，但早期研究受实验技术不灵敏、特异性不好的限制（比如细菌培养的技术），其发病机理至今未能全面揭示。近年来随着分子技术的发展检测微生物日益容易，核酸检测技术快速替换了传统的微生物学表型实验。本文展示了近期开展欧洲幼虫腐臭病及其病原体 M. plutonius方面的常用技术。

Keywords: honey bee, European foulbrood, *Melissococcus plutonius*, brood disease, methodologies, *BEEBOOK,* COLOSS

Footnote: Please cite this paper as: FORSGREN, E; BUDGE, G E; CHARRIÈRE, J-D; HORNITZKY, M A Z (2013) Standard methods for European foulbrood research. In *V Dietemann; J D Ellis, P Neumann (Eds) The COLOSS BEEBOOK: Volume II: Standard methods for Apis mellifera pest and pathogen research. Journal of Apicultural Research* 52(1): http://dx.doi.org/10.3896/IBRA.1.52.1.12

1. Introduction

1.1. Background

European foulbrood (EFB) is a bacterial brood disease caused by the Gram positive bacterium *Melissococcus plutonius*. EFB is listed in the *OIE Terrestrial Animal Health Code* (2011), but, unlike American foulbrood; it is not notifiable in all countries. The disease occurs in honey bees throughout the world, and may cause serious losses of brood and colony collapse. In many areas, the disease is endemic with occasional seasonal outbreaks, but, in a few countries, the scenario is different. In Switzerland, the incidence of EFB has increased dramatically since the late 1990s; it has become the most widespread bacterial brood disease in the UK, and Norway reported a regional outbreak of EFB during 2010 after a 30 year period of absence. Geographically, the disease appears to vary in severity from being relatively benign in some areas but increasingly severe in others (Wilkins *et al.*, 2007; Dahle *et al.*, 2011; Grangier, 2011; Arai *et al.*, 2012). Virulence tests on individual larvae using exposure bioassays (see section 9), shows that *M. plutonius* strains collected in different geographic European locations vary in their ability to cause larval mortality (Charrière *et al.*, 2011).

1.2. Disease symptoms

The field diagnosis of EFB is based on the visual inspection of brood combs and detection of diseased larvae (see section 4). The general symptoms observed in a colony suffering from EFB are irregular capping of the brood; capped and uncapped cells being found scattered irregularly over the brood frame (known as pepper pot brood). The youngest larvae that die from the infection cover the bottom of the cell and are almost transparent, with visible trachea. Older larvae die malpositioned and flaccid in their cells; twisted around the walls or stretched out lengthways (Fig. 1). The colour of affected larvae

Fig. 1. Malpositioned and discoloured larvae in brood with symptoms of European foulbrood. Photo by Eva Forsgren

changes from pearly white to pale yellow, often accompanied by a loss in segmentation. More advanced symptoms can manifest as further colour changes to brown and greyish black (Fig. 1), sometimes ultimately leaving a dark scale (Fig. 2) that is more malleable than those typically found with American foulbrood (AFB).

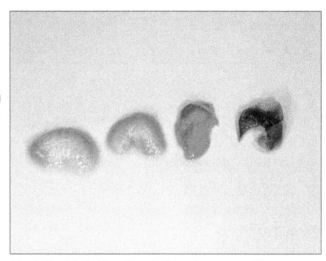

Fig. 2. The infected larva loses its internal pressure and becomes flaccid, ultimately leaving a dark scale. Photo by Kaspar Ruoff

1.3. Secondary bacteria

Several other bacteria such as: *Enterococcus faecalis; Achromobacter euridice; Paenibacillus alvei* and *Brevibacillus laterosporus* may be associated with EFB (Forsgren, 2010). Although the presence of *P. alvei* - like spores of *E. faecalis* has been considered presumptive evidence of European foulbrood, the role of such secondary bacterial invaders in disease development has been poorly investigated. *A. euridice* is frequently isolated in mixed culture with *M. plutonius* and EFB symptoms in larvae may be more easily induced with inoculate containing *M. plutonius* in combination with *A. euridice* or *P. alvei* (Bailey, 1957). However, a more recent study from Switzerland showed that the simultaneous or 3 days delayed inoculation of *P. alvei* had no influence on the virulence of M. *plutonius* in individual larvae (Charrière *et al.*, 2011; see section 9.2). This paper will focus solely on techniques for diagnosis and research of the causative agent of EFB, *M. plutonius*.

1.4. Diagnosis

Symptoms of EFB may easily be confused with other diseases or abnormalities in the brood, making diagnosis difficult. The diagnosis in the field can be further verified by microscopic examination of brood smear preparations (see section 6; Hornitzky and Wilson, 1989; Hornitzky and Smith, 1998), and a field test kit (see section 7.2) for the detection of *M. plutonius* in larval extracts is also available (Tomkies *et al.*, 2009). Analysing pooled samples of bees from the

Table 1. Reference strains of *M. plutonius*.

species	strain no.	other designation	source
Melissococcus plutonius	LMG 20360	= ATCC 35311	Honey bee larvae, UK
		= CIP 104052	
		= LMG 15058	
		= LMG 19520	
		= LMG 20206	
		= LMG 21267	
		= NCDO 2443	
	NCIMB 702439		Honey bee larvae, India
	NCIMB 702440		Honey bee larvae, Brazil
	NCIMB 702441		Honey bee larvae, Tanzania
	NCIMB 702442		Honey bee larvae, Australia
	NCIMB 702443		Honey bee larvae, UK

brood nest by PCR (see section 8) may be an alternative or complement to visual inspection (Roetschi *et al.*, 2008), although false negatives may sometimes occur (Budge *et al.*, 2010). Sensitive detection methods are required to ensure the absence of the bacterium from bee products and for the confirmation of the visual diagnosis made in the field or for research purposes. Pure isolates of *M. plutonius* may sometimes be desirable for various research purposes (see section 5). There are selective media for the cultivation of *M. plutonius* (Bailey, 1957; Bailey, 1983; Bailey and Collins, 1982; Hornitzky and Wilson, 1989; Hornitzky and Karlovskis, 1989), but to culture the bacterium can be difficult and there is some evidence that *M. plutonius* samples from different regions have a differential response to culturing (Allen and Ball, 1993; Arai *et al.*, 2012). Immunology-based tests such as enzyme linked immuno-sorbent assay (ELISA) (Pinnock and Featherstone, 1984) have been published and used for the detection and quantification of *M. plutonius* (see section 7.1), but DNA amplification using the polymerase chain reaction (PCR) provides lower thresholds of detection than ELISA, and has been successfully used for the detection of *M. plutonius* since the late 1990s (see section 8).

This paper therefore aims to present selected protocols useful for diagnosis and research on the honey bee brood disease European foulbrood.

2. Bio-safety recommendations

It is important to appreciate that there is no internationally accepted biohazard classification for the handling of *M. plutonius* for diagnostic or research purposes. Restrictions on the handling of the bacterium and diseased material vary significantly between countries, and any laboratory should check national policies for guidance before handling material. In those countries where *M. plutonius* is not notifiable, there is still a strong need for precautionary measures to reduce the risk of

infection, and further guidance on bio-safety recommendations can be found in the American foulbrood paper of the *BEEBOOK* (de Graaf *et al.*, 2013).

3. Reference strains of *Melissococcus plutonius*

Type or reference strains of *M. plutonius* are available at culture collections, such as the Belgian Co-ordinated Collections of Microorganisms, BCCM/LMG (Table 1). A type strain (or a prototype strain) is a nomenclatural type of a species or a subspecies. Authors who propose new bacterial names are supposed to deposit type strains in two publicly accessible recognized culture collections in two different countries. Type strains are also useful in validation work.

4. Sampling

Methods can be grouped into those looking to confirm the presence of the disease by testing an individual symptomatic larva for *M. plutonius,* and those that hope to confirm the presence of *M. plutonius* in asymptomatic material, such as in bulk samples of adult bees or disease-free colonies in proximity to disease. When considering the latter, it is important to consider the within-hive distribution of the pathogen and also how the sample size may affect the power of the subsequent test. For example, when assuming a hive population of 50,000 individuals, sampling 5 adult bees provides a 95% confidence of detecting a pathogen with a minimum prevalence of 50%, whereas sampling 60 adult bees increases the power of the testing regime to enable a more meaningful minimum pathogen detection of 5% prevalence. No single study has provided the necessary detail to make definitive recommendations on sample location and size, therefore this chapter concentrates on summarizing the sampling methods and

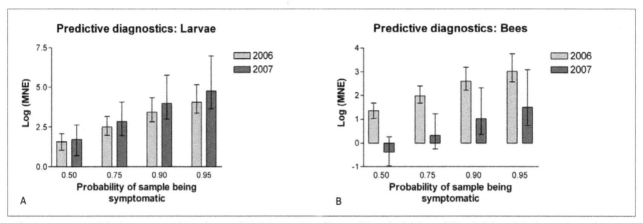

Fig. 3. Estimates of the log amount of *M. plutonius* (MNE) in samples of larvae (A) and adult bees (B) probability of the honey bee colony being symptomatic for EFB. Data from 2006 are from Budge et al. (2010), and those from 2007 previously unpublished findings from follow up work. The plots represent estimates and corresponding 95% confidence limits from a generalized linear model constructed using qPCR data (see Budge *et al.*, 2010).

knowledge to date (See also de Graaf *et al.*, 2013; Human *et al.*, 2013). Storage temperature is not crucial. All sample types can be refrigerated for several hours and stored at -20ºC for longer periods.

4.1. Brood

Upon visual inspection in the field, large pieces of symptomatic brood may be cut out and sent to the laboratory (a piece of 10 x 10 cm cuts through the metal wires of the brood frame) for further examination and confirmation of the diagnosis. Correct sampling of brood is important because even within the same brood frame, *M. plutonius* is mainly found in larvae with visual disease symptoms (Forsgren *et al.*, 2005). Alternatively, and in cases of lower severity, diseased larvae can be smeared on a microscope slide and submitted to the laboratory (see section 6). *M. plutonius* can survive for over three years on such slide preparations (Bailey, 1960), and 6 years within Lateral Flow Devices (Budge, unpublished data), and so culturing often remains a viable option many years after diagnosis.

It is possible to identify the presence of *M. plutonius* in the absence of disease symptoms by collecting bulk samples of 100 larvae, taken at random from across the brood nest and subjecting the samples to qPCR (Budge *et al.*, 2010; see also the molecular methods paper of the *BEEBOOK* (Evans *et al.*, 2013). This method provided robust quantification of *M. plutonius* and is a potentially useful tool to help predict the risk of a colony either prior to disease development or in the absence of an inspection to confirm disease (Fig. 3).

4.2. Adults

Analysis of worker bees indicates that individuals from the brood nest contain more bacteria than bees from flight entrances; therefore, it was suggested that samples of bees are preferably collected from the brood nest (Roetschi *et al.*, 2008). This result has not been replicated by others, where the amount of *M. plutonius* in foragers equalled that found in nurse bees (Budge, unpublished data). Pooled samples of

100 bees have been used for DNA extraction (Roetschi *et al.*, 2008; Budge *et al.*, 2010); however, adult bees show more variation in the amount of *M. plutonius* detected than samples of larvae (Fig. 3).

4.3. Honey and pollen

Brood nest honey, bulk honey and pollen have to some extent been used to confirm the presence of bacteria using culture methods (see section 5) and PCR (see section 8) in both diseased and healthy looking colonies (Hornitzky and Smith, 1998, McKee *et al.*, 2003). For further sampling instructions see the American foulbrood paper of the *BEEBOOK* (De Graaf *et al.*, 2013).

5. Cultivation of *M. plutonius*

For many experiments, it is imperative to use bacterial cultures in which all cells are genetically identical. Since all cells in a colony develop from one single cell, a single isolated colony of *M. plutonius* is an excellent source of a genetically pure bacterial stock. In order to propagate any bacterium, it is necessary to provide the appropriate biochemical and biophysical environment to encourage bacterial growth. The biochemical or nutritional environment is provided as a culture medium based on special needs for particular bacteria, and can be used for isolation and maintenance of bacterial cultures. Bacterial culture media can be classified based on consistency. Liquid media are sometimes referred to as "broths" where bacteria grow uniformly, and tend to be used when a large quantity of bacteria have to be grown. Moreover, liquid media can be used to obtain a viable bacterial count i.e. to physically quantify the amount of organism present (see section 9.1.1). Any liquid media can be solidified by the addition of agar (e.g. Oxoid Technical Agar No. 1) at a concentration of 1-3%. Although *M. plutonius* can be isolated from honey and diseased brood by cultivation, bacterial culture methods seem to be very insensitive detecting less than 0.2% of the bacterial cells (Djordjevic *et al.*, 1998; Hornitzky and Smith, 1998).

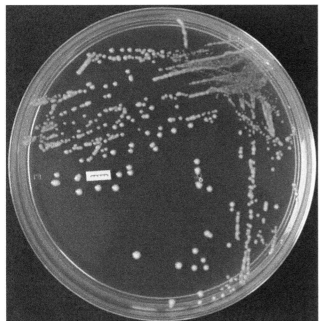

Fig 4a. Agar plate (basal medium) with colonies of *M. plutonius*. The yellow bar represents 5 mm.

Photo by Lena Lundgren and Karl-Erik Johansson

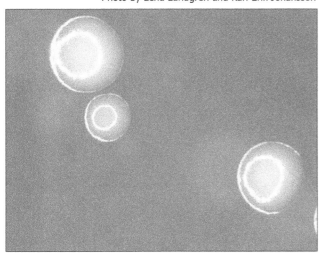

Fig 4b. Colony morphology of *M. plutonius* on basal medium. The bar represents 1 mm.   Photo by Eva Forsgren

5.1. Basal medium (modified from Bailey, 1957)

1. Dissolve the following components in approximately 800 ml deionized water:-
   - 10 g of yeast extract
   - 10 g of glucose
   - 10 g of starch
   - 0.25 g L-cysteine
   - 20 g of agar
2. Add 100 ml of 1M KH_2PO_4 (pH 6.7)
3. Adjust the pH to 6.6 using 2.5 M KOH.

4. Adjust the final volume to 1000 ml.
5. Sterilize by autoclaving at 115°C for 15 minutes.

Optional; in order to prevent growth of secondary bacteria, filter-sterilized nalidixic acid (dissolved in 0.1 M NaOH may be added to a final concentration of 3 µg per ml after autoclaving. For liquid cultures, the starch may be replaced with saccharose, making the medium clear. This will facilitate when checking the turbidity or the cloudiness of the cell suspension, e.g. to see if there is any bacterial growth.

6. Incubate the plates for 7 days at 35°C anaerobically. Colonies about 1 mm in diameter will appear after 4-7 days (Figs 4 a, b), and can be further confirmed by staining, LFI or PCR (see sections 6.1, 7.2 and 8). Single bacterial colonies can be screened using real-time PCR / conventional PCR by simply touching a small (10 µl) tip directly onto the colony of interest, touch the tip onto some agar to sub-culture if required, before placing the tip directly into the PCR master-mix (see section 8).

5.2. M110 agar (from BCCM/LMG bacteria catalogue)
5.2.1. Agar base

- 2.5 g of peptone (Oxoid L37)
- 10 g of glucose
- 2 g of soluble starch
- 2.5 g of yeast extract (Oxoid L21)
- 5 g of neopeptone (Difco 0119)
- 2 g of trypticase (BBL211921)
- 50 ml of 1 M phosphate buffer (pH 6.7) = 49.7 ml of 1M K_2HPO_4 + 50.3 ml of 1M KH_2PO_4.

1. Mix all ingredients and make up to 1000 ml with distilled water.
2. Adjust to pH 7.2 with 5M KOH.
3. Add agar (Difco or Oxoid No 1) to a 1.5% final concentration. For 250 ml amounts of agar base, weigh 3.75 g of agar into the 250 ml Duran flasks, for 500 ml, 7.5 gram. Dispense the correct amount of made up and pH corrected broth into the Duran flask.
4. Mix well before dispensing broth into Duran flasks with agar, the soluble starch settles out quite quickly.
5. Sterilize by autoclaving at 115°C for 15 minutes. Higher temperature autoclaving will tend to caramelize the agar (make it darker).
6. Add cysteine hydrochloride H_2O to a final concentration of 0.025%.

NOTE: Cysteine hydrochloride reduces the oxygen content in the agar more efficiently if added just before the agar is to be used. If the agar is made up in 250 or 500 ml amounts, then autoclaved, this base can be kept for quite a long time (2-3 months or longer), but if the cysteine is added before autoclaving, the base will only keep a week at the most.

5.2.2. Agar plates

1. Melt the agar base in a steamer and cool to 46°C.
2. Add 625 µl of a freshly made 0.2 µm filter sterilized 10% solution of cysteine hydrochloride (does NOT keep at all) to 250 ml of the agar base, or 1.25 ml for 500 ml of base. Make sure the agar is mixed well, but not bubbly as soluble starch settles out.
3. Pour immediately into petri-dishes.
4. Cool and dry in the lamina flow for 20 minutes.
5. Use the plates as soon as possible.
   It is not advised to keep unused plates to use another day.

5.3. Anaerobic incubation

1. Seal in an anaerobic jar as soon as possible with an anaerobic indicator (Oxoid BR055B) and Oxoid AnaeroGen anaerobic generator sachet (appropriate size for anaerobic jar volume. AnaeroGen AN025A for 2.5 l jar, AnaeroGen AN035A for 3.5 l jar).
2. Incubate for about a week at 35°C.
   The anaerobic indicator should go colourless from pink after a few hours incubation. If it does not, the jar has failed to achieve anaerobic conditions.

5.4. Medium for long-term storage

Freezing is an effective long-term storage of *M. plutonius* as well as other bacterial isolates. Broth cultures are mixed with chemicals such as glycerol or DMSO to limit damage upon freezing. It is possible to use proprietary cryopreservation kits which can even specialize in the preservation of anaerobic organisms (e.g. PROTECT system from Thermo Fisher Scientific). Using such proprietary kits, pure cultures of *M. plutonius* are added to a tube containing cryopreservation liquid and ceramic beads. After temperature controlled freezing, cultures can be recovered by removing a single ceramic bead and plating directly onto selective media. Results to date indicate 100% successful recovery after 6 months storage at -80°C (Budge, unpublished data).

6. Microscopy

The laboratory diagnosis of European foulbrood is based on the identification of *M. plutonius* in affected brood. One method for the identification *M. plutonius* is the microscopy of smears prepared from diseased brood.

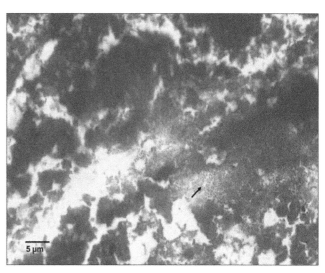

Fig 5a. Early infection- only *Melissocoocus plutonius*. Arrow indicates a mass of coccoid/lanceolate *M. plutonius* organisms.

Photo by Michael Hornitzky

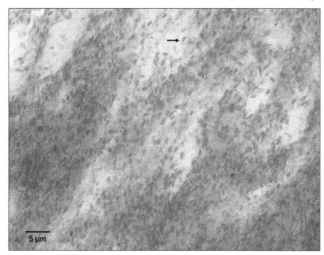

Fig 5b. Infiltration of secondary invader *Paenibacillus alvei*. Arrow indicates one of the many vegetative *P. alvei* cells.

Photo by Michael Hornitzky

6.1. Carbol fuchsin staining

Prepare the carbol fuchsin stain by mixing the following 2 solutions.
Solution A: 0.2 g basic fuchsin and 10 ml 95% ethanol.
Solution B: 5 g phenol and 90 ml distilled water.
Procedure:

1. Select larvae and/or pupae showing signs of European foulbrood and place them on a microscope slide.
2. Using a swab or stick, pulp the larvae together and spread over the slide pushing any excess off one end, to leave a thin smear. Allow the smears to dry before processing.
3. Heat fix by flaming the slide over a burner a few times.
4. Flood with 0.2% carbol fuchsin for 30 seconds.
5. Wash off the stain and either air dry or gently blot dry before microscopic examination at 1000 times magnification.

A diagnosis of European foulbrood is made if examination revealed *M.*

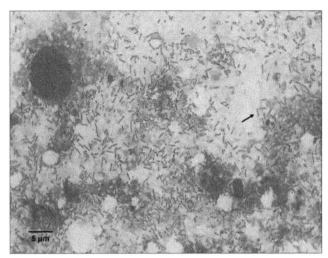

Fig 5c. Proliferation of *P. alvei* spores to the virtual exclusion of *M. plutonius*. Arrow indicates one of the many *P. alvei* spores.

Photo by Michael Hornitzky

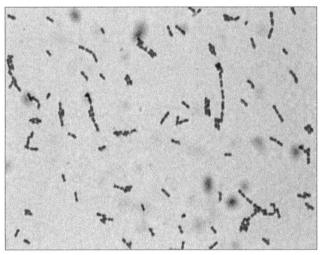

Fig. 6. Gram staining of *Melissococcus plutonius*. The coccoid-shaped bacteria forming pairs or even chains are clearly visible.

Photo by Lena Lundgren and Karl-Erik Johansson

plutonius-like organisms. Organisms are considered to be *M. plutonius* if they are lanceolate cocci, approximately 0.5 x 1.0 μm. *E. faecalis* is very like *M. plutonius* in appearance and has frequently been confused as being the causative agent (Bailey and Gibbs, 1962; Hornitzky and Wilson, 1989) (Figs 5 a, b, c).

An alternative to the 0.2% carbol fuchsin stain is the Gram stain, useful mainly when the Gram positive feature of *M. plutonius* needs to be confirmed.

6.2. Gram staining

Gram-staining is a four part procedure which uses certain dyes to make a bacterial cell stand out against its background. The reagents you will need are:

- Crystal violet (the primary stain)
- Iodine solution (the mordant)
- Decolourizer (ethanol + acetone)
- Safranin (the counter stain)

Procedure:

1. Mount and heat fix the specimen (about 6 times through the flame).
2. Flood (cover completely) the entire slide with crystal violet.
3. Let the crystal violet stand for about 60 seconds.
4. Flood your slide with the iodine solution.
5. Let it stand for 60 seconds.
6. Rinse the slide with water for 5 seconds and immediately proceed to next step.
7. Rinse the slide with decolourizer for 20-60 seconds.
8. Rinse the slide carefully with water for about 5 seconds.
9. Apply the counter stain, safranin, by flooding the slide with the dye.
10. Let it stand for about 10-15 seconds.
11. Rinse with water for 5 seconds.
12. Dry the slide with paper or allow it to air dry.
13. View it under the microscope at 1,000 times magnification (Fig. 6).

7. Immunology-based methods

Various laboratory techniques based on the use of antibodies to visualize or distinguish between microorganisms exist. The key component in any of the vast array of methods used is the antibody. Polyclonal antibodies against *M. plutonius* can be prepared by injection of washed cultures of *M. plutonius* into rabbits either by intravenous injections (Bailey and Gibbs, 1962) or by a single intramuscular injection of 1ml *M. plutonius* suspension mixed with an equal volume of Freund´s incomplete adjuvant (OIE, 2008). Monoclonal antibodies can be prepared by injecting mice as described by Tomkies *et al.* (2009).

7.1. ELISA (Enzyme Linked Immuno Sorbent Assay)

The enzyme-linked immunosorbent assay (ELISA) is a common serological test for particular antigens or antibodies. There are two forms of the test: i. the direct ELISA employs antibodies to detect presence of a particular antigen in a samples and; ii. the indirect ELISA is usually used to detect specific antibodies in a specimen such as blood serum. However, the indirect ELISA method can also be applied for detection of antigens as described in section 7.1.3. The ELISA method described by Pinnock and Featherstone (1984) is unable to detect bacterial levels less than 10^5 cells per ml.

7.1.2. Sample processing

Individual or pooled samples of bees, larvae or pupae (sampled and stored at -20ºC) can be crushed in phosphate buffered saline, PBS, pH 7.4 (for the recipe, see Table 1 of the cell culture paper of the *BEEBOOK* (Genersch *et al.*, 2013)) , the homogenate centrifuged for 10 sec at 10,000 g and the supernatant stored at -20°C or used directly in an ELISA.

7.1.3. Indirect ELISA

The reagents needed to perform the ELISA are:

- Bicarbonate / carbonate coating buffer, 100 mM, pH 9.6.
- Phosphate buffered saline, PBS, pH 7.4.
- Blocking solution (PBS with 1-2% BSA).
- Washing buffer (PBS with 0.05% Tween 20).
- Primary antibody (rabbit, chicken, mouse).
- Peroxidase or alkaline phosphatase-conjugated secondary antibody (anti-rabbit, anti-chicken; anti-mouse).
- Substrate for peroxidase alkaline phosphatase-conjugated secondary antibody (e.g. TMB (3,3´,5,5´- tetramethylbenzidine)).
- Stop solution (0.5 M H_2SO_4).
- Microtiter plates.
- Microtiter plate reader.

Many different types of enzymes can be used for detection. Peroxidase-conjugated secondary antibodies and TMB (3,3´,5,5´- tetramethylbenzidine) are commonly used and accessible.

Procedure:

1. Dilute the bee homogenates in coating buffer. The total protein concentration should not exceed 20 µg per ml.
2. Coat the wells of a microtiter plate with 100 µl per well of the antigen dilution.
3. Cover the plate using an adhesive plastic.
4. Incubate for 2 hours at room temperature or at 4°C over night.
5. Remove the coating buffer.
6. Wash the plates two times filling the wells with washing buffer.
7. Block the remaining protein-binding sites by adding 200 µl blocking solution to the wells.
8. Incubate for 2 hours at room temperature or at 4°C over night.
9. Wash the plate two times with washing solution.
10. Add 100 µl of the *M. plutonius* specific antibody diluted in blocking solution.
11. The optimal dilution should be determined using a dilution assay.
12. Cover the plate and incubate for 2 hours at room temperature.
13. Wash the plate three times with PBS.
14. Add 100 µl of the secondary, conjugated antibody diluted according to the manufacturer´s instruction.
15. Cover the plate.
16. Incubate for 1 hour at room temperature.
17. Wash four times with washing solution.
18. Dispense 100 µl per well of the substrate solution.
19. Incubate for 15 min in room temperature (dark).
20. Add equal volume of the stop solution (2 M H_2SO_4).
21. Read the optical density at 450 nm using a plate reader. Compare the density reads of unknown samples against standards (*e.g.* suspensions of known concentrations of *M. plutonius*). To ensure accuracy, include standards and at least one blank sample to each plate.

7.2. Lateral flow immunoassay (LFI)

A commercially available lateral flow device for the detection of *M. plutonius* using specific monoclonal antibodies is available. The kit was designed primarily for the confirmation of disease symptoms in the field, but may also be used in the laboratory (Tomkies *et al.,* 2009). The kit is produced by Vita (Europe) Ltd and the protocols available at: http://www.vita-europe.com. Using LFIs gives an instant result (meaning no time delay between disease suspicion and treatment), no expensive equipment required and is relatively cheap compared to posting the samples to the laboratory for diagnosis. However, it works only on single larvae and requires field knowledge to select the correct/infected larvae in a brood sample (see section 1.2).

8. PCR-based methods

Detection of infectious microorganisms has been revolutionized by the polymerase chain reaction (PCR), and has increasingly been described as the "gold standard" for detecting some microbes. Theoretically, a single target DNA molecule is sufficient for detection, making PCR one of the most sensitive biological techniques ever described. For a more generic overview on PCR-based methods and other molecular biology methodologies used in *A. mellifera* research see the molecular methods paper of the *BEEBOOK* (Evans *et al.*, 2013).

8.1. Processing

Samples can be homogenized with glass beads in mechanical 'bead mills', in mesh bags (*e.g.* Bioreba, Neogen) using a grinding pestle, a stomacher (e.g. Seward Ltd UK) or in microfuge tubes with a micropestle. The choice depends on sample size and type. Individuals can be extracted in the manufacturer´s buffer directly, but for bulk samples a primary extract may be necessary (see section 8.2.1).

Table 2. PCR-based methods for the detection of *M. plutonius*.

publication	primers	Sequence (5'-3')	Size (bp)	method and target
Govan et al., 1998	Primer 1 Primer 2	GAAGAGGAGTTAAAAGGCGC TTATCTCTAAGGCGTTCAAAGG	831	PCR, 16S rRNA gene
Djordjevic et al., 1998	MP1 MP2 MP3	CTTTGAACGCCTTAGAGA ATCATCTGTCCCACCTTA TTAACCTCGCGGTCTTGCGTCTCTC	486 276	hemi-nested PCR, 16S rRNA gene
Roetschi et al., 2008	MelissoF MelissoR Probe	CAGCTAGTCGGTTTGGTTCC TTGGCTGTAGATAGAATTGACAAT FAM-CTTGGTTGGTCGTTGAC-MBGNFQ	79	real-time PCR, sodA gene
Budge et al., 2010	EFBFor EFBRev2 Probe	TGTTGTTAGAGAAGAATAGGGGAA CGTGGCTTTCTGGTTAGA FAM-AGAGTAACTGTTTTCCTCGTGACGGT-TAMRA	69	real-time PCR, 16S rRNA gene

8.2. DNA extraction

Cellulose-based affinity columns such as QIAGEN, or generic
equivalents are most practical for obtaining clean DNA preparations.
They are reliable and yield good quality DNA. Magnetic bead-based
purification also works well (e.g. Budge *et al.*, 2010). Since samples of
adult bees contain more secondary metabolites and phenolics than
larvae, including a QiaShredder in the protocol will yield purer nucleic
acid (DNeasy® Plant Mini Kit) and prevent inhibition of the PCR
reaction. This is also recommended when extracting bacterial DNA
from honey. The columns can be used for manual DNA extraction or
in a QiaCube® (QIAGEN) for automated extraction. There are two
options when considering extraction controls for the quantification of
M. plutonius in honey bee samples. First, it is possible to monitor
extraction efficiency using a honey bee reference gene (e.g. 18S;
Budge *et al.*, 2010). Alternatively extraction failures or PCR
amplification inhibition can be monitored by amending the sample
with a known amount of *Staphylococcus aureus* before extraction
(Grangier, 2011). It is also recommended to include a negative
extraction control (e.g. water) to check for possible contamination
during the extraction process (Bustin *et al.*, 2009). For further
information on nucleic acid extraction see the molecular methods
paper of the *BEEBOOK* (Evans *et al.*, 2013).

8.2.1. Adults
Procedure:
1. Place adult bees in filter grinding bag (Neogen™, Bioreba).
2. Add 0.5 ml grinding buffer (e.g. GITC[1]) per bee.
3. Crush the bees.
4. Transfer 1.5 ml of the supernatant to a 2 ml Eppendorf tube.
   OR; include a "crude" centrifugation step for bigger volumes.
5. Centrifuge at 2,000 g for 10 minutes.
6. Transfer 1.5 ml to an Eppendorf tube.
7. Centrifuge at 20,000 g for 2 minutes.
8. Discard the supernatant.
9. Resuspend the pellet in the manufacturer's lysis buffer
   (DNeasy® Plant Mini Kit, QIAGEN).

10. For manual DNA-extraction: Use DNeasy® Plant Mini Kit
    (QIAGEN). Follow the protocol for plant tissue (Mini Protocol).
    For automated DNA extraction using a QiaCube® (QIAGEN);
    follow the purification of total DNA from plant tissue standard
    protocol.
11. Use the DNA templates directly in a PCR or store in −20ºC
    until needed.
    [1]GITC = for 100 ml, add 50 g guanidine thiocyanate, 50 ml
    nuclease free water, 5.3 ml 1M Tris-Cl (pH 7.6), 5.3 ml 0.2 M
    EDTA. Stir until completely solved and store at 4ºC.

8.2.2. Larvae / pupae
Procedure:
1. Place the larva / pupa in an Eppendorf tube.
2. Add 0.5 ml grinding buffer (e.g. GITC).
3. Ground with a micropestle.
4. Centrifuge for 10 min at 7,500 g.
5. Discard the supernatant.
6. Resuspend the pellet in 180 µl enzymatic lysis buffer
   (DNeasy® Blood and Tissue kit, QIAGEN).
7. Use the Qiacube and the DNeasy® Blood and Tissue kit
   protocol for enzymatic lysis of Gram + bacteria for automated
   purification.
8. Use the DNA templates directly in a PCR or store in −20ºC
   until needed.

8.2.3. Honey
Procedure:
1. Heat 5 ml of honey to 40ºC.
2. Mix thoroughly with an equal volume of PBS.
3. Centrifuge at 27,000 g for 20 minutes.
4. Discard the supernatant.
5. Resuspend the pellet in the manufacturer's lysis buffer
   (DNeasy® Plant Mini Kit, QIAGEN).
   Follow the protocol for plant tissue (Mini Protocol).
6. Use the DNA templates directly in a PCR or store in -20ºC
   until needed.

8.3. PCR

When PCR is used solely for detecting the presence or absence of a specific DNA signature, it is referred to as qualitative PCR (yes or no answer). The qualitative PCR detects only the end product whereas the real-time PCR detects the amplicon as it accumulates and determines the number of new DNA molecules formed in each reaction. The amount of the target molecule can be quantified (qPCR) either relatively or as absolute values or numbers (for further general information see the molecular methods paper of the *BEEBOOK* (Evans *et al.*, 2013)).

Four protocols for the detection and quantification of *M. plutonius* using PCR have been published to date (Table 2). Two protocols for qualitative PCR; one for detection in diseased larvae (Govan *et al.*, 1998) and a hemi-nested PCR assay (Djordjevic *et al.*, 1998).The latter method was further developed for the detection of *M. plutonius* in larvae, adult bees, honey and pollen (McKee *et al.*, 2003; see section 8.3.1). The results obtained indicate: 1. that the PCR assay is far more sensitive than culture; 2. that not all the *M. plutonius* detected is viable or amenable to culturing; and 3. that honey samples may be a useful tool for detecting sources of *M. plutonius.*

Real-time PCR assays for the quantification (qPCR) of *M. plutonius* (Roetschi *et al.*, 2008; Budge *et al.*, 2010) have been used to analyse pooled samples of brood nest workers from several colonies within an apiary as a suggested alternative to routine visual brood control (Roetschi *et al.*, 2008). However, more recent results suggest the amount of *M. plutonius* in adult bees provides a less stable estimate of the likelihood of finding disease than using larvae (Budge *et al.*, 2011). The qPCR method can also be used to attribute a risk of EFB infection to collected samples measured as probability of the sample showing clinical symptoms and providing a trigger for later inspection of apiaries at risk (Budge *et al.*, 2010; Grangier, 2010). This may provide a definitive diagnosis of EFB, based on a combination of the presence of clinical disease and the confirmed presence of *M. plutonius*. However, in some territories, the costs of such preliminary screening using real-time PCR may not be economically viable (Grangier, 2011).

8.3.1. Qualitative PCR

Procedure (after McKee *et al.*, 2003):

1. Genomic DNA (5-30 ng) is amplified using a thermal cycler in a 50 μl reaction comprising:
   - 4 mM $MgCl_2$,
   - 200 μM of each deoxyribonucleotide triphosphate,
   - 100 ng of primers MP1 and MP2 (Table 2),
   - 5 μl of 10 x PCR buffer (100 mM tris-HCl, pH 8.3; 500 mM KCL),

   - 2-5 μg of *Taq* DNA polymerase.

2. Conditions for amplification consist of:
   - initial denaturation at 95° C for 2 min,
   - 40 cycles of denaturation (95° C, 30 s),
   - primer annealing (61° C, 15 s),
   - primer extension (72° C, 60 s),
   - final extension cycle (72° C, 5 min).

3. Amplification products are analysed by electrophoresis (55 V, 1.5 h) through 1.0-1.5% (wt / vol) agarose containing ethidium bromide. A 486 bp PCR product is produced from primers MP1 and MP2. To ensure test specificity; a second PCR following the same protocol (using primers MP1 and MP3) is conducted, and a specific 276 bp hemi-nested product is amplified from the 486 bp template.

8.3.2. Quantitative PCR, qPCR

Procedure (after Budge *et al.*, 2010):

1. Genomic DNA is amplified in a 25 μl reaction comprising:
   - 1 x buffer A (Applied Biosystems),
   - 0.025 U/μl AmpliTaq Gold,
   - 0.2 mM each dNTP,
   - 5.5 mM $MgCl_2$,
   - 300 nM of each primer,
   - 100 nM probe,
   - 10 μl of nucleic acid extract.

2. PCR reactions are carried out in duplicate or triplicate wells and plates cycled using generic system conditions:
   - 95ºC for 10 min,
   - 40 cycles of 60ºC for 1 min,
   - 95ºC for 15 sec.

   in a 7900 Sequence Detection System (Applied Biosystems; Branchburg, New Jersey, USA) or equivalent with real-time data collection.

3. Quantification of *M. plutonius* in each sample can be achieved using the standard curve method (Anon, 1997) with assay EFBFor/EFBRev2/EFBProbe (*M. plutonius* 16S; Budge *et al.*, 2010; Table 2) as the target and assay AJ307465-955F/1016R/975T (*A. mellifera* 18S; Budge *et al.*, 2010) as the reference assay.

4. As fluorescence increases in the presence of the target, the change in fluorescence (DRn) enters an exponential phase. The quantification cycle (Cq) is defined as midway through the exponential phase of this amplification curve (Bustin *et al.*, 2009). It is often required to manually move the threshold of measurement manually to intercept midway through the

exponential phase of the amplification curve and obtain an appropriate Cq.

5. To account for variation in extraction efficiency between samples, the result can be expressed as a ratio of the number of *M. plutonius* and *A. mellifera* cells.

9. Exposure bioassays using *in vitro* rearing of larvae

Bioassays can be used to determine the biological activity of a substance by its effects on a test organism. Differences in virulence of a pathogen are best analysed in exposure bioassays, and such methods involving *in vitro* rearing of honey bee larvae (see the *in vitro* rearing paper of the *BEEBOOK* (Crailsheim *et al.*, 2013)) have been used for both *Paenibacillus larvae* and *M. plutonius* (McKee *et al.*, 2004; Genersch *et al.*, 2005; Giersch *et al.*, 2009). Virulence tests using this technique show that *M. plutonius* strains collected in different geographic places in Europe present important variations in the mortality rate and how fast the larvae die (Charrière *et al.*, 2011).

Three common measurement results can be obtained from exposure bioassays: the dose (LD_{50}) or concentration (LC_{50}) of the pathogen it takes to kill 50% of the hosts tested, and the time (LT_{50}) required for killing 50% of infected individuals. For the purpose of determining the LD_{50} or LC_{50}, a reliable estimation of the concentration of bacterial cells used in the exposure bioassay is crucial.

9.1. Estimating the concentration of bacteria

The plate (viable) count method is an indirect measurement of bacterial cell density as it only detects live (or cultivable) bacteria whereas the microscopic (total) count includes all bacterial cells, cultivable or not (see the miscellaneous methods paper of the *BEEBOOK* (Human *et al.*, 2013)).

9.1.1. Plate count

The plate count method means diluting bacteria with a diluent solution (*e.g.* sterile saline) until the bacteria are dilute enough to count accurately when spread on a plate. The assumption is that each viable bacterial cell will develop into a single colony. Bacterial cell numbers need to be reduced by dilution, because more than 200 colonies on a standard 9 cm plate are likely to produce colonies too close to each other to be distinguished as distinct colony-forming units (CFUs).

The materials needed to perform a plate count are:

- Sterile 0.9% NaCl (sterile saline)
- Sterile tubes, tips and spreaders
- Agar plates (three per sample)

Procedure:

1. Make a ten-fold dilution serial dilution of your bacterial culture (broth). Dilute the suspension to a dilution factor of 10^{-6} (a million-fold dilution).

2. Spread out aliquots using a sterile bacterial spreader (0.1 ml) of each dilution onto 3 agar plates.

3. Incubate the plates for 4-7 days as previously described in section 5.3.

4. Count the number of bacterial colonies that appear on each of the plates that has between 30 and 200 colonies.

   Any plate which has more than 200 colonies is designated as "too numerous to count". Plates with fewer than 30 colonies do not have enough individuals to be statistically acceptable.

5. To compute the estimated number of bacteria on the surface that you tested, use the following formula: $B = N/d$

   where: B = number of bacteria; N = average number of colonies counted on three plates; d = dilution factor.

Example: Plate 1: 56 CFU; Plate 2: 75 CFU; Plate 3: 63 CFU; Average (N) = 64.7; Dilution (d) = 1/1,000; B = (64.7 x 1,000) = 64,700 bacteria in 0.1 ml, 647,000 bacteria per ml.

9.1.2. Total or microscopic count

A direct microscopic or total count is the enumeration of bacteria found within a demarcated region of a slide, a counting chamber. The slide is placed under a microscope, preferably with phase contrast. For counting bacteria, an oil immersion lens is usually required (1000 x magnification). For the procedure description refer to the section on hemocytometer counting in the miscellaneous methods paper of the *BEEBOOK* (Human *et al.*, 2013).

9.2. Protocol for inducing EFB infection in honey bee larvae reared in vitro

A protocol for inducing EFB involves grafting an individual larva (less than 24 hours old) into a single well in a micro-titer plate (for detailed protocols see the *in vitro* rearing paper of the *BEEBOOK* (Crailsheim *et al.*, 2013)). Older larvae may also become infected but are less susceptible. Each larva is fed 10 µl of larval diet (Crailsheim *et al.,* 2013) containing a defined number of *M. plutonius* cells (*e.g.* 500,000; see section 9.1.2). From 72 hours post grafting, the larvae are examined for mortality and fed uninfected feed daily, following the feeding regime recommended in Crailsheim *et al.* (2012). The mortality of the larvae can be evaluated using a microscope or by eye. Dead larvae are distinguished by the lack of respiration and loss of body elasticity.

The ability of *M. plutonius* to produce symptoms in the absence of secondary bacteria such as *P. alvei* seems to differ regionally. In Australia, feeding only *M. plutonius* has been demonstrated not to produce the typical clinical signs of EFB, but in Europe, *M. plutonius* was capable of inducing significant mortality in isolation (Charrière *et al.*, 2011). When infecting with *M. plutonius* in combination with *P. alvei*, the larval colour changes to a greyish brown rather than a yellowish colour and the gut content of infected larvae turns watery rather than pasty. Infecting larvae with *M. plutonius* and subsequently feeding *P. alvei* (*e.g.* 60,000 spores in 10 μl larval diet) after 72 hours may produce signs typical of that seen in field cases of EFB (Giersch *et al.*, 2010). The simultaneous or 3 days delayed inoculation of *P. alvei* has, however, been demonstrated not to influence the virulence of some European strains of *M. plutonius* (Charrière *et al.*, 2011). The feeding of *P. alvei* in addition to *M. plutonius* has no influence on larval mortality as such, but may be important for the presence of all the typical EFB-symptoms, and the saprophyte *P. alvei* is probably important for the presence of some of the clinical symptoms in the field.

10. Measuring susceptibility / resistance to antibiotics of *Melissococcus plutonius*

Due to the fastidious culture requirements and slow growing nature of *M. plutonius*, measuring antibiotic susceptibility of this organism using traditional techniques such as a disc diffusion assay, which is a test that uses antibiotic - impregnated discs to determine whether particular bacteria are susceptible to specific antibiotics, is not possible. Oxytetracycline hydrochloride (OTC) is the antibiotic of choice for the treatment of EFB. However, only two reports of the sensitivity of *M. plutonius* to this antibiotic have been published (Waite *et al.*, 2003; Hornitzky and Smith, 1999) and both these studies indicated that all strains tested were sensitive to OTC. In both studies an agar plate method was used. This involves incorporating antibiotic at decreasing concentrations into culture plates (see section 5) of EFB culture medium, to determine the lowest concentration at which growth would occur. This methodology would be suitable for testing the susceptibility of *M. plutonius* to other antibiotics.

11. Conclusions

The pathogenic mechanisms of EFB are poorly understood, and the factors and timescales leading to overt symptomatology remain enigmatic. Molecular tools will open new possibilities for the identification of putative virulence factors in both the bacterium as well as the host in order to unravel some of the pathogenic mechanisms. To date, there are no published methods for genotyping and molecular differentiation of *M. plutonius* strains, but the nucleotide sequence of the bacterial genome was recently deposited in the DNA Database of Japan under accession no. AP012200 and AP012201 (Okomura *et al.*, 2011), and it is likely that new molecular methods such as genotyping will be developed in the near future. Moreover, research fields and methods already in use for research on *P. larvae* such as selection of reference genes, quantifying and knocking down gene expression (see designated parts in the American foulbrood and molecular protocols papers of the *BEEBOOK* (deGraaf *et al.*, 2013; Evans *et al.*, 2013)) could be adapted to *M. plutonius* and EFB research. Moreover, new technologies may also be useful tools to study interactions between secondary bacteria and the causative agent and to fully understand their role in symptomatology.

Molecular diagnostic methods such as PCR are also widely employed for EFB diagnosis. The PCR method is user-friendly and theoretically, a single target DNA molecule is sufficient for detection, making it one of the most sensitive biological techniques ever described. Considering this, we might ask whether a positive PCR result is always biologically relevant. Low levels of *M. plutonius* can be found in apiaries where no symptoms of disease are present and the PCR will also detect non-viable bacterial cells. However, it is clear that *M. plutonius* is still below the level of detection in honey bee colonies located in some geographical areas (Budge *et al.*, 2010). Future work should help understand whether this observation is due to the genetics of the honey bees from these areas, unfavourable meteorological conditions, lower apiary density, gut microbiota unfavourable to disease development, or simply down to an absence of movement of the causative organism.

Infectivity tests causing disease at the colony level using both cultured *M. plutonius* and extracts from diseased larvae were carried out during the 1930s (Tarr, 1936) and the 1960s (Bailey, 1960; Bailey, 1963; Bailey and Locher, 1968), but not much has been published since. This is an area of research where new information can be obtained by a combination of colony level infection experiments and modern diagnostic methods. Such advances would benefit from cross country collaborations, where advanced diagnostics from one country may complement field trials in another country where there may be less stringent rules governing EFB control.

12. References

ANON (2012) *User Bulletin #2: ABI PRISM 7700 Sequence Detection*, December 11, updated 10/2001:1-36.

ARAI, R; TOMINAGA, K; WU, M; OKURA, M; ITO, K; OKAMURA, N; ONISHI, H; OSAKI, M; SUGIMURA, Y; YOSHIYAMA, M; TAKAMATSU, D (2012) Diversity of *Melissococcus plutonius* from honey bee larvae in Japan and experimental reproduction of European foulbrood with cultured atypical isolates. *PLoS ONE* 7 (3): e33708. http://dx.doi.org/10.1371/journal.pone.0033708

BAILEY, L (1957) The isolation and cultural characteristics of *Streptococcus pluton* and further observations on "*Bacterium eurydice*". *Journal of General Microbiology* 17(2): 39-48.

BAILEY, L (1960) The epizootiology of European foulbrood of the larval honey bee, *Apis mellifera* Linneaus. *Journal of Insect Pathology* 2: 67-83.

BAILEY, L (1963) The pathogenicity for honey bee larvae of microorganisms associated with European foulbrood. *Journal of Insect Pathology* 5: 198-205.

BAILEY, L (1983) *Melissococcus pluton*, the cause of European foulbrood of honey bees (*Apis* spp.). *Journal of Applied Bacteriology* 55(1): 65-69.

BAILEY, L; COLLINS, M D (1982) Reclassification of '*Streptococcus plutori*' (White) in a new genus *Melissococcus*, as *Melissococcus pluton* nom. rev.; comb. nov. *Journal of Applied Bacteriology* 53 (2): 215-217.

BAILEY, L; GIBBS, A J (1962) Cultural characters of *Streptococcus pluton* and its differentiation from associated *Enterococci*. *Journal of General Microbiology* 28: 385-391.

BAILEY, L; LOCHER, N (1968) Experiments on the etiology of European foulbrood of the honey bee. *Journal of Apicultural Research* 7: 103-107.

BUDGE, G E; JONES, B; POWELL, M; ANDERSON, L; LAURENSSON, L; PIETRAVALLE, S; MARRIS, G; HAYNES, E; THWAITES, R; BEW, J; WILKINS, S; BROWN, M A (2011) Recent advances in our understanding of European Foulbrood in England and Wales. In *Proceedings of the COLOSS Workshop; The future of brood disease research – guidelines, methods and development, Copenhagen, Denmark, 10-12 April, 2011*, p 7.

BUDGE, G E; BARRETT, B; JONES, B; PIETRAVALLE, S; MARRIS, G; CHANTAWANNAKUL, P; THWAITES, R; HALL, J; CUTHBERTSON, A G S; BROWN, M A (2010) The occurrence of *Melissococcus plutonius* in healthy colonies of *Apis mellifera* and the efficacy of European foulbrood control measures. *Journal of Invertebrate Pathology* 105: 164-170. http://dx.doi.org/10.1016/j.jip.2010.06.004

BUSTIN, S A; BENES, V; GARSON, J A; HELLEMANS, J; HUGGETT, J; KUBISTA, M; MUELLER, R; NOLAN, T; PFAFFL, M W; SHIPLEY, G L; VANDESOMPELE, J; WITTWER, C T (2009) The MIQE guidelines: Minimum information for publication of quantitative real-time PCR experiments. *Clinical Chemistry* 55: 611-622. http://dx.doi.org/10.1373/clinchem.2008.112797

CHARRIÈRE, J-D; KILCHENMANN, V; ROETSCHI, A (2011) Virulence of different *Melissococcus plutonius* strains on larvae tested by an *in vitro* larval test; In *Proceedings of the 42nd International Apicultural Congress, Buenos Aires 2011*: p 158.

CRAILSHEIM, K; BRODSCHNEIDER, R; AUPINEL, P; BEHRENS, D; GENERSCH, E; VOLLMANN, J; RIESSBERGER-GALLÉ, U (2013) Standard methods for artificial rearing of *Apis mellifera* larvae. In *V Dietemann; J D Ellis; P Neumann (Eds) The COLOSS BEEBOOK, Volume I: standard methods for* Apis mellifera *research. Journal of Apicultural Research* 52(1): http://dx.doi.org/10.3896/IBRA.1.52.1.05

DAHLE, B; SØRUM, H; WEIDEMANN, J E (2011) European foulbrood in Norway: How to deal with a major outbreak after 30 years absence. In *Proceedings of the COLOSS Workshop: The future of brood disease research – guidelines, methods and development, Copenhagen, Denmark, 10-12 April, 2011*, p 8.

DE GRAAF, D C; ALIPPI, A M; ANTÚNEZ, K; ARONSTEIN, K A; BUDGE, G; DE KOKER, D; DE SMET, L; DINGMAN, D W; EVANS, J D; FOSTER, L J; FÜNFHAUS, A; GARCIA-GONZALEZ, E; GREGORC, A; HUMAN, H; MURRAY, K D; NGUYEN, B K; POPPINGA, L; SPIVAK, M; VANENGELSDORP, D; WILKINS, S; GENERSCH, E (2013) Standard methods for American foulbrood research. In *V Dietemann; J D Ellis; P Neumann (Eds) The COLOSS* BEEBOOK, *Volume II: standard methods for* Apis mellifera *pest and pathogen research. Journal of Apicultural Research* 52(1): http://dx.doi.org/10.3896/IBRA.1.52.1.11

DJORDJEVIC, S P; NOONE, K; SMITH, L; HORNITZKY, M A Z (1998) Development of a hemi-nested PCR assay for the specific detection of *Melissococcus pluton*. *Journal of Apicultural Research* 37: 165-174.

EVANS, J D; CHEN, Y P; CORNMAN, R S; DE LA RUA, P; FORET, S; FOSTER, L; GENERSCH, E; GISDER, S; JAROSCH, A; KUCHARSKI, R; LOPEZ, D; LUN, C M; MORITZ, R F A; MALESZKA, R; MUÑOZ, I; PINTO, M A; SCHWARZ, R S (2013) Standard methodologies for molecular research in *Apis mellifera*. In *V Dietemann; J D Ellis; P Neumann (Eds) The COLOSS* BEEBOOK, *Volume I: standard methods for* Apis mellifera *research. Journal of Apicultural Research* 52(4): http://dx.doi.org/10.3896/IBRA.1.52.4.11

FORSGREN, E (2010) European foulbrood in honey bees. *Journal of Invertebrate Pathology* 103(Suppl. 1): S5-S9. http://dx.doi.org/10.1016/j.jip.2009.06.016

FORSGEN, E; LUNDHAGEN, A C; IMDORF, A; FRIES, I (2005) Distribution of *Melissococcus plutonius* in honey bee colonies with and without symptoms of European foulbrood. *Microbial Ecology* 50: 369-374. http://dx.doi.org/10.1007/s00248-004-0188-2

GENERSCH, E; ASHIRALIEVA, A; FRIES, I (2005) Strain and genotype-specific differences in virulence of *Paenibacillus larvae* subsp. *larvae*, a bacterial pathogen causing American foulbrood disease in honey bees. *Applied and Environmental Microbiology* 71: 7551-7555. http://dx.doi.org/10.1128/AEM.71.11.7551-7555.2005

GENERSCH, E; GISDER, S; HEDTKE, K; HUNTER, W B; MÖCKEL, N; MÜLLER, U (2013) Standard methods for cell cultures in *Apis mellifera* research. In *V Dietemann; J D Ellis; P Neumann (Eds) The COLOSS* BEEBOOK, *Volume I: standard methods for* Apis mellifera *research. Journal of Apicultural Research* 52(1): http://dx.doi.org/10.3896/IBRA.1.52.1.02

GOVAN, V A; BROZEL, V; ALLSOPP, M H; DAVIDSON, S (1998) A PCR detection method for rapid identification of *Melissococcus pluton* in honey bee larvae *Applied and Environmental Microbiology* 64 (5): 1983-1985.

GRANGIER, V (2011) Early detection of European foulbrood using real-time PCR. PhD thesis, Vetsuisse faculty, University of Bern, Switzerland.

HORNITZKY, M A Z; SMITH, L (1999) The sensitivity of Australian *Melissococcus pluton* isolates to oxytetracycline hydrochloride. *Australian Journal of Experimental Agriculture* 39(7): 881-883.

HORNITZKY, M A Z; WILSON; S (1989) A system for the diagnosis of the major bacterial brood diseases. *Journal of Apicultural Research* 28: 191-195.

HUMAN, H; BRODSCHNEIDER, R; DIETEMANN, V; DIVELY, G; ELLIS, J; FORSGREN, E; FRIES, I; HATJINA, F; HU, F-L; JAFFÉ, R; KÖHLER, A; PIRK, C W W; ROSE, R; STRAUSS, U; TANNER, G; TARPY, D R; VAN DER STEEN, J J M; VEJSNÆS, F; WILLIAMS, G R; ZHENG, H-Q (2013) Miscellaneous standard methods for *Apis mellifera* research. In *V Dietemann; J D Ellis; P Neumann (Eds) The COLOSS BEEBOOK, Volume I: standard methods for* Apis mellifera *research. Journal of Apicultural Research* 52(4): http://dx.doi.org/10.3896/IBRA.1.52.4.10

McKEE, B A; DJORDJEVIC, S P; GOODMAN, R D; HORNITZKY, M A Z (2003) The detection of *Melissococcus pluton* in honey bees (*Apis mellifera*) and their products using a hemi-nested PCR. *Apidologie* 34: 19-27. http://dx.doi.org/10.1051/apido: 2002047

McKEE, B A; GOODMAN, R D; HORNITZKY, M A Z (2004) The transmission of European foulbrood (*Melissococcus plutonius*) to artificially reared honey bee larvae (*Apis mellifera*). *Journal of Apicultural Research* 43: 93-100.

OIE (2008) Chapter 9.3. European foulbrood of the honey bees. In *OIE Manual of diagnostic tests and vaccines for terrestrial animals (mammals, birds and bees) (6th Ed.), Vol. 1*. OIE; Paris, France. pp 405-409.

OIE (2011) Chapter 9.3. European foulbrood. In *OIE Terrestrial animal health code, vol. 2 (20th Ed.)*. OIE; France. pp 507-508.

OKOMURA, K; ARAI, R; OKURA, T; TAKAMATSU, D; OSAKI, M; MIYOSHI-AKIYAMA, T (2011) Complete genome sequence of *Melissococcus plutonius* ATCC 35311. *Journal of Bacteriology* 193: 4029-4030. http://dx.doi.org/10.1128/JB.05151-11

PINNOCK, D E; FEATHERSTONE, N E (1984) Detection and quantification of *Melissococcus pluton* infection in honey bee colonies by means of enzyme-linked immunosorbent assay. *Journal of Apicultural Research* 23: 168-170.

ROETSCHI, A; BERTHOUD, H; KUHN, R; IMDORF, A (2008) Infection rate based on quantitative real-time PCR of *Melissococcus plutonius*, the causal agent of European foulbrood, in honey bee colonies before and after apiary sanitation. *Apidologie* 39: 362-371. http://dx.doi.org/10.1051/apido: 200819

TARR, H L A (1936) Studies on European foulbrood of bees II. The production of the disease experimentally. *Annals of Applied Biology* 23: 558-584.

TOMKIES, V; FLINT, J; JOHNSSON, G; WAITE, R; WILKINS, S; DANKS, C; WATKINS, M; CUTHBERTSON, A G S; CARPANA, E; MARRIS, G; BUDGE, G; BROWN, M A (2009) Development and validation of a novel field test kit for European foulbrood. *Apidologie* 40(1): 63-72. http://dx.doi.org/10.1051/apido:2008060

WAITE, R; JACKSON, S; THOMPSON, H (2003) Preliminary investigations into possible resistance to oxytetracycline in *Melissococcus plutonius*, a pathogen of honey bee larvae. *Letters in Applied Microbiology* 36: 20-24. http://dx.doi.org/10.1046/j.1472-765X.2003.01254.x

WILKINS, S; BROWN, M A, CUTHBERTSON, A G S (2007) The incidence of honey bee pests and diseases in England and Wales. *Pest Management Science* 63: 1062-1068. http://dx.doi.org/10.1002/ps.1461

Journal of Apicultural Research 52(1): (2013)
DOI 10.3896/IBRA.1.52.1.13

REVIEW ARTICLE

Standard methods for fungal brood disease research

Annette Bruun Jensen[1]*, Kathrine Aronstein[2], José Manuel Flores[3], Svjetlana Vojvodic[4], María Alejandra Palacio[5] and Marla Spivak[6]

[1]Department of Plant and Environmental Sciences, University of Copenhagen, Thorvaldsensvej 40, 1817 Frederiksberg C, Denmark.
[2]Honey Bee Research Unit, USDA-ARS, 2413 E. Hwy. 83, Weslaco, TX 78596, USA.
[3]Department of Zoology, University of Córdoba, Campus Universitario de Rabanales (Ed. C-1), 14071, Córdoba, Spain.
[4]Center for Insect Science, University of Arizona, 1041 E. Lowell Street, PO Box 210106, Tucson, AZ 85721-0106, USA.
[5]Unidad Integrada INTA – Facultad de Ciencias Ags, Universidad Nacional de Mar del Plata, CC 276,7600 Balcarce, Argentina.
[6]Department of Entomology, University of Minnesota, St. Paul, Minnesota 55108, USA.

Received 1 May 2012, accepted subject to revision 17 July 2012, accepted for publication 12 September 2012.

*Corresponding author: Email: abj@life.ku.dk

Summary

Chalkbrood and stonebrood are two fungal diseases associated with honey bee brood. Chalkbrood, caused by *Ascosphaera apis,* is a common and widespread disease that can result in severe reduction of emerging worker bees and thus overall colony productivity. Stonebrood is caused by *Aspergillus* spp. that are rarely observed, so the impact on colony health is not very well understood. A major concern with the presence of *Aspergillus* in honey bees is the production of airborne conidia, which can lead to allergic bronchopulmonary aspergillosis, pulmonary aspergilloma, or even invasive aspergillosis in lung tissues upon inhalation by humans. In the current chapter we describe the honey bee disease symptoms of these fungal pathogens. In addition, we provide research methodologies and protocols for isolating and culturing, *in vivo* and *in vitro* assays that are commonly used to study these host pathogen interactions. We give guidelines on the preferred methods used in current research and the application of molecular techniques. We have added photographs, drawings and illustrations to assist bee-extension personnel and bee scientists in the control of these two diseases.

Métodos estándar para la investigación de enfermedades fúngicas de la cría

Resumen

La ascosferiosis (o cría yesificada) y cría de piedra son dos enfermedades fúngicas asociadas con la cría de la abeja melífera. La ascosferiosis, causada por el hongo *Ascosphaera apis,* es una enfermedad común de amplia distribución que puede resultar en una severa disminución en el número de obreras emergentes y por lo tanto afecta la productividad de la colonia. La cría de piedra es causada por *Aspergillus* spp. los cuales son raramente observados y su impacto en la salud de la colonia no está esclarecido. El mayor problema con la presencia de *Aspergillus* en abejas es la producción de conidios aéreos, que pueden llevar a aspergillosis broncopulmonares alérgicas, aspergilomas pulmonares o aun aspergilosis invasivas en tejidos pulmonares en humanos luego de inhalarlos. En este capítulo describimos los síntomas de las enfermedades producidas por estos patógenos fúngicos. Además proveemos metodologías de investigación y protocolos para su aislamiento y cultivo, ensayos in vivo e in vitro que son comúnmente usados para estudiar las interacciones de estos patógenos con su hospedero. Damos una guía sobre los métodos preferidos utilizados en la investigación actual y la aplicación de técnicas de moleculares. Hemos añadido fotografías, dibujos e ilustraciones para ayudar al personal.

Footnote: Please cite this paper as: JENSEN, A B; ARONSTEIN, K; FLORES, J M; VOJVODIC, S; PALACIO, M A; SPIVAK, M (2012) Standard methods for fungal brood disease research. In *V Dietemann; J D Ellis, P Neumann (Eds) The COLOSS BEEBOOK: Volume II: Standard methods for* Apis mellifera *pest and pathogen research. Journal of Apicultural Research* 52(1): http://dx.doi.org/10.3896/IBRA.1.52.1.13

幼虫真菌病研究的标准方法

白垩病和曲霉病是蜜蜂幼虫相关的真菌病，白垩病由蜜蜂球囊菌*Ascosphaera apis*引起，分布较广，可造成工蜂羽化数量和蜂群生产力的急剧下降。蜜蜂曲霉病由*Aspergillus* spp.引起，由于较少观察到，所以其对蜂群健康状况的影响研究的不够深入。蜜蜂中*Aspergillus*引起之所以引起关注，是因为风媒分生孢子，这种孢子可通过吸入引起人肺组织的多种疾病，如过敏性支气管曲菌病，肺曲菌球病，甚至侵袭性曲霉病。本章我们描述了这些真菌病原体的在蜂群中的疾病症状。此外，还提供了开展寄主／病原体相互作用研究中常用的病原菌分离及培养技术、体内和体外测定技术。并给出了最新研究中用到的首选方法以及分子技术应用方面的指导方针。增加了照片、插图和注释以帮助推广人员和研究者控制这两种疾病。

Keywords: *Ascosphaera*, Ascosphaerales, *Aspergillus*, chalkbrood, diagnostics, methods, recommendations, stonebrood, *COLOSS, BEEBOOK*, honey bee

Introduction

Insect pathogenic fungi can be found throughout the fungal kingdom (Humber, 2008), all being capable of invading their hosts and overcoming their immune systems. Two fungal genera (*Ascosphaera* and *Aspergillus*) are known to infect honey bee brood, causing chalkbrood and stonebrood diseases. Both are ascomycetes within the Eurotiomycetes. The fungus causing chalkbrood in honey bees has a narrow host range and a unique infection route, it relies solely on sexual reproduction and has many host-specific adaptations. Therefore many methods known from common insect pathogenic fungi are not easily adopted to study chalkbrood.

In contrast, the fungi causing stonebrood are facultative pathogens with a broad host range, they produce asexual conidia and their infection biology resemble many well-known insect pathogenic fungi, like *Beauveria* and *Metarhizium*; so several standard insect pathological methods can be directly transferred to this system. In the current chapter, we compile, discuss and provide detailed protocols for various methods to assist beekeepers and bee scientists entering this area of research.

In addition to fungal pathogens causing chalkbrood and stonebrood, two species of microsporidia, *Nosema apis* and *Nosema ceranae*, are known to infect adult honey bees. Although recently suggested to be fungi at the very base of the fungal tree (James *et al.*, 2006), these intracellular pathogens have a very different biology and are considered in a separate paper in the *BEEBOOK* (Fries *et al.*, 2013).

1. Chalkbrood

1.1. introduction

Chalkbrood, the most common fungal bee brood disease, is caused primarily by the fungus *Ascosphaera apis* (Maassen ex Claussen) Olive and Spiltoir (Spiltoir, 1955). It was recognized in the honey bee in the early 20th Century (Maassen, 1913). The field diagnosis of chalkbrood is based on visual detection of diseased, mummified brood, commonly known as "chalkbrood mummies". Chalkbrood can reduce colony productivity by lowering the number of newly emerged bees, and in some cases may lead to colony losses. The disease is found infecting honey bee brood in most regions of the world, including warm and dry climates. Clinical symptoms of chalkbrood often appear for only a short time, typically under cold and damp weather conditions (Aronstein and Murray, 2010).

The genus *Ascosphaera spp.* comprises species that are adapted to eusocial and solitary bees and their habitats. Some of the species are saprophytic, growing on nest materials, such as stored food, faecal matter, and nest debris; others have evolved as opportunistic and / or obligate bee pathogens (Bissett, 1988). In honey bees *Ascosphaera major* and *Arrhenosphaera cranae* (both belonging to Ascospheraceae) have been reported only a few times from chalkbrood infected colonies, but Koch's Postulates have never been demonstrated for them, so they might be secondary invaders.

Ascosphaera apis primarily infects honey bee brood by entering the host through the gut lining. In an infected larva, hyphae penetrate the gut wall, and mycelium develops inside the body cavity. After a few days, mycelium breaks out of the posterior end of the larva leaving the head unaffected (Maurizio, 1934). Fruiting bodies with the new ascospores are formed on aerial hyphae outside the dead larvae. The ascospores are a result of sexual reproduction of fungal mycelia with the opposite sex, unlike most other insect pathogenic fungi where asexually conidia are the infective units, as e.g. seen in *Aspergillus*. Ascospores of *A. apis* (hereafter referred to as spores) are believed to be adapted to the harsh gut environment of the host. *Ascosphaera apis* infect the host larvae if ingested, unlike other insect pathogenic fungi which mainly infect through the external cuticle. Spores of *A. apis* can stay dormant and viable for years, but upon exposure to CO_2 the spores becomes activated, resulting in spore swelling and subsequently germ tube formation that extends to form hyphae (Bamford and Heath, 1989). Presumably, CO_2 produced by the larval tissues accumulates in the closed hindgut of the larvae aiding in spore germination (Heath and Gaze, 1987). However additional factors that may be involved in spore germination within the host still remain to be discovered. In addition, it has been shown that chilling of the brood below the optimal rearing temperature increases the number of diseased larvae (Vojvodic *et al.*, 2011a; Flores *et al.*, 1996a). All these specific adaptations have to be taken into account when working with *A. apis*, and specific protocols have been developed.

1.2. Diagnostics and qualitative detection

1.2.1 Morphological description

1.2.1.1. Macroscopic diagnosis

The typical symptoms observed in a colony affected by chalkbrood are irregular wax cappings over the brood and uncapped cells scattered over the brood frames (Fig. 1). The cell capping may also have small holes or appear slightly flattened. Chalkbrood mummies can often be seen in the combs, at the hive entrance or found on the bottom board (Fig. 2). Observation of combs may reveal different stages of the disease; fresh larval cadavers covered with white cotton-like mycelium and desiccated mummies that appear as white, dark or a combination of white and dark solid clumps. White desiccated mummies look like small pieces of chalk giving rise to the name of the disease and dark mummies are coloured by fungal fruiting bodies.

Diagnosis in the field is generally based on the presences of chalkbrood mummies (as described above). Following field diagnosis, a microscopic examination is usually required to confirm the presence of spore cysts in the samples using the microscope slide smear technique. The spores can be mounted on a microscope slide with a drop of distilled water and observed at 100-400 x magnification.

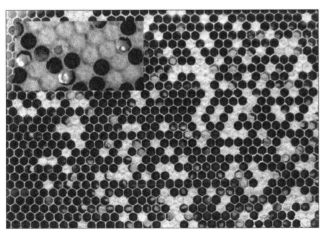

Fig. 1. A brood frame from a honey bee colony with clinical symptoms of chalkbrood and a close up insert of fresh mummies. In many cells chalkbrood mummies have been partly removed by the worker bees.
Photo: F Padilla, J M Flores and A B Jensen

1.2.1.2. Microscopic diagnosis

Ascosphaera apis has septate hyphae (2.5 – 8.0 µm in diameter) which show pronounced dichotomous branching (Spiltoir, 1955; Skou,

Fig. 2. Dark and white chalkbrood mummies. The black mummies contain millions of new infective spores.   Photo: A B Jensen

1988). As mentioned above, this is a heterothallic fungus with two mating types and only when + and - strains are grown in close proximity does the formation of fruiting bodies occur. The fruiting bodies are spherical spore cysts (47 – 140 µm in diameter; Fig. 3a) which contain numerous spore balls (9-19 µm, Fig. 3b) composed of hyaline spores (2.7 – 3.5 x 1.4 – 1.8 µm, Fig. 3c) (Skou, 1972; Bissett, 1988).

1.2.1.3. Biological diagnosis

Identification of the vegetative stages of the fungus can be done by a mating test with two reference strains (AFSEF 7405 and ARSEF 7406). Production of spore cysts with one of the reference strains would prove the identify of *A. apis* mating. The mating test is described under paragraph *1.4.1.4. Production and quality of inoculums*.

1.2.2. Molecular methods

The Polymerase Chain Reaction (PCR) has increasingly been used for detection of microorganisms. The internal transcribed spacer (ITS) region of the nuclear ribosomal repeat unit is the locus most often used for molecular identification of fungal species (Nilsson *et al.*, 2008) and it is now accepted as the general fungal barcode marker (Schoch *et al.*, 2012). Almost no variation between *A. apis* strain was detected in the ITS region (Anderson *et al.*, 1998; Jensen *et al.*, 2012) and several *A. apis* species specific primers have been designed targeting the ITS (Table 1). Irrespective of which primers are used, the presence of a band from PCR amplification indicates the presence of *A. apis* DNA.

DNA can be extracted with standard kits like DNeasy® Plant Mini Kit (Qiagen), Ultra Clean plant DNA isolation kits (Mo Bio Laboratories) or PrepMan Ultra reagent (Applied Biosystems) using the manufacturers' protocols (for other protocols of DNA extractions consult the *BEEBOOK* paper on molecular methods (Evans *et al.*, 2013)

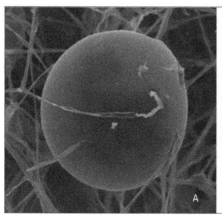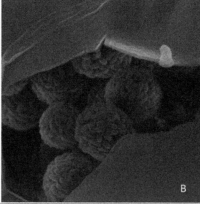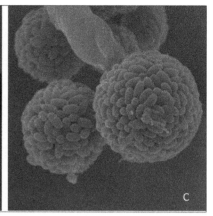

Fig. 3. Scanning electron microscope pictures of *Ascosphaera apis* fruiting body. **A.** Spherical fruiting body; **B.** Cracked fruiting body with sporeballs; **C.** Sporeballs with multiple ascospores.   Photos: J Bresciani and A B Jensen

1.2.2.1 PCR for species identification

Genomic DNA can be amplified in a 25 µl reaction containing:

1. 2.5 µl Taq polymerase buffer (100 mM tris-HCl, pH 8.3; 500 mM KCL)
2. 2 U of *Taq* DNA polymerase
3. 1.5-2.5 mM MgCl₂
4. 250 µM of each deoxyribonucleotide triphosphate
5. 0.25 µM of an forward and reverse primer (Table 1)
6. 1 µl of DNA

Reaction conditions are as follows:

1. initial denaturing for 10 min at 94°C
2. 30 cycles of 45 s denaturing at 94°C
3. 45 s annealing at 62-65°C
4. 1 min extension at 72°C
5. final extension for 5 min at 72°C.

Results are visualised using Gel Doc (BioRad) or any other DNA imaging systems. We would recommend that the primer pair 3-F1 and 3-R1 (Table 1) (James and Skinner, 2005) be used as a standard for identification of *A. apis*. Both primers are *A. apis* specific as opposed to the primer pair in Murray *et al.* (2005). Furthermore, 3-F1/3-R1 amplifies a longer PCR product that could be very useful for sequencing efforts (James and Skinner, 2005). The PCR protocol described above can be optimised depending on the type of equipment and reagents used in the laboratory. We would recommend sequencing the PCR product as a quality control, the first time the protocol is implemented in a new laboratory (for more information on sequencing consult the *BEEBOOK* paper on molecular methods (Evans *et al.*, 2013).

1.3. Quantitative detection

1.3.1. Quantify the level of fungal infection in bee colonies

Different methods have been used to assess the level of chalkbrood infection in the honey bee hive, but none of the methods are completely reliable. The most direct method is to remove brood combs from the hives and count the number of mummies inside capped and uncapped cells. These values can be added to the number of mummies found on the bottom board of the hive. It is however important to add an entrance hive trap like a pollen trap (see the *BEEBOOK* paper on miscellaneous methods (Human *et al.*, 2013) or a screened bottom board (Flores, unpublished data), to hinder the bees from removing mummies from the hive.

Assessing chalkbrood infection by counting mummies has important drawbacks: it does not include the number of mummies inside capped cells, unless all capped cells are opened and inspected; and it does not include the mummies that are removed from the hive in small pieces by worker bees which are thus not collected by the entrance traps. Furthermore, these counting methods are heavily influenced by environmental conditions. High nectar flow induces higher removal of mummies (Thompson, 1964; Momot and Rothenbuhler, 1971) and during a nectar flow more mummies may be recovered in the traps, leading to possible error in the estimation of disease prevalence.

For the above reasons, we propose that the level of colony infection be evaluated by chilling the brood to 25°C (Flores *et al.*, 1996a). The number of mummified larvae after chilling allows for an assessment of the degree of potential infection in bee colonies. This method has also been used to compare the degree of infection among

Table 1. List of *Ascosphaera apis* specific primers. * Primer pair recommended to be used as standard.

Primer name	Annealing temp. (°C)	Primer sequence	Citation
3-F1*	62	TGTCTGTGCGGCTAGGTG	James and Skinner, 2005
3-F2	62	GGGTTCTCGCGAGCCTG	James and Skinner, 2005
3-R1*	62	CCACTAGAAGTAAATGATGGTTAGA	James and Skinner, 2005
AscoF3	64	GCACTCCCACCCTTGTCTA	Murray *et al.*, 2005
AapisR3	64	CCCACTAGAAGTAAATGATGGTTA	Murray *et al.*, 2005

colonies before and after a treatment or as a routine colony inspection (Flores *et al.*, 2001; 2004a). To minimize the effects of brood removal by workers, it is important to place the brood comb in an incubator, as described in paragraph 1.3.1.1.

1.3.1.1. Procedure for quantifying the level of colony infection:

1. Remove one brood comb from the hive containing 5th instar larvae that have not yet been sealed (i.e., the larvae are close to being capped with wax)
2. Mark on a plastic transparency the area with these unsealed larvae (this step is only important if the frame also contains capped brood)
3. Return the brood comb to the colony
4. Remove the brood comb from the hive a maximum of 20 hours later
5. Cut a piece of comb containing at least 100 recently capped cells (capped within the last 20 hours). These larvae can be identified using the plastic transparency. (It is best to remove unsealed brood from the comb before placed in the incubator)
6. Place one or more pieces of comb with capped cells in an incubator at 25ºC at approximately 65% humidity for 5 days. The chilling will ensure disease development. An open water bottle placed in the incubator will provide sufficient humidity
7. After 5 days, open all capped cells and record the results
8. Repeat the experiment three times to obtain individuals from enough patrilines and cover the variation in individual susceptibility to the fungus.

1.3.2. Molecular methods qPCR

In recent years, real-time PCR (RT-PCR) has been used to identify and quantify viral, bacterial and microsporidial pathogens in the honey bee (see the *BEEBOOK* papers on nosema (Fries *et al.*, 2013), European foulbrood (Forsgren *et al.*, 2013), viruses (de Miranda *et al.*, 2013) and molecular methods (Evans *et al.*, 2013)). There are, however, no methods yet available to quantify *A. apis* spores or the *A. apis* hyphal biomass.

1.4. Production and quality of inoculums

1.4.1. Isolation techniques

1.4.1.1. Growth media

In culture, *A. apis* isolates can grow on many different media such as potato-dextrose agar (PDA), yeast-glucose-starch agar (YGPSA), and Sabouraud dextrose agar) (Bailey, 1981; Heath, 1982; Anderson and Gibson, 1998; Hornitzky, 2001), but they grow very well on media with high sugar content like MY-20 (Udagawa and Horie, 1974).

MY-20:
1. Dissolve the following components in approximately 750 ml deionised water:

- 4.0 g peptone
- 3.0 g yeast extract
- 200.0 g glucose
- 20.0 g agar
2. Adjust the final volume to 1000 ml.
3. Autoclave at 115°C for 15 minutes.

1.4.1.2. Isolation of A. apis strains

Ascosphaera apis can be isolated from fresh mummies or dry mummies collected directly from brood frames or mummies from the bottom board or hive entrance. Both white and dark mummies can be used, but it is generally easier with white mummies, since the latter will often readily produce spore cysts following incubation when + and - strains come into contact.

1. Surface sterilize the mummy in 10% sodium hypochlorite for 10 min
2. Rinse the mummy twice in sterile distilled water for 2 min
3. Cut the mummy into smaller pieces, place the pieces on an agar plate (SDA, PDA,YGPSA or MY20)
4. Incubate the plate in dark at 30-34°C
5. Hyphal growth is visible usually within 2–4 days

Addition of antibacterial compounds (e.g. chloramphenicol, cycloheximide, streptomycin or dodine) in the media can be advantageous to prevent bacterial growth. *Ascosphaera apis* growth will often be inhibited, but once transferred back to non-antibiotic media it can be quickly recovered.

1.4.1.3. Hyphal tip isolation of A. apis strains

A single honey bee larva can ingest a large number of fungal spores, thus to ensure that the isolate is composed of a single strain, hyphal tip isolation is preferred (Fig. 4).

1. Place a dissecting microscope in a sterile hood and wipe with 70% ethanol
2. Use a fresh *A. apis* isolate (approx. 4 cm in diameter)
3. Find a single hypha at the edge of the culture under the microscope, and cut a hyphal tip just before the last branching point using a scalpel and a minutien pin
4. Transfer the tip to a new culture plate
5. Incubate the plate in darkness at 30-34°C

1.4.1.4. Mating test

Once a single isolate is obtained, the mating type can be determined using a sexual compatibility test. We recommend that the two reference strains (AFSEF 7405 and ARSEF 7406) be used in the mating tests.

1. Cut agar plugs from the border of well growing isolates

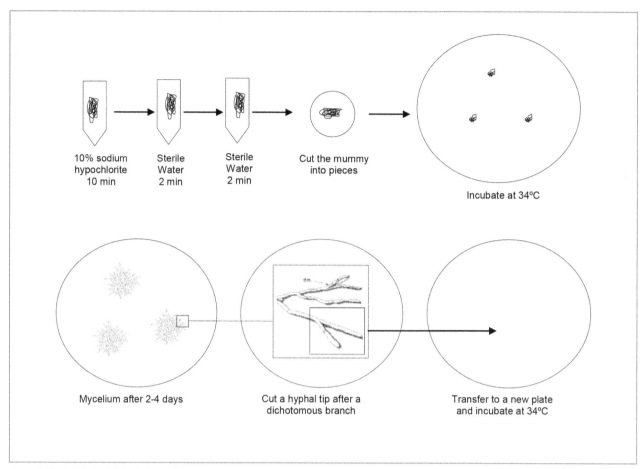

Fig. 4. Hyphal tip isolation of *Ascosphaera apis*.

2. Place the plugs of the new isolate between agar plugs of the two reference strains with a distance of approximately 4 cm
3. Incubate the plate in darkness at 30-34°C
4. After 5-7 days the presence of dark spore cysts (Fig. 5) normally can be observed as a black zone where hyphae of two strains with opposite mating types meet, indicating successful reproduction
5. If the test isolates are sexually compatible with the reference + strain, the other is the − mating type, or vice versa

On a standard agar plate (diameter 9 cm) 4 new isolates can be tested simultaneously (Fig. 5).

1.4.2. PCR for strain differentiation

Genetic variation is crucial for strain differentiation, and the arms race in host-pathogen relationships has been recorded with genetically different *A. apis* strains (Vojvodic *et al.*, 2011b). The PCR fingerprinting method, using BOX, REP, and ERIC as random primers (Reynaldi *et al.*, 2003), has been used to identify *A. apis* isolates, and microsatellite markers have been developed (Rehner and Evans, 2009). Recently several sequences of intergenetic regions or introns were screened. Three loci were found to be highly variable and can be used in identifying and differentiating *A. apis* strains (Jensen *et al.*, 2012)

(Table 2). The advantages of DNA sequences of polymorphic loci with specific primers over the PCR fingerprinting and microsatellite markers is that the sequences produced in one study can easily be compared with those in another study.

1.4.2.1. PCR

Genomic DNA can be amplified in a 50 µl reaction containing:

1. 10 µl of 5x HF buffer (1.5 mM MgCL$_2$)
2. 1 U of Phusion® High-Fidelity DNA Polymerase (New England Biolabs, Inc)
3. 200 µM of each deoxyribonucleotide triphosphate
4. 1 µM forward and reverse primer (Table 2)
5. 1 µl of DNA

Reaction conditions:

1. initial denaturing for 30 s
2. denaturing at 98°C
3. 10 touchdown cycles: 98°C for 30 s, 70-60°C (decrease of 1° C per cycle) for 30 s, and 72°C for 30 s,
4. 30 cycles of 98°C for 30 s, 60°C for 30 s, and 7 °C for 30 s;
5. 10 min extension at 72°C.

Table 2. List of qRT-PCR primers developed for targeting *A. apis* transcripts. The first three primer sets target genes that are involved in fungal mating and reproduction. *Actin* can be used as a reference gene.

Primer name	Annealing temp. (°C)	Primer sequence	Citation
Mat1-2-1 F	62	AAAATACCAAGGCCACCGA	Aronstein *et al.*, 2007
Mat1-2-1 R	62	GGAGCATATTGGTAATTTGG	Aronstein *et al.*, 2007
Ste11-like F	62	GGGAAGATTGCCAGGCC	Aronstein *et al.*, 2007
Ste11-like R	62	CAAACTTGTAGTCCGGATG	Aronstein *et al.*, 2007
Htf F	62	AAAATCCCAAGGCCTCGTA	Aronstein *et al.*, 2007
Htf R	62	CTGGTAGCGGTAGTCAGG	Aronstein *et al.*, 2007
Actin_Aapis F	58	CATGATTGGTATGGGTCAG	Aronstein *et al.*, 2007
Actin_Aapis R	58	CGTTGAAGGTCTCGAAGAC	Aronstein *et al.*, 2007

Amplified products can be analysed on a 1.5% (wt/vol) agarose gel stained with ethidium bromide or EZ-vision dye. Alternatively, PCR results could be visualised using DNA imaging. Sharp bands can be sequenced directly (further advice on sequencing and analyses can be found in the *BEEBOOK* paper on molecular methods (Evans *et al.*, 2013). Note: Alternatively another polymerase can be used and the touch down step can be replaced by adding five normal cycles; however, it will require some optimization.

1.4.3. Preservation of in vitro cultures

An *A. apis* isolate starts to show signs of aging after 30 days of growth (Ruffinengo *et al.*, 2000) and thus requires monthly transfer. However, frequent culture transfers are expensive, time consuming and increase the risk of contamination. In addition, *A. apis* isolates, as seen for many other microbes, can lose virulence when kept in culture with repeated transfers, thus it is important to utilize a long term storage method soon after an isolate is obtained. Two different culture preservation methods have been successfully used:

cryopreservation (Jensen *et al.*, 2009a) and propagation on Integral Rice Kernels (IRK) (Palacio *et al.*, 2007). Contamination during storage seems to be less risky during cryopreservation at -80°C, however for larger quantities and for laboratories not equipped with cryopreservation capacity, the use of rice kernels is a great alterative.

1.4.3.1. Cryopreservation in glycerol at -80°C

1. Cut three plugs from the border of a growing agar colony and put the plugs in a cryogenic vial containing 1 ml 10% sterile glycerol
2. Place the vials in a Cryo 1°C Freezing container, "Mr. Frosty" (Nalgene Co; Rochester, NY). Mr. "Frosty" ensures a 1°C /min cooling rate
3. Set the "Mr. Frosty" at 5°C for 4 hours and then in -80°C for 24 hours
4. After 24 hours transfer the vials to standard freezer boxes
5. Store at -80°C until use
6. Upon use, thaw the vial in a 34°C water bath and place the hyphal plugs on new agar plates

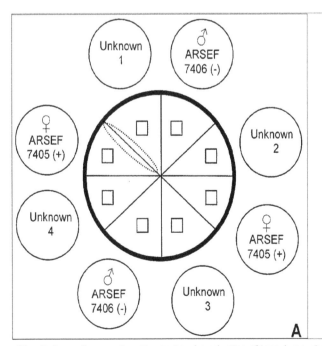

Fig. 5. *Ascosphaera apis* mating test and production of inoculums. **A.** Mating test with four isolates of unknown mating types. Unknown 1 mates with ARSEF 7405 (+) and therefore harbours the opposite mating type (-). **B.** Production of inoculums between two isolates of opposite mating types. Fruiting bodies with spores inside are formed where the hyphae meet (black zone). Photo: A B Jensen

1.4.3.2. Integral Rice Kernels (IRK) storage

1. Autoclave (1 atmosphere and 120°C for 15 min) 25 g moistened IRK (2:1 wt/vol) in Petri dishes (9-cm diameter)
2. Inoculate the IRK with 4-mm diameter plugs from the border of a growing agar colony
3. Incubate at 30°C for 15 days
4. Store the IRK at 19°C until use

1.4.4. Spore production

Ascosphaera apis spores are the only infective units of chalkbrood. It is essential to have a high quantity and quality of spores for experiments; e.g. for testing antifungal compounds or for breeding experiments. Spores can be retrieved from black mummies or produced *in vitro*. From a standardization point of view, the use of *in vitro* spores from reference strains are strongly recommended, but spores retrieved from black mummies might be chosen in certain situations. *Ascosphaera apis* will mate on various media but the yield of spores is increased on malt yeast with 20% dextrose agar (MY20) (Ruffinengo *et al.*, 2000). It is recommended that a fresh spore solution be made for every experiment.

1.4.4.1. Production and harvest of in vitro spores

1. Cut agar plugs from the border of well-growing isolates with different mating types
2. Place the plugs of the + and − strain a distance of approximately 4 cm on MY20 agar (see paragraph 1.4.1.1.)
3. Incubate the plate in the dark at 30-34°C at least 3 weeks to ensure production of mature spores
4. Use a scalpel to scrape off the spore cysts from the black mating stripes
5. Transfer them to a glass mortar with a ground-glass pestle (glass tissue grinder; Fig. 6a and b)
6. Add 5-10 µl sterile water and grind the suspension 1 min to release spores from the spore cysts and balls
7. Add 50 µl sterile water and grind again 1 min
8. Add 200 µl sterile water to the mortar
9. Pipette spores from the mortar into an 1.5 ml Eppendorf tube
10. Add additional 750 µl water to get a medium density of spores
11. Let large particles, unbroken sporocysts, spore clumps etc. in the suspension settle for 30 min
12. Pipette, from just below the surface, approximately 500 µl into a new tube, which should contain mostly released individual spores
13. Prepare a dilution series to 10^{-2} or 10^{-3} for counting and calculating the concentration of the undiluted suspension with a haemocytometer (refer to the *BEEBOOK* paper on miscellaneous methods (Human *et al.*, 2013).

1.4.4.2. Harvest of in vivo spores

1. Mix black mummies with 5 ml of sterile distilled water
2. Shake for 5 min
3. Homogenize the spore solution with a glass tissue grinder for 5 min
4. Centrifuge to spin-dry at 3500 rpm for 5 min
5. Discard the supernatant
6. Add 5 ml iodine-povidone (50% in distilled sterile water). Iodine-povidone is added to avoid bacterial contamination
7. Shake for 10 min
8. Centrifuge to spin-dry at 3500 rpm for 5 min
9. Discard the supernatant
10. Wash three times by adding 5 ml of sterile distilled water
11. Shake for 5 min
12. Spin-dry at 3500 rpm for 5 min
13. Add 1 ml water
14. Prepare a dilution series to 10^{-2} or 10^{-3} for counting and calculating the concentration of the undiluted suspension with a haemocytometer (refer to the *BEEBOOK* paper on miscellaneous methods (Human *et al.*, 2013).

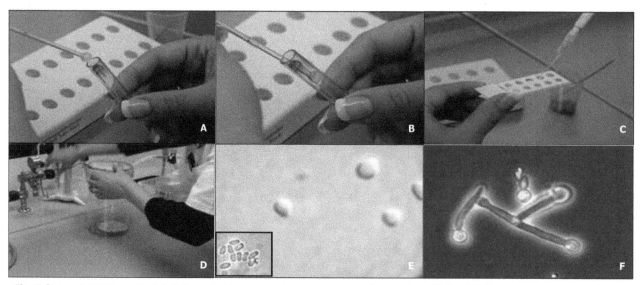

Fig. 6. Spore germination test: ***A & B.*** First grind the sporocysts to release spores from cysts and balls; ***C & D.*** incubate the spores in liquid medium and flush with CO_2 incubate the spores 32 hours at 34°C; ***E & F.*** check for spore enlargement.   Photos: A B Jensen

Spores from mummies can also be harvested using the *in vitro* harvest method described above. To get a better homogenization, scrape the spore cysts from the outside of the mummy before grinding. *In vivo* spores must be used the same day they are prepared, due to the presence of other microorganisms on the mummies even though the broad-spectrum bactericide iodine-povidone is used to avoid bacterial contamination in the spore solution preparation. Dry dark mummies with viable spores can be preserved for several years (at least 9 years) when stored dry in a closed bottle (Flores, pers. comm.)

1.4.5. Quality test of inoculums

Chalkbrood spores can contaminate all surfaces of the bee hive and accumulate in hive products. They are very resistant to the environment and can remain viable and infective for more than 15 years (Toumanoff, 1951). However, spore viability will decline over time and this will vary from isolate to isolate and from batch to batch. Thus it is important to check spore viability prior to an experiment.

1.4.5.1. Spore viability and germination test (Fig. 6)

1. Place a sterile Teflon coated slide in a sterile petri dish lined with wet filter paper
2. Prepare a concentration of 2.0×10^7 spores per ml
3. Mix 100 µl spore suspension with 500 µl GLEN (see recipe below)
4. Place 10 µl of the GLEN / spore mix onto the spot of the Teflon coated slides
5. Place the petri dish in an airtight container and flush with CO_2 30 sec to ensure a minimum of 10% CO_2
6. Incubate 32 hours at 34°C
7. Add a cover slip and count the spores directly on the Teflon slide under a microscope at 400 magnification
8. Evaluate 100 spores for enlargement or germ tube formation in three different fields of view on the slide

The AnaeroGen system (Oxoid; Basingstoke, UK) has been used with success in cases where the laboratory is not equipped with CO_2. A packet is placed together with the petri dish in a sealed jar and will generate 9 to 13% CO_2.

1.4.5.2. GLEN medium

1. Dissolve the following components in approximately 900 ml deionised water:
   - 4.000 g glucose-dextrose
   - 7.680 g natrium chloride
   - 6.500 g lactalbumin hydrolysate
   - 5.000 g yeast extract
   - 1.952 g MES-buffer (2-(N-morpholino)ethanesulfonic acid)
2. Adjust the pH to 7.0 using 1 M NaOH
3. Readjust the volume to 1000 ml
4. Sterilize by autoclaving at 121°C for 20 minutes.

1.4.6. Availability and recommended reference isolates

Microbial isolates including fungi are deposited and stored in several culture collections around the world. They are available for both commercial and scientific use. Isolates can be ordered from the culture collection for a fee that is usually lower for non-profit organizations. We would recommend that researchers deposit isolates that have been used in publications in one of these culture collections, so that they can be retrieved for future research.

Different strains of *A. apis* have shown variable virulence towards honey bees (Vojvodic *et al.*, 2011a) and may also vary in other biological traits. Thus to be able to compare different studies we recommend to always use AFSEF 7405 and ARSEF 7406 as reference isolates. These two isolates have been involved in the *A. apis* genome sequencing project (Qin *et al.*, 2006); they represent each of the two mating types and both can readily be retrieved from either the USDA-ARSEF or the ATCC culture collections.

Below is the list of the three most important culture collections we recommend for retrieving *A. apis* isolates:
USDA-ARSEF; ARS Collection of Entomopathogenic Fungi (7 isolates deposited, March 2012)
> http://www ars usda gov/Main/docs htm?docid=12125
ATCC; American Type Culture Collection (9 isolates deposited, March 2012)
> http://www lgcstandards-atcc org/
CBS; Centraalbureau voor Schimmelcultures (7 isolates deposited, March 2012)
> http://www cbs knaw nl/

1.5. Infection bioassays

A bioassay is a measurement of the effect of a substance on living organisms. Bioassays with *A. apis* can be conducted using *in vitro* reared larvae or directly in colonies. An *in vitro* bioassay has very important advantages over the whole colony bioassay; mainly, it excludes the effects of social immunity and allows for replicable experiments in controlled environment. However, to address questions related to hygienic behaviour, bioassays must be conducted directly within honey bee colonies.

1.5.1. Infection bioassays using in vitro rearing of larvae

Difficulties associated with controlling experimental conditions using honey bee colonies prompted the development of *in vivo* rearing conditions that allow rearing honey bee larvae in the laboratory and also a way to test the effects of pathogens, toxins and drugs. Refer to the *BEEBOOK* paper on larval rearing for a detailed protocol (Crailsheim *et al.*, 2013). Exposure bioassays of *in vitro* reared larvae have been used to test differences in virulence of various *A. apis* strains (Vojvodic *et al.*, 2011b), to compare the temperature response of chalkbrood and stonebrood (Vojvodic *et al.*, 2011a), to test susceptibility of various honey bee subspecies towards *A. apis* (Jensen *et al.*, 2009b), to test virulence of *A. apis* with or without presence of other

Ascosphaera species commonly associated with solitary bees (Vojvodic *et al.*, 2012) and to explore the host response to infection of *A. apis* at the molecular level (Aronstein *et al.*, 2010).

Ascosphaera apis spores have to be ingested by the larvae, hence they must be incorporated in the larval food. There are two ways spores can be administered - either the larvae can be fed a small quantity (5 µl) of spore contaminated diet, which they will ingest quickly, or the larvae can be fed spore-contaminated diet *ad libitum* for a certain period. The advantage of feeding small quantities is that the exact dose per larvae can be controlled, and if fed with different dosages an exact LD_{50} can be calculated (Jensen *et al.*, 2009b). Depending on the larval ages, additional feeding has to be done on the same day, approximately 2 hours later, after the first 5 µl is consumed. However if the aim is to produce a high numbers of infected individuals it can be more efficient to use surplus diet, as it is less time consuming.

All larval instars can be infected by *A. apis* (Jensen, unpublished), but it is recommended to use 2nd-4th instar larvae. When using 5th instar larvae, the *A. apis* spores have a limited time period in the gut for germination and penetration of the gut wall before the bee defecates, and 1st instar larvae are very fragile.

1.5.2. Infection bioassay of colonies

When it is necessary to provoke chalkbrood infection in bee colonies, it may be sufficient to infect a group of larvae, or it may be necessary to cause widespread disease in the colony. The procedure differs for each case. In the first case, the larvae are directly fed spore-contaminated food. In the second case, the entire colony is exposed to spores either by spraying, feeding sugar/honey with spores or feeding pollen patties with spores (Moffett *et al.*, 1978; Flores *et al.*, 2004a). Flores *et al.* (2004a) evaluated different ways to inoculate colonies with *A. apis* spores using pieces of comb containing bee brood that have been subjected to controlled chilling. A mummification of 90.63% of the brood was reached by spraying water containing spores over combs, and colonies fed spore-pollen mixtures reached 86.32% infection. The use of sugar syrup with spores only reached 60.13% mummification and proved to be less effective. We recommend spraying or feeding inoculated pollen since these methods have proven to be most effective.

1.5.2.1. Direct exposure of individual larvae

1. Prepare a spore solution of 2×10^8 spores/ml in sterile distilled water.
2. Mix (1:1, v:v) of the solution with honey food (50% honey : 50% distilled sterile water)
3. Remove sections of comb containing 5th instar larvae from the hive.
4. Supply each of the 5th instar larvae with 5 µl of contaminated food (equals a dose of 5×10^5 spores per larvae)

5. Place the food near the mouthpart. Use extreme care to avoid touching the larvae and confirm consumption by direct observation
6. Immediately after feeding, store the pieces of comb for 2 hours at 25ºC
7. Return the comb to the hive so that the bees will cap the cells containing the inoculated larvae
8. Remove the comb a maximum of 20 hours later and store it at 25ºC and approximately 65% humidity for 5 days. The chilling of the larvae will ensure disease development. An open water bottle placed in the incubator will provide sufficient humidity
9. After 5 days open all cell cappings and record the results. (See paragraph 3.1. Quantify infection degree in colonies)

1.5.2.2. Infection of colonies with a water spore suspension

1. Prepare a spore solution of 1.25×10^7 spores/ml in sterile distilled water. (See paragraph 1.4. Production and quality of inoculums)
2. Spray 10 ml on the each brood comb surface

This quantity has been used successfully on colonies with 8-10 combs (Flores *et al.*, 2004a).

1.5.2.3. Infection of colonies with spores in pollen

1. Prepare a spore solution of 1.25×10^7 spores/ml in sterile distilled water. (See paragraph 1.4.3. and 1.4.4. Production and quality of inoculums)
2. Add 100 ml spore solution to 150 g pollen
3. Distribute the mixture on the top of the combs

An alternative method, if spore concentration is not very important, is to homogenize 15 black mummies and mix with 150 g pollen.

1.6. Expression of fungal genes

Studies investigating host-pathogen interactions have taken a new dimension utilizing mRNA quantification (qRT-PCR). Here we describe qRT-PCR approach used for identification of *A. apis* mating type idiomorphs and quantification of *A. apis* transcripts in culture and in host tissue (Aronstein *et al.*, 2007).

Total RNA can be isolated using standard kits, TRIzol (Invitrogen; Carlsbad, California) and RNeasyR Mini Kit (Qiagen; Valencia CA). These reagents have been used successfully to make cDNA from mycelia or honey bee larvae (Aronstein *et al.*, 2007; 2010), for RNA extraction method, see the *BEEBOOK* paper on molecular methods (Evans *et al.*, 2013)

1.6.1. qRT-PCR for quantification of A. apis *transcripts*

1. cDNA can be amplified in a 20 µl reaction containing,
    1.1. 1.0 U of GoTaq® Flexi DNA polymerase (Promega Co; Madison, WI),
    1.2. 5 x GoTaq® Flexi buffer,

Table 3. List of qRT-PCR primers developed for targeting *A. apis* transcripts. The first three primer sets target genes that are involved in fungal mating and reproduction. *Actin* can be used as a reference gene.

Primer name	Annealing temp. (°C)	Primer sequence	Citation
Mat1-2-1 F	62	AAAATACCAAGGCCACCGA	Aronstein *et al.*, 2007
Mat1-2-1 R	62	GGAGCATATTGGTAATTTGG	Aronstein *et al.*, 2007
Ste11-like F	62	GGGAAGATTGCCAGGCC	Aronstein *et al.*, 2007
Ste11-like R	62	CAAACTTGTAGTCCGGATG	Aronstein *et al.*, 2007
Htf F	62	AAAATCCCAAGGCCTCGTA	Aronstein *et al.*, 2007
Htf R	62	CTGGTAGCGGTAGTCAGG	Aronstein *et al.*, 2007
Actin_Aapis F	58	CATGATTGGTATGGGTCAG	Aronstein *et al.*, 2007
Actin_Aapis R	58	CGTTGAAGGTCTCGAAGAC	Aronstein *et al.*, 2007

  1.3. 0.25 mM dNTP mix,
  1.4. 2.5 mM MgCl$_2$,
  1.5. 0.3 μM of each primer,
  1.6. 0.75 μl of a 1/1000 stock dilution of SYBR-Green (Invitrogen Corp),
  1.7. 1 μl of cDNA
2. Reaction conditions are as follows:
  2.1. Initial denaturing at 95° C for 3 min,
  2.2. 40 cycles at 95°C (20 s), 58-62°C (30 s) depending on the gene (Table 3) and 72°C (30 s).
  Negative controls (all reaction components except the DNA which is replaced with water) must be included in each run.
3. For standardization and normalization against housekeeping genes and data analysis, see the *BEEBOOK* paper on molecular methods (Evans *et al.*, 2013).

1.7. Hygienic behaviour

Hygienic behaviour is defined as the bees' ability to detect and remove diseased brood from the nest (Rothenbuhler, 1964). Hygienic behaviour was first described in the 1930s when researchers sought to determine the mechanism by which some honey bee colonies were resistant to American foulbrood (reviewed in Spivak and Gilliam, 1993). In the 1980s, it was shown that hygienic behaviour was also the primary mechanism of resistance to chalkbrood (Gilliam *et al.*, 1983), although resistance to this disease involves other factors as well, such as differences in the susceptibility of different colonies or even between patrilines within colonies (Invernizzi *et al.*, 2009; Jensen *et al.*, 2009b).

  Hygienic behaviour assays, involving killing brood by freezing or by piercing pupae with a pin (methods described in the *BEEBOOK* paper on queen rearing and selection (Büchler *et al.*, 2013)) are indirect, and record the proportion of dead brood removed by a colony after a particular amount of time. Most, but not all colonies, show a good correlation between removal of freeze-killed brood and

resistance to chalkbrood. However, researchers and beekeepers cannot assume that the ability of a colony to remove dead brood within a certain time will ensure colony-level resistance to chalkbrood. It is very important, especially for breeding purposes, to directly challenge colonies with *A. apis* in addition to the freeze-kill or pin-kill brood assay (See paragraph 1.5.1. Infection bioassay of colonies) and subsequent observation of the bees' response to the challenged brood (Palacio *et al.*, 2010).

1.8. Olfactory detection

Bees from hygienic colonies are particularly responsive to olfactory-based stimuli associated with diseased brood. All bees can perform uncapping and removal behaviours, but bees that detect abnormal brood odours at a low stimulus level may rapidly initiate uncapping behaviour, resulting in the removal of diseased brood before it becomes infectious (Wilson-Rich *et al.*, 2009). Individual bees from rapid-hygienic line breeds exhibited significantly increased sensitivity to the odour of chalkbrood disease at lower concentrations compared with bees from the slow-hygienic line, based on electrophysiological recordings of nerve impulses from the antennae (EAG), by proboscis-extension response conditioning (PER), and by isolation of volatiles from chalkbrood–infected larvae for use in field bioassays (see detailed methods in: Masterman *et al.*, 2000, 2001; Gramacho and Spivak, 2003; Spivak *et al.*, 2003; Swanson *et al.*, 2009; and general methods of EAG and collection of volatiles in the *BEEBOOK* paper on chemical ecology methods (Torto *et al.*, 2013) and of PER in the *BEEBOOK* paper on behavioural methods (Scheiner *et al.*, 2013)).

1.9. Inhibitory assays against chalkbrood

For both fundamental and applied research it can be important to test the inhibitory effects of certain chemicals, plant extracts, propolis, probiotic bacteria or hemolymph against chalkbrood. These substances can be tested for their direct effect on spore germination and on hyphal growth, or they can be used to test their effect on the ability of the fungus to infect individual bees *in vitro* or in a colony context.

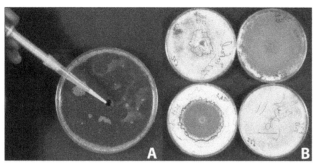

Fig. 7. Zone of inhibition assay of different compounds against *Ascosphaera apis*: **A.** MY-20 medium with spores spread on the entire surface and with a central hole for the test compound; **B.** The zone of inhibition was recorded daily by drawing on the Petri dish, a line over the border of the mycelium (each day a different colour) and the anti-fungal property was assessed as daily inhibition zone using an image analyser.   Photos: J M Flores

1.9.1. Inhibition of spore germination and hyphae (zone of inhibition)

1. Prepare MY20 medium (see paragraph 1.4.1.1.) in petri dishes (15 cm)
2. Spread 2 ml of a spore solution (9.0×10^7 spores/ml) on the surface of the medium, approximately 10^6 spores/cm^2
3. Make a central hole (7 mm) in the MY20 medium (Fig. 7A)
4. Place 0.5 ml of the test product into the hole
5. Incubate the cultures (30ºC, 12% CO_2, 65% relative humidity)
6. Measure the fungal growth (or the zone of inhibition) (Fig. 7B)

Measure the fungal growth (or the zone of inhibition) daily using a stereoscopic microscope magnification (X 20 to 40). Mycelium growth can be seen within the medium before emerging at the surface (Puerta *et al.*, 1990). An Image Analyzer can be used to measure the diameter of the zone of inhibition.

1.9.2. Inhibition in colonies

1. Assess the degree of infection in colonies before treatment, as described in paragraph 1.3.1. (Quantify infection degree in colonies)
2. Increase colony infection with *A. apis* as described in paragraph 1.5.2. (Infection bioassay of colonies)
3. Apply the test item to the colony
4. Assess the level of infection in the colony, as described in paragraph 1.3.1. (Quantify infection level in colonies) the next day after treatment and repeat the assessment depending of the aim of the experiment.

We propose an indirect assessment of *in vivo* treatments because quantifying the disease by counting mummies is very difficult and inaccurate. If the disease is increasing in the colony, the spore numbers will also increase; likewise, if the disease is decreasing, the spore number will be lower due to the natural behaviours of the bees

such as hygienic removal of larvae. The degree of infection would most probably be related to the dynamics of *A. apis* spore load in the colony. Evaluation of the level of infection in the colony before and after treatment is therefore recommended.

1.10. Minimizing chalkbrood in experimental colonies

The main techniques for chalkbrood control lie in management practices that reduce spore concentration in the colonies and that avoid stress (in particular chilling) of susceptible brood.

1.10.1. Breeding for resistance

Gilliam *et al.*, (1983) and Taber (1986) demonstrated that it is possible to select and breed honey bees for resistance to chalkbrood disease. Spivak and Reuter (2001) demonstrated that colonies selected for rapid removal of freeze-killed brood showed resistance to chalkbrood in field experiments. Palacio *et al.* (2000) observed that hygienic colonies had a lower frequency of brood diseases including chalkbrood. Commercial queen breeders in the US and Denmark have found that if they have "zero-tolerance" for chalkbrood; i.e. they never raise queens from a colony that has had clinical symptoms of chalkbrood and they simultaneously select for rapid hygienic behaviour then they get rather chalkbrood resistant lines (Spivak and Jensen, unpublished).

1.10.2. Management and treatment

There are a number of management's techniques that can be used to minimize the effects of chalkbrood in infected colonies:

1. Reduce volume of the brood chamber for the overwintering (Seal, 1957)
2. Enlarge colony entrance to aid ventilation (Gochnauer *et al.*, 1975)
3. Replace old combs (Betts, 1951)
4. Heat treatment of the wax (Flores *et al.*, 2005a)
5. Requeen affected colonies (Lunder, 1972)

To date, there is no effective chemical treatment against this disease. Nevertheless *in vitro* (Puerta *et al.*, 1990; Flores *et al.*, 1996b) and *in vivo* (Flores *et al.*, 2001) techniques have been used to evaluate chemical treatments for chalkbrood control.

2. Stonebrood

2.1. Introduction

Stonebrood is a very rare honey bee brood disease caused by several fungi from the genus *Aspergillus*. The disease was first described by Massen (1906) and has since then been found worldwide. *Aspergillus flavus* has most frequently been reported, followed by *Asp. fumigatus*, but also *Asp. niger* and other species can kill honey bees (Gilliam and Vandenberg, 1997). *Aspergillus* is able to infect the host through the gut if the spores are ingested, but also through the cuticle. Therefore, adults as well as larvae and pupae can become infected. In addition, most species of *Aspergillus* produce aflatoxins that have been suggested to be the primary cause of death in stonebrood infected honey bees (Burnside, 1930). However, a non-aflatoxin producing *Asp. flavus* strain has been observed to induce stonebrood symptoms equally well in *in vitro* reared honey bee larvae as afaltoxin producing strains (Vojvodic, unpublished).

Aspergillus spp. are cosmopolitan filamentous fungi often found in soil, where they thrive as saprophytes, but occasionally they do infect living hosts including, plants, insects and mammals. *Aspergillus* can infect human lungs, eyes, pharynx, skin and open wounds, but most commonly this has been observed in immune deprived individuals (Gefter, 1992; Germaud and Tuchais, 1995; Denning, 1998; Galimberti *et al.*, 1998; Garret *et al.*, 1999). In addition, the alfatoxins are carcinogenic if inhaled or ingested; therefore precautions need to be taken when stonebrood disease occurs in honey bees principally to protect beekeepers and consumers. In several countries stonebrood is a notifiable disease that has to be reported to the authorities if it occurs.

2.2. Biohazards

Working with fungi requires good microbiological practice and containment, irrespective of whether they possess a potential risk for the environment or human, since proliferation on the growth medium of contaminants always poses a potential risk. *Good microbiological practice* is in principle the handling of a microorganism in a "test tube" without any other organisms entering and contaminating it. *Containment* is in principle the handling of a microorganism with emphasis on safety of the laboratory worker and the environment. Good microbiological practice and containment involves:

Aseptic techniques
- Limit the open time of "test tube"
- Open tubes or plates only in bio-safe cabinet/bench to avoid
- contamination
- Only use sterile tools (e.g. pipette tips or loops)
- Avoid casual contact with the bench, fingers, or outside of the bottle
- Dispose or decontaminate tools immediately after use

Personal hygiene and dress
- Wash hands prior to and following manipulations
- Wear appropriate personal protective equipment (gloves, clean lab coats)
- Do not touch the skin, face, or unclean non-sterile surfaces
- Confine loose or long hair and keep fingernails short

Area cleanliness and organization
- Disinfect work area before and after work with 70% Ethanol
- Immediately clean spills, and then disinfect the work surface
- Keep only items important for the task in progress in the bench

- Plan and lay out work in a logical order so the work in the bench becomes efficient
- Minimize personnel traffic and unnecessary movements around the work area
- Routine cleaning of difficult-to-access areas to prevent build up of dust and debris

Aspergillus spp. produce airborne conidia which represent a potential risk for the experimenter, but no additional attention has to be taken as long as good microbiological practice and containment are followed. In particular, it is important that the cultures grown on agar plates are only opened in a sterile bench and that conidia are only handled outside the bench if they are in a liquid suspension.

If conducting a bioassay with *Aspergillus* spp. on *in vitro* reared larvae or in cages on adult honey bees, it is important that the assessment is done in a sterile bench or a fume hood once the *Aspergillus* start to sporulate. Dead infected bees can be removed with forceps to avoid production of numerous new conidia.

In several countries, permission from the authorities is required to work experimentally with stonebrood fungi, in particular if it is an outdoor experiment. If *Aspergillus* spp. are used for experiments in bee colonies we recommend using a mask and safety glasses for protection while conducting the experiments.

2.3. Diagnostics and qualitative detection

2.3.1 Morphological description

The typical symptoms observed in a colony affected by stonebrood are not very different from chalkbrood symptoms and includes irregular capping of the brood. Infected brood, also called "mummies", can be seen in the combs. Stonebrood mummies turn hard and they resemble small stones, not sponge-like as chalkbrood mummies. Stonebrood mummies are difficult to remove from the cells with forceps and removal by the worker bees is also difficult. Infected brood becomes covered with powdery yellow, brown, green or black fungal spores depending on the species. In some cases infected or deceased larvae looked dry, but they do not produce visible conidia within a 48 hrs after pathogen inoculation (Vojvodic, unpublished)

Stonebrood can be diagnosed by its gross symptoms, but positive identification requires its cultivation in the laboratory and subsequent microscopic examination. Structures of the conidiophores (spore forming structures) are very important for identification of *Aspergillus spp*. The conidiophores originate from a basal cell located on the supporting hyphae and terminate in a vesicle (Fig. 8). The morphology, colour and roughness of the conidiophores vary from species to species. Additionally, the position of the flask-shaped phialides (spore producing cells) on the vesicle is an important character. The phialides can cover the vesicle surface entirely ("radiate" head) or partially ("columnar" head) and the phialids can be attached to the vesicle directly (uniseriate) or attached via a supporting cell, called metula (biseriate). The phialides produce round conidia (2-5 μm in diameter)

Table 4. Microscopic characters of three *Aspergillus* species most often reported to cause stonebrood.

Species	Conidia colour	Conidiophores	Phialides	Vesicle
Asp. flavus	Yellow-green	Colourless. Rough	Uni/biseriate	Round, radiate head
Asp. fumigatus	Blue-green to grey	Short (<300μm), smooth, colourless-green	Uniseriate	Round, columnar head
Asp. niger	Black	Long, smooth, colourless or brown	Biseriate	Round, radiate head

that form radial chains (Fig. 8). The conidia of the different species can have different colours (Table 4) (See Fig. 9 for *Asp. flavus in vitro* infected larvae). We however recommend contacting a mycologist for correct species identification of the *Aspergillus* specimens.

2.3.2 Molecular methods

There is no single method (morphological, physiological or molecular) that can be used to recognize all of the approximately 250 *Aspergillus* species. Using a multi-locus approach will give a lot of information, but it also requires certain skills and equipment not always present in diagnostic laboratories (Geiser *et al.*, 2007). A two-step barcoding has been suggested for identification of *Aspergillus* species in a clinical setting. The ITS regions can be used for inter-section level identification and the β-tubulin for identification of individual species within the various *Aspergillus* sections (Balajee *et al.*, 2007) (see primers in Table 5).

Aspergillus DNA can, as for *Ascosphaera*, be extracted with standard kits (see also the *BEEBOOK* paper on molecular methods (Evans *et al.*, 2013)

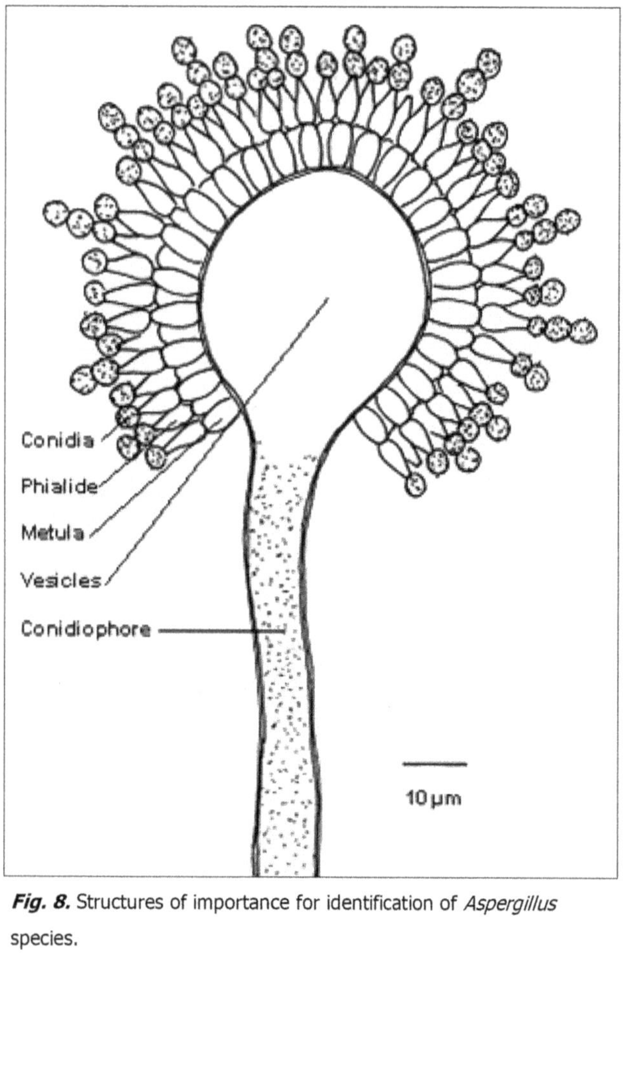

2.4. Production and quality of inoculums

2.4.1. Isolation techniques

Aspergillus grows readily on many different standard media e.g. SDA, PDA, but Czapek-Dox medium, which contain sucrose as carbon source and nitrate as the nitrogen source should be very suitable. Czapek-Dox medium with addition of yeast extract (5.0 g/l) is recommended (Frisvad, pers. comm.). All these three media can be purchased premixed and easily prepared. E.g. to prepare 1 l of SDA medium, suspend 65 g SDA (Merck) in 1 l of demineralized water and autoclave 15 min at 121°C.

2.4.1.1. Aspergillus *spp. can be isolated from sporulating mummies.*

1. Put a sterile microbial loop in an area with many conidia
2. Streak on agar plates
3. If the plates become contaminated with other microbes, repeat the procedure. Once a clean culture is established, proceed with single spore isolation described below.

2.4.1.2. Single spore isolation

1. Add 10 ml 0.05% Triton-X on the culture agar plate
2. Rub its surface gently with a sterile Drigalski spatula to loosen the conidia
3. Transfer the suspension conidia into a 15-50 ml sterile tube
4. Wash the suspension twice (to remove agar and hyphal fragments)
   4.1. Centrifuge 3 min at 7000 g for 3 min
   4.2. Discharge the supernatant and add 10 ml 0.005% Triton-X
   4.3. Centrifuge 3 min at 7000 g for 3 min
   4.4. Discharge the supernatant and add 10 ml 0.005% Triton-X
5. Prepare a serial dilution (remember to whirl mix before pipetting)

Conidia
Phialide
Metula
Vesicles
Conidiophore

10 μm

Fig. 8. Structures of importance for identification of *Aspergillus* species.

Table 5. List of primers used to be used in the two-step DNA barcoding of *Aspergillus* spp. for the ITS region and part of the β-tubulin gene.

Primer name	Annealing temp (°C)	Annealing temperature (°C)	Citation
ITS 1	55	TCCGTAGGTGAACCTGCGG	White *et al.*, 1990
ITS 4	55	TCCTCCGCTTATTGATATGC	White *et al.*, 1990
Bt2a (β-tubulin)	58	GGTAACCAAATCGGTGCTGCTTTC	Glass and Donaldson, 1995
Bt2b (β-tubulin)	58	ACCCTCAGTGTAGTGACCCTTGGC	Glass and Donaldson, 1995

6. Count the spore concentration in a haemocytometer (described in the *BEEBOOK* paper on miscellaneous methods (Human *et al.*, 2013))
7. Transfer 100 µl of a spore solution at 5 X 10² spore per ml to a new plate
8. Incubate at 25°C for two-four days
9. Transfer a single small colony to a new plate

To harvest conidia for experimental purposes the above procedure can be used (minus step 6-8). It is important to use Triton-X or another detergent to avoid spore clumping.

2.4.2. Preservation of *in vitro* cultures

Aspergillus has been observed to lose sporulation capacity and virulence after a couple of transfers on a standard medium (Scully and Bidochka, 2006). Therefore we recommend long-term storage of the isolates during the first transfers. Several long-term storage methods can be used: frozen in skim milk, refrigerated with silica gel, or freeze dried. It is possible to successfully recover various *Aspergillus* species (e.g. *Asp. flavus*, *Asp. parasiticus* and *Asp. niger*), stored in 10% glycerol at -80°C for more than a year, using the same method which we recommend for *A. apis* (see paragraph 1.4.3.1. Cryopreservation in glycerol at -80°C).

2.5. Quality test of inoculums

Viability testing of *Aspergillus* spp. can be performed using the standard insect pathology methods, which are briefly described below.

2.5.1. Spore viability and germination test

1. Transfer 10 µl of a spore solution at 5 x 10⁵ spore per ml to an agar plate
2. Spread the spore solution over the entire ager plate with a sterile Digralski spatula
3. Incubate at 25°C for 24 hour
4. Place three cover slips on the plate
5. Place the agar plate under the microscope and count the proportion of germinated spores out of 100 spores under each of the three cover slips

It is recommended to use only new fresh spore solution for each experiment because the shelf life of *Aspergillus* spores in a water solution is about 3 days (Vojvodic, unpublished).

2.5.2. Availability and recommended reference isolates

Aspergillus has been deposited in various fungal culture collections, such as USDA-ARSEF, ATCC and CBS (see link below). None of the isolates deposited so far have been isolated from honey bees, neither larvae nor adults, but from sources such as plant material, soil, vertebrates and invertebrates. USDA-ARSEF is a collection of Entomopathogenic fungi, thus the majority of isolated deposited there originate from insects or other arthropods. It is possible to retrieve *Aspergillus* isolates from hymenoptera; solitary megachilid bees (*Osmia lignaria* and *Megachile rotundata*) and formicine ants (*Anoplolepsis longipes* and *Solenopsis invicta*).

Limited work has been performed with stonebrood, thus it is difficult to recommend a specific reference strain. A reference strain can be chosen based on its pathobiological properties (to honey bees or at least Hymenoptera); as in Vojvodic *et al.* (2011a) where an *Asp. flavus* strain isolated from an infected honey bee larvae was used to infect *in vitro* reared honey bee larvae. This particular strain is unfortunately lost. However, the reference strain could be based on its geographical origin or type specimen.

USDA-ARSEF; ARS Collection of Entomopathogenic Fungi (83 isolates deposited, 15 *Asp. flavus*, 7 *Asp. niger* April 2012)

 http://www.ars.usda.gov/Main/docs.htm?docid=12125

ATCC; American Type Culture Collection (1618 isolates deposited, 180 *Asp. favus*, 111 *Asp. fumingatus*, 117 *Asp. niger*, April 2012)

 http://www.lgcstandards-atcc.org/

CBS; Centraalbureau voor Schimmelcultures (1055 isolates deposited, 47 *Asp. flavus*, 147 *Asp. fumigatus*, 7 *Asp. niger*, April 2012)

 http://www.cbs.knaw.nl/

Aspergillus is classified as an opportunistic human pathogen, and thus some countries will need an import permit and proof that the laboratory is accredited to handle it.

2.6. Infection bioassays

Exposure bioassays with *Aspergillus* spp. can potentially be carried out, using *in vitro* reared larvae (Fig. 9), caged bees or colonies, due to its ability to infect larvae, pupae and adults. A bioassay with *in vitro* reared larvae has been conducted showing that stonebrood and chalkbrood had opposite temperature-dependent responses in virulence. Chilling of chalkbrood exposed larvae increased the

Fig. 9. *In vitro* reared honey bee larvae infected with *Aspergillus flavus*: the bottom left cell contains larvae in an early stage of disease visible by the change in larval colour; bottom right cell contains later stage of the diseases with the visible fungal body and conidia protruding out of the larval cuticle. The two upper cells contain healthy larvae.   Photo: S Vojvodic

pathogen virulence; whereas chilling of stonebrood exposed larvae increased larval survival (Vojvodic *et al.*, 2011a). For infection of *in vitro* bee larvae with *Aspergillus spp.* the same precautions must be taken as described in paragraph 1.5.1. for *A. apis*.

Few experiments with caged bees have been conducted (Gilliam and Vandenberg, 1997). Infecting individual larvae while still in the brood frame is possible by placing the spore suspensions to be ingested in front of the larvae (Vojvodic unpublished).

3. Future perspectives

Chalkbrood and stonebrood diseases have been recognized for more than a century, but there is still much that remains to be discovered regarding these diseases and their impact on the general health status of honey bees. The numbers of studies of the two diseases reflects their frequency and abundance, with a magnitude difference in favour of chalkbrood.

The genome of *A. apis* was published in 2006 (Qin *et al.*, 2006), and although the full annotation is still lacking, it will be useful for future research investigations on the expression of genes important for the infection and virulence of *A. apis*. Chalkbrood is a stress related disease and a recent longitudinal cohort study based on monitoring data collected over six years indicated that colonies with high numbers of varroa mites in the same season or *Nosema ceranae* infection in the spring had significantly higher chances of chalkbrood

outbreaks (Hedtke *et al.*, 2011). Such a correlation needs experimental confirmation, but it elucidates the complexity of the host-pathogen-interaction in honey bee colonies. Research on the interaction with other pathogens and stressors, such as sublethal concentration of various chemicals, is also warranted.

Aspergillus research is mostly focused on human health and food spoilage due to aflatoxin contamination of grains. Stonebrood outbreaks are rarely observed in honey bee colonies, but should not be underestimated. Their rarity could be a result of honey bees removing the stonebrood infected individuals very quickly, an area of research that has not been previously investigated. Furthermore, the basic biology of stonebrood is still poorly understood, and several studies elucidating stonebrood etiology are still to be performed. *Aspergillus* spp. spores are present everywhere and a high virulence towards honey bee larvae have been shown. Even though stonebrood is rarely reported, it would be interesting to understand which factors and mechanisms might play a role in the establishment and resistance of this disease.

References

ANDERSON, D L; GIBSON, N L (1998) New species and isolates of spore-cyst fungi (Plectomycetes: Ascosphaerales) from Australia. *Australian Systematic Botany* 11: 53-72. http://dx.doi.org/10.1071/SB96026

ANDERSON, D L; GIBBS, A J; GIBSON, N L (1998) Identification and phylogeny of sporecyst fungi (*Ascosphaera* spp) using ribosomal DNA sequences. *Mycological Research* 102: 541–547. http://dx.doi.org/10.1017/S0953756297005261

ARONSTEIN, K A; MURRAY, K D; DE LEON, J; QIN, X; WEINSTOCK, G (2007) High mobility group (HMG-box) genes in the honey bee fungal pathogen *Ascosphaera apis*. *Mycologia* 99: 553–561. http://dx.doi.org/10.3852/mycologia.99.4.553

ARONSTEIN, K A; MURRAY, K D (2010) Chalkbrood disease in honey bees. *Journal of Invertebrate Pathology* 103: 20-29. http://dx.doi.org/10.1016/j.jip.2009.06.018

ARONSTEIN, K A; MURRAY, K D; SALDIVAR, E (2010) Transcriptional responses in Honey bee larvae infected with chalkbrood fungus. *BMC Genomics* 11: 391. http://dx.doi.org/10.1186/1471-2164-11-391

BAILEY, L (1981) *Honey bee pathology*. Academic Press; London, UK. pp. 40–44.

BAMFORD, S; HEATH, L A F (1989) The effects of temperature and pH on the germination of spores of the chalkbrood fungus, *Ascosphaera apis*. *Journal of Apicultural Research* 28: 36–40.

BALAJEE, S A; HOUBRAKEN, J; VERWEIJ, P E; HONG, S-B; YAGUCHI, T; VARGA, J; SAMSON, R A (2007). *Aspergillus* species identification in the clinical setting. *Studies in Mycology* 59: 39–46. http://dx.doi.org/10.3114/sim.2007.59.05

BETTS, A D (1951) *The diseases of bees: their signs, causes and treatment.* Hickmott: Camberley, UK.

BISSETT, J (1988) Contribution toward a monograph of the genus *Ascosphaera. Canadian Journal of Botany* 66: 2541–2560.

BURNSIDE, C E (1930) Fungous diseases of the honey bee. *US Department of Agriculture Technical Bulletin* 149.

BÜCHLER, R; ANDONOV, S; BIENEFELD, K; COSTA, C; HATJINA, F; KEZIC, N; KRYGER, P; SPIVAK, M; UZUNOV, A; WILDE, J (2013) Standard methods for rearing and selection of *Apis mellifera* queens. In *V Dietemann; J D Ellis; P Neumann (Eds) The COLOSS BEEBOOK, Volume I: standard methods for* Apis mellifera *research. Journal of Apicultural Research* 52(1): http://dx.doi.org/10.3896/IBRA.1.52.1.07

CRAILSHEIM, K; BRODSCHNEIDER, R; AUPINEL, P; BEHRENS, D; GENERSCH, E; VOLLMANN, J; RIESSBERGER-GALLÉ, U (2013) Standard methods for artificial rearing of *Apis mellifera* larvae. In *V Dietemann; J D Ellis; P Neumann (Eds) The COLOSS BEEBOOK, Volume I: standard methods for* Apis mellifera *research. Journal of Apicultural Research* 52(1): http://dx.doi.org/10.3896/IBRA.1.52.1.05

DE MIRANDA, J R; BAILEY, L; BALL, B V; BLANCHARD, P; BUDGE, G; CHEJANOVSKY, N; CHEN, Y-P; VAN DOOREMALEN, C; GAUTHIER, L; GENERSCH, E; DE GRAAF, D; KRAMER, M; RIBIÈRE, M; RYABOV, E; DE SMET, L VAN DER STEEN, J J M (2013) Standard methods for virus research in *Apis mellifera*. In *V Dietemann; J D Ellis; P Neumann (Eds) The COLOSS BEEBOOK, Volume II: standard methods for* Apis mellifera *pest and pathogen research. Journal of Apicultural Research* 52(4): http://dx.doi.org/10.3896/IBRA.1.52.4.22

DENNING, D W (1998) Invasive aspergillosis. *Clinical Infectious Diseases* 26: 781-803. http://dx.doi.org/10.1086/513943

EVANS, J D; CHEN, Y P; CORNMAN, R S; DE LA RUA, P; FORET, S; FOSTER, L; GENERSCH, E; GISDER, S; JAROSCH, A; KUCHARSKI, R; LOPEZ, D; LUN, C M; MORITZ, R F A; MALESZKA, R; MUÑOZ, I; PINTO, M A; SCHWARZ, R S (2013) Standard methodologies for molecular research in *Apis mellifera*. In *V Dietemann; J D Ellis; P Neumann (Eds) The COLOSS BEEBOOK, Volume I: standard methods for* Apis mellifera *research. Journal of Apicultural Research* 52(4): http://dx.doi.org/10.3896/IBRA.1.52.4.11

FLORES, J M; RUIZ, J A; RUZ, J M; PUERTA, F; BUSTOS, M; PADILLA, F; CAMPANO, F (1996a) Effect of temperature and humidity of sealed brood on chalkbrood development under controlled conditions. *Apidologie* 27: 185-192. http://dx.doi.org/10.1051/apido:19960401

FLORES, J M; RUÍZ, J A; RUZ, J M; PUERTA, F; BUSTOS, M; PADILLA, F; CAMPANO, F (1996b) Estudio *in vivo* e *in vitro* para el control de la ascosferiosis en *apis mellifera. Revista Iberoamericana de Micología* 13: 292-295.

FLORES, J M; PUERTA, F; GUTIÉRREZ, I; ARREBOLA, F (2001) Estudio de la eficacia del Apimicos-b® en el control y la prevención de la ascosferiosis en la abeja de la miel. *Revista Iberoamericana de Micología* 18: 187-190.

FLORES, J M; GUTIÉRREZ, I; PUERTA, F (2004a) A comparison of methods to experimentally induce chalkbrood disease in honey bees. *Spanish Journal of Agricultural Research* 2: 79-83.

FLORES, J M; GUTIÉRREZ, I; PUERTA, F (2004b) Oxytetracycline as a predisposing condition for chalkbrood in honey bee. *Veterinary Microbiology* 103: 195-199. http://dx.doi.org/10.1016/j.vetmic.2004.07.012

FLORES, J M; SPIVAK, M; GUTIÉRREZ, I (2005a) Spores of *Ascosphaera apis* contained in wax foundation can infect honey bee brood. *Veterinary Microbiology* 108: 141-144. http://dx.doi.org/10.1016/j.vetmic.2005.03.005

FORSGREN, E; BUDGE, G E; CHARRIÈRE, J-D; HORNITZKY, M A Z (2013) Standard methods for European foulbrood research. In *V Dietemann; J D Ellis, P Neumann (Eds) The COLOSS BEEBOOK: Volume II: Standard methods for* Apis mellifera *pest and pathogen research. Journal of Apicultural Research* 52(1): http://dx.doi.org/10.3896/IBRA.1.52.1.12

FRIES, I; CHAUZAT, M-P; CHEN, Y-P; DOUBLET, V; GENERSCH, E; GISDER, S; HIGES, M; MCMAHON, D P; MARTÍN-HERNÁNDEZ, R; NATSOPOULOU, M; PAXTON, R J; TANNER, G; WEBSTER, T C; WILLIAMS, G R (2013) Standard methods for nosema research. In *V Dietemann; J D Ellis, P Neumann (Eds) The COLOSS BEEBOOK: Volume II: Standard methods for* Apis mellifera *pest and pathogen research. Journal of Apicultural Research* 52(1): http://dx.doi.org/10.3896/IBRA.1.52.1.14

GALIMBERTI, R; KOWALCZUK, A; PARRA, I H; RAMOS M G; FLORES, V (1998) Cutaneous aspergillosis: a report of six cases. *British Journal of Dermatology* 139: 522-526. http://dx.doi.org/10.1046/j.1365-2133.1998.02424.x

GARRETT, D O; JOCHIMSEN, E; JARVIS, W (1999) Invasive *Aspergillus spp.* infections in rheumatology patients. *Journal of Rheumatology* 26: 146-149.

GEFTER, W B (1992) The spectrum of pulmonary aspergillosis. *Journal of Thoracic Imaging* 7: 56-74.

GERMAUD, P; TUCHAIS, E (1995) Allergic bronchopulmonary aspergillosis treated with itraconazole. *Chest* 107: 883. http://dx.doi.org/10.1378/chest.107.3.883

GEISER, D M; KLICH, M A; FRISVAD, J C; PETERSON, S W; VARGA, J; SAMSON, R A (2007) The current status of species recognition and identification in *Aspergillus. Studies in Mycology* 59: 1-10. http://dx.doi.org/10.3114/sim.2007.59.01

GILLIAM, M; TABER, S; RICHARDSON, G V (1983) Hygienic behaviour of honey bees in relation to chalkbrood disease. *Apidologie* 14: 29-39. http://dx.doi.org/10.1051/apido:19830103

GILLIAM, M; VANDENBERG, J (1997) Fungi. In *Morse, R A; Flottum, K (Eds). Honey bee pests, predators, & diseases (3rd Edition).* A I Root Co.; Medina, USA. pp. 81-110.

GLASS, N L; DONALDSON, G C (1995) Development of primer sets designed for use with the PCR to amplify conserved genes from filamentous ascomycetes. *Applied and Environmental Microbiology* 61: 1323–1330.

GOCHNAUER, T A; FURGALA, B; SHIMANUKI, H (1975) Diseases and enemies of the honey bee. In *Grout R (Ed.) The hive and the honey bee (3rd Ed.).* Dadant and Sons; Hamilton, USA. pp. 615-621.

GRAMACHO, K P; SPIVAK, M (2003) Differences in olfactory sensitivity and behavioural responses among honey bees bred for hygienic behaviour. *Behavioural Ecology and Sociobiology* 54: 472-479. http://dx.doi.org/10.1007/s00265-003-0643-y

HEATH, L A F (1982) Development of chalkbrood in a honey bee colony: a review. *Bee World* 63: 119-130.

HEATH, L A F; GAZE, B M (1987) Carbon dioxide activation of spores of the chalkbrood fungus *Ascosphaera apis. Journal of Apicultural Research* 26(4): 243–246.

HEDTKE, K; JENSEN, P M; JENSEN, A B; GENERSCH, E (2011) Evidence for emerging parasites and pathogens influencing outbreaks of stress-related diseases like chalkbrood. *Journal of Invertebrate Pathology* 108:167-73. http://dx.doi.org/10.1016/j.jip.2011.08.006

HORNITZKY, M (2001) *Literature review of chalkbrood.* A report for the RIRDC. Publication No. 01/150, Kingston, ACT, Australia.

HUMAN, H; BRODSCHNEIDER, R; DIETEMANN, V; DIVELY, G; ELLIS, J; FORSGREN, E; FRIES, I; HATJINA, F; HU, F-L; JAFFÉ, R; KÖHLER, A; PIRK, C W W; ROSE, R; STRAUSS, U; TANNER, G; TARPY, D R; VAN DER STEEN, J J M; VEJSNÆS, F; WILLIAMS, G R; ZHENG, H-Q (2013) Miscellaneous standard methods for *Apis mellifera* research. In *V Dietemann; J D Ellis; P Neumann (Eds) The COLOSS BEEBOOK, Volume I: standard methods for* Apis mellifera *research. Journal of Apicultural Research* 52(4): http://dx.doi.org/10.3896/IBRA.1.52.4.10

HUMBER, R A (2008) Evolution of entomopathogenicity in fungi. *Journal of Invertebrate Pathology* 98: 262–266. http://dx.doi.org/10.1016/j.jip.2008.02.017

INVERNIZZI, C; PENAGARICANO, F; TOMASCO, I H (2009) Intracolonial genetic variability in honey bee larval resistance to the chalkbrood and American foulbrood parasites. *Insectes Sociaux* 56: 233–240. http://dx.doi.org/10.1007/s00040-009-0016-2

JAMES, R R; SKINNER, J S (2005) PCR diagnostic methods for *Ascosphaera* infections in bees. *Journal of Invertebrate Pathology* 90: 98–103. http://dx.doi.org/10.1016/j.jip.2005.08.004

JAMES, T Y; KAUFF, F; SCHOCH, C; MATHENY, P B; HOFSTETTER, V; COX, C J; CELIO, G; GEUIDAN, C; FRAKER, E; MIADLIKOWSKA, J; LUMBSCH, H T; RAUHUT, A; REEB, V; ARNOLD, A E; AMROFT, A; STAJICH, J E; HOSAKA, K; SUNG, G -H; JOHNSON, D; O'ROURKE, B; CROCKETT, M; BINDER, M; CURTIS, J M; SLOT, J C; WANG, Z; WILSON, A W; SCHÜßLER, A; LONGCORE, J E; O'DONNELL, K; MOZLEY-STANDRIDGE, S;. ORTER, D; LETCHER, P M; POWELL, M J; TAYLOR, J W; WHITE, M M; GRIFFITH, G W; DAVIES, D R; HUMBER, R A; MORTON, J B; SUGIYAMA, J; ROSSMAN, A; ROGERS, J D; PFISTER, D H; HEWITT, D; HANSEN, K; HAMBLETON, S; SHOEMAKER, R A; KOHLMEYER, J; VOLKMANN-KOHLMEYER, B; SPOTTS, R A; SERDANI, M; CROUS, P W; HUGHES, K W; MATSUURA, K; LANGER, E; LANGER, G; UNTEREINER, W A; LÜCKING, R; BÜDEL, B; GEISER, D M; APTROOT, A; DIEDERICH, P; SCHMITT, I; SCHULTZ, M; YAHR, R; HIBBETT, D S; LUTZONI, F; MCLAUGHLIN, D J; SPATAFORA, J W; VILGALYS, R (2006) Reconstructing the early evolution of Fungi using a six-gene phylogeny. *Nature* 443: 818–822. http://dx.doi.org/10.1038/nature05110

JENSEN, A B; JAMES, R R; EILENBERG, J (2009a) Long-term storage of *Ascosphaera aggregata* and *A. apis*, pathogens of the leafcutting bee (*Megachile rotundata*) and the honey bee (*Apis mellifera*). *Journal of Invertebrate Pathology* 101: 157–160. http://dx.doi.org/10.1016/j.jip.2009.03.004

JENSEN, A B; PEDERSEN, B V; EILENBERG, J (2009b) Differential susceptibility across honey bee colonies in larval to chalkbrood resistance. *Apidologie* 40: 524-534. http://dx.doi.org/10.1051/apido/2009029

JENSEN, A B; WELKER, D L; KRYGER, P; JAMES, R R (2012) Polymorphic DNA sequences of the fungal honey bee pathogen *Ascosphaera apis. FEMS Microbiology Letters* 330: 17-22. http://dx.doi.org/10.1111/j.1574-6968.2012.02515.x

LUNDER, R (1972) Undersøkelse av kalkyngel i 1971 [Investigation on chalkbrood in 1971]; *Birøkteren* 88: 55–60.

MAASSEN, A (1913) Weitere Mitteilungen uber der seuchenhaften Brutkrankheiten der Bienen [Further communication on the epidemic brood disease of bees]. *Mitteilungen aus der Kaiserlichen Biologischen Anstalt fur Land- und Forstwirtscshaft* 14: 48-58.

MASTERMAN, R; SMITH, B H; SPIVAK M (2000) Brood odour discrimination abilities in hygienic honey bees (*Apis mellifera* L) using proboscis extension reflex conditioning. *Journal of Insect Behaviour* 13(1): 87-101. http://dx.doi.org/10.1023/A:1007767626594

MASTERMAN, R; ROSS, R; MESCE, K; SPIVAK, M (2001) Olfactory and behavioural response thresholds to odours of diseased brood differ between hygienic and non-hygienic honey bees (*Apis mellifera* L.). *Journal of Comparable Physiology A* 187:441-452. http://dx.doi.org/10.1007/s003590100216

MAURIZIO, A (1934) Uber die Kaltbrut (Pericystis-Mykose) der Bienen. *Archiv Bienenkunde* 15: 165–193.

MOFFETT, J O; WILSON, W T (1978) Feeding commercially purchased pollen containing mummies caused chalkbrood. *American Bee Journal* 118: 412-414.

MOMOT, J P; ROTHENBUHLER, W C (1971) Behaviour genetics of nest cleaning in honey bees. VI. Interactions of age and genotype of bees, and nectar flow. *Journal of Apicultural Research* 10: 11-21.

MURRAY, K D; ARONSTEIN, K A; JONES, W A (2005) A molecular diagnostic method for selected *Ascosphaera* species using PCR amplification of internal transcribed spacer regions of rDNA. *Journal of Apicultural Research* 44: 61-64.

NILSSON, H; KRISTIANSSON, E; RYBERG, M; HALLENBERG, N; LARSSON, K H (2008) Intraspecific ITS variability in the kingdom Fungi as expressed in the International Sequence Databases and its implications for molecular species identification. *Evolutionary Bioinformatics Online* 4: 193–201.

PALACIO, M A; FIGINI, E; RODRIGUEZ, E; RUFFINENGO, S; BEDASCARRASBURE, E; DELHOYO, M (2000) Changes in a population of *Apis mellifera* selected for its hygienic behaviour. *Apidologie* 31:471-478. http://dx.doi.org/10.1051/apido:2000139

PALACIO, M A; PEÑA, N; CLEMENTE, G; RUFFINENGO, S; ESCANDE, A (2007) Viability and pathogenicity of *Ascosphaera apis* preserved in integral rice culture. *Journal of Agricultural Research* 5(4): 481-486.

PALACIO, M A; RODRIGUEZ, E; GONCALVES, L; BEDASCARRASBURE, E; SPIVAK, M (2010) Hygienic behaviours of honey bees in response to brood experimentally pin-killed or infected with *Ascosphaera apis*. *Apidologie* 41: 602-612. http://dx.doi.org/10.1051/apido/2010022

PUERTA, F; FLORES, J M; TARÍN, R; BUSTOS, M; PADILLA, F, HERMOSO DE MENDOZA, M (1990) Actividad antifúngica de productos seleccionados contra *Ascosphaera Apis*. Estudios *in vitro*. *Revista Iberoamericana de Micología* 7: 103-106.

QIN, X; EVANS, J D; ARONSTEIN, K A; MURRAY, K D; WEINSTOCK, G M (2006) Genome sequences of the honey bee pathogens *Paenibacillus larvae* and *Ascosphaera apis*. *Insect Molecular Biology* 15(5): 715-718. http://dx.doi.org/10.1111/j.1365-2583.2006.00694.x

REHNER, S A; EVANS, J D (2009) Microsatellite loci for the fungus *Ascosphaera apis*: cause of honey bee chalkbrood disease. *Molecular Ecology Resources* 9: 855–858.

REYNALDI, F J; LOPEZ, A C; ALBO. G N; ALIPPI, A M (2003) Differentiation of *Ascosphaera apis* isolates by rep-PCR fingerprinting and determination of chalkbrood incidence in Argentinean honey samples. *Journal of Apicultural Research* 42: 68–76.

ROTHENBUHLER, W C (1964) Behaviour genetics of nest cleaning in honey bees. IV. Responses of F1 and backcross generations to disease-killed brood. *American Zoologist* 4(2): 111-123. http://dx.doi.org/10.1016/0003-3472(64)90082-X

RUFFINENGO, S; PEÑA, N I; CLEMENTE, G; PALACIO, M A; ESCANDE, R (2000) Suitability of culture media for the production of ascospores and maintenance of *Ascosphaera apis*. *Journal of Apicultural Research* 39(3/4): 143-148.

SCHEINER, R; ABRAMSON, C I; BRODSCHNEIDER, R; CRAILSHEIM, K; FARINA, W; FUCHS, S; GRÜNEWALD, B; HAHSHOLD, S; KARRER, M; KOENIGER, G; KOENIGER, N; MENZEL, R; MUJAGIC, S; RADSPIELER, G; SCHMICKLI, T; SCHNEIDER, C; SIEGEL, A J; SZOPEK, M; THENIUS, R (2013) Standard methods for behavioural studies of *Apis mellifera*. In *V Dietemann; J D Ellis; P Neumann (Eds) The COLOSS BEEBOOK, Volume I: standard methods for* Apis mellifera *research. Journal of Apicultural Research* 52(4): http://dx.doi.org/10.3896/IBRA.1.52.4.04

SCHOCH, C L; SEIFERT, K A; HUHNDORFC, S; ROBERT, V; SPOUGE, J L; LEVESQUEB, C A; CHEN, W; (FUNGAL BARCODING CONSORTIUM) (2012) Nuclear ribosomal internal transcribed spacer (ITS) region as a universal DNA barcode marker for Fungi. *Proceedings of the National Academy of Sciences* 109: 6241-6246. www.pnas.org/cgi/doi/10.1073/pnas.1117018109

SCULLY, L R; BIDOCHKA, M J (2006) A cysteine/methionine auxotroph of the opportunistic fungus *Aspergillus flavus* is associated with host-range restriction: a model for emerging diseases. *Microbiology* 152: 223–232. http://dx.doi.org/10.1099/mic.0.28452-0

SEAL, D W A (1957) Chalk brood disease of bees. *New Zealand Journal of Agriculture* 6: 562.

SKOU, J P (1972) Ascosphaerales. *Friesia* 10(1): 1–24.

SKOU, J P (1988) More details in support of the class Ascosphaeromycetes. *Mycotaxon* 31(1): 191–198.

SPILTOIR, C F (1955) Life cycle of *Ascosphaera apis* (*Pericystis apis*). *American Journal of Botany* 42(6): 501-508. http://dx.doi.org/10.2307/2438686

SPIVAK, M; GILLIAM, M (1993) Facultative expression of hygienic behaviour of honey bees in relation to disease resistance. *Journal of Apicultural Research* 32: 147-157.

SPIVAK, M; REUTER, G S (2001) Resistance to American foulbrood disease by honey bee colonies, *Apis mellifera*, bred for hygienic behaviour. *Apidologie* 32: 555-565. http://dx.doi.org/10.1051/apido:2001103

SPIVAK, M; MASTERMAN, R; ROSS, R; MESCE, K A (2003) Hygienic behaviour in the honey bee (*Apis mellifera* L.) and the modulatory role of octopamine. *Journal of Neurobiology* 55: 341–354. http://dx.doi.org/10.1002/neu.10219

SWANSON, J; TORTO, B; KELLS, S; MESCE, K; TUMLINSON, J; SPIVAK, M (2009) Odorants that induce hygienic behaviour in honey bees: Identification of volatile compounds in chalkbrood-infected honey bee larvae. *Journal of Chemical Ecology* 35: 1108-1116. http://dx.doi.org/10.1007/s10886-009-9683-8

TABER, S (1986) Breeding bees for resistance to chalkbrood disease. *American Bee Journal* 126: 823-825.

THOMPSON, V C (1964) Behaviour genetics of nest cleaning in honey bees. III. Effect of age of bees of a resistant line on their response to disease-killed brood. *Journal of Apicultural Research* 3: 25-30.

TORTO, B; CARROLL, M J; DUEHL, A; FOMBONG, A T; NAZZI, F., GOZANSKY, K T; SOROKER, V; TEAL, P E A (2013) Standard methods for chemical ecology research in *Apis mellifera*. In *V Dietemann; J D Ellis; P Neumann (Eds) The COLOSS BEEBOOK, Volume I: standard methods for* Apis mellifera *research. Journal of Apicultural Research* 52(4): http://dx.doi.org/10.3896/IBRA.1.52.4.06

TOUMANOFF, C; (1951) Les Maladies des Abeilles [The diseases of bees]. *Revue Française d'Apicculture numéro spécial* 68: 1-325.

UDAGAWA, S; HORIE, Y (1976) A new species of *Emericella*. *Mycotaxon* 4: 535-539.

VOJVODIC, S; BOOMSMA, J J; EILENBERG, J; JENSEN, A B (2012) Virulence of mixed fungal infections in honey bee brood. *Frontiers in Zoology* 9:5. http://dx.doi.org/10.1186/1742-9994-9-5

VOJVODIC, S; JENSEN, A B; JAMES, R R; BOOMSMA, J J; EILENBERG, J (2011a). Temperature dependent virulence of obligate and facultative fungal pathogens of honey bee brood. *Veterinary Microbiology* 149: 200-205. http://dx.doi.org/10.1016j.vetmic.2010.10.001

VOJVODIC, S; JENSEN, A B; MARKUSSEN, B; EILENBERG, J; BOOMSMA, J J (2011b) Genetic variation in virulence among chalkbrood strains infecting honey bees. *PLoS ONE* 6(9): e25035. http://dx.doi.org/10.1371/journal.pone.0025035

WHITE, T; BURNS, T; LEE, S; TAYLOR, J (1990) Amplification and direct sequencing of fungal ribosomal RNA genes for phylogenetics. In *Innis, M A; Gelfand, D H; Sninsky, J J; White, T J (Eds). PCR protocols. A guide to methods and applications.* Academic Press, Inc.; Sand Diego, USA. pp. 315–322.

WILSON-RICH, N; SPIVAK, M; FEFFERMAN, N H; STARKS, P T (2009) Genetic, individual, and group facilitation of disease resistance in insect societies. *Annual Review of Entomology* 54: 405-423. http://dx.doi.org/10.1146/annurev.ento.53.103106.093301

Journal of Apicultural Research 52(1): (2013)
DOI 10.3896/IBRA.1.52.1.14

REVIEW ARTICLE

Standard methods for *Nosema* research

IBRA

INTERNATIONAL BEE
RESEARCH ASSOCIATION

Ingemar Fries[1]*, Marie-Pierre Chauzat[2], Yan-Ping Chen[3], Vincent Doublet[4], Elke Genersch[6], Sebastian Gisder[6], Mariano Higes[7], Dino P McMahon[5], Raquel Martín-Hernández[7], Myrsini Natsopoulou[4], Robert J Paxton[4,5], Gina Tanner[8], Thomas C Webster[9] and Geoffrey R Williams[8]

[1]Department of Ecology, Swedish University of Agricultural Sciences, 75007 Uppsala, Sweden.
[2]Anses, Unit of honey bee pathology, 105 route des Chappes, BP 111, F06 902 Sophia Antipolis cedex, France.
[3]Bee Research Laboratory, USDA-ARS, Beltsville, MD, USA.
[4]Institute for Biology, University of Halle-Wittenberg, 06099 Halle (Saale), Germany.
[5]School of Biological Sciences, Queen's University Belfast, Belfast BT9 7BL, UK.
[6]Department for Molecular Microbiology and Bee Diseases, Institute for Bee Research Hohen Neuendorf, 16540 Hohen Neuendorf, Germany.
[7]Centro Apícola de Castilla-La Mancha, Laboratorio de Patología, San Martín s/n, 19180 Marchamalo, Spain.
[8]Swiss Bee Research Centre, Agroscope Liebefeld-Posieux Research Station ALP-HARAS, Bern, Switzerland.
[9]Land Grant Program, Kentucky State University, Frankfort, KY 40601, USA.

Received 10 April 2012, accepted subject to revision 14 June 2012, accepted for publication 29 October 2012.

*Corresponding author: Email: Ingemar.Fries@slu.se

Summary

Methods are described for working with *Nosema apis* and *Nosema ceranae* in the field and in the laboratory. For fieldwork, different sampling methods are described to determine colony level infections at a given point in time, but also for following the temporal infection dynamics. Suggestions are made for how to standardise field trials for evaluating treatments and disease impact. The laboratory methods described include different means for determining colony level and individual bee infection levels and methods for species determination, including light microscopy, electron microscopy, and molecular methods (PCR). Suggestions are made for how to standardise cage trials, and different inoculation methods for infecting bees are described, including control methods for spore viability. A cell culture system for *in vitro* rearing of *Nosema* spp. is described. Finally, how to conduct different types of experiments are described, including infectious dose, dose effects, course of infection and longevity tests.

Métodos estándar para la investigación sobre *Nosema*

Resumen

Se describen procedimientos para trabajar con *Nosema apis* y *Nosema ceranae* en el campo y en el laboratorio. Para el trabajo de campo, se describen diferentes métodos de muestreo para determinar infecciones al nivel de colonia en un momento determinado, y también para el seguimiento de la dinámica temporal de infección. Se hacen sugerencias para la forma de estandarizar los ensayos de campo para evaluar los tratamientos y el impacto de la enfermedad. Los métodos de laboratorio descritos incluyen diferentes formas de determinar los niveles de infección al nivel de colonia y de abeja individual, y los métodos para la determinación de las especies, incluyendo microscopía óptica, microscopía electrónica y métodos moleculares (PCR). Se hacen sugerencias para estandarizar los ensayos con cajas, y se describen diferentes métodos de inoculación para infectar abejas, incluyendo métodos de control para la viabilidad de las esporas. Se describe un sistema de cultivo celular para la cría *in vitro* de *Nosema* spp. Finalmente, se describe cómo llevar a cabo diferentes tipos de experimentos, incluyendo la dosis infecciosa, efectos de la dosis, curso de la infección y las pruebas de longevidad.

Footnote: Please cite this paper as: FRIES, I; CHAUZAT, M-P; CHEN, Y-P; DOUBLET, V; GENERSCH, E; GISDER, S; HIGES, M; MCMAHON, D P; MARTÍN-HERNÁNDEZ, R; NATSOPOULOU, M; PAXTON, R J; TANNER, G; WEBSTER, T C; WILLIAMS, G R (2013) Standard methods for nosema research. In *V Dietemann; J D Ellis, P Neumann (Eds) The COLOSS BEEBOOK: Volume II: Standard methods for* Apis mellifera *pest and pathogen research. Journal of Apicultural Research* 51(5): http://dx.doi.org/10.3896/IBRA.1.52.1.14

孢子虫研究的标准方法

本文描述了实验室及野外实验中关于孢子虫*Nosema apis* 和 *Nosema ceranae* 的研究方法。对于野外实验，本文列举了多种取样方法，用于研究特定时间段内，蜂群群体感染水平，以及随后开展长期感染规律的研究。同时也对如何标准化评估蜂场治疗效果、感病程度及选用哪些指标用于标准化评估提出了建议。实验室方法包括，确定蜂群感染水平及个体蜜蜂感染水平的方法以及测定孢子虫种类的方法，如光学显微镜法，电子显微镜法以及分子方法（PCR）。对于如何标准化蜂笼实验的各项指标提出了建议，并描述了感染蜜蜂的不同接种方法，包括孢子生存能力的对比法。描述了孢子虫的一种体外细胞培养体系。最后描述了如何进行不同类型的实验，包括感染剂量、剂量效能、感染过程以及寿命试验。

Keywords: *Nosema apis, Nosema ceranae,* field methods, laboratory methods, sampling methods, infection dynamics, infection level, microscopy, species identification, standardised cage trials, inoculation methods, spore viability, cell culture, infectious dose, dose effects, course of infection, longevity tests, honey bee, *BEEBOOK*, COLOSS

Table of Contents

Introduction

Since the description of *Nosema apis* in the early part of the last century (Zander, 1909), nosema disease, or nosemosis of honey bees has been regarded as a serious obstacle for profitable beekeeping in temperate climates (Fries, 1993). With the detection of the new parasite *Nosema ceranae*, originally described from the Asian honey bee *Apis cerana* (Fries *et al.*, 1996), in European honey bees (Higes *et al.*, 2006) the need for research in this field has become urgent. In particular, since early reports on the effects of this new parasite suggested a more severe impact on colony health compared to infections by *N. apis* (Higes *et al.*, 2008a). Following the COLOSS workshop "Nosema disease: lack of knowledge and work standardization" in Guadalajara (19-22 October, 2009) data were collected on the heterogeneity of methods used in *Nosema* research in different laboratories in Europe and in the USA. This survey showed the widely heterogeneous experimental conditions applied in the nine different participating laboratories. Even if common sense implies that some conditions should be applied in all cases, their costs and the availability of these analyses have to be taken into account. For example, one could be tempted to assert that virus presence has to be checked in colonies providing honey bees for *Nosema* experiments, but some laboratories are not equipped with virus diagnostics, and this could restrict these teams from performing experiments on *Nosema* without controlling for confounding viral infections. The level and extent of diagnosis should also be specified: which viruses should be studied, is viral detection sufficient, or should virus quantification be included?

Here we attempt to standardize study of the microsporidians *Nosema apis* and *Nosema ceranae*. *Nosema apis*, the historical microsporidian parasite of European honey bees, can decrease worker longevity and cause considerable winter colony losses (Fries, 1993), whilst *N. ceranae*, probably introduced into European honey bees from its Asian congener (*Apis cerana*) within the last few decades (Higes *et al.*, 2006; Martín-Hernández *et al.*, 2007; Klee *et al.*, 2007; Paxton *et al.*, 2007; Chen *et al.*, 2008; Williams *et al.*, 2008; Invernizzi *et al.*, 2009; Currie *et al.*, 2010; Botías *et al.*, 2011), is associated with colony depopulation and collapse in warmer areas of Europe (Higes *et al.*, 2008a), but not in northern parts of Europe (Gisder *et al.*, 2010a), in North America (Guzman-Nova, 2010; Williams *et al.*, 2010) or in South America (Invernizzi *et al.*, 2009). Yet because detection of *N. ceranae* in European honey bees coincided with recent large-scale honey bee colony losses throughout the world, data on the pathology and management of this parasite are of significant interest.

When working in the field with full-sized colonies, several considerations need to be made, such as where to sample, how often to sample and also the size of samples. We find, for example, that sample sizes are often too small to satisfy a statistically reasonable level of diagnostic precision (Fries *et al.*, 1984). Please refer to the section on statistics in the *BEEBOOK* paper on miscellaneous methods (Human *et al.*, 2013) to determine sample size.

Similarly, laboratory tests using bees in cages are often employed to investigate *Nosema* intra-host development (e.g. Higes *et al.*, 2007; 2010; Martín-Hernández *et al.*, 2009; Forsgren and Fries, 2010), effects of parasitism on host mortality (e.g. Paxton *et al.*, 2007), immunity (e.g. Antúnez *et al.*, 2009), and physiology (e.g. Dussaubat *et al.*, 2010; Mayack and Naug, 2009; Martín-Hernández *et al.*, 2011), as well as for testing the efficacy of potential control treatments (e.g. Maistrello *et al.*, 2008; Higes *et al.*, 2011). When designing cage experiments, researchers typically must control for a number of variables, ranging from selection of study subjects (e.g. parasite and host strains) to experimental environment (e.g. growth chamber conditions, food quality and quantity). Although decisions typically do not jeopardize the scientific rigor of a study, they may profoundly affect results, and may make comparisons with similar, independent studies difficult. An important consideration is that most current data on *Nosema* were collected from experiments with *N. apis*. The same research is now needed for *N. ceranae* in order to assess the similarities and differences between the two species.

Here we discuss some important factors that researchers must consider when studying the *Nosema*-honey bee system using field as well as laboratory cage experiments. This will allow researchers to make informed choices when developing experimental protocols and will increase confidence when comparing results among studies.

2. Method type

2.1. Field methods

2.1.1. Colony level infection

The degree of *Nosema* spp. infection in a colony has most commonly been described through the average number of spores per bee in a pooled sample (Cantwell, 1970; OIE, 2008). However, some studies suggest that the best way to determine the degree of infection is to estimate the proportion of infected bees in the colony (L'Arrivee, 1963; Doull, 1965; Higes *et al.*, 2008a; Botías *et al.*, 2012a). Nevertheless, there is a good correlation between the proportion of infected bees and the average number of spores in a pooled sample of bees (Fries *et al.*, 1984), but not in all cases (Higes *et al.*, 2008a). Evaluating the proportion of infected bees is much more laborious than to count the number of spores in a pooled sample, so pooled sampling will probably remain an important tool for quantifying infections in colonies. Because there is a wide variation in the numbers of spores found in individual bees in pooled samples, when the highest precision is needed, it may still be motivating to investigate the proportion of infected bees. The highest proportion of infected bees are found in foraging bees (Higes *et al.*, 2008a, b; Botías *et al.*, 2012a,b; Meana *et al.*, 2010; Smart and Sheppard, 2011), as is the greatest infection (spores per bee) (El-Shemy and Pickard, 1989). Recent studies suggest the importance of determining the proportion of infected house bees to establish the viability and impact from infection on colonies (Botías *et al.*, 2012a). Considerations regarding pooled sampling versus individual bee diagnosis are discussed in section 2.2.1.2.

2.1.1.1. Sampling

For the diagnosis and detection of *Nosema* spp.-infected colonies, the oldest honey bees should be the target population, since they are more frequently infected compared to younger bees (Meana *et al.*, 2010; Smart and Sheppard, 2011). Forager bees can be sampled outside the entrance. This method is useful in all areas during flight and foraging conditions. Caution should be taken to avoid collection of young bees performing their orientation flights. The time of the day of these flights could change in different geographic areas. Bees that conduct their orientation flights are easily recognized by the hovering behaviour in large numbers outside the entrance. When this behaviour is seen, attempts to sample for foragers should be avoided.

During non-flight conditions, old bees may still be the target population for diagnosis. To avoid sampling of newly emerged, uninfected bees, the samples can either be taken from peripheral combs in the brood area, without hatching bees, or in a super above a queen excluder. The variation introduced by not sampling flight bees is unfortunate, because it reduces the possibility for meaningful comparisons between sites. Nevertheless, when samples are taken during late autumn or winter or during the active season outside of foraging conditions (e.g. study of infection development across the year), bees from within the hive must be sampled.

To determine the degree of infection within colonies, both pooled sampling or individual sampling of bees can be used (see section 2.2.1.2). Samples should preferably be stored in the deep freezer until further processing.

2.1.1.2. Sample size

Please refer to the statistics section in the *BEEBOOK* paper on miscellaneous methods (Human *et al.*, 2013).

2.1.1.3. Clinical signs and symptoms of infection

The effects from *N. apis* infections on honey bee colonies have been extensively documented (Fries, 1993). This disease is characterized (in acute forms) by the trembling of honey bee workers or dead bees around the hive. Bees may also exhibit a dilated abdomen and brown faecal marks on the comb and the front of the hives are often found. Further, heavily infected colonies have a decrease in brood production and slow colony growth, particularly in spring (Bailey, 1955; OIE, 2008). Although *N. apis* is correlated with winter mortality of infected colonies, the disease also appears without causing losses of infected colonies (Bailey, 1955). Colony level symptoms produced by *N. ceranae* infections have been described to be different from *N. apis* symptoms (Higes *et al.*, 2008; 2009). A gradual depopulation of adult bees, higher autumn / winter colony mortality, and finally the queen

surrounded by only young bees have been observed in southern Europe (Higes *et al.*, 2006; 2008; 2009), whereas such severe symptoms or effects have not been described from more temperate climates (Gisder *et al.*, 2010a) nor from South (Invernizzi *et al.*, 2009) or North (Williams *et al.*, 2011; Guzman-Novoa *et al.*, 2010) America. Possibly, differences in honey bee subspecies, foraging conditions, agricultural practices, differences in hive management practices, or other abiotic or biotic factors may contribute to the variation in symptoms described for *N. ceranae* infections in different regions of the world. Further research efforts using standardized methodologies are most certainly warranted.

2.1.2. Colony level infection dynamics

The temporal dynamics of *N. apis* infections in temperate climates have been described by many authors (White, 1919; Borchert, 1928; Michailoff, 1928; Bailey, 1955; Furgala and Hyser, 1969; Furgala and Mussen, 1978), with a similar pattern in both northern and southern hemispheres (Doull and Cellier, 1961). In short, the typical infection exhibits low prevalence during the summer, a small peak in the autumn, and a slow rise of infection during the winter. In the spring, infections increase rapidly as brood rearing starts while flight possibilities are still limited. There are few data on the temporal dynamics on *N. apis* infections in tropical or sub-tropical climates, but the infection appears to be present but with low impact on colony fitness (Wilson and Nunamaker, 1983) and probably without the pronounced temporal dynamics observed in temperate climates (Fries *et al.*, 2003). For *N. ceranae*, few long-term studies have been performed on the temporal dynamics of this infection in honey bee colonies. Studies from central Spain suggest much less variation in infection prevalence over the season for *N. ceranae* compared to what has been described for *N. apis* (Higes *et al.*, 2008a). However, a clear seasonal effect on disease prevalence, with higher prevalence and infection levels in the early season has been documented in eastern USA (Traver and Fell, 2011; Traver *et al.*, 2012) and in untreated colonies in maritime Canada (Williams *et al.*, 2010; 2011). There is an urgent need for long-term studies of the temporal dynamics of *N. ceranae* infections under different climatic conditions.

2.1.2.1. Sampling period and sampling frequency

To gain information on how amounts and prevalence of infections change over time, it is necessary to monitor changes over more than one year. Samples should be taken with at least monthly intervals during the period when colonies can be opened without adverse effects on colony survival. For higher resolution of the temporal dynamics of infections, sampling must be bi-weekly or even weekly during periods of rapid change of infection prevalence, as described for *N. apis* during early spring. Refer to section 2.1.1.1. sampling honey bees.

2.1.3. Standardising field trials

2.1.3.1. Selection of colonies

A main problem with field trials using free flying honey bee colonies is the large natural variation in size, productivity or behaviour between colonies. In comparative field trials, it is therefore advisable to minimize such variations, and if considerable variation is expected, then increase the number of colonies involved. This can be achieved by using:

- Artificial swarms from healthy colonies
- Sister queens mated the same way or instrumentally inseminated (see *BEEBOOK* papers on queen rearing and selection (Büchler *et al.*, 2013) and on instrumental insemination (Cobey *et al.*, 2013))
- Adding a controlled degree of infection at the onset of the experiment

Adding a controlled degree of infection at the onset of the experiment can be achieved by adding a known number of bees with a documented degree of infection, either from naturally infected colonies, or from colonies where infections have been propagated for this purpose. Spores can also be sprayed onto combs and bees in sugar solution as is done with American foulbrood spores (see the *BEEBOOK* paper on American foulbrood (de Graaf *et al.*, 2013)).

Naturally infected colonies can also be used for comparative studies after careful documentation of infection prevalence (see section 2.1.1.).

2.1.3.2. Behaviour of infected bees

There is a need to study if the impact on behaviour and physiology described from infections with *N. apis* also occurs during infections with *N. ceranae*. In particular, we need to find out:

- Do honey bees infected by *N. ceranae* start their foraging activities at a younger age compared to uninfected bees, as is the case for *N. apis* infections (Hassanein, 1953)?
- Do infections with *N. ceranae* cause a faster physiological aging of the bees, as is the case for *N. apis* infections (Wang and Muller, 1970)?
- Do bees infected by *N. ceranae* feed the queen less frequently compared to uninfected bees, as is the case for *N. apis* infections (Wang and Muller, 1970)?
- Are queen bees infected by *N. ceranae* superseded, as is the case for *N. apis* infections (Farrar, 1947)?

2.1.3.3. Parameters to record

The type of data to be collected during field experiments will vary depending on the objective of the study. Field studies often involve observing colonies and the development of disease in relation to different treatments or management practices. It is thus advisable to

always report on certain standard parameters if possible, for interpretation of the data. Such parameters include:

- Intra-hive mortality (dead bee traps; see the *BEEBOOK* paper on miscellaneous methods (Human *et al.*, 2013))
- Colony size and population dynamics (number of bees and amount of brood, see the *BEEBOOK* paper on estimating colony strength (Delaplane *et al.*, 2013)
- Clinical signs of any diseases (see Volume II of the *BEEBOOK*)
- *Varroa* mite infestation (see the *BEEBOOK* paper on varroa (Dietemann *et al.*, 2013))
- Honey production
- Climatic conditions during the experiment
- Subspecies of honey bees used (see the *BEEBOOK* paper on characterising subspecies and ecotypes (Meixner *et al.*, 2013))
- Specific management practices

2.2. Laboratory methods

2.2.1. Colony and individual level infection

2.2.1.1. Spore counts

To determine the degree of infection of *Nosema* spp. in a sample it has been suggested that subjective judgement on an arbitrary infection scale can be used (Doull and Eckert, 1962; Gross and Ruttner, 1970). However, for research purposes, such estimations should not be employed if some degree of precision is required.

The easiest way to count *Nosema* spp. spores is to use a traditional haemocytometer as described in the miscellaneous methods paper of the *BEEBOOK* (Human *et al.*, 2013) and Cantwell (1970).

Although yeast and *Nosema* spores reflect light differently in regular light microscopy, the use of phase contrast microscopy will avoid any misidentification (see section 2.2.2.1).

2.2.1.2. Individual bees or pooled samples

To determine the degree of infection in colonies from pooled samples of bees, the required precision of the diagnosis should first be determined (see section 2.1.1.2).

1. Measure 1 ml of water per bee in the pooled sample.
2. Grind the required number of bees, their abdomen or ventriculus (see section 2.2.1.3.) thoroughly in water (1/3 of the measured amount) in a mortar with a pestle.
3. Add the remaining water and mix thoroughly.
4. Add a small droplet to a haemocytometer and allow spores to settle before counting (see the miscellaneous methods paper of the *BEEBOOK* (Human *et al.*, 2013).

By examining infection in each bee in a sample, the proportion of infected bees can be determined. For *N. apis,* this is a better measurement than the average spore count per bee in a pooled sample (pool of individual bees), if the objective is to measure the influence of infection on honey yield (Fries *et al.*, 1984). To determine the proportion of infected bees within a colony, individual bees must be examined.

1. Use 1 ml of water per bee. If other amounts of water are used it is necessary that the dilution factor is stated. Depending on the content of the bees' intestine with sometimes massive pollen amounts, dilution may be necessary to actually see the spores in light microscopy.

2. Grind each bee, their abdomen or ventriculus separately (see section 2.2.1.3.) thoroughly in water in a mortar with a pestle.

3. Look for the presence of spores in the macerate under a microscope (see section 2.2.2.1.).

The use of 96-well PCR plates with a single bee or abdomen or ventriculus per well for maceration can be useful for high throughput analysis. If using a 96-well plate, a system (manual or automatic) to guarantee that every bee or abdomen is completely macerated to release the spores to be detected needs to be developed. Nevertheless, there is a highly significant and positive correlation between the proportion of infected bees and average spore counts per bee. Thus, under most circumstances, the less labour intensive use of pooled samples can be used, rather than determining the infection status of at least 59 individual bees (see section 2.1.1.2. and the statistics paper of the *BEEBOOK* (Human *et al.*, 2013)).

Use of dead bees from the floor board of live colonies should be avoided, since the correlation with the health status of the live bees is too low (Fries *et al.*, 1984).

It remains to be determined whether the relations described here are the same for *N. ceranae.*

2.2.1.3. Parts to examine

Although molecular evidence of *N. ceranae* DNA in other tissues than the ventriculus have been described, no spore production has been demonstrated outside the epithelial cells of the ventriculus for either of the two microsporidians concerned (Chen *et al.*, 2009; Bourgeois *et al.*, 2012). Thus, for spore counts, the ventriculus is ideal since the amount of surplus debris is low compared to using the entire abdomen or the whole bee in mash preparations.

Dissecting the ventriculus:
With some training, it is easy to pull the ventriculus out from CO_2 immobilized bees (see the section on anesthetising bees in the *BEEBOOK* paper on miscellaneous research methods (Human *et al.*, 2013)) using forceps.

1. Grip over the A7 abdominal dorsal and ventral segments with the forceps

2. Hold the abdomen in the other hand

3. Slowly pull apart; the posterior portion of the alimentary canal comes out, sometimes with, sometimes without, the honey sac attached to the anterior end of the ventriculus (Fig. 1).

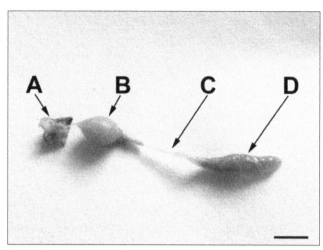

Fig. 1. Posterior section of the worker honey bee alimentary canal and sting apparatus: A = sting apparatus; B = rectum; C = small intestine; and D = ventriculus (midgut). Line = 2 mm (Dade, 2009).

It should be noted that this procedure is difficult to perform on bees that have been frozen and thawed.

Because of the labour involved, the use of the ventriculus for spore counts is generally more suitable for laboratory cage experiments. For field investigations of colonies, it is recommended to use samples of whole bees, or abdomens of adult bees.

2.2.2. Nosema species identification
2.2.2.1. Light microscopy (LM)

A compund microscope using 400 X magnification is sufficient for observing *Nosema* spp. spores in macerated bee preparations. Use of phase contrast light microscopy facilitates distinguishing spores of microsporidia from yeast or other particles.

Although the differences in spore size between *N. ceranae* and *N. apis* are not immediately apparent in light microscopy, there is a consistent difference. Spores of *N. ceranae* are clearly smaller compared to spores of *N. apis*. Fresh, unfixed spores of *N. apis* measure approximately 6 x 3 μm (Zander and Böttcher, 1984); whereas, fresh spores of *N. ceranae* measure approximately 4.7 x 2.7 μm (Fries *et al.*, 1996) (Fig. 2). Although there is a slight overlap, with the smallest *N. apis* spores being smaller than the largest *N. ceranae* spores, the average spore size of *N. apis* is approximately 1 μm larger in length (Fig. 2).

In contrast to spores of *N. apis*, the spores of *N. ceranae* are often slightly bent, and appear less uniform in shape compared to *N. apis* spores (Fries *et al.*, 1996; Fig. 2). Although the difference in the size of spores between these species is clear, it may still be difficult to detect the difference in routine diagnosis of infected bees using light microscopy. This is particularly true because mixed infections of both species can occur (Chen *et al.*, 2009), even in individual bees (Burgher-MacLellan *et al.*, 2010).

Because of their light refractive properties, spores of *Nosema* spp. are easily seen without contrast colouring in the light microscope in

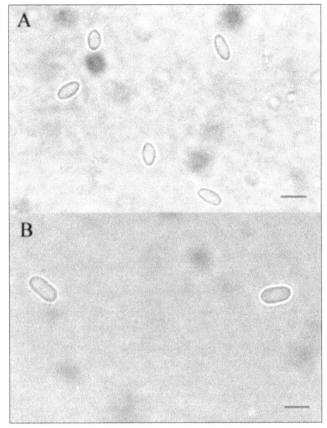

Fig. 2. Spores of *Nosema ceranae* (A) and *Nosema apis* (B) in light microscopy squash preparations. Bars = 5 mm. From Fries *et al.*, 2006.

water squash preparations at 200-400 X magnifications. Methanol-fixed smears contrast coloured using Giemsa staining is the standard technique for microsporidians and is described in section 2.2.2.1.1. However, the nuclei are not revealed because of the staining of the spore contents. Giemsa staining can be useful to visualize infection in tissue preparations. Another method to identify *Nosema* species with LM is to mount sections of material embedded for electron microscopy (section 2.2.2.2) for light microscopy after contrast colouring with toluidine blue (section 2.2.2.1.2). For a further range of colouring techniques for microsporidians see Vavra and Larsson (1999).

2.2.2.1.1. Giemsa staining

Giemsa's stain stock solution is obtained from commercial sources. The prepared staining solution is prepared fresh by diluting the stock solution 1:9 using PBS buffer (pH 7.4).

1. Apply smear onto microscope slide
2. Allow to air dry
3. Flood with 95% methanol for 5 minutes, then pour off
4. Flood with 10% buffered Giemsa (pH7.4) for 10 minutes
5. Wash carefully using tap water
6. Dry using filter paper

2.2.2.1.2. Toluidine staining

Toluidine staining (Fig. 3) is applied to semi thin sections of epoxy embedded tissue (section 2.2.2.2). The stained sections can be used to determine the area of interest for further processing, but also for LM observations. Toluidine stock solution is obtained from commercial sources.

The 1% toluidine blue and 2% borate in distilled water needed is obtained by mixing:

- 1 g toluidine blue O
- 2 g sodium borate (borax)
- 100 ml distilled water
1. Dissolve the sodium borate in the water
2. Add the toluidine blue powder
3. Stir until dissolved
4. Filter the stain solution (use syringe filter) before use

Note: The borax makes the stain alkaline so it will help penetrating to the epoxy sections

Staining process
1. Cut semi-thin sections at 0.5 - 1.0 μm
2. Transfer sections to a drop of distilled water on a glass slide
3. Dry sections on a glass slide on a slide warmer or over a 40 W lamp

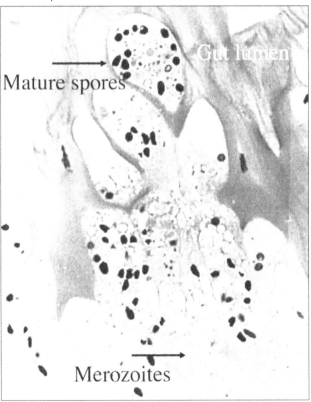

Fig. 3. Toluidine staining of a semi thin section of epoxy embedded ventricular tissue.

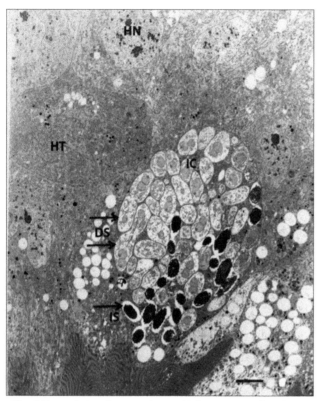

Fig. 4. Transmission electron micrograph of *N. ceranae* infected tissue: host nucleus (HN); healthy tissue (HT); infected cell (IC); dividing stages (DS); immature spore (IS). Bar= 10 mm.

4. Cover section with staining solution (with the heat source still on) for 1-2 minutes
5. Rinse off excess stain with distilled water and air dry

2.2.2.2. Transmission electron microscopy (TEM)

Several fixation procedures are available for studies of microsporidia infections. The following methodology has been widely used with good results for both *N. apis* (Fries, 1989; Fries *et al.*, 1992) and *N. ceranae* (Fries *et al*, 1996, Fig. 4):

1. Prefix tissue specimens for transmission electron microscopy using 4%. glutaraldehyde (v/v) in 0.067 M cacodylate buffer, pH 7.4, for 3 days to three weeks.
2. Keep material refrigerated (+7°C) during prefixation.
3. Wash in cacodylate buffer.
4. Post fix for 2 hours in 2% OsO4 (w/v) in 0.1 M S-colloidine buffer.
5. Dehydrate through ethanol series at room temperature:
   5.1. 5 min. in 30% EtOH,
   5.2. 5 min. in 50% EtOH,
   5.3. 5 min. in 75% EtOH,
   5.4. 5 min. in 95% EtOH,
   5.5. 3 X 10 min. in 100% EtOH,
   5.6. 3 X 10 min. in 100% propylene oxide.
6. Embed in epoxy resin (Agar 100) by routine procedures for electron microscopy.

The number of polar filament coils is one tool that helps to differentiate between species of *Nosema* (Burges *et al.*, 1974). In *N. ceranae*, the number of filament coils varies between 20 and 23 in mature spores (Fries *et al.*, 1996), whereas the number of polar filament coils in spores of *N. apis* is always larger and often more than 30 (Fries, 1989). The immature spores, where the filament is still developing, can be distinguished from mature spores on the less developed spore wall.

2.2.2.3. Molecular detection of Nosema spp. (N. apis , N. ceranae and N. bombi)

In addition to the microscopic techniques described above, various molecular methods have been developed for the detection and identification of *Nosema* spp. because such molecular methods are more sensitive and species specific (Klee *et al.*, 2006; 2007). As a consequence, it is possible to confirm the presence of *Nosema* (and other disease organisms) in bees using molecular techniques even when visual detection suggests its absence. Care must be exercised, though, in interpretation of results from molecular detection. *Nosema* may be detected molecularly in a bee, though very few spores are present in the bee and the pathogen has little or no impact on its host. In addition, vegetative forms of *Nosema* as well as spores can be detected using molecular methods, whereas only spores can be detected using visual methods. A further advantage of molecular methods, is that their extreme sensitivity of detection may provide insights into hitherto unknown modes of transmission of pathogens.

The molecular techniques developed for detection of *Nosema* spp. in bees (i.e. *N. apis*, *Nosema bombi*, found to date only in bumble bees, and *N. ceranae*) are usually PCR-based (i.e. uniplex or multiplex PCR, PCR-RFLP, qPCR; see the *BEEBOOK* paper on molecular methods (Evans *et al.*, 2013), and a wide range of species-specific PCR primer sets for these *Nosema* species can be found in the literature (Table 1). A test of the specificity and detection limits of nine of these primer sets suggests that some of them may lack specificity or exhibit low sensitivity (Erler *et al.*, 2011). In addition, the use of different molecular methods or conditions across laboratories can lead to inconsistencies. For this reason, it is recommended that PCR-based screening protocols be optimized and adjusted to fit each individual laboratory's conditions, research and monitoring questions. To allow comparisons across laboratories, we recommend analysis of the same homogenates of infected bees in each laboratory to account for differences in sensitivity, or for threshold sensitivities of detection to be reported per laboratory in terms of minimum number of spores per bee that can be detected by molecular markers.

Whilst most primers are designed for conventional PCR, real time PCR (qPCR or quantitative PCR, see the *BEEBOOK* paper on molecular methods (Evans *et al.*, 2013)) primers and protocols for quantification of *N. ceranae* and *N. apis* have also been developed (Table 1), including primers that quantify both species in one reaction (Martín-Hernández *et al.,* 2007). As for standard PCR, primer sets for

Table 1. List of primer sets available for the detection of *Nosema* spp. in bees by PCR.

Name	Source		Primer sequence (5'-3')	locus	use	N.a.	N.b.	N.c.
218MITOC	Martín-Hernández *et al.* 2007	fwd rev	CGGCGACGATGTGATATGAAAATATTAA CCCGGTCATTCTCAAACAAAAAACCG	SSU rRNA	qPCR			218-219
321APIS	Martín-Hernández *et al.* 2007	fwd rev	GGGGGCATGTCTTTGACGTACTATGTA GGGGGGCGTTTAAAATGTGAAACAACTATG	SSU rRNA	qPCR	321		
BOMBICAR	Plischuk *et al.* 2009	fwd rev	GGCCCATGCATGTTTTTGAAGATTATTAT CTACACTTTAACGTAGTTATCTGCGG	SSU rRNA	PCR		101	
ITS	Klee *et al.* 2006	fwd rev	GATATAAGTCGTAACATGGTTGCT CATCGTTATGGTATCCTATTGATC	ITS region	PCR	120	120	120
N.b.a	Erler *et al.* 2011	fwd rev	TGCGGCTTAATTTGACTC GGGTAATGACATACAAACAAAC	SSU rRNA/ITS	PCR		511	
Nbombi-SSU-J	Klee *et al.* 2006	fwd rev	CCATGCATGTTTTTGAAGATTATTAT CATATATTTTTAAAATATGAAACAATAA	SSU rRNA	PCR		323	
NOS	Higes *et al.* 2006	fwd rev	TGCCGACGATGTGATATGAG CACAGCATCCATTGAAAACG	SSU rRNA	PCR	240		252
NosA	Webster *et al.* 2004	fwd rev	CCGACGATGTGATATGAGATG CACTATTATCATCCTCAGATCATA	SSU rRNA	PCR	209		
SSU-res	Klee *et al.* 2007	fwd rev	GCCTGACGTAGACGCTATTC GTATTACCGCGGCTGCTGG	SSU rRNA	PCR	402	402	402
NaFor	Forsgren and Fries 2010	fwd(a)	CTAGTATATTTGAATATTTGTTTACAATGG [b]	LSU rRNA	qPCR	278		
NcFor		fwd(c)	TATTGTAGAGAGGTGGGAGATT					316
UnivRev		Urev	GTCGCTATGATCGCTTGCC					
Nosema	Chen *et al.* 2008	fwd rev	GGCAGTTATGGGAAGTAACA GGTCGTCACATTTCATCTCT	SSU-rRNA	generic	208		212
N. ceranae	Chen *et al.* 2008	fwd rev	CGGATAAAAGAGTCCGTTACC TGAGCAGGGTTCTAGGGAT	SSU-rRNA	PCR			250[a]
N. apis	Chen *et al.* 2008	fwd rev	CCATTGCCGGATAAGAGAGT CACGCATTGCTGCATCATTGAC	SSU-rRNA	PCR	401[a]		
Nos-16S	Stevanovic *et al.* 2011	fwd rev	CGTAGACGCTATTCCCTAAGATT CTCCCAACTATACAGTACACCTCATA	SSU rRNA	PCR	488		488
Mnceranae-F	This report	fwd	CGTTAAAGTGTAGATAAGATGTT	SSU rRNA	PCR			
Mnapis-F		fwd	GCATGTCTTTGACGTACTATG					143
Mnbombi-F		fwd	TTTATTTTATGTRYACMGCAG				171	
Muniv-R		Urev	GACTTAGTAGCCGTCTCTC				224	
SSUrRNA-f1b	Tay *et al* 2005	Ufwd	CACCAGGTTGATTCTGCCT	SSU rRNA	generic		ca.	
SSUrRNA-r1b		Urev	TGTTCGTCCAGTCAGGGTCGTCA					

ITS: internal transcribed spacer region; SSU: small subunit rRNA (16S rRNA); *N.a.*: *Nosema apis*; *N.b.*: *Nosema bombi*; *N.c.*: *Nosema ceranae*.
[a]Fragment size could not be verified. [b]Sequence modified to complement original GenBank entry U97150.
Use: PCR, standard PCR (for the detection of different *Nosema* spp.); qPCR, for quantitative or real time PCR (for the quantification of different *Nosema* spp.) and standard PCR (for the detection of different *Nosema* spp.); generic, primers amplify all known *Nosema* spp. or all Microsporidia without differentiating among species.

qPCR need to be tested in each laboratory for sensitivity and reliability (see also Bourgeois *et al.*, 2010; Burgher-MacLellan *et al.*, 2010; Hamiduzzaman *et al.*, 2010; Traver and Fell, 2012 and the molecular paper of *BEEBOOK* by Evans *et al.*, 2013).

Here we report the use of a multiplex PCR-based method that is able to detect and differentiate simultaneously the three *Nosema* species of high prevalence in European bee populations (*N. apis*, *N. bombi* and *N. ceranae*) using genomic DNA. Microsporidia from genera other than *Nosema* are known from bees, where they can be very abundant (see Paxton *et al.*, 1997; Li *et al.*, 2012). To capture all of these Microsporidia, it is advisable to PCR amplify microsporidian DNA using 'generic' primers (see Table 1) and then sequence PCR products, which can be laborious and expensive. Here we present a much faster and cheaper method in which we combined multiple primers based on the 16S ribosomal rRNA gene into a single reaction to simultaneously detect *N. apis*, *N. bombi* and *N. ceranae* in bees.

2.2.2.3.1. *Nosema DNA extraction*

DNA extraction can be performed from specific tissue (e.g. ventriculus, rectum, fat body), subdivided bee sections (e.g. metasoma), whole bees, or homogenates from pooled samples (see the *BEEBOOK* paper on bee anatomy and dissection (Carreck *et al.*, 2013)).

Nosema DNA extraction from bee homogenates:

1. Crush fresh or flash frozen tissue to generate a homogeneous homogenate of bee/bee guts. For example,
   1.1. place a maximum of 30 bees in a filter grinding bag (e.g. extraction bags from BIOREBA AG, Switzerland).
   1.2. Add 0.5 ml (DNAase/RNAase free) ddH$_2$O per bee.
   1.3. homogenize the mixture using a homogenizer (e.g. Homex6 from BIOREBA AG; Switzerland).
   1.4. Flash-freeze in liquid nitrogen is possible prior to homogenization to aid in mechanically breaking open cells.
   1.5. Crush the sample.

Without access to a robot, one can use a pestle to crush the bee tissue.

2. Transfer 100 µl of the liquid homogenate into a microcentrifuge tube and centrifuge for 3 min at 16,100 g to precipitate the microsporidia and other cellular material.
3. Discard the supernatant.
4. Freeze the pellet by using liquid nitrogen.
5. Crush using a pestle until pulverized (in order to break open *Nosema* spore walls).
6. Repeat steps 4 and 5 two or three times so that *Nosema* DNA goes into solution.
7. Use the DNeasy® Plant Mini kit protocol (Qiagen) following the Mini protocol for plant tissue to extract DNA from the homogenate.

Other non-proprietary DNA extraction protocols (e.g. those using phenol/chloroform; chelex resin; Tay *et al.* 2005) were used with poor success, possibly because bee guts contain plant secondary compounds and tissue may be in a state of decay due to poor preservation. Research on other extraction techniques (e.g. using CTAB) is needed to provide cheaper yet efficient methods for the extraction of *Nosema* DNA from bees.

8. Complete the final elution step in 100 µl of 0.01M Tris (pH 7.5) buffer.

The same Qiagen protocol can also be implemented in the QiaCube (Qiagen) for automated DNA extraction.

2.2.2.3.2. Multiplex PCR for detection of *Nosema* and differentiation between *N. apis*, *N. bombi* and *N. ceranae*

For multiplex PCR amplification of partial 16S rRNA (= SSU rRNA) gene fragments, we recommend the following primer combination, though others from Table 1 (standard or qPCR primers) may be more suitable for different purposes and in different laboratories:

Primers were designed based on alignment of all available sequence data in GenBank of the 16S rRNA gene from *N. apis*, *N. bombi* and *N. ceranae*.

Mnapis-F   forward primer: 5'-GCATGTCTTTGACGTACTATG-3'
Mnbombi-F   forward primer: 5'-TTTATTTTATGTRYACMGCAG-3'
Mnceranae-F   forward primer: 5'- CGTTAAAGTGTAGATAAGATGTT-3'
Muniv-R: reverse primer: 5'- GACTTAGTAGCCGTCTCTC-3'

Note that the Mnbombi-F primer contains variable sites to account for the sequence diversity observed for this species.

PCR product size:
- for *N. ceranae*: 143 bp
- for *N. bombi*: 171 bp
- for *N. apis*: 224 bp

2.2.2.3.2.1. PCR reaction mix
- 1 µl of DNA extract (ca. 1 ng)
- 0.5 U of GoTaq® polymerase (Promega)
- 2x GoTaq® reaction buffer (3mM $MgCl_2$ final concentration, Promega)
- 0.3 mM of each dNTP (dNTP mix from Promega)
- 0.4 µM of Mnceranae F
- 0.4 µM of MnapisF
- 0.5 µM of Mnbombi-F
- 0.5 µM of Muniv-R
- H_2O as required, to make up to a 10 µl total volume.

Amplification is carried out on a thermocycler (e.g. TProfessional Biometra) using the following conditions:

1. Initial denaturation step of 95°C for 2 min,
2. 35 cycles of (95°C for 30 s, 55°C for 30s and 72°C for 60 s),
3. Final extension step of 72°C for 5 min.

Each laboratory might have to optimise *de novo* primers, protocols and PCR conditions.

2.2.2.3.2.2. Visualization
1. Resolve the amplification products
2. Visualize in a QIAxcel electrophoresis system with a QIAxcel DNA high resolution kit (QIAGEN)
3. Analyse using the QIAxcel ScreenGel software (v1.0.0.0).

The resolution method is:
- OM700 (3-5 bp),
- QX DNA Size Marker: 25 bp-450 bp,
- QX Alignment Marker: 15 bp/400 bp.

Typical results are presented in Fig. 5.

As a cheaper alternative, PCR products can be resolved in a 1-2% agarose gel with suitable size marker and then visualised by staining with ethidium bromide and photographing on a u/v transilluminator (see the *BEEBOOK* paper on molecular methods (Evans *et al.*, 2013).

2.2.2.3.2.3. Controls
Attention needs to be given to the use of controls when undertaking DNA extractions and PCR amplifications to avoid false positives (detection of a band of the appropriate size when a *Nosema* species was not present) and false negatives (absence of a band of the appropriate size because of poor extraction or poor PCR set-up despite a *Nosema* species being present). We recommend using newly emerged honey bees as 'negative controls' because such bees are not infected with *N. apis* or *N. ceranae* at emergence although for

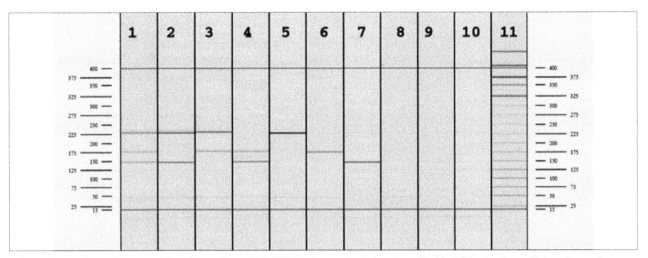

Fig. 5. PCR products of *Nosema*-infected bees from which DNA has been extracted and amplified by PCR using the multiplex primer mix Mnapis F, Mnbombi F, Mnceranae F and Muniv-R to reveal *N. apis*, *N. bombi* and *N. ceranae*: 1. *N. apis* + *N. bombi* + *N. ceranae*; 2. *N. apis* + *N. ceranae*; 3. *N. apis* + *N. bombi*; 4. *N. ceranae* + *N. bombi*; 5. *N. apis*; 6. *N. bombi*; 7. *N. ceranae*; 8. Control, no template; 9. *Apis mellifera* negative control; 10. *Bombus* spp. negative control; 11. size ladder.

N. ceranae vertical transmission of infection has recently been suggested (Traver and Fell, 2012). Greater care needs to be given to negative controls for *N. bombi* in bumble bees as this microsporidian may be transmitted transovarially from queen to offspring (Rutrecht and Brown, 2008). For positive controls, it is best to use adult bees that have been experimentally infected with spores and to confirm visually that spores are present in the homogenate before DNA extraction.

2.2.2.3.3. Realtime PCR for quantification of N. apis and N. ceranae

Molecular quantification of *N. apis* and *N. ceranae* allows both relative quantification of the two *Nosema* species or their absolute quantification per bee or per sample, which can be of considerable interest for studies of the interactions between these two pathogens. However, molecular quantification requires use of a real-time PCR machine; these machines are currently relatively expensive and they require calibration with dilution series of *N. apis* and *N. ceranae* to generate accurate quantification.

Several realtime PCR machines are on the market, each employing different fluorophore chemistries for molecular quantification. Here we present a method based on Forsgen and Fries (2010) that uses a BioRad MiniOpticon real-time PCR machine and EvaGreen chemistry for quantification as it has functioned well in several laboratories.

2.2.2.3.3.1. PCR reaction mix

PCRs use a 20 µl final volume and should be set up as follows for the quantification of either *N. apis* or *N. ceranae* using primers from Forsgen and Fries (2010) in Table 1. Set up one mastermix as below for quantification of *N. apis* for each sample and set up another

mastermix as below for quantification of *N. ceranae* for each sample, using either NaFor or NcFor respectively.

- SsoFast™ EvaGreen® Supermix (BioRad)   1 x
- NaFor or NcFor primer   0.4 µM
- UnivRev primer   0.4 µM
- H_2O 6.4 µl
- Sample DNA extract   2 µl (ca. 2 ng DNA)

2.2.2.3.3.2. Quantification

Amplification and data acquisition are carried out in a MiniOpticon® (Bio-Rad) real-time PCR machine using the following program:

1. Initial enzyme activation step:
   a. 98°C for 15 min
2. Followed by 40 cycles of:
   a. denaturation at 98°C for 5 sec
   b. annealing/extension at 63°C for 10 sec
   c. melt curve analysis from 65-95°C (in 0.5°C increments) 10 sec/step

Specificity and the absence of non-specific amplification are determined based on the melting temperature (*T*m) of the amplified products (see also molecular methods paper of the *BEEBOOK* (Evans *et al.*, 2013)).

2.2.2.3.3.3. Controls

For controls, the criteria and warnings provided previously for standard PCR for the detection of *Nosema* spp. are applicable (section 2.2.2.3.2.3). For quantification, *N. apis* and *N. ceranae* recombinant amplicons should be used as external standards. Set up a standard curve using serial dilutions of recombinant target DNA fragments ranging from 10^{-2} to 10^{-8} and include them as quantification standards in every PCR run (see also the *BEEBOOK* paper on molecular methods (Evans *et al.*, 2013)).

2.2.3. Standardising cage trials

This section contains information on methodologies specifically for performing laboratory investigations of the adult European honey bee-*Nosema* system. Additional details on general methodologies for maintaining European honey bees in the laboratory are described by Williams *et al.* (2013) in the laboratory cages paper of the *BEEBOOK*.

2.2.3.1. Source of bees

General details on selecting bees and colonies for experiments can be found in the *BEEBOOK* papers by Williams *et al.* (2013) and Human *et al.* (2013). Details specific to *Nosema* investigations are discussed here. See the *BEEBOOK* paper on maintaining bees in cages (Williams *et al.*, 2013) for information on how to obtain honey bees for experiments.

For every experiment state:

1. The subspecies of European honey bee used because of there may be genetic variation for disease resistance (Evans and Spivak, 2010).
2. The time of year the experiment was performed because bee physiology can differ seasonally, in particular between summer and winter (Fluri *et al.*, 1977), but also when brood rearing declines (Amdam *et al.*, 2009).

Further considerations:

1. Cage replicates should be performed during the same season to allow for easier comparison of data.
2. Bees from multiple colonies should be used to ensure that the particular question being asked is relevant to honey bees in general (see Human *et al.* (2013) and Williams *et al.* (2012)).
3. Use a sufficient number of replicate cages per treatment (see Human *et al.* (2013) and Williams *et al.* (2012)).

Bees from all colonies should be homogenized among cages to eliminate effect of 'colony', thereby leaving only 'cage' as a random factor.

2.2.3.1.1. Source colonies

Colonies should be 'healthy', and quantitatively and qualitatively assessed, as described above and elsewhere in the *BEEBOOK* (the paper on estimating colony strength (Delaplane *et al.*, 2013)), to demonstrate zero or low infection of major pathogens or parasites, including *N. apis*, *N. ceranae* (this paper), and the mite *Varroa destructor* (see the *BEEBOOK* paper on varroa (Dietemann *et al.*, 2013)). Because *Nosema* interacts with pesticides (Alaux *et al.*, 2010a), colonies likely to be exposed to high levels of agro-chemicals, such as

those pollinating agricultural crops, should not be used. *Nosema* also interacts with viruses (Bailey *et al.*, 1983; Costa *et al.*, 2011); but because colonies with asymptomatic viral infections are nearly ubiquitously distributed (e.g. Tentcheva *et al.*, 2004; Williams *et al.*, 2009), colonies without symptomatic individuals will suffice. If possible, titre levels of common honey bee viruses, in particular deformed wing and black queen cell viruses, should be quantitatively assayed in bees from potential source colonies, as described by de Miranda *et al.* (2013) in the virus paper of the *BEEBOOK*.

2.2.3.1.2. Age of bees

The age when honey bees are inoculated with *Nosema* spores may also influence parasite development and virulence due to changes in honey bee immune response (Amdam *et al.*, 2005) and morphology (Rutrecht *et al.*, 2007) as bees grow older. However, no such influence has been studied for *Nosema* spp. to our knowledge, and handling of very young bees for spore inoculation may damage the bees and reduce longevity. It is generally advisable to initiate inoculations when the bees are a few days old (2-5 days).

2.2.3.2. Type of cages

Numerous cage designs exist for maintaining honey bees in the laboratory and for performing experiments (see Williams *et al.* (2013) in the *BEEBOOK*). Despite this diversity, it is necessary that cages meet basic criteria described by Williams *et al.* (2013).

- Vital for *Nosema* studies, cages must be used once and discarded, or sterilized if used multiple times, to prevent contamination by *Nosema* spores. Multiple-use cages should be made from materials that are easily autoclaveable such as stainless steel and glass because spores of *N. ceranae* in PBS can only be confidently destroyed by exposure to 121°C for 30 minutes (Fenoy *et al.*, 2009). Dry sterilisation of wooden cages (*i.e.*, 80°C for 1 hr.) destroys viability of *N. apis* spores (Cantwell and Shimanuki, 1969); this method is also probably effective for *N. ceranae*, although it remains to be verified.

- To reduce risk of contamination by *Nosema* spores among individuals differentially treated and maintained in the same incubator, cages should be placed sufficiently apart. If screens or holes are used to provide ventilation, they should face in opposite directions.

Additional materials, such as comb (e.g. Czekońska, 2007) and plastic strips for releasing queen mandibular pheromone (e.g. Alaux *et al.*, 2010a), are sometimes used during laboratory experiments. Although queen mandibular pheromone probably promotes honey bee health and reduces stress of the caged individuals, its effect on *Nosema* development is not understood and therefore its use should be avoided until our knowledge is improved.

2.2.3.3. Type of food

Because diet can affect honey bee longevity (Schmidt *et al.*, 1987), immune response to *Nosema* infection (Alaux *et al.*, 2010b), as well as spore development (Porrini *et al.*, 2011), it is important to carefully consider food provided to experimental bees. Generally, researchers should attempt to maintain their honey bees as healthy as possible.

Honey bees are capable of surviving for long periods of time on 50% (weight/volume) sucrose solution (Barker and Lehner, 1978); however, to ensure normal development of internal organs and glands (Pernal and Currie, 2000), as well as proper immune response (Alaux *et al.*, 2010b), supplementing a strict carbohydrate diet with protein, and even nutrients such as vitamins and minerals, is recommended for maintaining honey bees in the laboratory. Bee-collected pollen provides an adequate medium for providing protein and nutrients (Brodschneider and Crailsheim, 2010); however, such material may be contaminated with *Nosema* spores (Higes *et al.*, 2008b) or pesticides (Pettis *et al.*, 2012). Additionally, it is possible that pollen may stimulate *N. ceranae* development by promoting bee health (Porrini *et al.*, 2011).

Therefore, in addition to *ad libitum* 50% (weight/volume) sucrose solution (i.e. 100 g table sugar dissolved in 100 ml tap water), individuals should be provided with an easily accessible source of multi-floral, radiation sterilised, bee-collected pollen *ad libitum* as described by Williams *et al.* (2013) and Human *et al.* (2013) in the *BEEBOOK*.

Further studies are needed to investigate the effects of commercial pollen substitutes on *Nosema* development and individual bee health before they should be considered as a replacement for bee collected pollen.

To sterilize for *N. apis* spores, pollen can be exposed to $\geq 0.2 \times 10^6$ rads gamma radiation from cobalt-60 (Katznelson and Robb, 1962) or heat treated at 49°C for 24 hours (Cantwell and Shimanuki, 1969). On the other hand, very little is known about the factors responsible for making *N. ceranae* spores non-viable in bee products. *N. ceranae* will lose viability during freezing (Forsgren and Fries, 2010), but it is more resistant to heat than its congener (Fenoy *et al.*, 2009). It is likely that temperatures and/or exposures higher than required for *N. apis* spore destruction will also render *N. ceranae* spores non-viable. As a result, a combination of heating and freezing pollen may be possible to develop as an alternative to radiation for sterilising bee-collected pollen of *Nosema* spores. Specific protocols need to be developed for this purpose because to date we only know that one week of freezing kills approximately 80% of *N. ceranae* spores (Fries, 2010), and we do not know what temperatures will reduce the nutritive value of pollen.

2.2.3.4. Incubation conditions

Researchers should attempt to maintain their experimental bees in optimal conditions for both host and parasite, and should consider the possible effects of growth chamber conditions, as environmental conditions can have a large influence on both host susceptibility and parasite virulence (e.g. Kraus and Velthuis, 1997; Ferguson and Read, 2002; McMullan and Brown, 2005). Although few data exist on the effects of specific temperatures on *Nosema* parasitism in honey bees, it is clear that intra-host development of both parasites in European honey bees, as well as spore viability, can be affected by temperature (e.g. Malone *et al.*, 2001; Fenoy *et al.*, 2009; Martín-Hernández *et al.*, 2009; Fries, 2010; Higes *et al.*, 2010).

Adult workers should be maintained in complete darkness at 30°C and approximately 60-70% RH in a growth chamber or incubator with adequate ventilation. A data logger should be used to record both temperature and relative humidity within the incubator during the course of each experiment. These data will ensure adequate conditions were maintained, and may explain deviations from expected results during changes in incubator conditions as a result of mechanical problems or changes in ambient conditions. See the *BEEBOOK* chapter on maintaining workers in cages (Williams *et al.*, 2013) for more details.

2.2.4. Inoculation methods

2.2.4.1. Spore source

As with other parasites (Ferguson and Read, 2002), including *N. bombi* in bumble bees (*Bombus* spp.) (Tay *et al.*, 2005) and bacterial diseases in honey bees (Genersh *et al.*, 2005; Charriere *et al*, 2011), it is possible that genetic variants detected in both *N. apis* and *N. ceranae* (Williams *et al.*, 2008; Chaimanee *et al.*, 2010; Sagastume *et al.*, 2011) may at least partially explain differences in host susceptibility and parasite virulence. Future studies should seek to identify genes responsible for *Nosema* epidemics (Chen and Huang, 2010). Based on differences in reported pathology of *Nosema* species around the world, researchers should state the region and country spores originated from. It is recommended that spores be sourced from multiple colonies.

When creating inoculums, it is important to use fresh spores because their viability is lost over time when spore suspensions are stored. In particular, this is true for *N. ceranae* isolates, which rapidly lose viability in the refrigerator and almost completely lose infection capacity after freezing of spores (Fenoy *et al.*, 2009; Fries, 2010). See section 2.2.5. for viability test procedures. Group feeding of caged bees using crushed bees infected with the respective *Nosema* spp. in sugar solution is a good way to propagate spores for experiments (see sections 2.2.4.4. and 2.2.4.5. for details on individual and group feeding, respectively). After 10-12 days, spore-inoculated bees from cages maintained in appropriate conditions (section 2.2.3.4) can be extracted and the ventriculi used for preparations of spore suspensions according to methods described previously (section 2.2.1 and 2.2.4.2). Molecular assays previously described (section 2.2.2.3) should be employed to ensure the proper species of spore is used for inoculations because mixed infections occur, even in naturally-infected individual bees (Burgher-MacLellan *et al.*, 2010). Prepared spore suspensions can be used the next day when kept in sugar solution in the refrigerator (section 2.2.4.2.3).

2.2.4.2. Spore suspension

A suspension of spores obtained as described previously (e.g. 1 infected bee ventriculus in 1 ml water; section 2.2.1.) can be filtered using sterile 74 μm sized mesh to remove large pieces of host materials. After filtration, 1-5 mM pH 9.0-buffered ammonium chloride (NH₄Cl) can be added to the filtrate to inhibit spore germination.

However, sometimes a high purity of spores is needed. Purification can remove unwanted host tissues and microbial contaminants that may confound experimental data, and can also facilitate accurate microscopic counting.

2.2.4.2.1. Purification

The purpose of purifying a spore suspension is to remove unwanted host tissues and microbial contaminants that may confound experimental data. A high purity of spores can facilitate accurate microscopic counting and is an important quality assurance parameter in pathological studies. A number of purification methods describe below give high purity preparations, and the choice of purification method depends upon the specific requirements and applications of spore suspension. For *in vivo* feeding assays, it is sufficient to purify spores using filtration and centrifugation techniques. Alternatively, methods of triangulation or density gradient purification, in addition to initial filtration, are recommended for experiments that require a very high level of spore purity. Note that amount of water added to resuspend the pellet will affect the spore concentration and the final concentration of the inoculum must therefore be checked using a haemocytometer (see section 2.2.1.1. and the *BEEBOOK* paper on miscellaneous methods (Human *et al.*, 2013).

2.2.4.2.1.1. Centrifugation

1. Centrifuge spore suspension at 5,000 G for five minutes to produce a pellet of spores.
2. Discard supernatant containing tissue debris that is lighter than spores.
3. Resuspend the pelleted spores in distilled water by vortexing for five seconds.

Repeat the centrifugation 2-3 times to wash spores and to create a *Nosema* spore suspension with over 85% purity level.

2.2.4.2.1.2. Triangulation

1. Centrifuge spore suspension at 300 G for five minutes to form a spore pellet which contains two strata, the upper supernatant fluid and the lower stationary phase containing most of spores.
2. Transfer supernatant fluid to another tube using a pipette, and resuspend lower stationary phase using sterile water.

3. Centrifuge supernatant for five minutes at 300 G to pellet the spores, and again transfer supernatant fluid to another tube and resuspend the lower stationary phase.
4. Repeat procedure three times.
5. Combine resuspended lower strata created from each centrifugation to yield a spore suspension with a purity greater than 99% (Cole, 1970).

2.2.4.2.1.3. Density gradient

The most commonly used density gradient substrates used for microsporidian spore purification are sucrose, sodium chloride, cesium chloride, and two silica colloids: Ludox HS40 and Percoll. The density gradient can be continuous or discontinuous (i.e. stratified), heaviest at the bottom and lightest at the top, thus allowing purified particles to be located at specific regions of the density gradient. Among all density gradient media, Percoll is frequently used for *Nosema* spore purification because it offers many advantages, including: 1. Ease of preparation; 2. Low viscosity, permitting rapid sedimentation at low speed centrifugation; 3. Low osmolarity and no toxicity; 4. Excellent stability under any autoclaving sterilization conditions (Fig. 6).

Purification procedure with Percoll substrate

1. Gently overlay the *Nosema* spore suspension on a discontinuous Percoll (Sigma-Aldrich; St. Louis, USA) gradient consisting of 10 ml each of 25, 50, 75, and 95% Percoll solution from top to bottom, respectively.
2. Centrifuge the column at 10,000 G for 20 minutes at 4°C.
3. Discard the supernatant.
4. Re-suspend the pellet in sterile water.
5. Repeat the process 3 times.
6. Overlay the spore suspension on 20 ml 100 % Percoll solution.
7. Centrifuging the column at 3,000 G for 2-3 minutes to pellet the lighter cellular debris and leave the spores on the top of Percoll solution.
8. Dilute the 100% Percoll solution by adding the equal amount of water and pellet the spores by centrifugation at 10,000 G for 15 minutes.
9. Re-suspend the spore pellet in 2-5 ml sterile water to produce a spore suspension with a purity of greater than 99% (Chen *et al.*, 2009).

2.2.4.2.2. Spore suspension concentration

Dilution of the purified spore suspension to the desired concentration can be performed using the formula: $C_1V_1 = C_2V_2$ where: C_1 = initial concentration; V_1 = initial volume; C_2 = final concentration; V_2 = final volume.

For example, if you want to feed 10,000 *Nosema* spores in 10 μl 50% (weight/volume) sucrose solution, you must, as determined by proportions, create a final spore concentration of 1,000,000 spores / ml

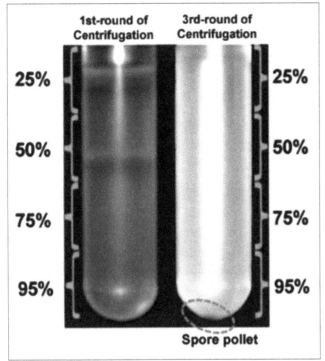

1st-round of Centrifugation | **3rd-round of Centrifugation**

25% | 25%
50% | 50%
75% | 75%
95% | 95%

Spore pollet

Fig. 6. The Percoll gradient will be constructed by layering 95%, 75%, 50% and 25% Percoll solution from bottom to top in a 50 ml ultracentrifuge tube. The spore suspension will be overlayed onto the gradient and centrifuged. After centrifugation, the supernatant will be removed and the spore pellet will be suspended in sterile water. After two or three times purification on Percoll gradient, the resulting pellet will contain very pure spores.

of 50% (w/v) sucrose solution. To create 10 ml of this final spore suspension in 50% sucrose solution when your initial spore suspension is: Example 1. 10,000,000 spores/ml water, then 1 ml of initial spore suspension must be added to 9 ml 55.6% (w/v) sucrose solution. See below for specific details:

Calculating initial volume of spore-water suspension needed
1. $C_1V_1 = C_2V_2$
2. $V_1 = C_2V_2/C_1$
3. $V_1 = 1,000,000$ spores per ml X 10 ml / 10,000,000 spores per ml
4. $V_1 = 1$ ml

Calculating initial concentration of spore-free sucrose solution needed*
1. $C_1V_1 = C_2V_2$
2. $C_1 = C_2V_2 / V_1$
3. $C_1 = 50$ g sucrose per 100 ml water X 10 ml final sucrose solution / 9 initial sucrose solution
4. 55.56 g sucrose per 100 ml water

*An initial concentration of spore-free sucrose solution of 50% is generally adequate when a low volume of spore suspension is added

to create a final spore suspension in approximately 50% sucrose solution.

Example 2. 12,456,000 spores/ml water, then 0.8 ml of initial spore suspension must be added to 9.2 ml 54.3% (w/v) sucrose solution.

Control suspensions to infected control bees should be created using uninfected individuals from the same colonies the spores were sourced from.

2.2.4.2.3. Storage of stock suspension

It should be emphasised that for infection experiments, newly made up spore suspensions should be used, prepared the same day or the day before the experiment and kept in the refrigerator. For *N. apis*, storage of spores for later preparations of new suspensions can be made by freezing adult infected bees or already prepared spore suspensions. For *N. ceranae* the situation is drastically different since the spores rapidly loose viability if frozen (Fenoy *et al.*, 2009; Fries, 2010). A protocol needs to be developed whereby infective spores of *N. ceranae* can be reliably stored over longer periods without serious loss of viability.

2.2.4.3. Handling of bees

Bee caste, activity level, and personal comfort level will dictate what devices are used to manipulate bees. Most importantly, bees should not be crushed, so particular attention must be placed on ensuring sensitive and more fragile parts of the bee such as the abdomen, antennae, eyes, and mouthparts, are not disturbed or damaged. A honey bee that is damaged or has stung should be replaced.

In most cases, use of anaesthesia to handle honey bees is not required, and generally not recommended because of the difficulty of ensuring a consistent dosage is given to bees and because of possible effects on the behaviour, physiology, and development of insects (Czekonska, 2007; Ribbands, 1950; Rueppell *et al.*, 2010). If anaesthesia is used, a consistent dosage should be provided to all experimental bees. Further details on handling and anaesthetising bees are provided in the *BEEBOOK* chapter on miscellaneous methods Human *et al.* (2013).

2.2.4.4. Individual feeding

Many *Nosema* laboratory experiments individually feed bees with an inoculum to ensure that bees are exposed to a known quantity of spores (e.g. Higes *et al.*, 2007; Paxton *et al.*, 2007; Maistrello *et al.*, 2008; Forsgren and Fries, 2010). The volume of the inoculum can be 5 to 10 ml, amounts that are readily consumed by bees. Individual feeding of spores produces significantly lower variation in response infection level compared to group feeding of spores (Furgala and Maunder, 1961) and should be preferred for most experimental purposes until further investigations comparing group versus individually fed honey bees occurs. As with other laboratory procedures concerning *Nosema*, it is important to minimize chances of contamination by setting up sterile feeding stations unique to each treatment group. To individually inoculate honey bees with *Nosema* spores, refer to the *BEEBOOK* chapter on maintaining bees *in vitro* (Williams *et al.*, 2013).

Specifically for *Nosema* studies, experimental bees should be starved in hoarding cages for two to four hours by removing all feeding devices to ensure that the entire inoculum is ingested quickly (e.g. Fries *et al.*, 1992; Malone and Stefanovic, 1999; Higes *et al.*, 2007; Maistrello *et al.*, 2008). Bees destined for each treatment group should be maintained in separate hoarding cages to avoid the possibility of contamination during the removal of individuals for feeding. Because feeding can require considerable time depending on the number of bees to be inoculated and the number of persons available to feed spores, it may be necessary to house bees for each treatment group in multiple cages that can be starved in different time blocks to ensure that all individuals are without food for a similar period of time. Pre-trials will determine how much time is required for feeding, but generally it should take one to two minutes per bee. Additionally, and depending on the time required to feed all experimental bees, it may be prudent to rotate feeding amongst treatment groups so that, for example, not all bees from one treatment group are fed in the morning whereas bees from a second treatment group are fed in the afternoon. See previously discussed subsections within section 2.2.4., as well as section 2.2.7., for creating and choosing specific spore suspensions.

Once inoculated, bees can be placed into a 1.5 ml microcentrifuge tube with a breathing hole at the end in suitable growth chamber conditions in detail below for 20 to 30 minutes to ensure that spores are not transferred among bees (see Williams *et al.*, 2012) (Kellner and Jacobs, 1978; Verbeke *et al.*, 1984). Afterwards, the bee can be placed in an appropriate treatment hoarding cage. The provided description can be applied to inoculating all castes. However, drones and queens can be transferred to appropriate treatment cages immediately after feeding because they are only recipients during trophallaxis (Crailsheim, 1998). Future studies should determine whether this is required as some debate exists regarding the frequency that newly emerged bees in cages engage in trophallaxis (Crailsheim, 1998).

2.2.4.5. Group feeding

In contrast to inoculating individual bees with *Nosema* spores as described above, inoculation can also occur by providing a group of caged bees with a spore suspension in one or more common feeding devices that may be fed on *ad libitum*. This method allows for individuals to be inoculated relatively quickly, and without the logistical and time constraints associated with individual feeding. Although not well studied, the primary disadvantage of group feeding of *Nosema* spores is the greater variance of *Nosema* intensity among

caged individuals because of unequal distribution of the inoculum among individuals over time (Furgala and Maunder, 1961). Preliminary studies suggest group feeding is as effective at infecting caged bees as individual inoculation (Tanner *et al.*, 2012). Greater cage replicates per treatment group may be required if group feeding of inoculum is used. Group inoculation of a spore suspension containing 10,000 and 33,300 spores per bee is sufficient to infect caged individuals (Webster, 1994; Pettis *et al.*, 2012).

Further details on group feeding are provided in the *BEEBOOK* chapter on maintaining bees in cages (Williams *et al.*, 2013). For *Nosema*, for example, to mass inoculate 100 honey bees with approximately 10,000 *Nosema* spores per bee:

1. Provide 1,000,000 spores in 4 ml 50% (w/v) sucrose solution to guarantee that the entire volume will be consumed within approximately 24 hours.

This short time period will ensure a similar initial inoculation period for all individuals, and help prevent bacterial degradation of unconsumed spores from occurring.

2. Top up the feeder with 50% (w/v) sucrose solution when the inoculation solution is close to empty so that the caged honey bees do not go without food.

To ensure all spores are ingested, small volumes can be regularly added throughout the day until one is confident that most spores have been consumed.

When cultivating *Nosema* spores, live bees can be killed 10 to 14 days post-inoculation, approximately when a full *N. apis* infection is reached (Fries, 1988), using methods described by Human *et al.* (2013) in the *BEEBOOK*.

2.2.5. Viability control of Nosema spores

The viability of spores to be tested is rarely checked, although it is a crucial point to ensure the reliability and reproducibility of the results. In particular regarding *N. ceranae*, it can be considered as a key point because of the high sensibility to cold temperatures of *N. ceranae*, spores (Forsgren and Fries, 2010). *N. apis* spores also loose infectivity after freezing, but in a matter of years (Bailey, 1972) rather than a week as shown for *N. ceranae* (Fries, 2010).

Spore viability can be tested using three methods: colouration and infectivity tests (*in vivo* and *in vitro*). The advantage of colouration is that suspensions can be checked before use, *in vivo* tests are useful for confirmation of spore viability used for infection experiments.

2.2.5.1. Colouration test for spore viability

1. Add 50 ml of spores in H_2O at a concentration of 5×10^5 spores / ml to 1 mM of Sytox green (Molecular Probes, Inc.).
2. Incubate for 20 minutes at room temperature.
3. Centrifuge for 3 minutes at 1600 x g and discard the supernatant.
4. Homogenize the pellet 1.5.
5. in H_2O and centrifuge for 3 minutes at 1600 x g.

6. Discard the supernatant.

7. Add 100 ml 4′,6-diamidino-2-phenylindole (DAPI) at 2 mg/ml.

8. Incubate for 30 minutes in room temperature.

9. Centrifuge for 3 minutes at 1600 x g and discard the supernatant.

10. Homogenize the pellet in H_2O and centrifuge for 3 minutes at 1600 x g and discard the supernatant.

11. Add 50 ml H_2O and apply aliquots if 15 ml onto glass slides.

12. Allow to dry in room temperature and view under oil in a fluorescent microscope.

Dead spores are identified as yellow-green ovals through the 470- to 490-nm excitation wavelength filter, and living spores are coloured with turquoise ovals through the 395- to 415-nm excitation wavelength filter. To differentiate extruded spores not visible by either Sytox green or DAPI staining, white-light microscopy, where extruded polar filaments can be seen (Fenoy *et al.*, 2009). The colouration test for viability has not yet been evaluated with the ultimate test - *in vivo* tests in live bees.

2.2.5.2. In vivo test for spore viability

Experimentally, the viability of *N. apis* spores can be assessed by feeding suspensions with different spore concentrations to groups of newly emerged adult honey bees (see section 2.2.7.1). Honey bees exposed to spores can be collected 10 days post infection when the infection appears to be almost fully developed for both *Nosema* spp. (Higes *et al.*, 2007; Martin-Hernandez *et al.*, 2011; Forsgren and Fries, 2010) and the ventriculus examined for the presence of spores. If spores are seen in the light microscope, the bee is recorded as infected (Malone *et al.*, 2001) and the viability of the spores in the suspension used can be calculated based on the ID_{50} and the ID_{100} obtained (see section 2.2.7.1).

2.2.5.3. In vitro test for spore viability

In vitro germination of spores of both *Nosema* spp. can also be triggered by a procedure that mimics the natural conditions for the germination of environmental spores. Since the germination is the first step in the infection process, this test gives data on spore infectivity ability (Gisder *et al.* 2010b) but the *in vitro* test for spore viability has not yet been evaluated with the ultimate test - *in vivo* tests in live bees. See section 2.2.6. for details on *in vitro* rearing of *Nosema*.

2.2.6. In vitro rearing of Nosema spp.: cell culture systems

Cell and tissue cultures are indispensable for the propagation and study of obligate intracellular pathogens like viruses and microsporidia. Bee pathogens comprise viruses and microsporidia as obligate intracellular parasites. However, studies on cellular and molecular aspects of pathogen-host interactions of these pathogens

and their target cells have been hampered in the past by the complete lack of permanent bee cell lines. Recently, protocols for the prolonged although limited maintenance of primary honey bee cells have been described (Bergem *et al.*, 2006; Hunter, 2010), but these cells have not been used for infection experiments. Several hundreds of non-honey bee insect cell lines are commercially available (Lynn, 2007; van Oers and Lynn, 2010) and have been proven to be valuable tools for elucidating attachment, entry, and replication of several intracellular insect pathogens and for analysing cellular reactions towards infection (Smagghe *et al.*, 2009; van Oers and Lynn, 2010). However, these cell lines were considered unsuitable for the study of bee pathogens due to the assumed host specificity of bee pathogenic viruses and microsporidia. We here describe both, the infection of primary ventricular cells established from honey bee pupae as well as the infection of commercially available insect cell lines established from several lepidoptera with *N. ceranae* and *N. apis*. General techniques for cell cultures are described by Genersch *et al.* (2013) in the *BEEBOOK*.

2.2.6.1 Infection of primary honey bee ventricular cells

1. Isolate pupal gut cells from 10 days old pupae as described in detail in the *BEEBOOK* paper on cell culture techniques (Genersch *et al.*, 2013).

2. Transfer approximately 5E+7 *Nosema* spp. spores in AE-buffer (Qiagen) into chamber slides (2 well glass slide, VWR).

3. Air dry for 3 hours at room temperature.

4. Initialize spore-germination with 50 µl 0.1M Sucrose in PBS-buffer (BDH, Laboratory Supplies).

5. Immediately resuspend 500 µl primary cell suspension in fresh L15 medium (1.49% L-15, 0.4% Glucose, 0.25% Fructose, 0.33% Prolin, 3% Sucrose, all w/v, pH 7.2; for a recipe see Table 2) with the germinating spores.

6. Incubate the spore-cell suspension at 33°C for 20 min.

7. Add 1 ml of fresh and pre-warmed (37°C) BM3 medium (L 15 medium + 0.075% Pipes, 3% inactivated FCS, 1.2% Yeastolate, 10% antimycotic/antibiotic solution from Sigma-Aldrich, pH 6.7; for a recipe see Table 2).

8. Long time incubation is performed in a cooling incubator (Thermo Fisher) at 33°C for 144 hours.

9. Remove the supernatant by aspiration.

10. Fix cells with 500 µl of 4% formalin solution (Roth) for 24 hours.

11. Identify infected cells by microscopic analysis (Figs 7A & B).

Although the infection of primary pupal cells can be achieved, this approach is time consuming, does not easily lead to reproducible results, and is accompanied by the problem of seasonal dependency because sufficient numbers of pupae are only available during the brood rearing period.

Table 2. Recipes for media used for infection of cultured primary honey bee ventricular cells with *Nosema* spp.

BM 3 medium pH 6.7	1000 ml L 15 medium + 0.75 g Pipes, 30 ml FCS (inactivated), 12 g Yeastolate
L 15 medium pH 7.2	14.9 g L-15 powder, 4.0 g glucose, 2.5 g fructose, 3.3 g prolin, 30 g sucrose, ad 1000 ml bidest

2.2.6.2. Infection of heterologous lepidopteran cell lines

Most of the commercially available, permanent insect cell lines are derived from Lepidoptera. Several lepidopteran cell lines are described to support propagation of homologous microsporidia (where the source species of the cell line is the original host), as well as microsporidia originally infecting other hosts (Jaronski, 1984).

Likewise, several lepidopteran cell lines have proved to be susceptible to *N. apis* and *N. ceranae* infection. Susceptibility could recently be demonstrated for the following cell lines (Gisder *et al.*, 2010b): MB-L2 (*Mamestra brassicae*), Sf-158 and Sf-21 (*Spodoptera frugiperda*), SPC-BM-36 (*Bombyx mori*), and IPL-LD-65Y (*Lymantria dispar*), and BTI-Tn-5B1-4 (*Trichoplusia ni*) which can all be obtained through national cell culture collections together with protocols how to maintain and passage the cell lines. It is recommended to maintain the cell lines for routine culture in 75 cm² cell culture flasks (e.g. Roth) in a cooling incubator (e.g. Thermo Fisher) at 27°C, and supply them with their individual medium composition (Table 3).

Approximately 2E+05 cells per ml should be used to establish the next passage (Table 3).

For infection of these insect cell lines with germinating *Nosema* spp. spores (after Gisder *et al.*, 2010b):

1. Prepare a pre-culture
2. Incubate the cells to their exponential growth phase.
3. Harvest the cells growing in exponential phase

Table 3. Maintained insect cell lines which can be infected with *Nosema* spp. FCS = fetal calf serum; w/o = without.

cell line	cell growth	Medium	FCS	time of passage
MB-L2	mostly adherent	Insect-Xpress (Lonza)	w/o	every 5 days
Sf-158	adherent	Insect-Xpress (Lonza)	5%	every 6 days
Sf-21	adherent	Insect-Xpress (Lonza)	5%	every 6 days
SPC-BM-36	mostly adherent	TC-100 (Invitrogen)	12%	every 7 days
IPL-LD-65Y	suspension	TC-100 (Invitrogen)	11%	every 7 days
BTI-Tn-5B1-4	adherent	Sf-900 II (Invitrogen)	w/o	every 5 days

4. Centrifuge at 210xg for 5 min. (Eppendorf 5810 R, rotor F34-6-38).
5. Remove medium by aspiration.
6. Wash the cell pellet twice with 1 ml of freshly prepared 0.1 M sucrose in PBS-buffer.
7. Dilute the cells to a final concentration of 2E+06 cells per ml in 0.1 M sucrose solution.
8. Transfer approximately 5E+07 freshly prepared spores (see above), diluted in AE-buffer (Qiagen), were into a chamber slide (VWR, 4 chambers each slide) for infection.
9. Air dry for 3 hours at room temperature.
10. Initialize infection by adding 100 µl 0.1 M sucrose in PBS buffer to the dried spores, so that spore germination is triggered.
11. Immediately add 50 µl of cell suspension (2.5E+05 cells) to the germinating spores.
12. Resuspend the spore-cell suspension (150 µl) thoroughly.
13. Incubate at room temperature for 5 minutes.
14. Add 350 µl medium with 250 µg ml⁻¹ penicillin/streptomycin (Roth) and 125 µl antimycotic/antibiotic solution (Sigma-Aldrich) to the spore-cell suspension to a final volume of 500µl.
15. Incubate the cells at 27°C up to 10 days.

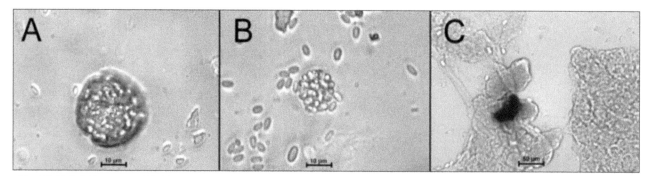

Fig. 7 A-C. Infection assay with *Nosema* spp. and primary pupal gut cells: A. infected primary ventricular cell; B. dissolving primary cell releasing new spores; C. *in situ*-hybridization (ISH) of infected ventricular cells, infected cells are stained blue. Specific hybridization was performed with 16S rRNA (SSU) probes coupled with digoxigenin and colour reaction to detect hybridized probes. Bars (A, B) represent 10 µm and bar in C represents 50 µm.

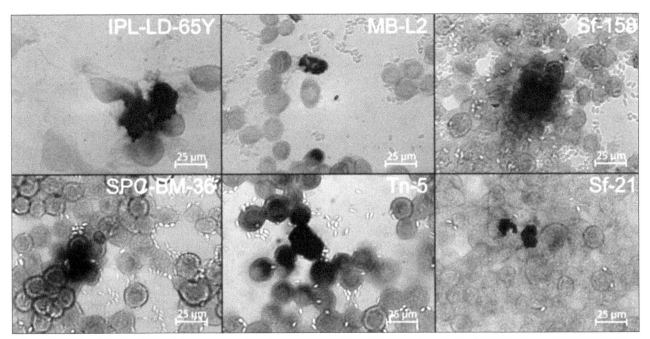

Fig. 8. *In situ*-hybridization (ISH) of *Nosema* spp. infected lepidopteran cell lines 72 hours post infection. Infected cells are stained dark blue. Specific hybridization was performed with 16S rRNA (SSU) probes coupled with digoxigenin and subsequent colour reaction to detect hybridized probes.

For *in situ*-hybridization (ISH) or fluorescence *in situ*-hybridization (FISH) medium was aspirated and cells were fixed on glass slides with 4% formalin (e.g. Roth) up to 24 hours. (Fig. 7c) ISH or FISH were performed according to recently published protocols (Gisder *et al.*, 2010b; Yue *et al.*, 2007, see also the *BEEBOOK* paper on molecular methods (Evans *et al.*, 2013)).

Six lepidopteran cell lines could be infected successfully (Fig. 8) as demonstrated by in situ-hybridization performed 72 hours post infection. For a detailed analysis of the life cycle of *Nosema* spp. in infected cells, IPL-LD-65Y cells were chosen and analysed by fluorescence in situ-hybridization (FISH, see the *BEEBOOK* paper on molecular methods (Evans *et al.*, 2013)). Although these newly developed cell culture models are valuable tools for studying pathogen-host interactions on cellular and molecular level (Troemel, 2011), the protocol does not yet allow the continuous propagation of *Nosema* spp. in cell culture and, hence, is not yet suitable to replace infection of bees for the production of spore suspensions.

2.2.7. Experiment type

Prior to any experiment with induced *Nosema* spp. infections, the infection level in the source colonies for bees to be used should be assessed prior to experimentation (see section 2.2.1). This is advisable also when bees for infection experiments are hatched in incubators.

2.2.7.1. Determination of infectious dose

The infectious dose (the dose that infects half of inoculated bees (ID_{50}) or 100% (ID_{100}) can be calculated relatively precisely when experimental bees are individually inoculated as described in section 2.2.4.4. and incubated in hoarding cages. The proportion of infected bees given different doses is then examined after 7-10 days. Usually, inoculums are prepared from a stock suspension of spores that is subsequently diluted (section 2.2.4.2.2). The final quantity of spores given to bees is therefore theoretical (= calculated). Under some circumstances and when the protocol is adapted to the goal of the study, the technique of group feeding can be used and gives reliable results (Pettis *et al.*, 2012; Tanner *et al.* (2012)). When the bees are mass fed, the quantity of spores given to each bee is an average count calculated from the overall quantity distributed to a given number of bees, but it should be noted, that group feeding is likely to yield an uneven distribution of spore dose among bees (section 2.2.4.5.). Similar to pesticide studies, where the lethal dose to 50% (LD_{50}) or 100% (LD_{100}) of tested insects are used to describe toxicity, the infectious dose can be described as the dose that produces infection in 50% (ID_{50}) or 100% of exposed insects (ID_{100}). For *N. apis*, ID_{50} has been determined to roughly 100 spores per bee (Fries, 1988), with 10,000 spores producing infection in all exposed honey bees (Czekonska, 2007; Fries, 1988). Recently, similar experiments were performed that compared *N. apis* and *N. ceranae*. Results revealed a slightly lower ID_{50} for *N. ceranae* (Forsgren and Fries, 2010). It should be noted that little is known about possible variations in infectivity between parasite isolates and/or different levels of susceptibility between host strains and, thus, infectious dose experiments need to be repeated using different bees and different spore sources.

2.2.7.1.1. Study of dose effects

Although the doses of 10,000 and 33,000 spores per bee have been shown to produce infection in all exposed bees, it is common to use much higher doses, even an order of magnitude higher or more, to ensure infection (Alaux *et al.*, 2010a; Alaux *et al.*, 2010b, Paxton *et al.*, 2007; Malone and Stefanovic, 1999; Higes *et al.*, 2007; Porrini *et al.*, 2010; Webster *et al.*, 2004; Woyciechowski and Moron, 2009). Depending on the type of experiment, it is often best not to use higher spore doses than is necessary to produce infection in individual bees, because when high spore doses are used, non-germinated spores may be retrieved and counted as spores produced from infection.

Honey bee queens become infected by both *N. apis* and *N. ceranae*. As with workers, a range of different doses have been used to study effects of infection on queens individually fed, although again, more spores were provided than were required (Alaux *et al.* 2010c; Webster, 2008). Using individual feeding, queens and worker bees have become infected using similar spore doses (Webster *et al.*, 2004), but the infectious doses for queens have never been established for either of the two microsporidian infections discussed.

2.2.7.1.2. Effects of different infection doses

It is necessary to test serial quantities of spores if the interest of the experiment is to investigate the effect of infection dose. In this case, for more precision, honey bees should always be individually fed the spore suspensions. The quantities of spores given to bees could increase by a factor of 10 (Forsgren and Fries, 2010; Malone *et al.*, 2001), but other increments can also be used (Martin-Hernandez *et al.*, 2011).

2.2.7.2. Course of infection in individual bees

The course of infection is checked in individual bees or in a group of tested bees. The time lap between spore counts has been highly variable between different authors and the goal of the experiment (Alaux *et al.*, 2010a; Alaux *et al.*, 2010b; Czekonska, 2007; Forsgren and Fries, 2010; Malone and Stefanovic, 1999; Paxton *et al.*, 2007). Generally speaking, a greater sample size and increased frequency of sampling will allow for more confidence in the data. Since individual feeding of honey bees is time consuming, the interval between sampling and the number of bees investigated must, nevertheless, be limited. Because *N. apis* spores are not produced from new infections until at least 3 days post infection (Fries, 1988 ; Fries *et al.*, 1996; Forsgren and Fries, 2010), with the first spores of *N. ceranae* produced slightly later (Forsgren and Fries, 2010), sampling should be

initiated no earlier than 4 days post infection. With a 2 day time interval between sampling, a relatively detailed data set on spore development can be accomplished. Because of variations in spore development between different bees, the sample size should never be below 3-4 bees per cage and treatment, and for statistical reasons, the more bees used, the better (see the statistics section in the *BEEBOOK* paper on miscellaneous methods (Human *et al.*, 2013)).

2.2.7.3. Longevity of infected bees

There is a surprising lack of data on the actual longevity of infected honey bees, probably because the mortality is rarely assessed *per se*. When the mortality takes place 10 to 14 days or more after the exposure to the spores, the researchers usually decide to sacrifice the tested honey bees. For *N. ceranae*, 100% mortality of infected bees have been reported within this time frame (Higes *et al.*, 2007; Martin-Hernandez, *et al.*, 2011), but significantly lower mortality rates have also been observed (Dussaubat *et al.*, 2010; Forsgren and Fries, 2010; Paxton *et al.*, 2007; Porrini *et al.*, 2010). There is a profound need for more mortality data from *Nosema* spp. infections, to better understand the impact from infection on colony viability. Possibly, cage experiments are less well suited for such tests because cage effects on longevity cannot be excluded, even if proper controls are used. An alternative would be to use marked bees, with and without infection, and introduce them to small functioning colonies and then study the rate at which such honey bees disappear compared to uninfected bees (see the section on statistics for survival analyses in the *BEEBOOK* paper on miscellaneous methods (Human *et al.*, 2013)).

Only a few studies have investigated the longevity of queens when infected with *N. apis* spores. Young infected queens that were allowed to lay eggs all died about 50 days after the onset of oviposition, with an average age when found dead or removed by the bees of about 25 days (Loskotova *et al.*, 1980). The impact from infection on queen as well as colony performance also need further studies to assess the impact on colony performance from these infections.

3. Future perspectives

The *Nosema* spp. parasites in honey bees still remain largely enigmatic. The described field symptoms differ between the parasites (Fries *et al.*, 2006), as do the seasonal prevalence (Higes *et al.*, 2008a, b). These observations suggest that the main mode of transmission between bees could differ between *N. apis* and *N. ceranae*. Faecal deposition within bee hives is associated with *N. apis* infections, but this is not the case with *N. ceranae* (Fries *et al.*, 2006). The main mode of parasite transmission for *N. apis* is believed to be a faecal-oral route through soiled comb (Bailey, 1953). There is a need to elucidate the main mode of transmission for *N. ceranae* to understand more of the epidemiology of this parasite. There is also a profound lack of data on differences in susceptibility of both parasites among different honey bee strains. Using standardized laboratory infectivity tests, it is probably possible to find differences in susceptibility to infection, both within and between strains of bees. With such information, breeding for resistance can be undertaken and genetic markers for resistance could possibly be located for genetic marker aided selection for disease resistance. As with differences in resistance in the host, it is likely that different isolates of the parasites differ in infectivity as well as in virulence. Although such differences have never been documented, along with differences in host susceptibility, they could complicate interpretations of experiments and possibly explain some contradictory results published on parasite virulence.

Ring tests among laboratories would be very useful for the scientific community but remain to be organized and funded.

The use of cell culture for studying the *Nosema* parasites is still in its infancy. With further developments, where continuous propagation of *Nosema* spp. in cell cultures becomes possible, new insights into infection biology may be gained. Also, there is a need to develop a reliable method for the long term storage of infective *N. ceranae* spores.

Lastly, a further complication when studying the *Nosema* parasites in honey bees is the associated virus infections. Infection with *N. apis* is associated with three unrelated viral infections; black queen cell virus (BQCV), bee virus Y (BVY) and filamentous virus (FV). A combination of *N. apis* and BQCV is distinctly more harmful than either infection alone, infection with BVY adds to *N. apis* virulence, whereas no such influence is seen with FV infections (Bailey *et al.*, 1983). Interestingly, *N. ceranae* infections have been shown to be negatively correlated with deformed wing virus (DWV) (Costa *et al.*, 2011).

4. Acknowledgements

The 7th framework project BEE DOC and the Ricola foundation supported this work period.

5. References

ALAUX, C; BRUNET, J L; DUSSAUBAT, C; MONDET, F; TCHAMITCHAN, S; COUSIN, M; BRILLARD, J; BALDY, A; BELZUNCES, L P; LE CONTE, Y (2010a) Interactions between *Nosema* microspores and a neonicotinoid weaken honey bees (*Apis mellifera*). *Environmental Microbiology* 12: 774-782. http://dx.doi.org/10.1111/j.1462-2920.2009.02123.x

ALAUX, C; DUCLOZ, F; CRAUSER, D ; LE CONTE, Y (2010b) Diet effects on honey bee immunocompetence. *Biology Letters* 6(4): 562-565. http://dx.doi.org/10.1098/rsbl.2009.0986

ALAUX, C; FOLSCHWEILLER, M; MCDONNELL, C; BESLAY, D; COUSIN, M; DUSSAUBAT, C; BRUNET, J L; LE CONTE, Y (2010c) Pathological effects of the microsporidium *Nosema ceranae* on honey bee queen physiology (*Apis mellifera*). *Journal of Invertebrate Pathology* 106(3): 80-385. http://dx.doi.org/10.1016/j.jip.2010.12.005

ALLEN, M D (1959) Respiration rates of worker honey bees of different ages and at different temperatures. *Journal of Experimental Biology* 36(1): 92-101.

AMDAM, G V; AASE, A L T O; SEEHUUS, S-C; FONDRK, M K; NORBERG, K; HARFELDER, K (2005) Social reversal of immunsenescence in honey bee workers. *Experimental Gerontology* 40(12): 939-947. http://dx.doi.org/10.1016/j.exger.2005.08.004

AMDAM, G V; RUEPPELL, O; FONDRK, M K; PAGE, R E; NELSON, C M (2009). The nurse's load: early-life exposure to brood-rearing affects behaviour and lifespan in honey bees (*Apis mellifera*). *Experimental Gerontology* 44(6-7): 467-471. http://dx.doi.org/10.1016/j.exger.2009.02.013

ANTÚNEZ, K; MARTIN-HERNANDEZ, R; PRIETO, L; MEANA, A; ZUNINO P; HIGES M (2009) Immune suppression in the honey bee (*Apis mellifera*) following infection by *Nosema ceranae* (Microsporidia). *Environmental Microbiology* 11(9): 2284-2290. http://dx.doi.org/10.1111/j.1462-2920.2009.01953.x

BAILEY, L (1953) The transmission of nosema disease. *Bee World* 34: 171-172.

BAILEY, L (1955) The epidemiology and control of *Nosema* disease of the honey-bee. *Annals of Applied Biology* 43(1-2): 379-389.

BAILEY, L (1972) The preservation of infective microsporidan spores. *Journal of Invertebrate Pathology* 20(3): 252-254. http://dx.doi.org/10.1016/0022-2011(72)90152-8

BAILEY, L; BALL, B V; PERRY, J N (1983) Association of viruses with 2 protozoal pathogens of the honey bee. *Annals of Applied Biology* 103(1): 3-20. http://dx.doi.org/10.1111/j.1744-7348.1983.tb02735.x

BARKER, R J; LEHNER, Y (1978) Laboratory comparison of high fructose corn syrup, grape syrup, honey, and sucrose syrup as maintenance food for caged honey bees. *Apidologie* 9(2): 111-116. http://dx.doi.org/10.1051/apido:19780203

BERGEM, M; NORBERG, N; AAMODT, R A (2006) Long-term maintenance of *in vitro* cultured honey bee (*Apis mellifera*) embryonic cells. *BMC Developmental Biology* 6(Article 17). http://dx.doi.org/10.1186/1471-213X-6-17

BOURGOIS, A L; BEARMAN, L D; HOLLOWAY, B; RINDERER, T E (2012) External and internal detection of *Nosema ceranae* on honey bees using real-time PCR. *Journal of Invertebrate Pathology* 109(3): 323-325. http://dx.doi.org/10.1016/j.jip.2012.01.002

BOURGEOIS, L A; RINDERER, T E; BEAMAN, L D; DANKA, R G (2010) Genetic detection and quantification of *Nosema apis* and *N. ceranae* in the honey bee. *Journal of Invertebrate Pathology* 103(1): 53-58. http://dx.doi.org/10.1016/j.jip.2009.10.009

BORCHERT, A (1928) Beiträge zur Kenntnis der Bienen Parasiten *Nosema apis. Archive für Bienenkunde* 9: 115-178.

BOTÍAS, C; MARTÍN-HERNÁNDEZ, R; BARRIOS, L; GARRIDO-BAILÓN, E; NANETTI, A; MEANA, A; HIGES, M (2012b) *Nosema* spp. parasitization decreases the effectiveness of acaricide strips (Apivar®) in treating varroosis of honey bee (*Apis mellifera iberiensis*) colonies. *Environmental Microbiology Reports* 4(1): 57-65. http://dx.doi.org/10.1111/j.1758-2229.2011.00299.x.v

BOTÍAS, C; MARTÍN-HERNÁNDEZ, R; DIAS, J; GARCÍA-PALENCIA, P; MATABUENA, M; JUARRANZ, A; BARRIOS, L; MEANA, A; NANETTI, A; HIGES, M (2012a) The effect of induced queen replacement on *Nosema* spp. infection in honey bee (*Apis mellifera iberiensis*) colonies. *Environmental Microbiology* 14(4): 845-859. http://dx.doi.org/10.1111/j.1462-2920.2011.02647.x

BOTÍAS, C; MARTÍN-HERNÁNDEZ, R; GARRIDO-BAILÓN, E; GONZÁLEZ PORTO, A, MARTÍNEZ-SALVADOR, A, DE LA RÚA, P, MEANA, A; HIGES, M (2011) The growing prevalence of *Nosema ceranae* in honey bees in Spain, an emerging problem for the last decade. *Research in Veterinary Sciences* 93(1): 150-155. http://dx.doi.org/10.1016/j.rvsc.2011.08.002

BOURGOIS, A L; BEARMAN, L D; HOLLOWAY, B; RINDERER, T E (2012) External and internal detection of *Nosema ceranae* on honey bees using real-time PCR. *Journal of Invertebrate Pathology* 109(3): 323-325. http://dx.doi.org/10.1016/j.jip.2012.01.002

BOURGEOIS, L A; RINDERER, T E; BEAMAN, L D; DANKA, R G (2010) Genetic detection and quantification of *Nosema apis* and *N. ceranae* in the honey bee. *Journal of Invertebrate Pathology* 103(1): 53-58. http://dx.doi.org/10.1016/j.jip.2009.10.009

BRODSCHNEIDER, R; CRAILSHEIM, K (2010) Nutrition and health in honey bees. *Apidologie* 41(3): 278-294. http://dx.doi.org/10.1051/apido/2010012

BÜCHLER, R; ANDONOV, S; BIENEFELD, K; COSTA, C; HATJINA, F; KEZIC, N; KRYGER, P; SPIVAK, M; UZUNOV, A; WILDE, J (2013) Standard methods for rearing and selection of *Apis mellifera* queens. In *V Dietemann; J D Ellis; P Neumann (Eds) The COLOSS BEEBOOK, Volume I: standard methods for* Apis mellifera *research. Journal of Apicultural Research* 52(1): http://dx.doi.org/10.3896/IBRA.1.52.1.07

BURGHER-MACLELLAN, K L; WILLIAMS, G R; SHUTLER, D; ROGERS, R E L; MACKENZIE, K (2010) Optimization of duplex real-time PCR with melting curve analysis for detection of microsporidian parasites *Nosema apis* and *Nosema ceranae* in *Apis mellifera*. *Canadian Entomologist* 142(3): 271-283. http://dx.doi.org/10.4039/n10-010

BURGES, H D; CANNING, E U; HULLS, I K (1974) Ultrastructure of *Nosema oryzaephili* and the taxonomic value of the polar filament. *Journal of Invertebrate Pathology* 23(2): 135-139. http://dx.doi.org/10.1016/0022-2011(74)90176-1

CANTWELL, G E (1970) Standard methods for counting nosema spores. *American Bee Journal* 110(6): 222-223.

CANTWELL, G E, SHIMANUKI, H (1969) Heat treatment as a means of eliminating nosema and increasing production. *American Bee Journal* 109(2): 52-54.

CARRECK, N L; ANDREE, M; BRENT, C S; COX-FOSTER, D; DADE, H A; ELLIS, J D; HATJINA, F; VANENGELSDORP, D (2013) Standard methods for *Apis mellifera* anatomy and dissection. In *V Dietemann; J D Ellis; P Neumann (Eds) The COLOSS BEEBOOK, Volume I: standard methods for* Apis mellifera *research. Journal of Apicultural Research* 52(4): http://dx.doi.org/10.3896/IBRA.1.52.4.03

CHAIMANEE, V; WARRIT, N; CHANTAWANNAKUL, P (2011) Infections of *Nosema ceranae* in four different honey bee species. *Journal of Invertebrate Pathology* 105(3): 207-210. http://dx.doi.org/10.1016/j.jip.2011.05.012

CHARRIÈRE, J-D; KILCHENMANN, V; RÖTSCHI, A (2011) Virulence of different *Melissococcus plutonius* strains on larvae tested by an *in vitro* larval test; In *Proceedings of the 42nd International Apicultural Congress, Buenos Aires, Argentina.* p 158.

CHEN, Y; EVANS, J D; MURPHY, C; GUTELL, R; ZUKER, M; GUNDENSEN-RINDAL, D; PETTIS, J S (2009) Morphological, molecular, and phylogenetic characterization of *Nosema ceranae*, a microsporidian parasite isolated from the European honey bee, *Apis mellifera. Journal of Eukaryotic Microbiology* 56(2): 142-147. http://dx.doi.org/10.1111/j.1550-7408.2008.00374.x

CHEN, Y; EVANS, J D; SMITH, I B; PETTIS, J S (2008) *Nosema ceranae* is a long-present and widespread microsporidian infection of the European honey bee (*Apis mellifera*) in the United States. *Journal of Invertebrate Pathology* 97(2): 186-188. http://dx.doi.org/10.1016/j.jip.2007.07.010

CHEN, Y P; HUANG, Z Y (2010) *Nosema ceranae*, a newly identified pathogen of *Apis mellifera* in the USA and Asia. *Apidologie* 41(3): 364-374. http://dx.doi.org/10.1051/apido/2010021

COBEY, S W; TARPY, D R ; WOYKE, J (2013) Standard methods for instrumental insemination of *Apis mellifera* queens. In *V Dietemann; J D Ellis; P Neumann (Eds) The COLOSS BEEBOOK, Volume I: standard methods for* Apis mellifera *research. Journal of Apicultural Research* 52(4): http://dx.doi.org/10.3896/IBRA.1.52.4.09

COLE, R J (1970) Application of triangulation method to purification of *Nosema* spores from insect tissues. *Journal of Invertebrate Pathology* 15(2): 193-195. http://dx.doi.org/10.1016/0022-2011(70)90233-8

COLTON, T (1974) *Statistics in medicine*. Little, Brown and Company; Boston, USA. 372 pp.

COSTA, C; TANNER, G; LODESANI, M; MAISTRELLO, L; NEUMANN, P (2011) Negative correlation between *Nosema ceranae* spore loads and deformed wing virus infection levels in adult honey bee workers. *Journal of Invertebrate Pathology* 108(3): 224-225. http://dx.doi.org/10.1016/j.jip.2011.08.012

CRAILSHEIM, K (1991) Interadult feeding of jelly in honey bee (*Apis mellifera* L) colonies. *Journal of Comparative Physiology B* 161(1): 55-60. http://dx.doi.org/10.1007/BF00258746

CRAILSHEIM, K (1998) Trophallactic interactions in the adult honey bee (*Apis mellifera* L.). *Apidologie* 29(1-2): 97-112. http://dx.doi.org/10.1051/apido:19980106

CURRIE, R W; PERNAL, S F; GUZMAN-NOVOA, E (2010) Honey bee colony losses in Canada. *Journal of Apicultural Research* 49(1): 104-106. http://dx.doi.org/10.3896/IBRA.1.49.1.18

CZEKONSKA, K (2007) Influence of carbon dioxide on *Nosema apis* infection of honey bees (*Apis mellifera*). *Journal of Invertebrate Pathology* 95(2): 84-86. http://dx.doi.org/10.1016/j.jip.2007.02.001

DADE, H A (2009) *Anatomy and dissection of the honey bee* (Revised edition). International Bee Research Association; Cardiff, UK. 196 pp.

DE GRAAF, D C; ALIPPI, A M; ANTÚNEZ, K; ARONSTEIN, K A; BUDGE, G; DE KOKER, D; DE SMET, L; DINGMAN, D W; EVANS, J D; FOSTER, L J; FÜNFHAUS, A; GARCIA-GONZALEZ, E; GREGORC, A; HUMAN, H; MURRAY, K D; NGUYEN, B K; POPPINGA, L; SPIVAK, M; VANENGELSDORP, D; WILKINS, S; GENERSCH, E (2013) Standard methods for American foulbrood research. In *V Dietemann; J D Ellis; P Neumann (Eds) The COLOSS BEEBOOK, Volume II: standard methods for* Apis mellifera *pest and pathogen research. Journal of Apicultural Research* 52(1): http://dx.doi.org/10.3896/IBRA.1.52.1.11

DELAPLANE, K S; VAN DER STEEN, J; GUZMAN, E (2013) Standard methods for estimating strength parameters of *Apis mellifera* colonies. In *V Dietemann; J D Ellis; P Neumann (Eds) The COLOSS BEEBOOK, Volume I: standard methods for* Apis mellifera *research. Journal of Apicultural Research* 52(1): http://dx.doi.org/10.3896/IBRA.1.52.1.03

DE MIRANDA, J R; BAILEY, L; BALL, B V; BLANCHARD, P; BUDGE, G; CHEJANOVSKY, N; CHEN, Y-P; VAN DOOREMALEN, C; GAUTHIER, L; GENERSCH, E; DE GRAAF, D; KRAMER, M; RIBIÈRE, M; RYABOV, E; DE SMET, L VAN DER STEEN, J J M (2013) Standard methods for virus research in *Apis mellifera*. In *V Dietemann; J D Ellis; P Neumann (Eds) The COLOSS BEEBOOK, Volume II: standard methods for* Apis mellifera *pest and pathogen research. Journal of Apicultural Research* 52(4): http://dx.doi.org/10.3896/IBRA.1.52.4.22

DIETEMANN, V; NAZZI, F; MARTIN, S J; ANDERSON, D; LOCKE, B; DELAPLANE, K S; WAUQUIEZ, Q; TANNAHILL, C; ZEGELMANN, B; ROSENKRANZ, P; ELLIS, J D (2013) Standard methods for varroa research. In *V Dietemann; J D Ellis; P Neumann (Eds) The COLOSS BEEBOOK, Volume II: standard methods for* Apis mellifera *pest and pathogen research. Journal of Apicultural Research* 52(1): http://dx.doi.org/10.3896/IBRA.1.52.1.09

DOULL, K M (1965) The effects of time of day and method of sampling on the determination of nosema disease in bee hives. *Journal of Invertebrate Pathology* 7(1): 1-4. http://dx.doi.org/10.1016/0022-2011(65)90143-6

DOULL, K M, CELLIER, K M (1961) A survey of the incidence of *Nosema* disease (*Nosema apis* Zander) in honey bees in South Australia. *Journal of Insect Pathology* 3(3): 280-288.

DOULL, K M; ECKERT, J E (1962) A survey of the incidence of *Nosema* disease in California. *Journal of Economic Entomology* 55(3): 313-317.

DUSSAUBAT, C; MAISONNASSE, A; ALAUX, C; TCHAMITCHAN, S; BRUNET, J L; PLETTNER, E; BELZUNCES, L P; LE CONTE, Y (2010) *Nosema* spp. Infection alters pheromone production in honey bees (*Apis mellifera*). *Journal of Chemical Ecology* 36(5): 522-525. http://dx.doi.org/10.1007/s10886-010-9786-2

EL-SHEMY, A A M; PICKARD, R S (1989) *Nosema apis* Zander infection levels in honey bees of known age. *Journal of Apicultural Research* 28(2): 101-106.

ERLER, S; LOMMATZSCH, S; LATTORFF, H (2012) Comparative analysis of detection limits and specificity of molecular diagnostic markers for three pathogens (Microsporidia, *Nosema* spp.) in the key pollinators *Apis mellifera* and *Bombus terrestris*. *Parasitology Research* 110(4): 1403–1410. http://dx.doi.org/10.1007/s00436-011-2640-9

EVANS, J D; SPIVAK, M (2010) Socialized medicine: individual and communal disease barriers in honey bees. *Journal of Invertebrate Pathology* 103(S): S62-S72. http://dx.doi.org/10.1016/j.jip.2009.06.019

EVANS, J D; CHEN, Y P; CORNMAN, R S; DE LA RUA, P; FORET, S; FOSTER, L; GENERSCH, E; GISDER, S; JAROSCH, A; KUCHARSKI, R; LOPEZ, D; LUN, C M; MORITZ, R F A; MALESZKA, R; MUÑOZ, I; PINTO, M A; SCHWARZ, R S (2013) Standard methodologies for molecular research in *Apis mellifera*. In *V Dietemann; J D Ellis; P Neumann (Eds) The COLOSS BEEBOOK, Volume I: standard methods for* Apis mellifera *research. Journal of Apicultural Research* 52(4): http://dx.doi.org/10.3896/IBRA.1.52.4.11

FARRAR, C L (1947) *Nosema* losses in package bees as related to queen supersedure and honey yields. *Journal of Economic Entomology* 40(3): 333-338.

FENOY, S; RUEDA, C; HIGES, M; MARTÍN-HERNÁNDEZ, R; DEL AGUILA, C (2009) High-level resistance of *Nosema ceranae*, a parasite of the honey bee, to temperature and desiccation. *Applied and Environmental Microbiology* 75(21): 6886-6889. http://dx.doi.org/10.1128/AEM.01025-09

FERGUSON, H M; READ, A F (2002) Genetic and environmental determinants of malaria parasite virulence in mosquitoes. *Proceedings of the Royal Society London* B 269(1497): 1217-1224. http://dx.doi.org/10.1098/rspb.2002.2023

FLURI, P (1977) Juvenile hormone, vitellogenin and haemocyte composition in winter worker honey bees (*Apis mellifera*). *Experientia* 33(9): 1240-1241. http://dx.doi.org/10.1007/BF01922354

FORSGREN, E; FRIES, I (2010) Comparative virulence of *Nosema ceranae* and *Nosema apis* in individual European honey bees. *Veterinary Parasitology* 170(3-4): 212-217. http://dx.doi.org/10.1016/j.vetpar.2010.02.010

FRIES, I (1988) Infectivity and multiplication of *Nosema apis* Z. in the ventriculus of the honey bee. *Apidologie* 19(3): 319-328. http://dx.doi.org/10.1051/apido:19880310

FRIES, I (1989) Observations on the development and transmission of *Nosema apis* Z. in the ventriculus of the honey bee. *Journal of Apicultural Research* 28(2): 107-117.

FRIES I (1993) *Nosema apis* - a parasite in the honey bee colony. *Bee World* 74(1): 5–19.

FRIES, I (2010) *Nosema ceranae* in European honey bees (*Apis melllifera*). *Journal of Invertebrate Pathology* 103(S): S73-S79. http://dx.doi.org/10.1016/j.jip.2009.06.017

FRIES, I; EKBOHM, G; VILLUMSTAD, E (1984) *Nosema apis*, sampling techniques and honey yield. *Journal of Apicultural Research* 23(2): 102-105.

FRIES, I; FENG, F; SILVA, A; DA SLEMENDA, S B; PIENIAZEK, N J (1996) *Nosema ceranae* n. sp. (Microspora, Nosematidae), morphological and molecular characterization of a microsporidian parasite of the Asian honey bee *Apis cerana* (Hymenoptera, Apidae). *European Journal of Protistology* 32(3): 356-365.

FRIES, I; GRANADOS, R R; MORSE, R A (1992) Intracellular germination of spores of *Nosema apis* Z. *Apidologie* 23(1): 61-71. http://dx.doi.org/10.1051/apido:19920107

FRIES, I; MARTIN, R; MEANA, A; GARCIA-PALENCIA, P; HIGES, M (2006) Natural infections of *Nosema ceranae* in European honey bees. *Journal of Apicultural Research* 45(4): 230-233.

FURGALA, B; HYSER, R A (1969) Minnesota *Nosema* survey. *American Bee Journal* 109(12): 460-461.

FURGALA, B, MAUNDER, M J (1961) A simple method of feeding *Nosema apis* inoculum to individual honey bees. *Bee World* 42(10): 249-252.

FURGALA, B; MUSSEN, E C (1978) Protozoa. In *Morse, R A; Nowogrodzki, R (Eds) Honey bee pests, predators, and diseases.* Cornell University Press; Ithaca, USA. pp. 48-63.

GENERSCH, E; GISDER, S; HEDTKE, K; HUNTER, W B; MÖCKEL, N; MÜLLER, U (2013) Standard methods for cell cultures in *Apis mellifera* research. In *V Dietemann; J D Ellis; P Neumann (Eds) The COLOSS BEEBOOK, Volume I: standard methods for* Apis mellifera *research. Journal of Apicultural Research* 52(1): http://dx.doi.org/10.3896/IBRA.1.52.1.02

GISDER, S; HEDTKE, K; MÖCKEL, N; FRIELITZ, M C; LINDE, A; GENERSCH, E (2010a) Five-year cohort study of *Nosema* spp. in Germany: climate shape virulence and assertiveness of *Nosema ceranae*? *Applied and Environmental Microbiology* 76(9): 3032-3038. http://dx.doi.org/10.1128/AEM.03097-09

GISDER, S; MÖCKEL, N; LINDE, A; GENERSCH, E (2010b) A cell culture model for *Nosema ceranae* and *Nosema apis* allows new insights into the life cycle of these important honey bee-pathogenic microsporidia. *Environmental Microbiology.* 13(2): 404-413. http://dx.doi.org/10.1111/j.1462-2920.2010.02346.x

GROSS, K P; RUTTNER, F (1970) Entwickelt *Nosema apis* Zander eine Resistenz gegenüber dem Antibiotikum Fumidil B? *Apidologie* 1 (4): 401-422. http://dx.doi.org/10.1051/apido:19700403

GRÜNEWALD, B; WERSING, A (2008) An ionotropic GABA receptor in cultured mushroom body Kenyon cells of the honey bee and its modulation by intracellular calcium. *Journal of Comparative Physiology A Neuroethology Sensory Neural and Behavioural Physiology* 194(4): 329-340. http://dx.doi.org/10.1007/s00359-007-0308-9

GUZMAN-NOVOA, E; CORREA-BENITEZ, A; ECCLES, L; CALVETE, Y; MCGOWAN, J; KELLY, P G (2010) *Varroa destructor* is the main culprit for the death and reduced populations of overwintered honey bee (*Apis mellifera*) colonies in Ontario, Canada. *Apidologie* 41(4): 443-450. http://dx.doi.org/10.1051/apido/2009076

HAMIDUZZAMAN, M M; GUZMAN-NOVOA, E; GOODWIN, P H (2010) A multiplex PCR assay to diagnose and quantify *Nosema* infections in honey bees (*Apis mellifera*). *Journal of Invertebrate Pathology* 105(2): 151-155. http://dx.doi.org/10.1016/j.jip.2010.06.001

HASSANEIN, M H (1953) The influence of infection with *Nosema apis* on the activities and longevity of worker honey bees. *Annals of Applied Biology* 40(2): 418-423. http://dx.doi.org/10.1111/j.1744-7348.1953.tb01093.x

HIGES, M; GARCÍA-PALENCIA, P; BOTÍAS, C; MEANA, A; MARTÍN-HERNÁNDEZ, R (2010) The differential development of microsporidia infecting worker honey bee (*Apis mellifera*) at increasing incubation temperature. *Environmental and Microbiology Reports* 2(6): 745-748. http://dx.doi.org/10.1111/j.1758-2229.2010.00170x

HIGES, M; GARCÍA-PALENCIA, P; MARTÍN-HERNÁNDEZ, R; MEANA, A (2007) Experimental infection of *Apis mellifera* honey bees with *Nosema ceranae* (Microsporidia). *Journal of Invertebrate Pathology* 94(3): 211-217. http://dx.doi.org/10.1016/j.jip.2006.11.001

HIGES, M; MARTIN-HERNANDEZ, R; BOTIAS, C; BAILON, E G; GONZALES-PORTO, A V; BARRIOS, L; DEL NOZAL, M J; BERNAL, J L; JIMENEZ, J J; PALENCIA, P G; MEANA, A (2008a) How natural infection by *Nosema ceranae* causes honey bee colony collapse. *Environmental Microbiology* 10(10): 2659–2669. http://dx.doi.org/10.1111/j.1462-2920.2008.01687.x

HIGES, M; MARTÍN-HERNÁNDEZ, R; GARRIDO-BAILÓN, E; GARCÍA-PALENCIA, P; MEANA, A (2008b) Detection of infective *Nosema ceranae* (Microsporidia) spores in corbicular pollen of forager honey bees. *Journal of Invertebrate Pathology* 97(1): 76-78. http://dx.doi.org/10.1016/j.jip.2007.06.002

HIGES, M; MARTIN-HERNANDEZ, R; MEANA, A (2006) *Nosema ceranae*, a new microsporidian parasite in honey bees in Europe. *Journal of Invertebrate Pathology* 92(2): 93-95. http://dx.doi.org/10.1016/j.jip.2006.02.005

HIGES, M; NOZAL, M J; ALVARO, A; BARRIOS, L; MEANA, A; MARTIN-HERNANDEZ, R; BERNAL, J L; BERNAL, J (2011) The stability and effectiveness of fumagillin in controlling *Nosema ceranae* (Microsporidia) infection in honey bees (*Apis mellifera*) under laboratory and field conditions. *Apidologie* 42(4): 364-377. http://dx.doi.org/10.1007/s13592-011-0003-2

HIGES, M; PILAR, G P; MARTIN-HERNANDEZ, R; ARANZAZU, M (2007) Experimental infection of *Apis mellifera* honey bees with *Nosema ceranae* (Microsporidia). *Journal of Invertebrate Pathology* 94(3): 211-217. http://dx.doi.org/10.1016/j.jip.2006.11.001

HUMAN, H; BRODSCHNEIDER, R; DIETEMANN, V; DIVELY, G; ELLIS, J; FORSGREN, E; FRIES, I; HATJINA, F; HU, F-L; JAFFÉ, R; KÖHLER, A; PIRK, C W W; ROSE, R; STRAUSS, U; TANNER, G; VAN DER STEEN, J J M; VEJSNÆS, F; WILLIAMS, G R; ZHENG, H-Q (2013) Miscellaneous standard methods for *Apis mellifera* research. In V Dietemann; J D Ellis; P Neumann (Eds) The COLOSS BEEBOOK, Volume I: standard methods for Apis mellifera research. *Journal of Apicultural Research* 52(4): http://dx.doi.org/10.3896/IBRA.1.52.4.10

HUNTER, W B (2010) Medium for development of bee cell cultures (*Apis mellifera*: Hymenoptera: Apidae). In Vitro *Cell Developmental Biology Animal* 46(2): 83-86. http://dx.doi.org/10.1007/s11626-009-9246-x

INVERNIZZI, C; ABUD, C; TOMASCO, I H; HARRIET, J; RAMALLO, G; CAMPA, J; KATZ, H; GARDIOL, G; MENDOZA, Y (2009) Presence of *Nosema ceranae* in honey bees (*Apis mellifera*) in Uruguay. *Journal of Invertebrate Pathology* 101(2): 150-153. http://dx.doi.org/10.1016/j.jip.2009.03.006

JARONSKI, S T (1984) Microsporidia in cell culture. *Advances in Cell Culture* 3: 183-229.

KATZNELSON, H; ROBB, J A (1962) The use of gamma radiation from cobalt-60 in the control of diseases of the honey bee and the sterilization of honey. *Canadian Journal of Microbiology* 8(2): 175-179.

KELLNER, N; JACOBS, F J (1978) In hoeveel tijd bereiken de sporen van *Nosema apis* Zander de ventriculus van de honingbij (*Apis mellifera* L.)? *Vlaams Diergeneeskundig Tijdschrift* 47(3): 252-259.

KLEE, J; BESANA, A M; GENERSCH, E; GISDER, S; NANETTI, A; TAM, D Q; CHINH, T X; PUERTA, F; RUZ, J M; KRYGER, P; MESSAGE, D; HATJINA, F; KORPELA, S; FRIES, I; PAXTON, R J (2007) Widespread dispersal of the microsporidian *Nosema ceranae*, an emergent pathogen of the western honey bee, *Apis mellifera*. *Journal of Invertebrate Pathology* 96(1): 1-10. http://dx.doi.org/10.1016/j.jip.2007.02.014

KLEE, J; TAY, W T; PAXTON, R J (2006) Specific and sensitive detection of *Nosema bombi* (Microsporidia : Nosematidae) in bumble bees (*Bombus* spp.; Hymenoptera : Apidae) by PCR of partial rRNA gene sequences. *Journal of Invertebrate Pathology* 91(2): 98-104. http://dx.doi.org/10.1016/j.jip.005.10.012

KRAUS, B; VELTHUIS, H H W (1997) High humidity in the honey bee (*Apis mellifera* L.) brood nests limits reproduction of the parasitic mite *Varroa jacobsoni* Oud. *Naturwissenschaften* 84(5): 217-218. http://dx.doi.org/10.1007/s001140050382

KREISSL, S; BICKER, G (1992) Dissociated neurons of the pupal honey bee brain in cell culture. *Journal of Neurocytology* 21(8): 545-556. http://dx.doi.org/10.1007/BF01187116

L'ARRIVEE, J C M (1963) Comparison of composite versus individual bee sampling for *Nosema apis* Zander. *Journal of Insect Pathology* 5(4): 349-355.

LI, J; CHEN, W; WU, J; PENG, W; AN, J; SCHMID-HEMPEL, P; SCHMID-HEMPEL, R (2012) Diversity of *Nosema* associated with bumble bees (*Bombus* spp.) from China. *International Journal of Parasitology* 42(1):49-61. http://dx.doi.org/10.1016/j.ijpara.2011.10.005

LINDER, A (1947) Über das Auswerten zahlenmässiger Angaben in der Bienenkunde. Beihefte zur *Schweizerischen Bienen-Zeitung* 2: 77-138.

LYNN, D E (2007) Available lepidopteran insect cell lines. In *Murhammer, D W (Ed.) Methods in molecular biology 388*. Humana Press Inc.; Totowa, NJ, USA. pp. 117-138.

LOSKOTOVA, J; PEROUTKA, M; VESLEY, V (1980) *Nosema* disease of honey bee queens (*Apis mellifica* L.). *Apidologie* 11(2): 153-161. http://dx.doi.org/10.1051/apido:19800205

MAISTRELLO, L; LODESANI, M; COSTA, C; LEONARDI, F; MARANI, G; CALDON, M; MUTINELLI, F; GRANATO, A (2008) Screening of natural compounds for the control of nosema disease in honey bees. *Apidologie* 39(4): 436-445. http://dx.doi.org/10.1051/apido:2008022

MALONE, L A; GATEHOUSE, H S; TREGIDGA, E L (2001) Effects of time, temperature, and honey on *Nosema apis* (Microsporidia: Nodematidae) a parasite of the honey bee, *Apis mellifera* (Hymenoptera: Apidae). *Journal of Invertebrate Pathology* 77(4): 258-268. http://dx.doi.org/10.1006/jipa.2001.5028

MALONE, L A; STEVANOVIC, D (1999) Comparison of the responses of two races of honey bees to infection with *Nosema apis* Zander. *Apidologie* 30(5): 375-382. http://dx.doi.org/10.1051/apido:19990503

MARTÍN-HERNÁNDEZ, R; BOTIAS, C; BARRIOS, L; MARTINEZ-SALVADOR, A; MEANA, A; MAYACK, C; HIGES, M (2011) Comparison of the energetic stress associated with experimental *Nosema ceranae* and *Nosema apis* infection of honey bees (*Apis mellifera*). *Parasitology Research* 109(3): 605-612. http://dx.doi.org/10.1007/s00436-011-2292-9

MARTÍN-HERNÁNDEZ, R; MEANA, A; GARCÍA-PALENCIA, P; MARÍN, P; BOTÍAS, C; GARRIDO-BAILÓN, E; BARRIOS, L; HIGES, M (2009). Effect of temperature on the biotic potential of honey bee microsporidia. *Applied and Environmental Microbiology* 75(8): 2554-2557. http://dx.doi.org/10.1128/AEM.02908-08

MARTÍN-HERNÁNDEZ, R; MEANA, A; PRIETO, L; MARTINEZ-SALVADOR, A; GARRIDO-BAILÓN, E; HIGES, M (2007) Outcome of colonization of *Apis mellifera* by *Nosema ceranae*. *Applied and Environmental Microbiology* 73(20): 6331-6338. http://dx.doi.org/10.1128/AEM.00270-07

MAYACK, C; NAUG, D (2009) Energetic stress in the honey bee *Apis mellifera* from *Nosema ceranae* infection. *Journal of Invertebrate Pathology* 100(3): 185-188. http://dx.doi.org/10.1016/j.jip.2008.12.001

MCMULLAN, J B; BROWN, M J F (2005) Brood pupation temperature affects the susceptibility of honey bees (*Apis mellifera*) to infestation by tracheal mites (*Acarapis woodi*). *Apidologie* 36(1): 97-105. http://dx.doi.org/10.1051/apido:2004073

MEANA, A; MARTÍN-HERNÁNDEZ, R; HIGES, M (2010) The reliability of spore counts to diagnose *Nosema ceranae* infections in honey bees. *Journal of Apicultural Research* 49(2): 212-214.

MEIXNER, M D; PINTO, M A; BOUGA, M; KRYGER, P; IVANOVA, E; FUCHS, S (2013) Standard methods for characterising subspecies and ecotypes of *Apis mellifera*. In *V Dietemann; J D Ellis; P Neumann (Eds) The COLOSS BEEBOOK, Volume I: standard methods for Apis mellifera research. Journal of Apicultural Research* 52(4): http://dx.doi.org/10.3896/IBRA.1.52.4.05

MICHAILOFF, A S (1928) Statistische Untersuchungen über Nosema an der Tulaer Versuchsstation für Bienenzucht. *Archive für Bienenkunde* 9: 89-114.

OFFICE INTERNATIONAL DES EPIZOOTIES (OIE) (2008) Nosemosis of honey bees. [WWW document].URL http://www.oie.int/eng/normes/mmanual/2008/pdf/2.02.04_NOSEMOSIS.pdf

PAXTON, R J; FRIES, I; PIENIAZEK, N J; TENGÖ, J (1997) High incidence of infection of an undescribed microsporidium (Microspora) in the communal bee *Andrena scotica* (Hymenoptera, Andrenidae) in Sweden. *Apidologie* 28: 129-141. http://dx.doi.org/10.1051/apido:19970304

PAXTON, R J; KLEE, J; KORPELA, S; FRIES, I (2007) *Nosema ceranae* has infected *Apis mellifera* in Europe since at least 1998 and may be more virulent than *Nosema apis*. *Apidologie* 38(6): 558-565. http://dx.doi.org/10.1051/apido:2007037

PERNAL, S F; CURRIE, R W (2000) Pollen quality of fresh and 1-year-old single pollen diets for worker honey bees (*Apis mellifera* L.). *Apidologie* 31(3): 387-409. http://dx.doi.org/10.1051/apido:2000130

PETTIS, J S; VANENGELSDORP, D; JOHNSON, J; DIVELY, G (2012) Pesticide exposure in honey bees results in increased levels of the gut pathogen *Nosema*. *Naturwissenschaften* 99(2): 153–158. http://dx.doi.org/10.1007/s00114-011-0881-1

PORRINI, M P; AUDISIO, M C; SABATE, D C; IBARGUREN, C; MEDICI, S K; SARLO, E G; GARRIDO, P M; EGUARAS, M J (2010) Effect of bacterial metabolites on microsporidian *Nosema ceranae* and on its host *Apis mellifera*. *Parasitology Research* 107(2): 381-388. http://dx.doi.org/10.1007/s00436-010-1875-1

PORRINI, M P; SARLO, E G; MEDICI, S K; GARRIDO, P M; PORRINI, D P; DAMIANI, N; EGUARAS, M J (2011) *Nosema ceranae* development in *Apis mellifera*: influence of diet and infective inoculum. *Journal of Apicultural Research* 50(1): 35-41. http://dx.doi.org/10.3896/IBRA.1.50.1.04

RIBBANDS, C R (1950) Changes in the behaviour of honey bees following their recovery from anaesthesia. *Journal of Experimental Biology* 27(3-4): 302-310.

RUEPPELL, O; HAYWORTH, M K; ROSS, N P (2010) Altruistic self-removal of health-compromised honey bee workers from their hive. *Journal of Evolutionary Biology* 23(7): 1538-1546. http://dx.doi.org/10.1111/j.1420-9101.2010.02022.x

RUTRECHT, S T; BROWN, M J F (2008) Within colony dynamics of *Nosema bombi* infections: disease establishment, epidemiology and potential vertical transmission. *Apidologie* 39(5): 504-514. http://dx.doi.org/10.1051/apido:2008031

RUTRECHT, S T; KLEE, J; BROWN, M J F (2007) Horizontal transmission success of *Nosema bombi* to its adult bumble bee hosts: effects of dosage, spore source and host age. *Parasitology* 134: 1719-1726.

SAGASTUME, S; DEL ÁGUILA, C; MARTÍN-HERNÁNDEZ, R; HIGES, M; HENRIQUES-GIL, N (2011) Polymorphism and recombination for rDNA in the putatively asexual microsporidian *Nosema ceranae*, a pathogen of honey bees. *Environmental Microbiology* 13(1): 84-95. http://dx.doi.org/10.1111/j.1462-2920.2010.02311.x

SCHMIDT, J O; THOENES, S C; LEVIN, M D (1987) Survival of honey bees, *Apis mellifera* (Hymenoptera: Apidae), fed various pollen sources. *Annals of the Entomological Society of America* 80(2): 176-183.

SMAGGHE, G; GOODMAN, C L; STANLEY, D (2009) Insect cell culture and applications to research and pest management. In Vitro *Cell Developmental Biology Animal* 45(3-4): 93-105. http://dx.doi.org/10.1007/s11626-009-9181-x

SMART, M D; SHEPPARD, W S (2011) *Nosema ceranae* in age cohorts of the western honey bee (*Apis mellifera*). *Journal of Invertebrate Pathology* 109(1): 148-151. http://dx.doi.org/10.1016/j.jip.2011.09.009.

TAY, W T; O'MAHONEY, E M; PAXTON, R J (2005) Complete rRNA gene sequences reveal that the microsporidium *Nosema bombi* infects diverse bumble bee (*Bombus* spp.) hosts and contains multiple polymorphic sites. *Journal of Eukaryotic Microbiology* 52 (6): 505-513. http://dx.doi.org/10.1111/j.550-7408.2005.00057.x

TANNER, G; WILLIAMS, G R; MEHMANN, M; NEUMANN, P (2012) Comparison of mass versus individual inoculation of worker honey bees with *Nosema ceranae*. In *Proceedings of the 8th COLOSS Conference / MC meeting FA0803. Halle-Saale, Germany, 1-3 September 2012.* p 59.

TENTCHEVA, D; GAUTHIER, L; ZAPPULLA, N; DAINAT, B; COUSSERANS, F; COLIN, M E; BERGOIN, M (2004) Prevalence and seasonal variations of six bee viruses in *Apis mellifera* L. and *Varroa destructor* mite populations in France. *Applied and Environmental Microbiology* 70(12): 7185-7191. http://dx.doi.org/0.1128/AEM.70.12.7185-7191.2004

TRAVER, B E; FELL, R D (2011) Prevalence and infection intensity of *Nosema* in honey bee (*Apis mellifera* L.) colonies in Virginia. *Journal of Invertebrate Pathology* 107(1): 43-49. http://dx.doi.org/10.1016/j.jip.2011.02.003

TRAVER, B E; FELL, R D (2012) Low natural levels of *Nosema ceranae* in *Apis mellifera* queens. *Journal of Invertebrate Pathology* 110(3): 408-410. http://dx.doi.org/10.1016/j.jip.2012.04.001

TRAVER, B E; WILLIAMS, M R; FELL, R D (2012) Comparison of within hive sampling and seasonal activity of *Nosema ceranae* in honey bee colonies. *Journal of Invertebrate Pathology* 109(2): 187-193. http://dx.doi.org/10.1016/j.jip.2011.11.001

TROEMEL, E R (2011) New models for microsporidiosis: infections in zebrafish, *C. elegans* and honey bee. *PLoS Pathology* 7(2): e1001243. http://dx.doi.org/10.1371/journal.ppat.1001243

VAN OERS, M M; LYNN, D E (2010) *Insect cell culture*. eLS. Wiley online Library. http://dx.doi.org/10.1002/9780470015902.a0002574.pub2

VAVRA, J; LARSSON, R (1999) Structure of the microsporidia. In *Wittner, M; Weiss* [initials?] *(Eds) The microsporidia and microsporidiosis*. ASM Press; Washington D C; USA. pp. 7-84.

VERBEKE, M; JACOBS, F J; DE RYCKE, P H (1984) Passage of various particles through the ventriculus in the honey bee (*Apis mellifera* L.). *American Bee Journal* 123(6): 468-470.

WANG, D-I, MOELLER, F E (1970) The division of labor and queen attendance behaviour of nosema-infected worker honey bees. *Journal of Economic Entomology* 63(5): 1539-1541.

WEBSTER, T C (1994) Fumagillin effects on *Nosema apis* and honey bees (Hymenoptera: Apidae). *Journal of Economic Entomology* 87 (3): 601-604.

WEBSTER, T C (2008) *Nosema apis* infection in honey bee *(Apis mellifera)* queens. *Journal of Apicultural Research* 47(1): 53-57.

WEBSTER, T C; POMPER, K W; HUNT, G; THACKER, E M; JONES, S C (2004) *Nosema apis* infection in worker and queen *Apis mellifera*. *Apidologie* 35(1): 49-54. http://dx.doi.org/10.1051/apido:2003063

WHITE, G F (1919) *Nosema* disease. *US Department of Agriculture Bulletin* 780.

WILLIAMS, G R; SHAFER SHAFER, A B A; ROGERS, R E L; SHUTLER, D; STEWART, D T (2008) First detection of *Nosema ceranae*, a microsporidian parasite of European honey bees (*Apis mellifera*), in Canada and central USA. *Journal of Invertebrate Pathology* 97 (2): 189-192. http://dx.doi.org/10.1016/j.jip.2007.08.005

WILLIAMS, G R; ALAUX, C; COSTA, C; CSÁKI, T; DOUBLET, V; EISENHARDT, D; FRIES, I; KUHN, R; MCMAHON, D P; MEDRZYCKI, P; MURRAY, T E; NATSOPOULOU, M E; NEUMANN, P; OLIVER, R; PAXTON, R J; PERNAL, S F; SHUTLER, D; TANNER, G; VAN DER STEEN, J J M; BRODSCHNEIDER, R (2013) Standard methods for maintaining adult *Apis mellifera* in cages under *in vitro* laboratory conditions. In *V Dietemann; J D Ellis; P Neumann (Eds) The COLOSS BEEBOOK, Volume I: standard methods for* Apis mellifera *research. Journal of Apicultural Research* 52(1): http://dx.doi.org/10.3896/IBRA.1.52.1.04

WILLIAMS, G R; ROGERS, R E L; KALKSTEIN, A L; TAYLOR, B A; SHUTLER, D; OSTIGUY, N (2009) Deformed wing virus in western honey bees (*Apis mellifera*) from Atlantic Canada and the first description of an overtly-infected emerging queen. *Journal of Invertebrate Pathology* 101(1): 77-79. http://dx.doi.org/10.1016/j.jip.2009.01.004

WILLIAMS, G R; SHUTLER, D; ROGERS, R E L (2010) Effects at Nearctic north-temperate latitudes of indoor versus outdoor overwintering on the microsporidium *Nosema ceranae* and western honey bees (*Apis mellifera*). *Journal of Invertebrate Pathology* 104(1): 4-7. http://dx.doi.org/10.1016/j.jip.2010.01.009

WILLIAMS, G R; SHUTLER, D; LITTLE, C M; BURGHER-MACLELLAN, K L; ROGERS, R E L (2011) The microsporidian *Nosema ceranae*, the antibiotic Fumagilin-B®, and western honey bee (*Apis mellifera*) colony strength. *Apidologie* 41(1): 15-22. http://dx.doi.org/10.1051/apido/20100230.

WILSON, W T; NUNAMAKER, R A (1983) The incidence of *Nosema apis* in honey bees in Mexico. *Bee World* 64(3): 132-136.

WOYCIECHOWSKI, M; MORON, D (2009) Life expectancy and onset of foraging in the honey bee (*Apis mellifera*). *Insectes Sociaux* 56 (2): 193-201. http://dx.doi.org/10.1007/s00040-009-0012-6

YUE, C; SCHRÖDER, M; GISDER, S; GENERSCH, E (2007) Vertical-transmission routes for deformed wing virus of honey bees (*Apis mellifera*). *Journal of General Virology* 88: 2329-2336. http://dx.doi.org/10.1099/vir.0.83101-0

ZANDER, E (1909) Tierische Parasiten als Krankenheitserreger bei der Biene. *Münchener Bienenzeitung* 31, 196-204.

ZANDER, E; BÖTTCHER, F K (1984) *Krankheiten der Biene, 7:e. edition.* Verlag Eugen Ulmer, Stuttgart, Germany. 408 pp.

Journal of Apicultural Research 52(4): (2013)
DOI 10.3896/IBRA.1.52.4.22

© IBRA 2013

REVIEW ARTICLE

Standard methods for virus research in *Apis mellifera*

INTERNATIONAL BEE
RESEARCH ASSOCIATION

Joachim R de Miranda[1*], **Lesley Bailey**[2], **Brenda V Ball**[2], **Philippe Blanchard**[3], **Giles E Budge**[4],
Nor Chejanovsky[5], **Yan-Ping Chen**[6], **Laurent Gauthier**[7], **Elke Genersch**[8], **Dirk C de Graaf**[9],
Magali Ribière[3], **Eugene Ryabov**[10], **Lina De Smet**[9] and **Jozef J M van der Steen**[11]

[1]Department of Ecology, Swedish University of Agricultural Sciences, PO Box 7044, Uppsala, 750-07 Sweden.
[2]AgroEcology Department, Rothamsted Research, Harpenden, Hertfordshire, AL5 2JQ, UK.
[3]Anses, Sophia-Antipolis Laboratory, Bee Diseases Unit, BP 111, Sophia Antipolis, 06902 France.
[4]National Bee Unit, Food and Environment Research Agency, Sand Hutton, York, YO41 1LZ, UK.
[5]Agricultural Research Organization, The Volcani Center, PO Box 6, Bet Dagan, 50250 Israel.
[6]USDA-ARS, Bee Research Lab, BARC-E Bldg 306, Beltsville, MD 20705, USA.
[7]Swiss Bee Research Center, Agroscope Liebefeld-Posieux (ALP), Schwarzenburgstr. 161, Bern, CH-3003 Switzerland.
[8]Institute for Bee Research, Friedrich-Engels-Str. 32, Hohen Neuendorf, 16540 Germany.
[9]Laboratory of Zoophysiology, Department of Physiology, Ghent University, Ghent, B-9000 Belgium.
[10]School of Life Sciences, University of Warwick, Coventry CV4 7AL, UK.
[11]Plant Research International, Wageningen University and Research Centre, Business Unit Biointeractions and Plant Health,
Box 16, Wageningen, 6700AA The Netherlands.

Received 27 April 2012, accepted subject to revision 19 July 2012, accepted for publication 20 February 2013.

*Corresponding author: Email: joachim.de.miranda@slu.se

Summary

Honey bee virus research is an enormously broad area, ranging from subcellular molecular biology through physiology and behaviour, to individual and colony-level symptoms, transmission and epidemiology. The research methods used in virology are therefore equally diverse. This article covers those methods that are very particular to virological research in bees, with numerous cross-referrals to other *BEEBOOK* papers on more general methods, used in virology as well as other research. At the root of these methods is the realization that viruses at their most primary level inhabit a molecular, subcellular world, which they manipulate and interact with, to produce all higher order phenomena associated with virus infection and disease. Secondly, that viruses operate in an exponential world, while the host operates in a linear world and that much of the understanding and management of viruses hinges on reconciling these fundamental mathematical differences between virus and host. The article concentrates heavily on virus propagation and methods for detection, with minor excursions into surveying, sampling management and background information on the many viruses found in bees.

Métodos estándar para la investigación de virus en *Apis mellifera*

Resumen

La investigación de los virus de la abeja de la miel es un área sumamente amplia, que abarca desde la biología molecular subcelular hasta la fisiología y el comportamiento, desde síntomas al nivel de individuo hasta al nivel de la colmena, transmisión y epidemiología. Los métodos de investigación en virología son, por tanto, diversos. Este artículo incluye aquellos métodos específicos de la investigación virológica en las abejas, con numerosas referencias cruzadas con otros artículos del BEEBOOK y otros más generales, usados tanto en virología como en otras disciplinas. La base de estos métodos es la comprensión de los virus en su nivel primario de hábitat molecular, ambiente subcelular, que manipulan y con el que interactúan, para producir otros fenómenos de orden superior asociados a la infección del virus y la enfermedad. En segundo lugar, estos virus actúan en un mundo exponencial, mientras que los hospedadores actúan en un mundo lineal y gran parte del entendimiento y manejo de los virus depende de los fundamentos matemáticos de las diferencias entre el virus y el hospedador. El artículo se centra principalmente en la propagación de virus y en los métodos para su detección, con inclusiones menores en su estudio, el manejo del muestreo y la información general sobre los numerosos virus que se encuentran en las abejas.

Footnote: Please cite this paper as: DE MIRANDA, J R; BAILEY, L; BALL, B V; BLANCHARD, P; BUDGE, G; CHEJANOVSKY, N; CHEN, Y-P; GAUTHIER, L; GENERSCH, E; DE GRAAF, D; RIBIÈRE, M; RYABOV, E; DE SMET, L; VAN DER STEEN, J J M (2013) Standard methods for virus research in *Apis mellifera*. In *V Dietemann; J D Ellis; P Neumann (Eds) The COLOSS BEEBOOK, Volume II: standard methods for* Apis mellifera *pest and pathogen research. Journal of Apicultural Research* 52(4): http://dx.doi.org/10.3896/IBRA.1.52.4.22

西方蜜蜂病毒研究的标准方法

摘要

蜜蜂病毒研究是一个非常广阔的领域,涉及亚细胞分子生物学、生理学和行为学、个体和蜂群症状、传播和流行病学。因此病毒学研究中用到的方法种类繁多。本文涵盖了蜜蜂病毒学研究中特有的一些方法。其中涉及大量在病毒学和其它研究中都有用到的方法,在 *BEEBOOK* 关于普通方法的章节中已有介绍。这些方法的根源是认识到病毒归根到底生活于一个分子和亚细胞世界,它们操控和作用于这一环境,以产生和病毒感染与疾病相关的更高级别的现象。其次,病毒在指数世界运行,而宿主在线性世界运行,理解和控制病毒很大程度上依赖于协调病毒和宿主之间的这些基本数学差异。本文重点针对病毒复制和病毒检测方法,也提及调查、取样操作及蜜蜂上发现的一些病毒的背景信息。

Keywords: COLOSS *BEEBOOK*, honey bee, virus, sampling, purification, assays, standardisation

Table of Contents

Table of Contents cont'd

1. Introduction

1.1. Honey bee viruses

There are currently about 24 viruses identified in honey bees, whose physical and biological properties are described in Table 1 and Table 2 respectively. Most of these were discovered by Bill Bailey, Brenda Ball and colleagues at Rothamsted Experimental Station, UK during the 1960s-1980s (Bailey and Ball, 1991; Ribière *et al.,* 2008). More recent additions have come mostly from mass sequencing of RNA and DNA from whole bee extracts (Fujiyuki *et al.,* 2004; Cox-Foster *et al.,* 2007; Cornman *et al.,* 2010; Runckel *et al.,* 2011), and it may well be that there is overlap between the traditionally described viruses and these newly described viral sequences. Several viruses are also closely enough related to be regarded as members of a single species complex (DWV/VDV-1/EBV; ABPV/KBV/IAPV; SBV/TSBV; BVX/BVY and LSV-1/LSV-2), reducing the total to around 16-18 truly unique viruses.

Although some viruses produce recognizable symptoms at sufficiently elevated titres, honey bee viruses generally persist naturally in honey bee populations at low levels, without causing overt symptoms, using a variety of transmission routes (Fig. 1; Table 2). Symptoms are, however, still the principal method by which diseases are diagnosed in the apiary. The advantages of symptom-based diagnosis are that it is robust, simple, fast and cheap and for some diseases accurate. The major disadvantages are that:

- many virus infections do not present visible symptoms at all times
- not all life stages present symptoms
- often different viruses produce similar symptoms (*e.g.* paralysis)
- a single virus may present different symptoms (*e.g.* CBPV)
- symptoms can be confounded if multiple virus infections are present

All viruses are asymptomatic at lower levels of infection and most shorten the life span of bees to varying degrees. The diagnostic symptoms for the major virus diseases have been described in detail by Bailey and Ball (1991) and can be summarized as follows:

1.1.1. Acute bee paralysis virus /Kashmir bee virus /Israeli acute paralysis virus

Acute bee paralysis virus (ABPV), Kashmir bee virus (KBV) and Israeli acute paralysis virus (IAPV) are three closely related viruses (de Miranda *et al.,* 2010a) that are largely symptomless, but they can be lethal at individual and colony level (Allen and Ball, 1995; Todd *et al.,* 2007), particularly when transmitted by *Varroa destructor* (Ball, 1985; 1989; Ball and Allen, 1988) which is an active vector of these viruses (Chen *et al.,* 2004a; Shen *et al.,* 2005a; 2005b; DiPrisco *et al.,* 2011). These viruses are characterized by the ability to kill both pupae (after injection; Bailey, 1967; Bailey and Ball, 1991) and adult bees (after injection or feeding: Maori *et al.,* 2007a; 2009; Hunter *et al.,* 2010)

very rapidly; 3-5 days after inoculation with sufficient virion loads. This exerts a strong negative selection pressure on the transmission by varroa, since infected pupae fail to complete development, preventing the release of infectious mites from the pupal cells (Sumpter and Martin, 2004). The association of these viruses with varroa infestation is therefore unstable and much influenced by the presence of other viruses that are better adapted to transmission by varroa.

1.1.2. Black queen cell virus

The main symptoms for black queen cell virus (BQCV) consist of blackened cell walls of sealed queen cells, containing dead pro-pupae (Bailey and Ball, 1991; Leat *et al.,* 2000). Diseased larvae have a pale yellow appearance and tough sac-like skin, much like sacbrood. The virus is present in adult bees but without obvious symptoms.

1.1.3. Aphid lethal paralysis virus & Big Sioux River virus

Aphid lethal paralysis virus (ALPV) is a common intestinal dicistrovirus of several major agricultural aphid pests, associated with aphid population declines (van Munster *et al.,* 2002 Laubscher and von Wechmar, 1992; 1993). Big Sioux River virus (BSRV) is closely related to *Rhopalosiphum padi* virus (RhPV; Moon *et al.,* 1998), another common intestinal Dicistrovirus that uses the plant vascular system to transmit horizontally between aphids (Gildow and D'Arcy, 1990). Both can be detected infrequently at very low background levels in adult honey bees throughout the year, with a sharp quantitative increase during late summer (Runckel *et al.,* 2011) when bees often feed on honeydew (aphid excreta) during low nectar flows. It is unclear therefore whether these viruses are incidental or truly infectious in bees. Either of these may be related to Berkeley bee picorna-like virus (BBPV; Lommel *et al.,* 1985), which has not yet been sequenced.

1.1.4. Deformed wing virus /kakugo virus /Varroa destructor virus-1 /Egypt bee virus

The symptoms for deformed wing virus (DWV) consist of bees with crumpled and/or vestigial wings and bloated abdomen and infected bees die soon after emergence. Asymptomatic bees can also be heavily infected, though with lower titres than symptomatic bees (Bowen-Walker *et al.,* 1999; Lanzi *et al.,* 2006; Tentcheva *et al.,* 2006). The virus is detected in all other life stages as well, but without obvious symptoms (Chen *et al.,* 2005a; 2005b; Yue and Genersch 2005; Lanzi *et al.,* 2006; Tentcheva *et al.,* 2006; Fievet *et al.,* 2006; Yue *et al.,* 2006; de Miranda and Genersch, 2010). 'kakugo' virus (KV; Fujiyuki *et al.,* 2004; 2006) and other strains of DWV (Terio *et al.,* 2008) have been associated with elevated aggression in bees, although naturally aggressive bee races are not more infected with DWV than gentle bee races (Rortais *et al.,* 2006). DWV also affects sensory response, learning and memory in adults (Iqbal and Müller, 2007).

Table 1. Summary of the physical properties, such as particle shape, size, capsid protein profile, genome type and length and taxonomy, of the currently known honey bee viruses. Adapted from Bailey and Ball (1991).

VIRUS		SHAPE	SIZE	CAPSID PROTEINS	NUCLEIC ACID	GENOME SIZE	TAXONOMY
		PHYSICAL PROPERTIES					
Acute bee paralysis virus	ABPV	icosahedral	30nm	35-9-33-24kDa	ssRNA	~9.5kb	Dicistroviridae
Kashmir bee virus	KBV	icosahedral	30nm	37-6-34-25kDa	ssRNA	~9.5kb	Dicistroviridae
Israeli acute paralysis virus	IAPV	icosahedral	30nm	35-7-33-26kDa	ssRNA	~9.5kb	Dicistroviridae
Black queen cell virus	BQCV	icosahedral	30nm	31-14-29-30kDa	ssRNA	~9.5kb	Dicistroviridae
Aphid lethal paralysis virus	ALPV	icosahedral	30nm	25-7-32-28kDa*	ssRNA	~10kb	Dicistroviridae
Big Sioux River virus	BSRV	icosahedral	30nm	28-5-29-30kDa	ssRNA	~10kb	Dicistroviridae
Deformed wing virus	DWV	icosahedral	30nm	32-2-44-28kDa	ssRNA	~10kb	Iflaviridae
Varroa destructor virus-1	VDV-1	icosahedral	30nm	32-2-46-28kDa	ssRNA	~10kb	Iflaviridae
Egypt bee virus	EBV	icosahedral	30nm	30-2-41-25kDa	ssRNA	?	Iflaviridae
Sacbrood virus	SBV	icosahedral	30nm	31-2-32-30kDa	ssRNA	~9kb	Iflaviridae
Thai/Chinese sacbrood virus	TSBV	icosahedral	30nm	31-2-32-30kDa	ssRNA	~9kb	Iflaviridae
Slow bee paralysis virus	SBPV	icosahedral	30nm	27-2-46-29kDa	ssRNA	~9.5kb	Iflaviridae
Chronic bee aralysis virus	CBPV	anisometric	30~60nm	23-(30/50/75?)kDa	ssRNA	~2.3kb/~3.7kb	unclassified
Chronic bee paralysis satellite virus	CBPSV	icosahedral	17nm	15kDa	ssRNA	(3x)~1.1kb	satellite
Cloudy wing virus	CWV	icosahedral	17nm	19kDa	ssRNA	~1.4kb	?
Bee virus-X	BVX	icosahedral	35nm	52kDa	ssRNA	?	?
Bee virus-Y	BVY	icosahedral	35nm	50kDa	ssRNA	?	?
Lake Sinai Virus-1	LSV-1	?	?	63kDa*	ssRNA	~5.5kb	unclassified
Lake Sinai Virus-2	LSV-2	?	?	57kDa*	ssRNA	~5.5kb	unclassified
Arkansas bee virus	ABV	icosahedral	30nm	43kDa	ssRNA	~5.6kb	?
Berkeley bee picorna-like virus	BBPV	icosahedral	30nm	37-?-35-32kDa	ssRNA	~9kb	?
Varroa destructor Macula-like virus	VdMLV	icosahedral	30nm	24kDa*	ssRNA	~7kb	Tymoviridae
Apis mellifera filamentous virus	AmFV	rod	150x450nm	12x(13~70kDa)	dsDNA	?	Baculoviridae
Apis iridescent virus	AIV	polyhedral	150nm	?	dsDNA	?	Iridoviridae
		* (genome predicted)		SDS-PAGE	(order in polyprotein)		

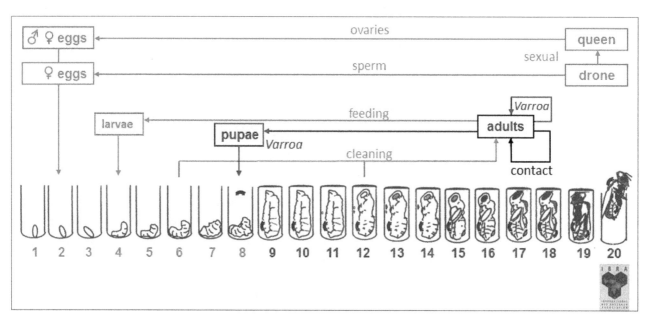

Fig. 1. Diagram describing the different possible transmission routes for honey bee viruses. Adapted from de Miranda *et al.* (2011).

Table 2. Summary of the current state of knowledge concerning biological properties of honey bee viruses, such as transmission routes, associations with other parasites/pathogens, principal life stages affected and seasonal incidences. Adapted from de Miranda *et al.* (2011).

VIRUS		TRANSMISSION							ASSOCIATION				LIFE STAGE				SEASON		
		HORIZONTAL				VERTICAL							INFECT/SYMPTOMS						
		ORAL-FECAL	CONTACT	AIR	VARROA	OVARIES	SEMEN	SPERM	VARROA	ACARAPIS	NOSEMA	MALPIGHAMOEBA	EGGS	LARVAE	PUPAE	ADULTS	SPRING	SUMMER	AUTUMN
Acute bee paralysis virus	ABPV	+	-	?	+	+	+	?	+	?	?	?	+/-	+/-	+/~	+/+	+	+++	++
Kashmir bee virus	KBV	+	-	?	+	+	~	?	+	?	?	?	+/-	+/-	+/+	+/+	+	++	+++
Israeli acute paralysis virus	IAPV	+	-	?	+	+	~	?	+	?	?	?	+/-	+/-	+/~	+/+	+	++	++
Black queen cell virus	BQCV	+	-	?	~	+	?	?	+	?	+	?	+/-	+/-	+/+	+/-	+	+++	+
Aphid lethal paralysis virus	ALPV	?	?	?	?	?	?	?	?	?	?	?	?/?	?/?	-/-	+/?	-	+++	-
Big Sioux River virus	BSRV	?	?	?	?	?	?	?	?	?	?	?	?/?	?/?	-/-	+/?	-	+++	++
Deformed wing virus	DWV	+	-	?	+	+	+	?	+	?	?	?	+/-	+/-	+/+	+/+	+	++	+++
Varroa destructor virus-1	VDV-1	+	-	?	+	+	+	?	+	?	?	?	+/-	+/-	+/+	+/+	+	++	+++
Egypt bee virus	EBV	?	?	?	?	?	?	?	?	?	?	?	?/?	?/?	+/?	+/~	?	?	?
Sacbrood virus	SBV	+	-	?	-	?	?	?	~	?	?	?	?/?	+/+	+/-	+/~	+++	++	+
Thai/Chinese sacbrood virus	TSBV	+	?	?	?	?	?	?	?	?	?	?	?/?	+/+	+/-	+/~	?	?	?
Slow bee paralysis virus	SBPV	+	-	?	+	?	?	?	+	?	?	?	?/?	+/-	+/-	+/+	+	+	+
Chronic bee paralysis virus	CBPV	+	+	?	-	?	?	?	~	~	?	?	~/-	+/-	+/-	+/+	++	++	+
Chronic bee paralysis satellite virus	CBPSV	?	?	?	?	?	?	?	?	?	?	?	?/?	?/?	?/?	+/?	+	+	+
Cloudy wing virus	CWV	?	~	~	-	?	?	?	~	?	?	?	-/-	~/-	~/-	+/+	+	+	+
Bee virus-X	BVX	+	?	?	?	?	?	?	?	?	-	+	-/-	-/-	-/-	+/+	+++	+	+
Bee virus-Y	BVY	+	?	?	?	?	?	?	?	?	+	-	-/-	-/-	-/-	+/+	+	+++	+
Lake Sinai Virus-1	LSV-1	?	?	?	?	?	?	?	?	?	?	?	?/?	?/?	+/?	+/?	++	+++	++
Lake Sinai Virus-2	LSV-2	?	?	?	?	?	?	?	?	?	?	?	?/?	?/?	~/?	+/?	+++	+	+
Arkansas bee virus	ABV	?	?	?	?	?	?	?	?	?	?	?	?/?	?/?	~/?	+/?	?	?	?
Berkeley bee picorna-like virus	BBPV	?	?	?	?	?	?	?	?	?	?	?	?/?	?/?	?/?	+/?	?	?	?
Varroa destructor Macula-like virus	VdMLV	?	?	?	+	?	?	?	+	?	?	?	?/?	?/?	+/?	+/?	+	++	+++
Apis mellifera filamentous virus	AmFV	+	?	?	?	?	?	?	?	?	+	?	-/-	-/-	-/-	+/+	+++	+	+
Apis iridescent virus	AIV	?	?	~	?	?	?	?	?	?	?	?	-/-	-/-	-/-	+/+	+	++	+
		+		(present)							~		(uncertain)		?		(unknown)		

Varroa destructor virus-1 (VDV-1) is genetically closely related to DWV but is reported to be more specific to *Varroa destructor* than to bees (Ongus, 2006). However, both viruses replicate in varroa mites as well as in honey bees (Ongus *et al.*, 2004; Yue and Genersch, 2005; Zioni *et al.*, 2011); both have been detected at high titres in different honey bee tissues (Zioni *et al.*, 2011; Gauthier *et al.*, 2011); both have been found in regions where *V. destructor* is absent (Martin *et al.*, 2012) and natural recombinants between them have been found (Moore *et al.*, 2011). VDV-1 and DWV therefore appear to co-exist in bees and mites as part of the same species-complex (de Miranda and Genersch, 2010; Moore *et al.*, 2011; Gauthier *et al.*, 2011; Martin *et al.*, 2012).

Egypt bee virus (EBV) is serologically related to DWV, but has no known symptoms in adults, pupae or larvae (Bailey *et al.*, 1979).

1.1.5. Sacbrood virus / Thai sacbrood virus

The clearest symptoms of sacbrood virus (SBV) appear a few days after capping, and consist of non-pupated pale yellow larvae, stretched on their backs with heads lifted up towards the cell opening, trapped in the unshed, saclike larval skin containing a clear, yellow-brown liquid. The virus is also present in adult bees, but without symptoms (Lee and Furgula, 1967; Bailey, 1968). Diseased larvae are most commonly seen in spring, but the disease normally clears quickly with rapid expansion. However, the Asian honey bee, *Apis cerana*, frequently suffers from lethal sacbrood epidemics caused by a closely related strain of SBV, variously called Thai sacbrood virus (TSBV), Chinese sacbrood virus (CSBV) or Korean sacbrood. The genetic differences of these strains with the SBV infecting *A. mellifera* are minimal. SBV-infected adults cease to attend brood or eat pollen, start foraging much sooner than normal, and only forage nectar, rarely pollen (Bailey and Fernando, 1972). These may be behavioural adaptations by *A. mellifera* to prevent sacbrood epidemics, since SBV is shed in the hypopharyngeal secretions fed to larvae and combined with pollen to make bee-bread (Bailey and Ball, 1991).

1.1.6. Slow bee paralysis virus

Slow bee paralysis virus (SBPV) is characterised by the paralysis of the front two pairs of legs of adult bees, a few days before dying, after inoculation by injection (Bailey and Woods, 1974). The virus is associated with, and transmitted by, *V. destructor* (Bailey and Ball, 1991; Denholm, 1999). Despite this association, SBPV is rarely detected in bee colonies (Bailey and Ball, 1991; de Miranda *et al.*, 2010b). SBPV can also be detected in larvae and pupae, but produces no symptoms in these.

1.1.7. Chronic bee paralysis virus / satellite virus

Chronic bee paralysis virus (CBPV) manifests itself in adult bees through two distinct set of symptoms. One set consists of trembling of the wings and bodies and a failure to fly, causing them to crawl in front of the hive in large masses. They often have partly spread, dislocated wings and bloated bodies as well. The other set of symptoms consists of hairless, greasy black bees caused by nibbling attacks from healthy bees in the colony. They soon also become flightless, tremble and die (Bailey, 1965; Bailey and Ball, 1991; Ribière *et al.*, 2010). The virus also infects the larval and pupal stages, can be detected in faecal material and is efficiently transmitted through contact and feeding (Bailey *et al.*, 1983b; Ribière *et al.*, 2010). CBPV is sometimes associated with a small "satellite" virus; chronic paralysis satellite virus (CBPSV; originally called chronic bee paralysis virus associate CBPVA), which has a unique genome and capsid protein to CBPV (Ribière *et al.*, 2010) and is of unknown significance to symptomatology (Bailey *et al.*, 1980; Ball *et al.*, 1985).

1.1.8. Cloudy wing virus

The symptoms for cloudy wing virus (CWV) consist of opaque wings of severely infected adult bees, with lower titres resulting in asymptomatic infected bees (Bailey *et al.*, 1980; Bailey and Ball, 1991; Carreck *et al.*, 2010). It cannot be propagated in larvae or pupae. It has an unpredictable incidence, no regular associations with other pathogens or pests. Like chronic bee paralysis satellite virus it has a small particle and very small genome, but they are serologically unrelated and their single capsid proteins are of different size (Table 1; Bailey *et al.*, 1980).

1.1.9. Bee virus X / Bee virus Y

Bee virus X (BVX) is largely symptomless in adult bees and does not multiply in larvae or pupae (Bailey and Ball, 1991). It is associated with the protozoan *Malpighamoeba mellificae* that causes dysentery in winter bees (Bailey *et al.*, 1983a). Bee virus Y (BVY) is serologically related to BVX and is similarly symptomless in adult bees, larvae or pupae. It is associated in adult bees with the dysentery inducing microsporidium *Nosema apis* (Bailey *et al.*, 1983a). Both viruses are common, BVY more so than BVX, with strong peaks in late winter for BVX and early summer for BVY (Table 2; Bailey and Ball, 1991).

1.1.10. Lake Sinai virus-1 / Lake Sinai virus-2

Lake Sinai virus-1 (LSV-1) and Lake Sinai virus-2 (LSV-2) are two closely related viruses that were identified in through a mass metagenomic sequencing survey of honey bee colonies in the USA (Runckel *et al.*, 2011). Their genome organization and sequences place them together with CBPV, in a unique family somewhere between the Nodaviridae and Tombusviridae. Both viruses are common and very abundant at peak incidence. LSV-1 is more common than LSV-2, present throughout the year with a peak in early summer. LSV-2 has a very sharp incidence and abundance peak in late winter with low incidence and abundance the rest of the year. These viruses have also been detected, with similar incidences and titres, in historical European honey bee samples. LSV-1 and LSV-2 have strong similarities in capsid and genome size, seasonal incidence, predominantly adult-based infection and absence of overt symptoms with Bee virus Y and Bee virus X respectively (Table 1 and 2), and may therefore be related.

1.1.11. Arkansas bee virus & Berkeley bee virus

Arkansas bee virus (ABV) and Berkeley bee picorna-like virus (BBPV) are two viruses first identified in the USA (Bailey and Woods, 1974; Lommel *et al.,* 1985; Bailey and Ball, 1991) of which very little is known other than that they often occur together. They have no known symptoms in adult bees or brood. BBPV has typical capsid and genome size characteristics of the Dicistro- and Iflaviruses.

1.1.12. *Apis mellifera* filamentous virus

Apis mellifera filamentous virus (AmFV) is a baculovirus-like DNA virus that has no physical symptoms. It renders the haemolymph of adult bees milky white with rod-shaped viral particles, when examined by electron microscopy (Clark, 1978; Bailey and Ball, 1991).

1.1.13. Apis iridescent virus

The symptoms for Apis iridescent virus (AIV) are similar to the adult flightless clustering symptoms of CBPV (Bailey *et al.,* 1976; Bailey and Ball, 1978). It is only known to occur in adult bees. A partial sequence of AIV has been published (Webby and Kalmakoff, 1999).

1.2. Definitions: pathogenicity *vs* virulence; incidence *vs* prevalence

The terms 'Infectivity', 'Pathogenicity', 'Virulence' and 'Transmissibility' are often used interchangeably, which has led to efforts to tighten and standardize their definition and adapt them to our improved understanding of host-pathogen interactions (Casadevall and Pirofski, 1999; 2001). The same is true for the terms 'Incidence' and 'Prevalence' in surveys and epidemiology. Here are their definitions:

- Prevalence: The proportion of a population that is infected, or diseased, at any one time.
- Incidence: This is the risk of new infection during a specified time. It is globally related to prevalence as a function of time: *prevalence = incidence* x *time*
- Infectivity: This refers to the ability of a microorganism to invade and replicate in a host tissue, whether the microbe is pathogenic or not.
- Pathogenicity: This is a qualitative trait, referring to the inherent, genetic capacity of a microorganism to cause disease, mediated by specific virulence factors. Whether or not it does so, is the result of the specific host-pathogen interactions.
- Virulence: This is a quantitative trait, representing the *extent* of the pathology caused by a microorganism. Virulence is therefore a trait expressing the interaction between a pathogen and its host. Its definition has been re-assessed recently (Casadevall and Pirofski, 1999; 2001), in view of the significant influence of the host's immunological condition on the extent of the damage (*i.e.* virulence) caused by a pathogen.

Virulence is usually correlated to the pathogen's capacity to multiply in the host (Casadevall and Pirofski, 2001) represented, for example, by the virion titre when symptoms appear (Figure 2) or the rate of multiplication. It can also be affected by host and environmental factors, such as the transmission route or type of tissue/life-stage infected. For example, a pathogen may be virulent when infecting one type of tissue and non-virulent when infecting a different tissue (Casadevall and Pirofski, 1999). Virulence is therefore dependent on the nature of the infection.

Virulence is also a relative trait, referring to the differences in the degree of pathology caused by strains of the same pathogen, or differences in the efficiency with which different strains can cause symptoms (Pirofski and Casadevall, 2012). For example, a pathogen strain that requires few particles to produce disease symptoms (strain-A in Figure 2) would be more virulent than a strain that requires many particles to produce the same symptoms (strain-B in Figure 2).

Since virulence is a quantitative measure, methods have been developed to quantify the relative contributions of different virulence factors to a phenotype (McClelland *et al.,* 2006).

- Transmissibility: This refers to the efficiency with which a pathogen is transmitted to naïve hosts. There are valid arguments that at epidemiological level, transmissibility could be considered a component of virulence (Figure 2). The relationship between transmission and virulence is a major topic in pathogen-host evolutionary theory (*e.g.* Ebert and Bull, 2003) and has been discussed within the context of honey bee colony structure (Fries and Camazine, 2001) and honey bee virus transmission (de Miranda and Genersch, 2010).

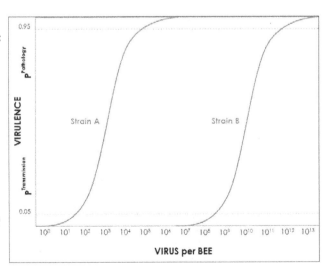

Fig. 2. Diagram describing the Log-linear relationship between virus concentration (X-axis) and virulence (Y-axis), represented by the degree/probability of Pathology ($P^{Pathology}$) or the efficiency/probability of Transmission ($P^{Transmission}$). Other variables can also be plotted on the Y-axis.   Image © J R de Miranda.

1.3. Virus replication and variation

Viruses have two main characteristics that are fundamental to the design, analysis and interpretation of virological experiments, surveys and assays. These are:

- The potential for rapid, exponential growth
- The potential for rapid evolution and high levels of molecular variability

Below we briefly discuss these two features and their impact on experimentation and data management.

1.3.1. The mathematics of virus replication and transmission

Viruses are obligatory cell parasites and as such are capable of rapid, exponential growth. This is particularly the case for viral replication within individual organisms. This means that the virus replication dynamics can range from linear (when the virus persists as a covert infection, with minimal replication) all the way to fully logarithmic (when the virus is growing exponentially) and back to linear again when the maximum virus load within diseased or dying organisms is reached, due to exhaustion of the resources for replication (Fig. 2).

The epidemiological spread between organisms is influenced by the transmission medium (air, water, vector), whose rules of dispersion are often not fully exponential. This also applies to other barriers to virus proliferation, such as tissue-specificity, interference, auto-interference, RNA silencing, and immune reactions which all can influence virus multiplication, shedding and dispersal. These restrictions can temper the logarithmic character of the quantitative virus data distribution, at the individual bee, colony or regional level.

What this means is that, from the design of experiments through to the analysis of the data, allowance has to be made for non-linear distributions of the data, ranging from fully logarithmic (pathogenic replication) through semi-exponential (epidemic proliferation) to near-linear (covert replication, dispersal). This can be addressed through transformations, thresholds or non-linear models, but it MUST be dealt with appropriately. Guidelines for this can be found in detail in the *BEEBOOK* paper on statistical methods (Pirk *et al.,* 2013), with aspects specific to virus research also covered in section 3; "Statistical Aspects" of this chapter.

1.3.2. Virus variability and evolution

The second major characteristic of viruses, particularly important when designing molecular assays, is the ease and speed with which they can generate and maintain large amounts of molecular variability. The virus encoded RNA dependant RNA polymerase (RdRp), which facilitates genome replication, lacks proofreading and repair mechanisms causing a high mutation rate in RNA viruses. Therefore a virus is not so much an individual entity with a fixed genome, but rather a large 'swarm' of closely related variants, recombinants and other genetic oddities that are transmitted between individuals as a unit. There are two forces that shape the genetic identity of such a 'mutant swarm' (or 'quasi-species' as it is officially known)

- Fierce *competition* between molecular variants for supremacy in a particular cell, host *etc.*
- Functional *co-operation* between variants, where a temporarily disfavoured variant can remain within the quasi-species by 'borrowing' essential functions such as replication and packaging from the locally dominant variant.

The functional co-operation is an adaptive super-feature of viruses, since it allows a wide range of genetic diversity to persist within a quasi-species across time, hosts and environments. The true adaptive strength of a virus lies therefore more in the diversity within the swarm than in the evolutionary abilities of any one strain.

The importance of this variability for experimentation is in the design of diagnostic assays for virus detection. Serological assays, such as ELISA, are generally not affected by this variability which is mainly expressed at the nucleic acid level. However, nucleic acid assays, especially those based on the Polymerase Chain Reaction (PCR), are often very sensitive to microvariation at the nucleotide level, where even a single base-pair difference can be exploited for specific diagnosis. It is here therefore that supreme care must be taken to ensure that the assays developed for detecting viruses, or virus strains, are designed accurately and conservatively, to avoid non-detection due to assay inadequacy (see also section 12; "Quality Control").

2. Virus surveys

2.1. Introduction

Bee diseases caused by viruses are significant threats to apiculture. Pathogen surveillance is an essential component of a structured (inter)national management strategy to contain or prevent epidemics of viral diseases in honey bee populations. Such surveillance is done both through questionnaires of beekeepers (see the *BEEBOOK* paper on surveys; van der Zee *et al.,* 2013) and through monitoring bee colonies for pathogen prevalence and amount.

Although some honey bee viruses, such as chronic bee paralysis virus (CBPV), deformed wing virus (DWV), black queen cell virus (BQCV), sacbrood bee virus (SBV) and cloudy wing virus (CWV), are capable of causing diseases with recognizable symptoms, most honey bee viruses usually persist and spread between colonies as covert infections without apparent symptoms in bees. Many other bee viruses such as bee virus X and bee virus Y (BVX; BVY) or filamentous virus (AmFV), either do not cause outward symptoms at all, or others, such as slow bee paralysis virus (SBPV) and acute bee paralysis virus (ABPV) only do so under laboratory conditions or produce vague, non-descript symptoms, such as 'early death' by Kashmir bee virus (KBV) and Israeli acute paralysis virus (IAPV) or 'clustering' associated with Apis iridescent virus (AIV). Many bee virus infections cannot therefore be identified through field observations, because the symptoms are non-existent, inconsistent or absent due to low titres. Until field-ready pathogen ID-kits become available, pathological analysis requires the transport of samples to a laboratory for analysis.

2.2. Types of survey

In epidemiology, there are two broad types of pathogen surveillance systems. Although they are often treated together in summaries and reviews, they are in fact radically different in purpose, strategy, methods and implications. Here we discuss these briefly as they apply to honey bee virus surveillance. For more detail information on how to manage these different approaches, see the *BEEBOOK* papers on epidemiology (vanEngelsdorp *et al.*, 2013) and surveys (van der Zee *et al.*, 2013).

2.2.1. Passive surveillance

This is the type of survey where beekeepers voluntarily send in suspect bee samples to a diagnostic laboratory for analysis. Most countries operate such a service and often the samples are of colonies that died suddenly, usually after winter. The survey is 'passive' since it only analyses material received from the public. It is analogous in human epidemiology to the entry reports of health care facilities for active disease cases. There is no statistically designed sampling strategy, samples will be biased by beekeeper interest, experience and knowledge and proximity to the diagnostic facilities. Samples will arrive in various states of decomposition making definitive diagnosis problematic (see section 4; 'Virus Sample Management'), the data will be heavily biased towards diseased colonies with high pathogen prevalence/titre and the results will be more relevant to the management of epidemics rather than to their prevention.

2.2.2. Active surveillance

Active surveillance schemes fill many of the gaps of passive surveillance. These are usually statistically designed sampling schemes to determine pathogen prevalence within the general bee population, irrespective of symptoms. The bee samples are alive when collected, making molecular detection uniform and reliable, and the data are an accurate representation of the complete pathogen presence within a region. Often samples are taken repeatedly from the same colonies throughout a season, which has to be taken into account during data analysis. A sub-category of active surveillance systems is 'sentinel' surveillance, *i.e.* a series of designated 'monitoring' colonies placed to catch certain pathogens before they reach a particular region.

3. Statistical aspects

3.1. Introduction

Whether experimenting or conducting surveys, data will be generated that will need to be analysed, and this requires the use of statistics. Fortunately, there are excellent statistical tools available right now for helping design experiments, determining the sample size needed to be able to make certain conclusions, for analysing the data and for modelling and prediction. It is highly recommended to include an expert statistician in the project right from the beginning, at the design stage. It will mean that the experiment is set up correctly, that sampling is as efficient as possible and that the data are analysed correctly. The most important statistical concepts and practices relevant to honey bee research and surveying are covered in the *BEEBOOK* paper on statistical methods (Pirk *et al.*, 2013), to which the reader is referred to for guidance.

3.2. Statistical distribution of virus data

The major additional point specifically relevant to honey bee virus research (and probably other pathogens) is that virus titres follow an Exponential (~Logarithmic) distribution, rather than a Normal distribution (Fig. 2; Gauthier *et al.*, 2007; Brunetto *et al.*, 2009; Yañez *et al.*, 2012; Locke *et al.*, 2012). Also, the prevalence of very rare viruses (pathogens, parasites) may follow a Poisson distribution rather than a Binomial distribution. These differences in how the primary virus data is statistically distributed affects the design of an experiment, the determination of sample sizes (Wolfe and Carlin, 1999), surveying strategy, the analysis of pooled samples and the management of the data produced. This subject is treated in more detail in the *BEEBOOK* paper on statistical methods (Pirk *et al.*, 2013).

3.2.1. Log-transformation

Exponentially distributed data (*i.e.* quantitative virus titre data) must be treated with a power transformation (Box and Cox, 1964; Bickle and Doksum, 1981), usually a log-transformation, before they can be used to estimate descriptive statistics (means, medians, variances etc.) or be used in parametric statistical analyses (ANOVAs, correlations, GLMs etc.). This can be done prior to statistical analysis, or can be incorporated as part of the statistical analysis.

3.2.2. Zero values

Since it is not possible to log-transform zero values, instances of non-detection (*i.e.* a 'zero' value) should be replaced by a non-zero constant value appropriate to the variable in question (Cox *et al.*, 2000). A logical constant value to use for replacing zero-values in quantitative virus data sets is one that is set just below the detection threshold for the virus in question (Yañez *et al.*, 2012; Locke *et al.*, 2012). This approach treats zero values as "below detection threshold" rather than as "absence of virus", which is usually also a more accurate description of the virus status of a sample, especially if the virus is known to be present within the wider bee population.

4. Virus sample management

4.1. Introduction

The aim of a survey or experiment is that the final data should as closely as possible represent the (virus) status of the bee or colony when the sample was taken, since ultimately the data interpretation, conclusions and recommendations will again refer to live bees. Sample management (from collection and field preservation, to transportation, short- and long-term storage, processing and finally analysis) is therefore crucial to the accurate interpretation of survey and experimental data. Honey bee sample management is covered in detail in the *BEEBOOK* paper on molecular methods (Evans *et al.*, 2013) with a reduced version given below. Most viruses have an RNA genome and much of the host's molecular response to virus infection is also at the RNA level. However, RNA is highly sensitive to degradation by nucleases and a major criterion for bee sample management is therefore to minimize this degradation (Chen *et al.*, 2007; Dainat *et al.*, 2011). The primary concern is not so much the virus particle (which is usually relatively robust), but rather the viral replicative RNA intermediates and host mRNAs.

4.2. What, where and when to sample

Different viruses have different infection patterns, life-stage/tissue preferences and seasonal prevalences (Table 2), and so the decision as to what bee stages to sample, when/how often to sample and where to sample depends to a large degree on the objective of the experiment/survey and on the virus studied. When the experiments/ surveys are detailed and specific, the sampling regimen should be designed to suit those specific aims. However, often the same samples will be analysed for multiple viruses, or the experiments/ surveys are more global in character, requiring a more consensual approach to sampling. Here are some considerations for making sampling decisions in these situations. See also the *BEEBOOK* paper on statistical methods (Pirk *et al.*, 2013) for decisions on how to determine the optimal sample size and pooled samples.

4.2.1. What to sample?

All viruses described to date can be detected in adult bees (Table 2). This is logical, since adult bees are central to most of the virus transmission pathways (Fig. 1), due to their high mobility, contact rate and diverse contact network. This makes adult bees the most suitable single bee stage for detecting all viruses. Within adult bees, the gut is a major site of accumulation for most viruses (and many other pathogens), and is thus the most suitable single tissue type for sampling.

4.2.2. Where to sample?

The difficulty with adult bees is that the virus titre can be influenced by the age and the tasks of the bee. It is therefore advisable to sample as much as possible the same age/task group throughout the experiment, in order to minimize the influence of such effects on the data (Van der Steen *et al.*, 2012). In practical terms, this means either sampling from the brood chamber (mostly young nurse bees), the honey supers (medium-age bees) or at the entrance (older foraging bees). The choice of age-class for sampling is less important than consistently sampling the same age-class. See also the *BEEBOOK* paper on statistical methods (Pirk *et al.*, 2013).

4.2.3. When to sample?

There are two considerations here. The first is the best time of day to collect a sample and the second how often to sample. The best time to sample is either on sunny days, during the afternoon, when bees are actively foraging and the adult population is most clearly subdivided according to tasks/age, or at the other extreme during cold rainy days, when there is no substructuring of the population and all bees are sampled randomly, irrespective of age class. This choice depends on the design and purpose of the sampling scheme.

The frequency of sampling depends on the type survey/ experiment conducted:

- For single virus geographic prevalence surveys, the best time of year for sample collection would be during the seasonal peak for the virus in question (Table 2).
- For multivirus-pathogen geographic prevalence surveys, and if only a single sample is collected, the best time would be autumn, when most viruses have a seasonal peak.
- For multivirus-pathogen surveys it is advisable to sample at least three times per season; in early spring when the colony is expanding, during peak productivity in summer and during late autumn when the colony is contracting, in order to catch the different pathogens at their peaks, observe seasonal variations in prevalence and identify possible associations between different pathogens-parasites.
- For colony-level experiments with repetitive sampling, it may be useful to take into account the natural turnover of the adult population when considering sampling frequency. During summer, both the brood stage and the adult stage last about three weeks. Sampling every three weeks therefore means that a completely new generation of adult bees is sampled each time, corresponding to the brood generation of the previous sampling point.

4.3. Sample collection

Methods for collecting different sample types (adults *vs* brood; whole bees *vs* extracted tissues), for sample preservation (chemical and/or temperature) and for transportation (hours *vs* days) are described in detail in the *BEEBOOK* paper on molecular methods (Evans *et al.,* 2013). Faeces may be a useful sample type for the non-destructive or repetitive sampling of the individual bees, such as queens (Hung, 2000). The general rules for sample collection are to get the samples to a freezer as quick as possible, and to keep dead, preserved or processed bee samples as cold as possible during transport.

4.3.1. Adult bees

Adult bees are usually collected from brood frames (young bees); honey frames (older bees); at the hive entrance (foragers) or in dead-bee traps (dead bees). Around 200 bees (about 20g, or a cup-full) are shaken or brushed into either a cage or ventilated box with food (live transport) or in a 1l plastic bag (cold transport). This is enough for both mite and virus analyses.

4.3.2. Pupae

Pupae are usually collected as a 10cm x 10cm section of sealed brood, placed in a suitably sized ventilated box (live transport) or in a plastic bag (cold transport). Individual pupae can be collected in microcentrifuge tubes or on collection cards (see section 4.4.5.).

4.3.3. Larvae

Larvae are usually collected in tubes or on collection cards and transported on ice, since they tend to crawl out of comb sections during live transport.

4.3.4. Eggs

Eggs can be collected as cut comb section, transported in a non-ventilated container to prevent dehydration, or individually in tubes or on collection cards.

4.3.5. Extracted guts

Adult bee guts can be collected by carefully pulling out the stinger plus last integument, and slowly drawing out the hind and midgut. Extracted guts can be transported in tubes or on collection cards, on ice.

4.3.6. Drone endophallus and semen

Drone endophali and/or semen can be collected by squeezing out the endophallus as described in the *BEEBOOK* article on artificial insemination (Cobey *et al.,* 2013). The endophallus or semen can be transported in tubes or on collection cards, on ice.

4.3.7. Faeces

Faeces can be collected destructively by removing the bee gut (see section 4.3.5.) and expelling the faeces, or non-destructively by placing the bee in a petri dish and waiting for defecation.

4.3.8. Dead colonies

Many virus experiments involve bee death as a parameter. Dead bee samples from such experiments should be treated like freshly killed material and frozen as soon as possible to minimize the effects of decay on RNA integrity.

Passive surveys also involve dead bee samples, in this case those sent in by beekeepers for *post-mortem* analysis of the cause of colony death. The RNA from such bees will most likely be degraded, which will affect the reliability of the data, especially of negative results (virus absence). How to manage such samples and data is covered in detail in the *BEEBOOK* paper on molecular methods (Evans *et al.,* 2013). Collect the more desiccated dead bees and transport in a ventilated cardboard box at ambient temperature.

4.4. Sample transport

The integrity of samples can stabilized during transport by:

4.4.1. Freezing

Freezing on-site is the gold standard for sample transport, but usually too expensive for mass surveys. There are several alternatives for different purposes and samples sizes (see the *BEEBOOK* paper on molecular techniques; Evans *et al.,* 2013), including the 'dry shipper' which can hold up to 1300 2-ml cryo-tubes below freezing for up to 3 weeks and is approved for international air-shipment.

4.4.2. Ice

Transport on ice is a cheap and practical substitute for freezing, if the samples can be (re)-frozen within 48 hours.

4.4.3. Live transport

Live transport is very practical and cheap, especially if the samples are sent by post. Obviously there is no RNA degradation due to bee death, but live transport may affect the expression of host genes, and possibly virus replication, which should be taken into account when planning experiments.

4.4.4. Chemical stabilizers

There are a number of chemicals that help prevent RNA degradation. For these to work, they have to penetrate the bee exoskeleton and get into the tissues. They are therefore more useful for extracted tissues, eggs and larvae and less useful for adult bees. There are two types of chemical preservation: salts (in solution or impregnated on collection cards) and organic solvents (usually alcohol). Solvents penetrate the exoskeleton better than salt solutions and are more suitable for adult bees. A large excess (>5-fold by weight) should be used to make sure the chemical's concentration in the tissues is high enough to inhibit the nucleases. Various solutions and how to use them are described in the *BEEBOOK* paper on molecular methods (Evans *et al.,* 2013).

4.4.5. Sample collection cards

FTA™ collection cards (Whatman) preserve tissues both by desiccation and chemical (salt) preservatives embedded in the filter paper (Becker *et al.,* 2004; Rensen *et al.,* 2005). They are ideal for soft tissues and for remote collecting situations where sample weight and ambient storage are important factors. They are very reliable, but pricey, only suitable for small sample sizes and processing can be messy. See the *BEEBOOK* paper on molecular methods (Evans *et al.,* 2013).

4.5. Long-term sample storage

The factors important for long-term storage and preservation are the same ones highlighted for sample transport, which are (in order of effectiveness): temperature, desiccation (lyophilisation) and chemical preservatives. See the *BEEBOOK* paper on molecular methods (Evans *et al.,* 2013) for details.

5. Virus propagation

5.1. Introduction

The methods for propagating and purifying honey bee viruses in bees have not changed much from those described by Bailey and Ball (1991). Many viruses, including all of the important 'picorna-like' viruses, can be propagated by injection in either pupae or adult bees. Pupal injections are easier to manage than injection into adult bees, from pretty much every aspect: injection, incubation, homogenization and purification. Some viruses can only be infected orally and/or are only infectious in adult bees. The propagation criteria for each virus are listed in Table 3. Infectivity tests are essentially more precise versions of the propagation protocols. Another way to propagate and purify honey bee viruses is through tissue culture. This removes the potential for contamination and the dependence on the bee season that comes with propagating in bees, and allows for large volume propagation. It also is a highly effective tool for detailed laboratory experimentation at cellular level, without the influence of bee and hive effects. Attempts at virus propagation in bee tissue culture have so far met with limited success. However, significant progress has recently been made with *Nosema* cultivation in commercial (Lepidopteran) insect cell lines (see the *BEEBOOK* paper on cell culture (Genersch *et al.,* 2013).

5.2. Starting material

Often the starting material for propagation is a previous virus preparation that has been checked for the absence of contaminating viruses and retained as a pure isolate. Virus preparations can, however, lose infectivity during prolonged storage. For example deformed wing virus (DWV) and its relatives kakugo virus (KV) and Varroa destructor virus-1 (VDV-1), are particularly sensitive to decay during storage. Since neither serological nor molecular assays can distinguish between degraded or intact virus particles, they are not

reliable methods for determining the infectivity of a preparation. Furthermore, such well-characterized and precious reference material is often in limited supply and highly valuable as a historical "reference" isolate for future experiments. Such reference material can be stored long-term either as freeze-dried bees/pupae/larvae, or as a (semi)purified virus solution in an appropriate virus purification buffer (Table 4) and stabilized by 50% glycerol. It is therefore advisable, particularly for infectivity tests but also for propagation, to first prepare a "working" inoculum, by injecting or feeding a small number of bees (larvae, pupae or adults; around 5-10 individuals) with a small amount of the pure reference material. This also serves as an infectivity test for the viability of the stored reference material. After incubating for the appropriate time, a crude extract should be prepared from the bees, the purity and virus concentration of this extract determined, and then this working extract should be used for large-scale propagation or infectivity tests within the next few of weeks.

5.3. Oral propagation

Oral propagation is relatively inefficient for most viruses, requiring high titre inoculums (10^6-10^{11} particles, depending on the virus; Table 3) to establish an infection (Bailey and Gibbs, 1964; Bailey and Ball, 1991).

5.3.1. Larvae

1. Mix purified virus of the appropriate minimum concentration (Table 3) with medium for *in-vitro* larval rearing and allocate this to the wells of a 48-well tissue culture plate (see the *BEEBOOK* paper on *in-vitro* larval rearing (Crailsheim *et al.,* 2013).

2. Transfer two-day old larvae to the wells, making sure there is enough virus so that each larva gets the minimum infectious dose.

3. Follow the procedures for *in-vitro* larval rearing (see the *BEEBOOK* paper on *in-vitro* larval rearing (Crailsheim *et al.,* 2013), transferring the larvae periodically to fresh food, either including or excluding further virus extract.

4. Include a series of control inoculations, using larval food medium without virus.

5.3.2. Adults

1. Mix purified virus with 60% sucrose to the desired concentration for infecting bees individually or in bulk.

2. Feed the virus-sucrose solution individually to newly emerged adult bees in 5-10 μl volumes, using a Pasteur or micro-pipette.

3. Immobilize the bees by either holding their wings or in a suitable restrainer, such as head-first in a 1.5 ml microcentrifuge tube with the bottom cut off (see the section

'standard methods for immobilising, terminating, and storing adult *Apis mellifera*' in the *BEEBOOK* paper on miscellaneous methods; Human *et al.,* 2013).

4.   Bulk-feeding of the virus-sucrose solution to adult bees is done either in hoarding cages, using disposable 15 ml plastic feeding tubes (see the *BEEBOOK* paper on maintaining adult *Apis mellifera* workers in cages; Williams *et al.,* 2013) or in whole colonies using internal or top feeders. In both cases it is important to calculate the amount of purified virus needed to ensure that each bee gets the minimum infectious dose (Table 3).

5.   Include a series of control inoculations, using sucrose solution without virus.

5.4. Injection propagation

Propagation by injection into pupae or adults is generally very efficient, requiring very low virus doses and concentrations to establish an infection (10^2-10^4 particles, depending on the virus). However, this high efficiency also makes propagation by injection susceptible to the amplification of any contaminating viruses, either those present in the injected inoculum or those present naturally within the bees. Extra care has to be taken therefore to confirm the purity of the propagation, after purification.

5.4.1. Pupae

1.   Lay the frame horizontally at a slight angle, bottom-to-top, under good light.

2.   Using needle-forceps, remove the wax capping from 10-50 cells containing white-eyed pupae, by cutting along the inside of the cell.

It is easiest to work from the bottom of the frame upwards, clearing room underneath for opening up the cells higher up the frame and picking up the pupae from behind.

3.   With blunt, curved forceps remove the top part of each cell, exposing the head of the pupa.

4.   Place the curved forceps underneath the head of the pupa, from the back, and carefully lift the pupa out of its cell.

It is critical to remove the white-eye pupae very carefully from the comb, to avoid damage.

5.   Collect the pupae in a plastic Petri dish containing a circular filter paper dampened with sterile water.

6.   For the purpose of propagation, you need:
   *   a 10~50 µl syringe,
   *   a thin needle (around 28G~30G)
   *   a semi-automatic volume dispenser control unit that can dispense 1~5 µl volumes.

The Hamilton Company produces both dispenser units and 10~50 µl syringes with Luer locks that fit disposable needles. The pupae can be injected by hand, which is faster (but less precise) than doing so under microscope, with moveable trays.

7.   Attach the syringe and control unit horizontally to a stand.

8.   Close the four fingers of your hand and carefully lay a white-eye pupa on its back in the groove between index and middle finger.

The abdomen of the pupa points towards the tip of the fingers and the head is supported by the tip of the thumb.

9.   Move the pupa towards the needle, inserting the needle at the narrowest angle possible under the skin of the pupa, on the lateral side, between the 2nd and 3rd integuments of the abdomen.

10.   Inject 1~5 µl of virus suspension using the control unit.

11.   Move the pupa backwards off the needle.

12.   Move the pupa carefully from the hand to a plastic, disposable tissue culture plate, using forceps to support the pupa underneath.

Use plates with matching lids, both to prevent cross-contamination between wells and to control the humidity.

13.   Incubate the plate at 30°C in a humidity-controlled incubator. If this is not available, place the plates in a closed plastic box containing moistened paper towels.

14.   Check the progress of the infection by monitoring the change in eye colour of the pupae.

Those pupae injected with virus that remain alive will change eye colour, but more slowly than those pupae injected with only buffer.

15.   Include a control series with buffer-only inoculations, and a control series without inoculation, just incubation of the pupae.

The best propagations are with those inoculum concentrations that keep the pupae alive for as long as possible, generating the highest propagation concentrations. Sometimes parts of the body become necrotic, either only the abdomen or only the thorax/head. This can happen when the virus concentration of the inoculum is too high, killing the tissue too quickly for efficient propagation. Too low a virus inoculums concentration will, however, increase the risk of amplifying unrelated covert infections already present in the pupa.

It is therefore advisable to first propagate a range of log-scale dilutions (1/10, 1/100, 1/1000 *etc.*) of the virus inoculum. Then for large-scale propagation, choose the highest inoculum concentration that does not necrotize the pupae before the 4th day of incubation.

Table 3. Summary of the protocols and conditions for the oral and injection propagation of the different honey bee viruses in larvae, pupae and adult bees. Non-viable propagation routes are marked with an 'x'. Absence of reliable information is marked with '?'. Adapted from Bailey and Ball (1991).

PROPAGATION						
VIRUS	**METHOD**	**STAGE**	**AGE**	**DOSE** (2ng $\approx 10^7$ particles)	**INCUBATION***	**NOTES**
Acute bee (ABPV)	ORAL	larvae	48 hour	?	?	
		adult	< 24 hour	> 10^{10} particles/bee	?	
	INJECTION	pupae	white eye	> 10^2 particles/bee	5 days	
		adult	any	> 10^2 particles/bee	5 days	
Kashmir bee virus (KBV)	ORAL	larvae	48 hour	?	?	
		adult	< 24 hour	> 10^7 particles/bee	5 days	
	INJECTION	pupae	white eye	> 10^2 particles/bee	3 days	
		adult	any	> 10^2 particles/bee	3 days	
Israeli acute (IAPV)	ORAL	larvae	48 hour	?	?	
		adult	< 24 hour	> 4ng pure virus/bee	6 days	
	INJECTION	pupae	white eye	> 10^2 particles/bee	3 days	
		adult	any	> 10^2 particles/bee	4 days	
Black queen cell virus (BQCV)	ORAL	larvae	48 hour	?	?	requires N. apis co-infection
		adult	< 24 hour	> 4mg crude extract/bee	40 days	
	INJECTION	pupae	white eye	> 10^3 particles/bee	5 days	
		adult	any			
Deformed wing Varroa destructor virus-1 (DWV & VDV-1)	ORAL	larvae	48 hour	> 10^9 genomes/bee	5 days	
		adult	< 24 hour	> 10^8 genomes/bee	?	
	INJECTION	pupae	white eye	> 10^2 genomes/bee	14 days	
		adult	any	> 10^7 genomes/bee	3 days	
Egypt bee virus (EBV)	ORAL	larvae	48 hour	?	?	
		adult	< 24 hour	?	?	
	INJECTION	pupae	white eye	> 10^2 particles/bee	8 days	
		adult	any	?	?	
Sacbrood virus (SBV)	ORAL	larvae	48 hour		7 days	
		adult	< 24 hour			
	INJECTION	pupae	white eye	> 10^3 particles/bee	5 days	
		adult	any			
Thai/Chinese Sacbrood virus (TSBV)	ORAL	larvae	48 hour		7 days	
		adult	< 24 hour			
	INJECTION	pupae	white eye	> 10^3 particles/bee	5 days	
		adult	any			
Slow bee paralysis virus (SBPV)	ORAL	larvae	48 hour	?	?	
		adult	< 24 hour	?	?	
	INJECTION	pupae	white eye	> 10^3 particles/bee	5 days	
		adult	any	> 10^3 particles/bee	12 days	
Chronic bee paralysis virus (CBPV)	ORAL	larvae	48 hour	> 10^{10} particles/bee	5 days	
		adult	< 24 hour	> 10^{10} particles/bee	? days	
	CONTACT	larvae	x	x	x	
		adult	< 48 hour	> 10^7 genomes/bee		
	INJECTION	pupae	white eye	> 10^2 particles/bee	5 days	
		adult	any	> 10^2 particles/bee	7 days	

Table 3. continued

VIRUS	METHOD	STAGE	AGE	DOSE (2ng ≈ 10^7 particles)	INCUBATION*	NOTES
Cloudy wing virus (CWV)	CONTACT	larvae	?	?	?	
		adult	?	?	?	
	INJECTION	pupae	x	x	x	
		adult	x	x	x	
Bee virus X (BVX)	ORAL	larvae	x	x	x	enhanced by M. mellificae co-infection
		adult	< 24 hour	> 4mg crude extract/bee	30 days	
	INJECTION	pupae	x	x	x	
		adult	x	x	x	
Bee virus Y (BVY)	ORAL	larvae	x	x	x	requires N. apis co-infection
		adult	< 24 hour	> 4mg crude extract/bee	30 days	
	INJECTION	pupae	x	x	x	
		adult	x	x	x	
Arkansas bee virus (ABV)	ORAL	larvae	48 hour	?	?	
		adult	< 24 hour	?	?	
	INJECTION	pupae	white eye	> 10^3 particles/bee	5 days	
		adult	any	> 10^3 particles/bee	21 days	
Filamentous virus (AmFV)	ORAL	larvae	?	?	?	enhanced by N. apis
		adult	< 24 hour	> 4mg crude extract/bee	30 days	
	INJECTION	pupae	x	x	x	
		adult	x	x	x	
Apis iridescent virus (AIV)	ORAL	larvae	48 hour	?	?	
		adult	< 24 hour	?	?	
	INJECTION	pupae	white eye	> 10^3 particles/bee	5 days	
		adult	any			
					* based on minimum	

5.4.2. Adults

Propagation in adult bees is best done with newly emerged bees, whose exoskeleton is still soft, making it easier for controlled injection. Using young bees also avoids any age-related variability in propagation and unintentional propagation of adult-acquired viruses.

1. Collect the bees in lots of 10-20 in queen cages, or similar containers.
2. Anaesthetize the bees for 1 minute with CO_2 from a pressurized cylinder.
3. Make sure to bubble the CO_2 gas through water to melt any CO_2 micro-particles, which can be very injurious to bees. See the section 'Standard methods for immobilising, terminating, and storing adult *Apis mellifera*' in the *BEEBOOK* paper on miscellaneous methods (Human *et al.,* 2013).
4. Proceed quickly to avoid excess anaesthesia for the bees and do not anaesthetize the bees more than once a day.

Bees sometimes die from excess anaesthesia. This is especially important for infectivity tests, where the death of a bee is often a recorded experimental variable.

5. Inject 1~5 µl of virus suspension between the 2nd and 3rd integuments of the bee using a similar controlled-volume syringe set-up as for pupal propagation (see section 5.4.1.).
6. Incubate the inoculated bees in a hoarding cage at 30°C in 60-70% relative humidity with sufficient sterilized food and water (see the *BEEBOOK* paper on maintaining adult *Apis mellifera* workers in cages (Williams *et al.,* 2013) for the appropriate amount of time for each virus (Table 3).
7. Include a control series with buffer-only inoculations, and a control series without inoculation, just incubation of the adult bees.

Table 4. Summary of the protocols for the purification of the different honey bee viruses. Adapted from Bailey and Ball (1991).

VIRUS	EXTRACTION BUFFER (2 ml/g tissue)	LOWSPEED CENTRIFUGE	HIGH SPEED CENTRFUGE	RESUSPENSION BUFFER (0.1 ml/g tissue)	SUCROSE GRADIENTS
Acute bee paralysis virus (ABPV)	0.01 M potassium phosphate (7.0) 0.02% diethyldithiocarbamate 0.1 volume ether 0.1 volume carbon tetrachloride	8000 g 10 minutes 15°C supernatant	75 000 g 3 hours 15°C retain pellet	0.01 M potassium phosphate (7.0) incubate 12-36 hours; 5°C low-speed centrifugation; 5°C	45 000 g 3 hours 4°C collect band
Kashmir bee virus (KBV & IAPV)	0.01 M potassium phosphate (7.0) 0.02% diethyldithiocarbamate 0.1 volume ether 0.1 volume carbon tetrachloride	8000 g 10 minutes 15°C supernatant	75 000 g 3 hours 15°C retain pellet	0.01 M potassium phosphate (7.0) incubate 12-36 hours; 5°C low-speed centrifugation; 5°C	45 000 g 3 hours 4°C collect band
Israeli acute paralysis (IAPV)	0.01M potassium phosphate (7.6) 0.2% Na-deoxycholate 2% BRIJ-58	10 000 g 20 minutes 15°C supernatant	100 000 g 3 hours 15°C retain pellet	0.01 M potassium phosphate (7.6) 0.6 g/ml CsCl * CsCl gradient centrifugation	100 000 g* 24 hours* 20°C collect band
Black queen cell virus (BQCV)	0.01 M potassium phosphate (7.0) 0.02% diethyldithiocarbamate 0.1 volume ether 0.1 volume CCl₄	8000 g 10 minutes 15°C supernatant	75 000 g 3 hours 15°C retain pellet	0.01 M potassium phosphate (7.0) incubate 12-36 hours; 5°C low-speed centrifugation; 5°C	45 000 g 3 hours 4°C collect band
Deformed wing virus & Varroa destructor virus-1 (DWV & VDV-1)	0.5 M potassium phosphate (8.0) 0.2% diethyldithiocarbamate 0.1 volume ether 0.1 volume carbon tetrachloride	8000 g 10 minutes 15°C supernatant	75 000 g 3 hours 15°C	0.5 M potassium phosphate (8.0) immediately to sucrose gradients	45 000 g 3 hours 4°C collect band
Egypt bee virus (EBV)	0.5M potassium phosphate (8.0) 2% ethylene diamine tetra acetic acid 1% ascorbic acid; 0.1 volume ether 0.1 volume carbon tetrachloride	8000 g 10 minutes 15°C supernatant	75 000 g 3 hours 15°C retain pellet	0.5 M potassium phosphate (8.0) immediately to sucrose gradients	45 000 g 3 hours 4°C collect band
Sacbrood virus (SBV)	0.01 M potassium phosphate (7.0) 0.02% diethyldithiocarbamate 0.1 volume ether 0.1 volume carbon tetrachloride	8000 g 10 minutes 15°C supernatant	75 000 g 3 hours 15°C retain pellet	0.01 M potassium phosphate (7.0) incubate 12-36 hours; 5°C low-speed centrifugation; 5°C	45 000 g 3 hours 4°C collect band
Thai/Chinese Sacbrood virus (TSBV)	0.5 M potassium phosphate (8.0) 0.2% diethyldithiocarbamate 0.02 M ethylene diamine tetra acetic acid 0.1 volumes ether; carbon	8000 g 10 minutes 15°C supernatant	75 000 g 3 hours 15°C retain pellet	0.5 M potassium phosphate (8.0) 0.02 M ethylene diamine tetra acetic acid	45 000 g 3 hours 4°C collect band
Slow bee paralysis virus (SBPV)	0.01 M potassium phosphate (7.0) 0.02% diethyldithiocarbamate 0.1 volume ether 0.1 volume carbon tetrachloride	8000 g 10 minutes 15°C supernatant	75 000 g 3 hours 15°C retain pellet	0.01 M potassium phosphate (7.0) incubate 12-36 hours; 5°C low-speed centrifugation; 5°C	45 000 g 3 hours 4°C collect band
Chronic bee paralysis virus (CBPV)	0.2 M potassium phosphate (7.5) 0.02% diethyldithiocarbamate 0.1 volume ether 0.1 volume carbon tetrachloride	3000 g 30 minutes 15°C supernatant	100 000 g 2 hours 15°C retain pellet	0.2 M potassium phosphate (7.5) 15°C	45 000 g 4.5 hours 15°C collect band

Table 4. cont.

VIRUS	EXTRACTION BUFFER (2 ml/g tissue)	LOW-SPEED CENTRIFUGE	HIGH-SPEED CENTRFUGE	RESUSPENSION BUFFER (0.1 ml/g tissue)	SUCROSE GRADENTS
Cloudy wing virus (CWV)	0.01 M potassium phosphate (7.0)	8000 g	. 75 000 g	0.01 M potassium phosphate (7.0)	45 000 g
	0.5 M potassium phosphate (8.0)	8000 g	75 000 g	0.5 M potassium phosphate (8.0)	45 000 g
	0.2% diethyldithiocarbamate	10 minutes	3.5 hours	immediately to sucrose gradients	4.5 hours
	0.1 volume ether	15°C	15°C		4°C
	0.1 volume carbon tetrachloride	supernatant	retain pellet		collect band
Bee virus X (BVX)	0.01 M potassium phosphate (7.0)	8000 g	75 000 g	0.01 M potassium phosphate (7.0)	45 000 g
	0.02% diethyldithiocarbamate	10 minutes	3 hours	equal volume 0.2M ammonium	3 hours
	0.1 volume ether	15°C	15 °C	low-speed & high-speed	4°C
	0.1 volume carbon tetrachloride	supernatant	retain pellet	resuspend 0.1 M ammonium acetate (7.0)	collect band
Bee virus Y (BVY)	0.01 M potassium phosphate (7.0)	8000 g	75 000 g	0.01 M potassium phosphate (7.0)	45 000 g
	0.02% diethyldithiocarbamate	10 minutes	3 hours	equal volume 0.2 M ammonium	3 hours
	0.1 volume ether	15°C	15°C	low-speed & high-speed	4°C
	0.1 volume carbon tetrachloride	supernatant	retain pellet	resuspend 0.1 M ammonium acetate (7.0)	collect band
Arkansas bee virus (ArkBV)	0.01 M potassium phosphate (7.0)	8000 g	75 000 g	0.01 M potassium phosphate (7.0)	45 000 g
	0.02% diethyldithiocarbamate	10 minutes	3 hours	incubate 12-36 hours; 5°C	3 hours
	0.1 volume ether	15°C	15°C	low-speed centrifugation; 5°C	4°C
	0.1 volume carbon tetrachloride	supernatant	retain pellet		collect band
Filamentous virus (AmFV)	0.01 M ammonium acetate (7.0)	150 g	30 000 g	0.1 M ammonium acetate (7.0)	10 000 g
	0.02% diethyldithiocarbamate	10 minutes	30 minutes	layer on 50% w/v sucrose	30 minutes
		15°C	15°C	centrifuge 75 000g, 3 hours, 15°C	5°C
		supernatant	retain pellet	resuspend 0.1 M ammonium acetate (7.0)	collect band
Apis iridescent virus (AIV)	0.01 M potassium phosphate (7.0)	150 g	30 000 g	0.01 M potassium phosphate (7.0)	10 000 g
	0.02% diethyldithiocarbamate	10 minutes	30 minutes	incubate 12-36 hours; 5°C	30 minutes
	0.1 volume carbon tetrachloride	15°C	15°C	low-speed centrifugation; 5°C	5°C
		supernatant	retain pellet		collect band

5.5. Tissue culture

The importance of a viable tissue-culture system for the purification and propagation of honey bee viruses has long been acknowledged. There are two possible approaches to such a system. One is to develop a reliable, immortal honey bee cell line for infection. Only recently has there been any significant progress towards this goal (Bergem *et al.*, 2006; Hunter, 2010; Kitagishi *et al.*, 2011; Gisder *et al.*, 2012). The other approach is to propagate honey bee viruses in existing commercial, heterologous insect cell lines. Many of the honey bee viruses naturally infect other insect hosts, such as other *Apis* spp., varroa and tropilaelaps parasitic mites, bumble bees, wasps, ants and a range of solitary pollinators (Bailey and Gibbs, 1964; de Miranda *et al.*, 2010a; de Miranda and Genersch, 2010; Ribière *et al.*, 2010; Dainat *et al.*, 2009; Singh *et al.*, 2010; Peng *et al.*, 2011; Li *et al.*, 2011; DiPrisco *et al.*, 2011; Zhang *et al.*, 2012; Evison *et al.*, 2012) and the replication-translation control regions of honey bee virus genomes are active in several commercial Lepidopteran and Dipteran

cell lines (Ongus *et al.*, 2006). Protocols for the establishment and maintenance of honey bee and commercial insect cell lines can be found in the *BEEBOOK* article on cell cultures (Genersch *et al.*, 2013).

5.5.1. Virus infection

The most common method of infecting tissue culture cells is through passive co-incubation of purified virus particles with the cells, allowing the natural processes of virus entry to establish an infection (Minor, 1985; Gantzer *et al.*, 1998; Rhodes *et al.*, 2011; Amdiouni *et al.*, 2012). However, the virus particles (or the naked viral RNA genome) can also be forced into the cells using electroporation, which involves a short high-voltage pulse of electricity to temporarily open up the cell membrane to allow foreign elements to enter the cell, or chemical-mediated transfection, where a combination of membrane-active ions and concentrating agents interact to encourage the uptake of the virus or nucleic acid into the cell (*e.g.* Boyer and Haenni, 1994;

Benjeddou *et al.,* 2002; Ongus *et al.,* 2006; Yunus *et al.,* 2010). Whichever virus transfection protocol is chosen, it is essential that the virus preparation is free of bacteria or fungi to prevent contamination of the tissue culture (Minor, 1985; Gantzer *et al.,* 1998). Bacteria, fungi and their spores can be effectively removed from a virus preparation using microfilters with appropriate pore size, depending on the size of the virus (Gantzer *et al.,* 1998; Rhodes *et al.,* 2011; Amdiouni *et al.,* 2012), with 0.2 µm suitable for purifying most small, enteric picorna-like viruses of around 30-60nm (Rhodes *et al.,* 2011; Amdiouni *et al.,* 2012), which includes most of the honey bee RNA viruses (Table 2).

5.6. Full-length infectious virus clones

A supremely powerful tool in RNA virus research is cloning full-length genomic sequences of the virus into bacterial vectors. The naked RNA transcribed from such clones is usually infectious when introduced into a suitable host (Yunus *et al.,* 2010), especially for positive-stranded RNA viruses (*i.e.* the majority of known bee viruses). Such clones can be manipulated by site-directed mutagenesis and recombination for functional analysis of different open reading frames or control regions. Reporter genes such as green fluorescent protein can be inserted to make fusion proteins with viral genes or to study promoter function in real-time (Ongus *et al.,* 2006) and of course they can function as a genetically pure source of infectious virus, rather than having to rely on biological propagation with the associated dangers of contamination with other viruses and the changing genetic constitution of the virus, through evolution.

Full-length viral cDNAs are also an important tool in studying the genetic complexity of virus populations, since they make it possible to identify complete sequences of individual viruses within the population, including natural recombinants between major variants (Palacios *et al.,* 2007; Moore *et al.,* 2011).

5.6.1. Full-length viral RNA synthesis strategies

Full-length cloning of viral genomes has been a common tool in virology since the 1980's (Taniguchi *et al.,* 1978; Lowry *et al.,* 1980; Racaniello and Baltimore, 1981), but it is often a long and tedious process, mostly due to the frequent instability of the full-length clones in bacteria (Boyer and Haenni, 1994). It is thought that cryptic bacterial promoters and secondary structures within the viral sequences encourage the bacteria to excise problematic viral regions from the plasmid clone. Full-length clones therefore have to be monitored constantly for possible deletions and re-arrangements as part of their maintenance. The process has been made easier by improvements in cloning techniques and the stability of the cloning vectors with respect to accepting and maintaining long (~10kb) inserts. There are several alternatives to cloning full-length genomes. Sometimes it is easier to clone the genome in several partial clones first, and then recombine these afterwards into a single full-length clone (Rodriguez *et al.,* 2006).

Another strategy, used for viruses that prove impossible to clone full-length, is to clone the genome in two halves, which are maintained independently and then recombined *in vitro* prior to RNA transcription using a suitably engineered restriction site (Jakab *et al.,* 1997). This improves the stability of the clones, the engineered restriction site can be used for recombining different virus strains for gene function analysis and also serves as a useful marker for tracing the virus through experiments.

A third alternative is to dispense with cloning altogether and generate infectious RNA transcripts directly off full-length PCR products that have a suitable recognition site for the T3, T7 or SP6 RNA polymerase incorporated into the full-length amplification primers, for transcript synthesis. This avoids the instability problems of cloned full-length clones but limits the extent to which the genomes can be manipulated genetically. This approach was successful for synthesizing full-length, infectious transcripts of BQCV after it proved impossible to successfully clone full-length BQCV (Benjeddou *et al.,* 2002).

5.6.2. Protocol

Here we describe a method for:

- Producing full-length PCR products of positive-stranded, ssRNA honey bee viruses, with a T7 RNA polymerase promoter site incorporated into the forward primer sequence (sections 5.6.2.1. – 5.6.2.2.)
- Cloning this product into a stable plasmid vector (section 5.6.2.3.)
- Confirming the integrity and character of the full-length clones (section 5.6.2.4.)
- Synthesizing full-length infectious RNA transcripts of the virus (sections 5.6.2.5. – 5.6.2.6.)

5.6.2.1. Full-length reverse transcription

Genomic RNA of most honey bee ssRNA viruses is approximately 10 kb long and contains highly structured 5'-terminal un-translated region with extended hairpin structures. Therefore, the first strand cDNA synthesis should be performed using a reverse transcriptase with a high optimum temperature (*e.g.* InVitrogen's Superscript III), so that RNA secondary structures are also transcribed.

1. Prepare purified virus particles using gradient centrifugation (see section 7; "Virus purification")
2. Extract viral RNA from the purified virus particles (see section 8.3.; "Nucleic acid extraction")
3. Combine in a single 200 µl thin wall tube:
    3.1. 1 µg virus RNA,
    3.2. 1 µl 2 µM "Reverse Primer" 5'-CGGTGTTTAAAC(T)$_{27}$(X)$_{32}$-3', where (X)$_{32}$ is a sequence complementary to the final 32 nucleotides at the 3' end of the virus genome to be cloned,
    3.3. 1 µl 10 mM dNTPs,
    3.4. Make the total volume 13 µl with RNase-free water.

4. Mix well by pipetting on ice.

5. Incubate at 65°C for 3 minutes in PCR heating block.

6. Transfer to ice and cool down for 1 minute.

7. Add the following:

 7.1. 4 µl 10x First Strand Buffer (supplied with Superscript III),

 7.2. 1 µl 0.1M DTT,

 7.3. 1 µl RNAse OUT recombinant RNAse inhibitor (Invitrogen).

8. Mix well and incubate at 52°C for 2 minutes in heating block.

9. Add 2 µl Superscript III reverse transcriptase (Invitrogen) and mix well.

10. Incubate at 52°C for 10 min.

11. Incubate at 55°C for 60 min.

12. Incubate at 70°C for 15 min.

13. Store in freezer as a template for full length cDNA amplification.

5.6.2.2. Amplifying full-length viral RNAs

One critical factor in the successful amplification of viable, full-length cDNAs is the use of a thermostable DNA polymerase with proof-reading capacity. One such high-fidelity, high processivity DNA polymerase is Phusion DNA polymerase (New England Biolabs). If the fragment is to be cloned into a plasmid vector lacking a T7 or T3 RNA polymerase promoter site (required for the synthesis of full-length infectious RNA copies), then such a site should be incorporated into the 5′ amplification primer (*e.g.* Benjedou *et al.,* 2002; Ongus *et al.,* 2006).

1. Combine following in the 200 µl thin wall PCR tube placed on ice:

 1.1. 2 µl of the first strand cDNA reaction,

 1.2. 35 µl sterile nuclease free water,

 1.3. 10 µl of 5x HF Phusion amplification buffer,

 1.4. 1 µl 10mM dNTP mixture,

 1.5. 0.5 µl 2 µM "Reverse primer" (section *5.6.2.1.*; step 3.2.)

 1.6. 0.5 µl 2 µM T7 RNA polymerase promoter-tagged "Forward primer" 5′-GCTATAATACGACTCACTATAGG(X)$_{20}$-3′ where (X)$_{20}$ are the first 20 nucleotides at the 5′ end of the virus genome

2. Mix well by pipetting on ice.

3. Add 2 µl (5U) Phusion DNA polymerase.

4. Mix by pipetting.

5. Place the tube in the thermocycler when the block is 90°C (Hot-start PCR).

6. Amplify with the following cycling programme:

 6.1. 98°C:1 min,

 6.2. 5x [98°C:15 sec – 52°C:60 sec – 72°C:7 min],

 6.3. 25x [98°C:15 sec – 55°C:60 sec – 72°C:7 min],

 6.4. 72°C: 7 min.

7. Purify the reaction products with PCR purification kit (Qiagen), eluting into 50 µl of water.

5.6.2.3. Cloning full-length viral RNAs

Cloning of long (~10 kbp) PCR products into plasmid vectors may be not very efficient. We recommend to using pCR-XL-TOPO cloning, which is vector specifically designed for cloning of large products. This vector requires 3′ terminal A overhangs in the PCR products, therefore the first stage is the addition of 3′ overhangs to the Phusion-generated blunt ends.

1. Mix the following:

 1.1. 50 µl purified PCR fragments,

 1.2. 6 µl 10x Taq polymerase buffer,

 1.3. 3 µl 10mM dNTPs,

 1.4. 1 µl Taq polymerase.

2. Incubate at 72°C for 10 min.

3. Separate the fragments by electrophoresis in 0.8% agarose gel in TAE buffer.

4. Stain the gel with crystal violet (see InVitrogen TOPO XL PCR cloning kit).

5. Excise the 10 kb full-length cDNA fragments.

6. Extract DNA using the Gel Purification reagents included in the TOPO XL PCR Cloning kit (Invitrogen).

7. Ligate fragments into the pCR XL TOPO vector by mixing:

 7.1. 4 µl of the purified product (approximately 10 to 50 ng)

 7.2. 1 µl of pCR XL TOPO vector.

8. Incubate reaction for 5 min at 25°C.

9. Stop reaction by addition of 6x TOPO cloning Stop Solution.

10. Mix for a few seconds at room temperature and place the reaction to ice.

11. Proceed immediately to transformation of the OneShot competent *E. coli* cells (Invitrogen), following the manufacturer's instructions.

5.6.2.4. Confirmation of full-length clones

1. Select the white colonies and transfer to a fresh plate.

2. Amplify insert DNA from all colonies using the PCR primers and protocols from Section 9.3.3.; "RT-(q)PCR – Protocols".

3. Separate PCR products on gel electrophoresis.

4. Isolate clones with inserts > 10kb.

5. Prepare plasmid DNA from full-length clones using a Qiagen plasmid purification kit and corresponding instructions.

6. Sequence the inserts in selected plasmids using a series of oligonucleotide primers that are conserved between all known variants of the virus.

7. Confirm identity of clones through comparing the cloned sequences with the published consensus virus sequences.

5.6.2.5. Synthesizing full-length viral RNA

The plasmid can be used as a template for *in vitro* transcription using T7 RNA polymerase.

1. Linearize plasmid downstream of the 3' poly-A sequence with *Pme*I restriction endonuclease.
2. Purify the linearized templates with a Qiagen plasmid purification kit.
3. Synthesize capped *in vitro* RNA transcripts using "mMESSAGEmMAchine" T7 kit (Ambion).

5.6.2.6. Confirmation of Full-length Viral RNA

The full-length nature and activity of the transcripts is confirmed by *in vitro* translation experiments (Green and Sambrook, 2012), and ultimately by infection of pupae through injection with the synthesized RNA transcripts (see section 5.4). The infectivity of such transcripts is confirmed by comparing virus titres between transcript-inoculated pupae with control-inoculated pupae, and by sequencing the new virus infection (Benjedou *et al.*, 2002).

6. Virus infectivity assays

Infectivity assays were used before sensitive molecular techniques were developed to detect low levels of virus in surveys (Bailey, 1976; Bailey *et al.*, 1981; 1983b). These assays take advantage of the fact that most bee viruses when injected into adult bees or pupae multiply rapidly to high titres that can subsequently be detected by serology (Dall, 1987). Dilution series of the extracts provide a measure of quantitation. Different viruses develop titre and kill pupae at different rates, which can be detected by the 'breaking' of the eye-colour development in white-eyed pupae (Anderson and Gibbs, 1988; 1989). This can provide an early indication of which virus is being multiplied. Although labour intensive, infectivity assays can rival the most sensitive molecular tests available (Denholm, 1999). One serious drawback of honey bee infectivity assays is that often unapparent viruses present at very low levels in the assay pupae can also be amplified, sometimes by the mere injection of buffer (Bailey, 1967; Anderson and Gibbs 1988; 1989). Several important bee viruses (ABPV, KBV and SBPV) were discovered this way, as a by-product of the propagation of CBPV, AIV and BVX respectively (Bailey *et al.*, 1963; Bailey and Milne, 1969; Bailey and Woods, 1974; 1977; Bailey and Ball, 1991), and the technique may yet prove useful for the discovery of other symptomless bee viruses.

In general terms, the procedures for virus infection infectivity assays are the same as for virus propagation. It is especially important is that the larvae and pupae are transferred to the incubation plate as carefully as possible, and that they are checked for vitality and survival before being used for infection experiments. Larvae and pupae should be checked under a stereo microscope for

damage and vitality. In both cases it is advisable to incubate them for 12-24 hours prior to conducting the assay, and remove larvae or pupae that show signs of necrosis or low vitality. The infectivity assays should also include a number of additional methodological controls (effects of transfer, incubation, manipulation, feeding etc.), to facilitate the interpretation of the data.

7. Virus purification

7.1. Introduction

The method given below for purifying honey bee viruses is simple but only suitable for non-enveloped viruses, which fortunately covers the vast majority of RNA viruses. However, this method is unsuitable for viruses containing membranes, such as *Apis mellifera* filamentous virus. There are a number of enveloped virus families that have insect-infecting members and there may be more enveloped honey bee viruses to be discovered. The various buffers, solvents and centrifugation conditions for purifying individual viruses are given in Table 4. Most of the buffers shown are the phosphate buffers developed by Bailey and Ball (1991), between 0.01 and 0.5M and of neutral pH (between 7.0 and 8.0). In most cases, TRIS.Cl buffers of similar molarity and pH will perform equally well. Similarly, chloroform can be substituted for the more toxic ether/carbon tetrachloride combination for extract clarification. Nonionic detergents such as Triton X100 (0.05%), BRIJ-58 (2%) and/or sodium deoxycholate (0.2%) are also common agents for lipid solubilisation and extract clarification during virus purification. 0.1 M ascorbic acid is a common alternative to DIECA as antioxidant. Conduct as much of the purification as possible at 4°C (on ice). With each purification step, there is a considerable loss of yield, particularly during the high-speed and gradient centrifugation steps. As much as 80% of the primary extract can be lost during purification. It is therefore important to consider how pure the virus preparation needs to be for your experiments. For infection experiments, purity may be less important while for developing a specific antiserum, high purity is essential. The high-speed and gradient centrifugation steps are excellent for separating the virus from other cellular contents and particles, but are not suitable for separating different virus species: they all have very similar densities.

7.2. Protocols

7.2.1. Primary extract

1. Mix 2 ml of extraction buffer (Table 4) per 1 g of bee tissue.
2. Prepare a primary virus extract by either:
   - Grinding the bee tissues in liquid nitrogen in a mortar-and-pestle.
   - Liquidizing in an automatic blender.
   - Using a large-volume bead mill.

(see also section 8.2.; "Sample Homogenisation")

3. Transfer to a solvent-resistant container.

4. Add 0.5 ml of chloroform or carbon tetra-chloride per 1 g of bee tissue.

5. Shake vigorously by hand.

6. Centrifuge at 8,000 *g* and 4°C for 15 minutes.

7. Carefully collect the supernatant.

8. Discard the organic phase.

9. Remove 10 μl for virus analysis by RT-qPCR or ELISA (see section 9.3. and the *BEEBOOK* article on molecular methods (Evans *et al.,* 2013) to determine the viral purity of the extract.

10. The crude extract at this stage is appropriate for long-term storage. Add glycerol to a final concentration of 50%, aliquot and store at -80°C.

7.2.2. High speed centrifugation

1. Centrifuge the supernatant at 75,000 *g* and 4°C for 3 hours.

2. Discard the supernatant.

3. Re-suspend the pellet 5 ml extraction buffer.

This is best achieved by storing the pellet with buffer overnight at 4°C, to loosen the pellet.

4. Next day vortex the pellet lightly.

5. Centrifuge at 8,000 *g* and 4°C for 15 minutes.

6. Retain the supernatant for gradient centrifugation.

7.2.3. Gradient Centrifugation

The purpose of gradient centrifugation is to concentrate the virus particles according to their specific density and thereby separate them from other cellular material with different density. For high purity requirements, where the virus needs to be separated from other particles with similar density (e.g. ribosomes), 'continuous' gradients are used. These have a gradual transition from high to low density so that each particle-type can band at its own specific density. For lower purity requirements 'discontinuous' gradients can be used. These have low density solution layered on top of high density solution, with a sharp interface between them where all material with a specific density between the high and low solutions concentrates. Discontinuous gradients are slightly easier to prepare and to fractionate, but continuous gradients are cleaner and more secure if the specific density of a virus is not known. Many different substances can be used for creating the density differential (sugars, salts, polyethylene glycol, synthetic polymers), each with their (dis) advantages, but for most virus purification purposes sucrose gradients are adequate. The most common alternative is CsCl (caesium chloride) gradients. These are easier to prepare and generally leave cleaner virus preparations, but require longer centrifugation. CsCl is also chaotropic, stripping virus particles from other cellular constituents, and is therefore not suitable for purifying enveloped viruses. For these, sucrose gradients should be used. Gradient centrifugation is best done in a high-speed, swing-out rotor. These

usually have 6 buckets, which should all be used during each centrifugation run, since the rotor is only fully balanced when each bucket is present in its correct position. This means that for each run, six centrifugation tubes should be prepared and balanced.

7.2.3.1. Sucrose gradients

1. Prepare four solutions, containing the appropriate virus extraction buffer (Table 4) and either 10%, 20%, 30% or 40% sucrose.

This is best done by mixing the appropriate amounts of 10x buffer, 60% sucrose solution and water.

2. Divide the total volume of a centrifuge tube by 5.

3. Add 1/5th volume of the 10% sucrose-buffer solution to every centrifugation tube.

Be accurate with the volumes, to avoid problems with balancing the tubes later-on.

4. Using a syringe with a long needle, layer 1/5th volume of the 20% sucrose-buffer solution underneath the 10% solution.

5. Repeat with 1/5th volumes of the 30% and 40% sucrose-buffer solutions.

6. Place the tubes in a freezer with minimum disturbance,.

7. Once completely frozen, take the tubes out and thaw completely.

The higher concentration solution will thaw earlier than the lower density solutions, causing the boundaries between the concentration layers to blur.

8. Repeat the freeze-thaw process twice more to extend this process, generating a continuous density gradient.

Do not freeze-thaw too often, or you will end up with a single-density solution.

9. For discontinuous gradients, layer 1/5th volume of the 40% sucrose-buffer solution underneath 3/5th volume of the 10% sucrose-buffer solution.

10. Layer 1/5th volume of virus extract carefully on top of the gradient.

11. Balance the tubes by weight to within 1 mg, using buffer solution.

12. Insert tubes carefully in the buckets and hang the buckets in the correct orientation in their appropriate place on the rotor.

13. Centrifuge in a swing-out rotor at the appropriate speed, time and temperature for the virus in question (see Table 4).

7.2.3.2. Caesium chloride gradients

CsCl gradients can be used either instead of sucrose gradients, or as an additional purification step after sucrose gradients. CsCl gradients are formed automatically during centrifugation, from a single-density solution (isopycnic, or self-forming, gradients). This is a property of the heavy salt.

1. Resuspend the virus in its buffer (Table 4).

2. Make the solution 1.37 g/ml CsCl; final concentration. This is the density where most picorna-like viruses will band.

3. Centrifuge 28,000 rpm for 16-24 hours at 15°C-20°C. Higher centrifugation speeds will create steeper gradients.

7.2.3.3. Fractionation

1. After centrifugation, carefully remove the centrifuge tubes from each bucket.

If there is a lot of virus then the virus particles can be seen as an iridescent band underneath a top-light (Figure 3).

Often two bands can be seen; a lighter band higher in the gradient and a more intense band lower in the gradient. These correspond to 'empty' and 'filled' (with RNA) particles respectively.

2. Remove the band(s) with a disposable syringe and needle. This is best done using a needle with a 'flat' end, rather than the 'angled' end. Slide the needle along the centrifuge wall to just below the band and draw up the band slowly into the syringe.

3. If no bands can be seen, either because of low virus amounts or because the centrifuge tube is opaque, then the gradient needs to be fractionated by removing 0.5 ml volumes at a time.

The best way is to use an automated fractionator, which removes the fractions from the bottom of the gradient. The manual alternative is to remove 0.5 ml fractions from the top of the gradient.

4. Analyse 10 μl of each fraction for the presence of virus, using either ELISA (see section 9.2.) or RT-(q)PCR (see section 9.3. and the *BEEBOOK* article on molecular methods; Evans *et al.,* 2013).

Since there will be some virus contaminating every fraction, qualitative RT-PCR will not be able to distinguish very well between high and low virus fractions.

5. Pool the 3-4 fractions containing the highest virus concentrations, giving a final volume of approximately 2 ml.

6. If necessary, the virus can be concentrated further by another high-speed centrifugation (Table 4), although this will reduce the yield.

Fig. 3. White translucent band containing DWV particles after CsCl density gradient centrifugation.   Image © E Ryabov.

8. Virus sample processing

8.1 Introduction

The primary processing of a sample is crucial for the uniformity of a diagnostic method and should be optimized for high yield and low variability (Bustin, 2000). There are two main stages: sample homogenization and nucleic acid extraction. This section is covered in detail in the *BEEBOOK* article on molecular methods (Evans *et al.,* 2013), with a reduced version given below.

8.2. Sample homogenisation

The most variable step in sample processing is sample homogenization. Different sample types require different homogenization methods (see below), but all should be optimized experimentally to ensure minimal variability between replicate samples.

8.2.1. Bead-mill homogenizers

These are mechanical shakers that homogenize samples with glass, ceramic or steel beads and give excellent and highly uniform homogenization of small samples (1-10 bees).

8.2.2. Blender

Similar homogenization uniformity to the beadmills, but for large samples (50-1000 bees). There is a risk of cross-contamination between samples, through re-use of the blender, which should be assessed.

8.2.3. Mortar and pestle

This method is suitable for medium-size samples (10-50 bees) but is much less uniform than beadmills or blenders, and also has a cross-contamination risk due to re-use of the equipment.

8.2.4. Mesh bags

Mesh bags are a disposable alternative to mortars, also for medium-size samples (10-50 bees). Homogenization uniformity is also moderate, but without the cross-contamination risk.

8.2.5. Micropestle

These are disposable pestles for manually grinding individual bee samples in microcentrifuge tubes. They are much inferior to bead-mills in terms of homogenization uniformity, for the same samples size. At best they should only be used for soft (brood) stages.

8.2.6. Robotic extraction

Robotic extraction stations are very reliable and consistent, but generally only suitable for small sample sizes and soft tissues. They are best used after bead-mill homogenization, as part of a semi-automated homogenization-extraction chain.

8.3. Nucleic acid extraction

A denaturing buffer should be used during homogenization to protect the nucleic acids from degradation. The most common denaturants used are: high concentrations of chaotropic (guanidine) salts, strong antioxidants (β-mercaptoethanol), detergents and/or organic solvents. The nucleic acid is purified from the buffer using either cheap, disposable affinity purification columns or even cheaper precipitation with ethanol, isopropanol or lithium chloride (RNA only). Both methods are reliable, though not particularly uniform (Tentcheva *et al.,* 2006). Affinity columns generally produce cleaner nucleic acid samples, due to the column washing steps. Precipitation can produce higher yields, since columns have a limited binding capacity, or extract volume, equivalent to ~¼ bee.

8.3.1. Protocol 1 – affinity column purification

The processing consists of making a primary homogenate from 1-30 bees and purifying RNA from 100 µl aliquots of the homogenate (equivalent to 20 mg bee tissue: the maximum loading capacity of one affinity column). The protocol is based on the Qiagen RNA purification columns. β-mercaptoethanol is toxic.

1. Prepare fresh GITC buffer:
   1.1. 5.25 M guanidinium thiocyanate (guanidine isothiocyanate),
   1.2. 50 mM TRIS.Cl (pH 6.4),
   1.3. 20 mM EDTA,
   1.4. 1.3% Triton X-100,
   1.5. 1% β-mercaptoethanol.
2. Place frozen bees in the homogenizer of choice.
3. Per bee add the following amount of GITC buffer:

Bee	Weight	Buffer	Total volume
Worker bee	120 mg	500 µl	600 µl
Drone	180 mg	700 µl	900 µl
Worker pupa	160 mg	650 µl	800 µl
Drone pupa	240 mg	1000 µl	1200 µl

4. Proceed according to the Qiagen Plant RNA extraction protocol using 100 µl extract as sample. The Qia-shredder option significantly increases yield and purity of nucleic acid (see Qiagen instructions booklet).
5. Elute in 100 µl nuclease-free water.
6. Determine nucleic acid concentration and purity (see section 8.4.; "Nucleic acid quality assessment").
7. Store as two separate 50 µl aliquots at -80°C, one for working with and one for storage.
8. Include a 'blank' extraction (*i.e.* an extraction of purified water) after every 24 bee samples, to make sure none of the extraction reagents have become contaminated.

8.3.2. Protocol 2 – TRIzol extraction and isopropanol precipitation

This protocol uses TRIzol®; a proprietary mixture of phenol and high concentration salt solution (InVitrogen). The RNA is recovered through isopropanol precipitation.

1. Homogenize bees directly at 4°C in TRIzol® reagent in a glass-walled blender or mortar and pestle. Use 1 ml reagent per bee (~120 mg tissue).
2. Add 0.5 ml chloroform per bee.
3. Shake hard for 1 minute.
4. Centrifuge 8,000 g; 15 minutes; 4°C.
5. Recover the upper (aqueous) layer containing the nucleic acids, discarding the lower red (organic) phase and the semi-solid, white interphase (containing proteins and lipids).
6. Add an equal volume of iso-propanol, mix and precipitate at -20°C for at least 15 minutes.
7. Centrifuge 8,000 g; 15 minutes; 4°C.
8. Remove iso-propanol supernatant.
9. Resuspend nucleic acid pellets in 100 µl RNase-free water.
10. Determine nucleic acid concentration and purity (see section 8.4.; "Nucleic acid quality assessment").
11. Store as two separate 50 µl aliquots at -80°C, one for working with and one for storage.

8.4. Nucleic acid quality assessment

A number of sophisticated methods are available to determine the quantity, quality and integrity of an RNA sample, as described in the *BEEBOOK* paper on molecular methods (Evans *et al.,* 2013). The minimum requirements are to determine the yield of the RNA, and its purity with respect to protein and phenolic metabolite contaminants. This can be determined by UV spectrophotometry (Green and Sambrook, 2012), through comparing peak absorbance at 260 nm (nucleic acids), 280 nm (proteins) and 230 nm (phenolic metabolites):

- A^{260} of 1.0   = 40 ng/µl ssRNA
                   = 37 ng/µl ssDNA
                   = 50 ng/µl dsDNA
- $A^{260}/A^{280} < 2.0$ indicates contamination with proteins.
- $A^{260}/A^{230} < 2.0$ indicates contamination with phenolics.

9. Virus detection

9.1. Introduction

There are numerous techniques available for detecting and quantifying viruses (de Miranda, 2008; see also the *BEEBOOK* paper on molecular methods (Evans *et al.,* 2013). Most of these detect only a small portion of the viral genome or the capsid proteins, and almost all require some sort of amplification, either of the target (most of the nucleic acid-based detection technologies) or the detection signal (most of the protein-based detection technologies). Both are important considerations to bear in mind when interpreting virus diagnostic data. Here we will only cover the most commonly used methods.

Secondly, despite the popular classification of molecular assays as either 'qualitative' (presence/absence) or 'quantitative' (concentration), ultimately all assays are quantitative: qualitative assays are simply quantitative assays with a detection threshold (a visible colour; a band on a gel; a fluorescence level; a C_q value; a statistical index). This is an important consideration, since there are many factors besides the initial virus amount that can influence whether or not an assay reaches a detection threshold, such as degradation of the sample, changes to storage-extraction procedures, assay deterioration *etc.*. Furthermore, the molecular and mathematical rules underpinning any assay are the same whether this assay is 'qualitative' or 'quantitative'. The only difference is that in 'quantitative' assays these rules are specifically acknowledged and accounted for, whereas in 'qualitative' assays they are often ignored. It is therefore advisable to approach any experiment or assay from a quantitative perspective first, and include the appropriate controls for threshold-conversion to 'qualitative' data, if this is desired.

9.2. Enzyme-Linked ImmunoSorbent Assay (ELISA)

There are many versions of the ELISA, using different blocking agents, primary/secondary antibodies, reporter enzymes and their specific colorimetric substrate solutions for detection and quantification (Harlow and Lane, 1988). They generally fall into one of two major categories:

9.2.1. Normal ELISA

In conventional ELISA, the sample is adsorbed directly into the wells, to be detected by the specific antibody. This antibody is either conjugated directly to an enzyme (Fig. 4A), usually either horse radish peroxidase or alkaline phosphatase, or more commonly is detected in a subsequent incubation by a commercial enzyme-conjugated protein that recognizes antibodies in general (Fig. 4B).

9.2.2. Sandwich ELISA

In "sandwich" ELISA, a modified version of the primary antibody is adsorbed to the well first, in order to 'capture' the virus particles after the sample is added. The captured virus particles are then detected as before, either with the reporting enzyme directly conjugated to the detecting antibody (Fig. 4C) or with an extra incubation using an antibody-detecting protein conjugated to the reporter enzyme (Fig. 4D). The sandwich ELISA is cleaner and much more sensitive than conventional ELISA, but has a less predictable relationship between virus concentration and signal (depending on which component in the assay is limiting).

The most common reporter enzyme systems are horseradish peroxidase (HRP) and alkaline phosphatase (AP). Both are relatively robust enzymes that can be conjugated to the primary or secondary antibody. They convert a colourless substrate into a coloured reaction product, such that the absorbance at a wavelength appropriate for the specific colour produced is proportional to the amount of enzyme activity present in the sample, which in turn is proportional to the amount of antibody captured by the sample, and thus also the amount of virus in the sample. The protocols below are generic ones for conventional ELISA and sandwich ELISA, based on the methods of Allen *et al.* (1986), using horseradish peroxidase as the reporter enzyme. See also Harlow and Lane (1988) for alternatives and more extensive laboratory protocols involving antibodies.

9.2.3. Protocols

9.2.3.1. Sample preparation

1. Mix phosphate-buffered saline (PBS):
    1.1. 0.8% NaCl,
    1.2. 0.14% $Na_2HPO_4.2H_2O$,
    1.3. 0.02% KH_2PO_4,
    1.4. 0.02% KCl,
    1.5. Adjust to pH 7.4.
2. Grind each bee in 1 ml PBS.
3. Add 300 µl chloroform.
4. Mix on a vortex.
5. Centrifuge for 3 minutes to clarify.

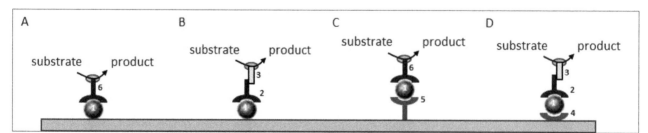

Fig. 4. Different types of ELISA, depending on whether the antigen[1] is adsorbed directly onto the assay well (A & B) or is captured by the F_{ab} fragment of a specific antibody[4] or the full antibody[5] (sandwich ELISA; C & D) and whether the detection system involves an enzyme reporter conjugated directly to the detecting antibody[6] (A & C) or a reporter conjugated to a generic antibody-recognizing protein[3] recognizing the detecting antibody[2] (B & D). Adapted from de Miranda (2008).

9.2.3.2. ELISA

1. Mix coating buffer (CB):
    1.1. 0.159% Na$_2$CO$_3$,
    1.2. 0.293% NaHCO$_3$,
    1.3. Adjust to pH 9.6.
2. Seed each well with 180 µl of CB.
3. Add 5-20 µl sample.
4. Incubate 3 hrs at ambient temperature, or overnight at 4°C.
5. Tip out and wash the wells 3X with PBS-T (= PBS containing 0.05% Tween20 detergent), shaking the ELISA plate dry each time.
6. Prepare a 1/2,000 - 1/5,000 dilution of the primary antibody in PBS-TPO:
    6.1. 2% Polyvynylpyrrolidone (PVP) mw 440000,
    6.2. 0.2% Bovine serum albumin (BSA),
    6.3. in PBS-T (fresh daily).
7. Add 200 µl of antibody/PBS-TPO to each well.
8. Incubate 3 hrs at ambient temperature, or overnight at 4°C.
9. Tip out fluid and wash the wells 3X with PBS-T, shaking the ELISA plate dry each time.
10. Prepare a ProteinA-HorseRadishPeroxidase (PrA-HRP) conjugate stock solution at 100 µg/ml.
11. Make a 1/2,000 – 1/5,000 dilution of PrA-HRP stock solution in PBS-TPO.
12. Add 200 µl to each well.
13. Incubate 3hrs at ambient temperature, or overnight at 4°C.
14. Tip out fluid and wash the wells 3X with PBS-T, shaking the ELISA plate dry each time.

9.2.3.3. Sandwich ELISA

1. Seed each well with 200 µl of a 1/2,000-1/5,000 dilution of the F$_{ab}$ fragment (Harlow and Lane, 1988) of the primary antibody in coating buffer (CB, see section 9.2.3.2.).
2. Incubate 3hrs at ambient temperature, or overnight at 4°C.
3. Tip out and wash the wells 3X with PBS-T, shaking the ELISA plate dry each time.
4. Add 180 µl PBS-TPO to each well.
5. Add 20 µl sample to each well.
6. Incubate 3hrs at ambient temperature, or overnight at 4°C.
7. Tip out fluid and wash the wells 3X with PBS-T, shaking the ELISA plate dry each time.
8. Add 200 µl 1/2,000 – 1/5,000 dilution of PrA-HRP stock solution in PBS-TPO to each well.
9. Incubate 3hrs at ambient temperature, or overnight at 4°C.
10. Tip out fluid and wash the wells 3X with PBS-T, shaking the ELISA plate dry each time.

9.2.3.4. Development

The development of the enzymatic reaction is the same for conventional and sandwich ELISA. This method is appropriate for horseradish peroxidase as a reporter enzyme. A different substrate is required if alkaline phosphatase is used as the reporter enzyme (Harlow and Lane, 1988), together with a different wavelength for determining the absorbance, but the overall procedure is the same.

1. Prepare TMB substrate solution:
    1.1. 100 ml water,
    1.2. 1 ml 10 mg/ml TMB (33'55'TetraMethylBenzidine) in DMSO (DiMethylSulfOxide),
    1.3. 10 ml 1M Na Acetate (pH 5.8 with 1.0M citric acid),
    1.4. 20 µl 30% H$_2$O$_2$.
2. Add 200 µl substrate solution to each well.
3. Let colour develop for 10-15 minutes .
4. Add 50 µl 3M H$_2$SO$_4$ to terminate the reaction.
5. **Immediately** read the absorbance at 450 nm (the termination reaction will continue to develop colour).

9.2.3.5. Controls

ELISA is a complex, multistep assay involving sensitive enzymatic reporters, which means that there are many opportunities for assay failure, either through false-positive or false-negative results. Enzymatic reporter systems, such as used by ELISA, are sensitive to any native enzymatic activity present in the sample (peroxidases, phosphatases). The initial coating step in a highly alkaline buffer abolishes most of such background activity, as does the specific capture of virus particles in sandwich ELISA and the washes with PBS. However, the user should be aware of the possibility of residual enzymatic activity in the samples, particularly if the substrate incubation step is extended to allow more colour to develop (for instance, when trying to detect very low amounts of virus). Secondly, either the enzyme or the substrate may be faulty, preventing colour development even though there has been antibody recognition of the sample. Alternatively, the primary or secondary antibody may fail, for a number of reasons. All ELISA assays should therefore have a number of controls to establish the correct functioning of the assay itself, thus validating the results from the samples.

*   Reporter-free negative control (quantification of background substrate absorbance).
*   Sample-free negative control (Absence of non-specific binding of antibodies/reporters to wells; test of blocking and washing efficacy).
*   Primary antibody-free negative control (Absence of non-specific binding of secondary antibody and/or reporters).
*   Secondary antibody/reporter-free negative control (quantification of background enzymatic activity in samples).

- Substrate-free negative control (quantification of background absorbance in the system).
- Primary antibody positive control (direct adsorption of antibody in CB: test for recognition of antibody by secondary antibody-reporter).
- Secondary antibody positive control (direct adsorption of secondary antibody-reporter in CB: test for functioning reporter enzyme).
- Purified virus positive control (correct recognition of virus by primary antibody; calibration standards).

9.3. RT-(q)PCR

The most common current methods for honey bee virus detection are based on Reverse Transcription Polymerase Chain reaction (RT-PCR), essentially the PCR amplification of cDNA. A detailed coverage of the principles and practices of PCR is found in Yuryev (2007) and the *BEEBOOK* paper on molecular methods (Evans *et al.*, 2013). A reduced version, including those elements specifically relevant to virus detection, is presented here.

9.3.1. Primer design

Designing RT-(q)PCR assays for detecting (honey bee) RNA viruses poses some unique challenges. The most critical components of a PCR assay are the two amplification primers. RNA viruses are genetically highly variable while PCR is very sensitive to nucleotide mismatches between primer and target, particularly at the 3′ primer termini where extension occurs (Onodera, 2007). A mismatch at the 3′ terminus of just one of the primers will result in non-amplification. Mismatches further away from the 3′ terminus have increasingly less influence on the success of amplification and generally only the last two 3′ nucleotides are critical for amplification specificity. The 3′ mismatch issue is therefore crucial to the specificity, accuracy, reliability and sensitivity of a PCR-based virus assay. Here we outline how to use this to our benefit, and how to avoid it when needed.

9.3.1.1. What do we want to detect?

The first decision is to establish precisely what the assay should detect and what it should not detect:

- For distinguishing closely related strains, locate the 3′ terminus of one primer at a position where the strains differ consistently. The other primer can be common for all strains (de Miranda *et al.*, 2010b).
- For detecting all potential variants within a virus species or complex, the primer sequences should be conserved between all known variants, so as to be able to detect both known and as-yet-unknown variants in the complex. Locate the primers at least 200 nucleotides apart, so that new variants can be identified by sequence analyses of the intervening region.
- Avoid locating the 3′ terminus of a primer on the 3rd base of a codon in the coding region of a virus genome, since these are

by far the most variable nucleotides in any virus genome (Grabensteiner *et al.*, 2001; Bakonyi *et al.*, 2002b; de Miranda *et al.*, 2004; Lanzi *et al.*, 2006; Olivier *et al.*, 2008; de Miranda *et al.*, 2010b).

- Use deoxyinosine as the 3′ nucleotide, which can pair with all nucleotides (Benjeddou *et al.*, 2001; Topley *et al.*, 2005), thus avoiding the 3′ mismatch problem altogether.

9.3.1.2. Where in the genome?

The genome of positive-strand RNA viruses is usually compact and efficiently coded, and there is normally no duplication of sequences within the genome. This facilitates the assay design enormously, since one can choose between many alternative assays on virological and assay performance-quantitation criteria, no matter where in the genome they are located, since they should all only amplify a single region of the genome.

9.3.1.3. Primer annealing temperature, length and composition

Both amplification primers should have similar melting temperature (T[m]), length and composition. It is useful to design all assays and primers around the same annealing temperature, so that a single cycling program can be used for all assays, and that different assays can be run concurrently with the same program, on the same plate. 56°C is a good, standard, robust target for the *in silico* estimated T[m] for primers.

9.3.1.4. Primer-dimer and other PCR artefacts

PCR is susceptible to qualitative and quantitative errors caused by the accidental, and highly efficient, amplification of short non-target PCR templates formed by fleeting complementarity of the primers with non-target templates, or among the primers themselves (SantaLucia, 2007; see the *BEEBOOK* paper on molecular methods; Evans *et al.*, 2013). Such artefacts can be identified by gel electrophoresis during assay optimization. The easiest solution to persistent PCR artefacts is to design new primers and test these experimentally (SantaLucia, 2007).

9.3.1.5. Fragment length

PCR assay design software packages usually design very short amplicons (< 100 nucleotides), with high amplification efficiency and short cycling times. However, amplicons up to 300 nucleotides amplify equally efficiently but are easier to separate from PCR artefacts, provide more room for designing probes and can be used to characterize new variants, through sequence analyses.

9.3.2. Detection and analysis of PCR products
9.3.2.1. "End-point" vs "real-time" detection

The PCR products can be detected after the PCR is completed, usually for "qualitative" analysis (presence/absence of product), either by (gel) electrophoresis or Melting Curve analysis. Detection can also be done after each cycle, as PCR proceeds, using laser optics (*i.e.* in 'real

-time'). The amount of initial target cDNA in a reaction can then be very accurately related to how many amplification cycles are required for a product to appear. This is the basis for "quantitative" PCR (qPCR), which is extremely accurate over a wide range of target concentrations (see the *BEEBOOK* paper on molecular methods; Evans *et al.*, 2013).

9.3.2.2. Cycles and thresholds

Continuous real-time detection also allows multiple detection thresholds (*i.e.* the cycle at which product appears) to be set for the same reaction, which can be related to different levels of risk for disease. For most practical (or even experimental) purposes, 35 cycles of amplification is sufficient. Beyond 35 cycles, the rapidly increasing risk of both false-positive and false-negative detection errors outweighs the marginal gains in sensitivity (see the *BEEBOOK* paper on molecular methods; Evans *et al.*, 2013).

9.3.2.3. Detection chemistry

There are many different detection chemistries available for following qPCR in real-time (de Miranda, 2008). The two most common are SYBR-green and similar DNA-binding dyes, and fluorophore-labelled hydrolysis (TaqMan™) probes. The merits of both systems are discussed in the *BEEBOOK* paper on molecular methods (Evans *et al.*, 2013). TaqMan™ assays are best suited for well-optimized, stable assays for widespread, routine diagnosis. SYBR-green assays are better suited for discovery, characterization of new strains and analysis of strain mixtures (Papin *et al.*, 2004). They are also much cheaper to design, adapt and optimize (Bustin and Nolan, 2004).

9.3.3. Assay optimization

Once a PCR assay has been designed, it should be optimized experimentally for annealing temperature (using annealing temperature gradients), primer concentration and cycling times (see the *BEEBOOK* paper on molecular techniques; Evans *et al.*, 2013). Optimization usually identifies the highest annealing temperature, the lowest primer concentrations and the shortest incubation time that consistently generates the right product, without artefacts, at a consistent amplification cycle (see section 12: "Quality control").

9.3.3.1. Reverse transcription

Reverse transcription is the most variable step in RT-PCR, whose efficiency is easily affected by inhibitors, reaction conditions (including primers) and even nucleic acid concentration (Ståhlberg *et al.*, 2004a; 2004b). To minimize this variability, the nucleic acid should be optimally prepared and a constant amount used in every reaction. Since PCR does not require large amounts of initial target, the RNA can be diluted to minimise the effects of any inhibitors. cDNA is best prepared with random hexamer primers that generate a bias-free cDNA copy of the entire RNA population, suitable for a multitude of analyses.

9.3.3.2. One-Step/Two-Step RT-PCR

Reverse transcription and PCR can be conducted in a single buffer, PCR following reverse transcription (One-Step RT-PCR) or in two separate reactions (Two-Step RT-PCR). The advantages of One-Step RT-PCR are speed and reduced contamination risk; the disadvantages are wasteful use of precious RNA and inability to control for differences in cDNA synthesis efficiency between reactions (Bustin, 2000; Bustin *et al.*, 2009). These (dis)advantages are reversed for Two-Step RT-PCR, with the additional advantage that the cDNA produced can be used for many other purposes as well. Two-Step RT-PCR also tends to be considerably more sensitive and more prone to artefacts, unless steps are taken to avoid this (see the *BEEBOOK* paper on molecular methods; Evans *et al.*, 2013).

9.3.4. Protocols

Numerous qualitative and quantitative RT-(q)PCR protocols have been published for honey bee viruses (Annex 1: http://www.ibra.org.uk/downloads/20130805/download), although few have been optimized experimentally. The European Reference Laboratory for bee diseases at ANSES (France) is in the process of designing fully optimized, validated RT-qPCR protocols for all bee viruses (see Blanchard *et al.*, 2012) for routine, standardised diagnostic use by accredited laboratories. For experimental purposes, existing published protocols can be used and optimized, many of which can be easily adapted to qPCR using SYBR-green dye detection. Alternatively, new protocols can be designed based on the following practical, robust protocols for Reverse Transcription, One-Step RT-qPCR and Two-Step RT-qPCR, suitable for either quantitative or qualitative analyses. These provide a useful basis for individual adaptation and optimization, using the guidelines given above and in the *BEEBOOK* paper on molecular techniques (Evans *et al.*, 2013).

9.3.4.1. Reverse transcription

The following is a robust, standard reverse transcription protocol for generating cDNA that is fully representative of the original RNA population:

1. Mix:
   1.1. 0.5 µg sample RNA template,
   1.2. 1 ng exogenous reference RNA (*e.g.* Ambion RNA250),
   1.3. 1 µl 50 ng/µl random hexamers,
   1.4. 1 µl 10mM dNTP,
   1.5. up to 12 µl RNAse free water.
2. Heat the mixture to 65°C for 5 min and chill quickly on ice.
3. Add:
   3.1. 4 µl 5X First-Strand Buffer,
   3.2. 2 µl 0.1 M DTT,
   3.3. 1 µl (200 units) of M-MLV RT.
4. Mix by pipetting gently up and down.
5. Centrifuge briefly to collect the contents at the bottom of the tube.

6. Incubate 10 min at 25°C.

7. Incubate 50 min at 37°C.

8. Inactivate the reaction by heating 15 min at 70°C.

9. Dilute the cDNA solution tenfold with nuclease-free water before using in PCR assays, to reduce the risk of PCR artefacts.

9.3.4.2. Two-Step RT-qPCR

The following is a robust, standard qPCR protocol for amplifying and quantifying cDNA templates < 300bp in length. The protocol is based on SYBR-green detection chemistry, with modifications for probe-based detection and qualitative PCR indicated:

1. Mix:

    1.1. 3 µl cDNA (pre-diluted 1/10, in nuclease-free water),

    1.2. 0.6 µl 10 µM Forward primer (0.3 µM final concentration),

    1.3. 0.6 µl 10 µM Reverse primer (0.3 µM final concentration),

    *[1.4. 0.4 µl*10 µM TaqMan™ probe* (0.2 µM final concentration*)],*

    1.5. x µl *TwoStep Buffer* + dNTP(0.2 mM final dNTP),

    1.6. y µl nuclease-free water,

    1.7. z µl Thermostable DNA polymerase mix,

    1.8. 20 µl total volume.

* Use appropriate buffer for either SYBR-green or TaqMan™ probe assays. dNTP is usually included in pre-optimized buffers. If not, add separately to 0.2 mM final concentration.

2. Incubate in real-time thermocycler:

    2.1. 5 min:95°C,

    2.2. 35 cycles [10 sec:95°C - 30 sec:58°C - read].

3. For SYBR-green assays, follow with Melting Curve analysis:

    3.1. 1 min:95°C,

    3.2. 1 min:55°C,

    3.3. 5 sec:0.5°C:read from 55°C to 95°C.

4. For qualitative PCR, a conventional thermocycler can be used and the products can be analysed by gel, capillary or chip-based electrophoresis.

9.3.4.3. One-Step RT-qPCR

The following is a robust, standard One-Step RT-qPCR protocol for amplifying and quantifying targets < 300bp in length, using SYBR-green detection chemistry, and starting with an RNA template:

1. Mix:

    1.1. 3 µl 5 ng/ µl RNA,

    1.2. 0.6 µl 10 µM Forward primer (0.3 µM final concentration),

    1.3. 0.6 µl 10 µM Reverse primer (0.3 µM final concentration),

    *[1.4. 0.4 µl*10 µM TaqMan™ probe* (0.2 µM final concentration*)],*

    1.5. x µl *OneStep Buffer* + dNTP(0.2 mM final dNTP),

    1.6. y µl nuclease-free water,

    1.7. z µl Reverse Transcriptase/Thermostable DNA polymerase mix,

    1.8. 20 µl total volume.

* Use appropriate buffer for either SYBR-green or TaqMan™ probe assays. dNTP is usually included in pre-optimized buffers. If not, add separately to 0.2 mM final concentration.

2. Incubate in real-time thermocycler:

    2.1. 15 min:50°C,

    2.2. 5 min:95°C,

    2.3. 35 cycles [10 sec:95°C - 30 sec:58°C - read].

3. For SYBR-green assays, follow with Melting Curve analysis:

    3.1. 1 min:95°C,

    3.2. 1 min:55°C,

    3.3. 5 sec:0.5°C:read from 55°C to 95°C.

4. For qualitative PCR, a conventional thermocycler can be used and the products can be analysed by gel, capillary or chip-based electrophoresis.

9.3.5. Quantitation controls

A number of controls are required for quantifying the amount of virus in a sample. These can be broadly divided into "external reference standards", which are used to quantify the absolute amount of target in each reaction, and "internal reference standards", which are used to correct the quantitative data for unique differences between individual samples. The *BEEBOOK* paper on molecular methods (Evans *et al.,* 2013) describes in detail the function, preparation and application of these standards.

9.3.5.1. External reference standards

These consist of dilution series of known concentrations of (cloned) target DNA or RNA, which is used to establish a calibration curve for converting RT-qPCR data to absolute amounts of target (genome copies) in a reaction (Pfaffl and Hageleit, 2001; Bustin *et al.,* 2009). External reference standards should be prepared for every target assayed, including all internal reference standards.

9.3.5.2. Internal reference standards

Unfortunately, external standards cannot correct for factors unique to each sample that affect the RT and/or PCR reactions, such as RNA quality and quantity, enzyme inhibitors, sample degradation, internal fluorescence *etc.* To correct for these factors, internal reference standards are used. These come in two forms:

"*Exogenous internal reference standard*", which is a pure, unrelated RNA of known concentration that is added to the RT mastermix prior to RT-qPCR (Tentcheva *et al.,* 2006). The amount used should be < 1% of the amount sample RNA, so as not to affect the RT-qPCR reaction efficiencies. These are used to calculate cDNA reaction efficiencies of individual samples (correcting for RT inhibitors).

"Endogenous internal reference standards" (commonly called 'housekeeping genes'), are relatively invariant host mRNAs present in every sample. These can be used to normalize quantitative data for differences between samples in RNA degradation or the presence of inhibitors (Bustin *et al.,* 2009; Radonić *et al.,* 2004) and to guard against 'false-negative' data (due to RNA degradation).

There are a couple of practical difficulties with endogenous internal reference standards. First, one can never be certain that they are truly invariant (Radonić *et al.,* 2004). The current recommendations are therefore to use an index of 3 or 4 endogenous reference standards for data correction (Bustin, 2000). Second, contaminating genomic DNA in an RNA sample can interfere with accurate quantification of the endogenous gene mRNA. This can be avoided by digesting the RNA sample with DNAse prior to RT-PCR, or more elegantly by designing intron-spanning primers for the endogenous reference gene (Bustin, 2000; Yañez *et al.,* 2012; Locke *et al.,* 2012), such that only cDNA to the mRNA can be amplified.

Internal reference standards are costly, since they are run for all samples. Their inclusion should therefore be evaluated in relation to their importance to the project. There are probably more relevant for fully-quantitative experiments and less for semi-quantitative surveys.

9.3.6. Multiple assays

With careful primer design (see section 9.3.1.) it should be possible to approach 100% correct detection (no false positive or false negative results) for most viruses with a single primer pair. This is, however, very much conditional on the natural variation and variability (*i.e.* the capacity to generate new variants) for each virus. There are valid arguments that PCR is perhaps too specific for the reliable detection of highly variable entities such as RNA viruses, even when employing several different primer sets (Gardner *et al.,* 2003). When the reliability of a primer set with respect to virus variability is in doubt, the best resolution is to employ several primer sets in parallel so that the failure of one set does not necessarily result in misdiagnosis. Multiple primer sets also allows one to estimate the rate of misdiagnosis by different primer sets due to virus variability (Chui *et al.,* 2005). Within the honey bee viruses, multiple primer sets may be needed for reliable diagnosis within the highly variable ABPV complex (de Miranda *et al.,* 2010a) and the slightly less variable DWV-VDV-1 complex (de Miranda and Genersch, 2010). Multiple assays are available for most honey bee viruses, and comparisons of multiple assays have been made for SBV (Grabensteiner *et al.,* 2001), ABPV (Bakonyi *et al., 2002a;* 2002b) and DWV (Genersch, 2005).

9.3.7. Multiplex RT-(q)PCR

Multiplex RT-PCR refers to the simultaneous amplification of several targets in the same reaction. The different end-products are usually identified by size, through (gel) electrophoresis. Several such qualitative multiplex protocols have been designed for honey bee viruses (Chen *et al.,* 2004b; Topley *et al.,* 2005; Grabensteiner *et al.,* 2007; Weinstein-Texiera *et al.,* 2008; Meeus *et al.,* 2010). Real-time qPCR can also be multiplexed, usually for the simultaneous amplification of a target and internal reference standards, by using TaqMan™ probes with different fluorophores.

The main reason for multiplexing is to save cost and time. However, multiplex PCR is less sensitive than uniplex PCR, more complex to optimize, more prone to artefacts and requires post-PCR fragment analysis, nullifying any gains in time and cost. Most importantly, the late amplification of low-abundance targets is strongly affected by the prior amplification of high-abundance targets, through the auto-inhibition of the PCR by the DNA it produces (SantaLucia, 2007). For these reasons, it is often more effective to use uniplex RT-PCR, even for large projects.

9.4. Microarrays

Multiplexing is far more effective through a microarray, which is an ordered array of hundreds of molecular probes specific for different target RNAs bound to a solid support, usually a slide. Most microarray technology has been developed for nucleic acid probes, although protein-based arrays are also being developed (Sage, 2004). The hybridization of RNA target sequences to these probes to these probes can be detected by a variety of methods (de Miranda, 2008), including PCR and sequencing. Numerous honey bee microarrays have been designed, including honey bee immune gene-pathogen arrays (Evans, 2006; Runckel *et al.,* 2011) and a honey bee virus array (Glover *et al.,* 2011). Microarrays are being superseded for research purposes by high-throughput sequencing technologies, but retain a future in routine screening applications, due to their adaptability and high multiplexing capacity (Glover *et al.,* 2011). See also the microarray section in the *BEEBOOK* paper on molecular methods (Evans *et al.,* 2013).

10. Virus replication

10.1. Introduction

The detection of viral replication is crucial for differentiating between an active infection and just the presence of virus particles in a host, or between a mechanical (virus non-replicating) and a biological (virus replicating) vector of a virus. Evidence that a virus is actively infecting a host includes the presence of viral particles and structures within host cells, revealed by electron microscopy, preferably including a specific nucleic acid or serological probe to positively identify the virus. Another approach is to detect the non-structural proteins involved in virus replication, which for most positive-stranded RNA viruses are only produced after invasion and mark the start of a replication cycle. Negative-stranded RNA viruses often carry their replicative proteins within the particle. A related philosophy, which is more sensitive and accessible, is to specifically detect the replicative strand RNA of a virus. Most of the described bee viruses are single- and plus-strand RNA viruses which replicate through a negative-strand RNA intermediate serving as template for the generation of new viral plus-strand RNA genomes. The specific detection of viral negative strand RNAs can therefore serve as a marker of active replication of these RNA viruses in a certain host, tissue or cell type. Below are outlined two methods for strand-specific detection of RNA virus sequences.

10.2. Strand-specific RT-qPCR

One of the most popular methods in bee virology for detecting virus replication is the specific detection of negative strand viral RNA, using strand-specific Reverse Transcriptase-PCR (Peng *et al.*, 2012; DiPrisco *et al.*, 2011; Boncristiani *et al.*, 2009; Dainat *et al.*, 2009; Eyer *et al.*, 2009; Gisder *et al.*, 2009; Celle *et al.*, 2008), following its first application in bee virology by Yue and Genersch (2005). The procedure is illustrated in Fig. 5, including the most common cause of false-positive results (non-specific cDNA synthesis) and how best to avoid this (tagged-cDNA primer followed by tag-specific PCR). Theoretically, strand-specificity can be achieved by performing the reverse transcription reaction in the presence of only one primer specifically annealing with a unique region of the viral negative strand before amplifying the obtained cDNA by adding the second primer or a specific primer pair for PCR amplification. Unfortunately, strand-specific RT-PCR is highly susceptible to false positive results (Gunji *et al.*, 1994; Lanford *et al.*, 1995; Lanford *et al.*, 1994; McGuiness *et al.*, 1994; Craggs *et al.*, 2001; Peyrefitte *et al.*, 2003; Boncristiani *et al.*, 2009) due to:

- False-priming of the incorrect strand by the cDNA primer.
- Self-priming of positive-strand RNA in areas of complex secondary structures.
- Random priming by contaminating cellular nucleic acids.
- Incomplete inactivation of the reverse transcriptase, leaving residual activity during PCR amplification (which contains both negative and positive strand primers).

To overcome these drawbacks and improve the specificity of the assays, certain effective techniques have been developed that enhance strand-specificity. These include:

- Thermostable reverse transcriptases.
- Tagged-cDNA primers.
- Inactivation/removal of residual tagged-cDNA primers prior to PCR.
- Chemical blocking of free 3' ends before or after reverse transcription.

10.2.1. Thermostable reverse transcriptases

Thermostable reverse transcriptases, operating at temperatures up to 50-70°C, avoid much non-specific priming of the RNA through elevated reaction temperatures (Lanford *et al.*, 1995; Laskus *et al.*, 1998; Craggs *et al.*, 2001; Horsington and Zhang, 2007; Carrière *et al.*, 2007; Celle *et al.*, 2008). Thermostable reverse transcriptases need to be inactivated thoroughly prior to the PCR step, otherwise the reverse transcriptase has access to primers for both strands (thus nullifying the strand-specificity). Another strategy is to inactivate the (viral) RNA by alkaline treatment or digestion with RNase H (McGuiness *et al.*, 1994), thereby removing any target for reverse transcription during PCR.

When using thermostable reverse transcriptase, make sure that the virus-specific portion of the tagged cDNA primer has a theoretical T^m ~60°C, to ensure adequate priming at elevated temperatures.

10.2.2. Chemical blocking of RNA 3' ends

The free 3' ends of the RNA can be blocked with borohydride, so that the RNA cannot serve as a primer for cDNA synthesis by self-priming or random priming with small cellular RNAs. This means that only RNA primed with the strand-specific cDNA primers can be elongated by the reverse transcriptase (Gunji *et al.*, 1994). The protocol involves oxidation of the RNA free ends with sodium periodate ($NaIO_4$) followed by reduction with sodium tetrahydroborate ($NaBH_4$).

10.2.3. Tagged cDNA primers

To further improve strand-specific detection of viral RNA, tagged RT-PCR can be used (Craggs *et al.*, 2001). This method relies on a primer for cDNA synthesis which contains a tag sequence at the 5'-end that is unrelated to either virus or host. PCR amplification is then carried out with a primer consisting of only the tag sequence, together with a virus-specific upstream primer. This ensures that only cDNA's derived from the tagged cDNA primer are amplified, and not cDNAs from false-, self- or mis-priming events. It is therefore important to ensure that the chosen tag sequence does not show any homology with a known bee pathogen or invertebrate sequence, by checking the tag sequences against the nucleotide sequence databases available on the NCBI website, using BLAST (Altschul *et al.*, 1990). Tag sequences screened for use with honey bee viruses are shown in Table 5.

10.2.4. Removal of tagged-cDNA primers from the cDNA reaction

Since the purpose of using tagged-cDNA primers is to amplify only with the tag sequence, it is important to either remove or inactivate the original tagged-cDNA primers after the cDNA reaction, and prior to PCR. If not, then these tagged-cDNA primers (which contain virus-specific sequences) can participate in the PCR reaction, just like a

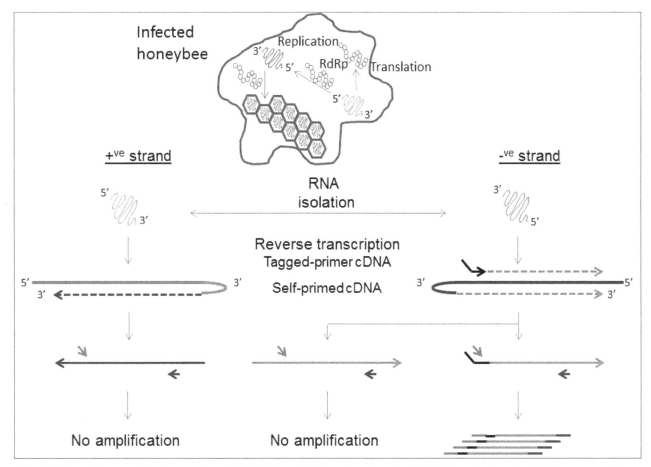

Fig. 5. Outline of the procedure for strand-specific RT-PCR amplification of negative-strand viral RNA, using tagged-cDNA primers to avoid amplification of non-target cDNAs. Only cDNA produced with tagged-cDNA primers, and amplified with the tag and a virus-specific primer will be amplified. RdRp refers to the viral RNA-dependent RNA polymerase. Image © JR de Miranda.

non-tagged virus-specific cDNA primer would. The presence of tagged -cDNA primers in the PCR reaction therefore permits the amplification of false-, self- or misprimed cDNAs of the 'wrong' strand, leading to an incorrect conclusion of strand-specificity (Craggs *et al.,* 2001; Peyrefitte *et al.,* 2003; Plaskon *et al.,* 2009; Boncristiani *et al.,* 2009).

Tagged-cDNA primers can be most easily removed from the cDNA reaction using commercial PCR/cDNA purification columns (Peyrefitte *et al.,* 2003). An alternative is to use biotinylated tagged-cDNA primers and then capture the tagged cDNA with Streptavidin-conjugated magnetic beads (Boncristiani *et al.,* 2009). Although highly effective at removing primers, the disadvantage of cDNA purification is that its DNA recovery efficiency of individual columns can be highly variable (Tentcheva *et al.,* 2006), leading to different types of error in (quantitative) detection and interpretation. This can be managed by adding a passive 'reference' DNA prior to cDNA purification (similar in concept to the "exogenous internal reference standards" used in RT-qPCR quantification, discussed in the *BEEBOOK* paper on molecular techniques; Evans *et al.,* 2013), which can be used to normalize the data again afterwards.

10.2.5. Inactivating tagged-cDNA primers in the cDNA reaction

It is also possible to inactivate the tagged-cDNA primer in the cDNA reaction, and so prevent it from participating in the PCR reaction. This can be done using exonuclease-I, which specifically digests only single-stranded DNA, *i.e.* the tagged-cDNA primer (Craggs *et al.,* 2001; Purcell *et al.,* 2006; Plaskon *et al.,* 2009; Lin *et al.,* 2009) or by phosphorylating the 3' end of the tagged-cDNA primer (making it impossible for the polymerases in PCR to synthesize DNA from this primer). These enzymatic reactions can be done right after the cDNA reaction, in the same reaction tube, after which the enzymes can be heat-inactivated prior to the PCR reaction. The exonuclease-I digestion has become the method of choice for quantitative strand-specific RT-qPCR (*e.g.* Purcell *et al.,* 2006; Plaskon *et al.,* 2009; Lin *et al.,* 2009; Runckel *et al.,* 2011). The main advantages of enzymatic inactivation of the tagged-cDNA primers, compared to the primer removal methods (see section 10.2.4.), is that enzymatic inactivation is much faster and cheaper, with fewer handling and contamination errors (no tube changes), and that it avoids the possible quantitation errors of the primer removal methods.

Table 5. Primers and tags used for detection of positive and negative strand honey bee virus RNAs by strand-specific RT-PCR.

TAG	VIRUS	STRAND	PRIMER	SEQUENCE	FUNCTION	REFERENCES
agcctgcgcaccgtgg	DWV	+ (pos)	tag-B23 F15	agcctgcgcaccgtggCCACCCAAATGCTAACTCTAAGCG TCCATCAGGTTCTCCAATAACGGA	tagged-cDNA virus-sense	Yue and Genersch, 2005 Gisder *et al.*, 2009 Dainat *et al.*, 2009
		- (neg)	tag-F15 B23	agcctgcgcaccgtggTCCATCAGGTTCTCCAATAACGGA CCACCCAAATGCTAACTCTAAGCG	tagged-cDNA virus-antisense	
	DWV	+ (pos)	tag-DWVas DWV-s	agcctgcgcaccgtggTCGACAATTTTCGGACATCA ATCAGCGCTTAGTGGAGGAA	tagged-cDNA virus-sense	Boncristiani *et al.*, 2009
		- (neg)	tag-DWVs DWV-as	agcctgcgcaccgtggATCAGCGCTTAGTGGAGGAA TCGACAATTTTCGGACATCA	tagged-cDNA virus-antisense	
	IAPV	+ (pos)	tag-IAPVas IAPV-s	agcctgcgcaccgtggCTTGCAAGATAAGAAAGGGGG GCGGAGAATATAAGGCTCAG	tagged-cDNA virus-sense	DiPrisco *et al.*, 2011
		- (neg)	tag-IAPVs IAPV-as	agcctgcgcaccgtggGCGGAGAATATAAGGCTCAG CTTGCAAGATAAGAAAGGGGG	tagged-cDNA virus-antisense	
	BQCV	+ (pos)	tag-BQCVas BQCV-s	agcctgcgcaccgtggGCAACAAGAAGAAACGTAAACCAC TCAGGTCGGAATAATCTCGA	tagged-cDNA virus-sense	Peng *et al.*, 2011
		- (neg)	tag-BQCVs BQCV-as	agcctgcgcaccgtggTCAGGTCGGAATAATCTCGA GCAACAAGAAGAAACGTAAACCAC	tagged-cDNA virus-antisense	
atcggaatcgcctagctt	CBPV	+ (pos)	tag-R23 R20	atcggaatcgcctagcttCCCAATGTCCAAGATGGAGT GCTTGATCTCCTCCTGCTTG	tagged-cDNA virus-sense	Celle *et al.*, 2008
		- (neg)	tag-R20 R23	atcggaatcgcctagcttGCTTGATCTCCTCCTGCTTG CCCAATGTCCAAGATGGAGT	tagged-cDNA virus-antisense	
ggccgtcatggtggcgaataa (Plaskon *et al.*, 2009)	LSV-1	+ (pos)	tag-LSVU-R1717 LSV1-F1434	ggccgtcatggtggcgaataaCCATATCATAAGTTGGCAAGTG CAGGTGCAGAGCAATTGGATTCA	tagged-cDNA virus-sense	Runckel *et al.*, 2011
		- (neg)	tag-LSV1-F1434 LSVU-R1717	ggccgtcatggtggcgaataaCAGGTGCAGAGCAATTGGATTCA CCATATCATAAGTTGGCAAGTG	tagged-cDNA virus-antisense	
	LSV-2	+ (pos)	tag-LSVU-R1717 LSV2-F1434	ggccgtcatggtggcgaataaCCATATCATAAGTTGGCAAGTG TAGGTGTCGGGCCATAGGGTTTG	tagged-cDNA virus-sense	
		- (neg)	tag-LSV2-F1434 LSVU-R1717	ggccgtcatggtggcgaataaTAGGTGTCGGGCCATAGGGTTTG CCATATCATAAGTTGGCAAGTG	tagged-cDNA virus-antisense	

10.2.6. Dilution

Another strategy to minimize the chance of illegitimate amplification of non-strand-specific cDNA molecules through the involvement of residual tagged-cDNA primer, is to dilute the cDNA reaction mixture 10-fold prior to PCR (Craggs *et al.*, 2001).

10.2.7. Strand-specific real-time RT-qPCR

Finally, a very effective way to manage the consequences of illegitimate priming events during cDNA synthesis is to use real-time qPCR for strand-specific detection (Purcell *et al.*, 2006; Gisder *et al.*, 2009; Boncristiani *et al.*, 2009; Plaskon *et al.*, 2009; Lin *et al.*, 2009; Zioni *et al.*, 2011). This allows all the PCR products arising from rare cDNAs generated by false-, self- or mis-priming events to be excluded from the data on quantitative grounds.

10.2.8. Protocols

All these conditions can easily be incorporated into a one-tube protocol combining the benefits of an RT-reaction at higher temperature, tagged primers, exonuclease-I digestion of the tag-cDNA primer and dilution of the cDNA, prior to real-time qPCR with a primer complementary to the tag and a virus-specific forward primer.

10.2.8.1. High temperature reverse transcription

Using an elevated temperature for the cDNA reaction significantly reduces mis-priming events, and thus the risk of falsely detecting the incorrect strand. Common alternatives are: SuperScript-III (50°C: Peyrefitte *et al.*, 2003; Purcell *et al.*, 2006; Plaskon *et al.*, 2009; Lin *et al.*, 2009; Runckel *et al.*, 2011), OmniScript/SensiScript (55°C: Yue and Genersch, 2005; Gisder *et al.*, 2009), Thermoscript (60°C: Carrière *et al.*, 2006; Horsington and Zhang, 2007) and r*Tth* reverse transcriptase (70°C: Lanford *et al.*, 1995; Laskus *et al.*, 1998; Craggs *et al.*, 2001; Celle *et al.*, 2009). The method below is a generic one, with individual adaptations for the different reverse transcription options:

1. Mix:
   1.1. 5 µl 50 ng/µl RNA,
   1.2. 1 µl 10 µM tagged-cDNA primer (0.5 µM final concentration),
   1.3. 2 µl nuclease-free water.
2. Heat 70°C for 5 min. Cool on ice for 2 min .
3.a. For SuperScript-III reactions, add:
   3.a.1. 10 µl 2x SuperScript-III buffer (containing 1 mM dNTP),
   3.a.2. 2 µl SuperScript-III/RNAseOUT mixture,
   3.a.3. Incubate 30 min at 50°C,

3.a.4. Inactivate 15 min at 95°C,

3.a.5. Cool reaction to room temperature, store on ice.

3.b. For OmniScript/SensiScript reactions, add:

3.b.1. 4 µl 5x Qiagen OneStep RT-PCR buffer,

3.b.2. 0.8 µl 10 mM dNTP (400 µM final concentration),

3.b.3. 0.8 µl Qiagen OneStep enzyme mix,

3.b.4. 5.4 µl nuclease-free water,

3.b.5. Incubate 30 min at 50°C. Go to section *10.2.8.4.*

3.c. For ThermoScript reactions, add:

3.c.1. 4 µl 5x ThermoScript buffer,

3.c.2. 2 µl 10 mM dNTP (1 mM final concentration),

3.c.3. 1 µl 0.1 M DTT (5 mM final concentration),

3.c.4. 1 µl 40 u/µl RNAseOut,

3.c.5. 1 µl 15 u/µl ThermoScript,

3.c.6. 4 µl nuclease-free water,

3.c.7. Incubate 30 min at 60°C,

3.c.8. Inactivate 15 min at 95°C,

3.c.9. Cool reaction to room temperature, store on ice.

3.d. For r*Tth* reactions, add:

3.d.1. 2 µl 10x r*Tth* buffer,

3.d.2. 0.4 µl 10 mM dNTP (200 µM final concentration),

3.d.3. 1 µl 10 mM MnCl$_2$ (1 mM final concentration),

3.d.4. 2 µl 2.5 u/µl r*Tth* reverse transcriptase,

3.d.5. 6.6 µl nuclease-free water,

3.d.6. Incubate 30 min at 70°C,

3.d.7. Add 2 µl 10x chelating buffer (to chelate the Mn^{+2}),

3.d.8. Inactivate 15 min at 98°C,

3.d.9. Cool reaction to room temperature, store on ice.

10.2.8.2. Exonuclease-I digestion of tagged primer

Exonuclease-I specifically digests only single-stranded DNA, in a 3'-5' direction, and thus inactivates unincorporated tagged-cDNA primer prior to PCR. This reduces ten-fold the chance of falsely detecting the incorrect strand (Craggs *et al.,* 2001) and is a common step in strand-specific RT-PCR (Purcell *et al.,* 2006; Lin *et al.,* 2009; Plaskon *et al.,* 2009; Runckel *et al.,* 2011).

1. Add to the cDNA reaction:

   10 u Exonuclease-I.

2. Incubate 30 min at 37°C; inactivate 15 min at 70°C.

3. Dilute cDNA reaction ten-fold, to 200 µl.

10.2.8.3. Column purification of cDNA

The cDNA can also be purified to remove unincorporated tagged-cDNA primer, using Qiagen affinity purification columns, and thus significantly reduce the chance of falsely detecting the incorrect strand through participation of residual tagged-cDNA primer in the early PCR reactions. This procedure is a common alternative to Exonuclease-I digestion (Peyrefitte *et al.,* 2003; Carrière *et al.,* 2007) and used in strand-specific detection of several honey bee viruses

(Boncristiani *et al.,* 2009; DiPrisco *et al.,* 2011; Peng *et al.,* 2012).

1. Follow Qiagen DNA affinity column purification protocol.

2. Elute the purified cDNA in 100 µl nuclease-free water.

10.2.8.4. OneStep PCR

Yue and Genersch (2005) developed a modified OneStep protocol for strand-specific RT-PCR that does not include specific steps to remove the tagged cDNA primer prior to PCR. Occasionally, weak bands are produced derived from non-strand-specific cDNA priming events (Gisder *et al.,* 2009).

1. For the OmniScript/SensiScript OneStep RT-PCR reactions, add:

   1.1. 0.5 µl 10 µM tag primer (0.25 µM final concentration),

   1.2. 0.5 µl 10 µM virus-specific primer (0.25 µM final concentration).

2. Incubate

   2.1. 15 min at 95°C,

   2.2. 35 cycles of [94°C:30 sec – 54.5°C:60 sec – 72°C:30 sec],

   2.3. 72°C:10 min.

10.2.8.5. Real-time qPCR

The SuperScript-III, Thermoscript and r*Tth* cDNA reactions all enter a separate (TwoStep) PCR protocol, which can be conveniently adapted to real-time qPCR, using a real-time qPCR kit containing SYBR-green:

1. Mix:

   1.1. 3 µl cDNA (column purified, or diluted 1/10),

   1.2. 0.4 µl 10 µM tag primer (0.2 µM final concentration),

   1.3. 0.4 µl 10 µM virus-specific primer (0.2 µM final concentration),

   1.4. 0.4 µl* 10 mM dNTP* (0.2 mM final concentration*),

   1.5. x µl Buffer + SYBR-green (as per manufacturer),

   1.6. y µl nuclease-free water,

   1.7. z µl Taq polymerase (as per manufacturer),

   1.8. 20 µl total volume.

(* dNTPs are often included in the optimized buffer)

2. Incubate in real-time thermocycler:

   2.1. 5 min:95°C,

   2.2. 35 cycles [10 sec:95°C - 30 sec:58°C - read].

3. For Melting Curve analysis of the products, incubate:

   3.1. 1 min:95°C,

   3.2. 1 min:55°C,

   3.3. 5 sec:0.5°C:read from 55°C to 95°C.

10.2.9. Controls

By now it should be evident that strand-specific RT-PCR should include a large number of controls, to account for the many ways by which an incorrect result can be generated. Most of these involve the reverse transcription reaction, since this is where most of the errors come from. The one essential control that should be run for every individual sample is:

- A primer-free cDNA reaction (proof that self-primed cDNA is not amplified).

Other controls that should be included at least once for the experiment are:

- A template-free cDNA reaction (absence of contamination of reagents/pipettes with target DNA).
- A reverse-transcriptase-free cDNA reaction (absence of reverse-transcriptase activity during PCR).
- An exonuclease-I-free cDNA reaction (disappearance of signal from mis-primed cDNA reactions).

The PCR step for all these controls should also include tagged-cDNA primer, equivalent to the estimated carry-over from the cDNA reaction, in addition to the regular concentrations of tag primer and virus-specific primer necessary for the PCR. Through this, the controls will contain the complete primer composition of the experimental reactions, which (as explained above) is an essential condition for excluding possible false positives.

Whether or not false-positive results during strand-specific RT-PCR presents a major problem also depends on the question to be answered. If the virus replication in a certain host, tissue or cell type is expected, then false-positive results are not a major factor. In contrast, if the absolute presence or absence of replication needs to be proven, then extreme care must be taken when conducting and interpreting the experiments.

10.3. Multiplex Ligation-dependent Probe Amplification (MLPA)

10.3.1. Introduction

Multiplex Ligation-dependent Probe Amplification (MLPA) technology is an amplification technique that allows simultaneous detection of up to 40 targets with the use of a single PCR primer pair. The procedure uses a series of paired oligonucleotides (half-probes), each pair specific for one target. The two half-probes; the Left Probe Oligo (LPO) and the Right Probe Oligo (RPO) lie adjacent to each other on the target genome so that they can be joined together by a ligation reaction, to produce an amplification probe (Fig. 6). In addition to a target-specific sequence, each of the half-probes contains one of two sequences recognized by a universal PCR primer, for probe amplification. Since these PCR primer sequences are common to all half-probe pairs, a single pair of PCR primers can amplify all target probes in a multiplex reaction. The half-probe pairs also contain a 'stuffer' fragment of variable length, allowing each amplified probe to be identified by its size, using (capillary) electrophoresis (Fig. 6). This technique was recently adopted to detect the most common honey

bee viruses including CBPV, DWV (KV & VDV-1), ABPV (IAPV & KBV), BQCV, SBPV, SBV (De Smet *et al.*, 2012). Because these are all RNA viruses, the MLPA is preceded by a reverse transcription of RNA into cDNA. Since the probes are strand-specific, this technique is highly suitable for the selective detection of either the positive-strand genomic viral RNA or the negative-strand virus replicative intermediate RNA, which is a marker for virus replication.

Since several targets are amplified at the same time, there will be competition between different targets for the amplification resources (primers, nucleotides, enzyme). This 'competitive' PCR allows for a measure of relative quantification between the targets, in the sense that the relative proportion of the targets after amplification should, if all targets amplify equally efficiently, reflect their initial proportions in the sample. By including one or more internal reference genes or exogenously added absolute quantification standards among the targets, the procedure can be made (semi-) quantitative.

10.3.2. Protocol

The reactions are performed in a thermocycler with heated lid (105°C) in 0.2 ml thin-walled PCR tubes. The specific MLPA reagents can be obtained from MRC-Holland. The various probes and oligonucleotides used in the honey bee virus MLPA are given in Table 6. It is recommended to use the wildtype MuMLV Reverse Transcriptase from Promega (M1701). The right probe oligos (RPO) are phosphorylated and should be synthesized as 'ultramer' grade.

10.3.2.1. Primer and probe mixtures

1. Prepare RT-primer mix:
   1.1. 5 mM each dNTP,
   1.2. 5 µM each RT primer (Table 6).
2. Prepare probe mix:
   1.33 nM of each half-probe (Table 6) in TE(8.0) buffer.

10.3.2.2. Reverse transcription

1. Mix on ice:
   1.1. 10~500 ng RNA,
   1.2. 1 µl SALSA RT buffer,
   1.3. 0.5 µl RT-primer mix,
   1.4. Sterile water to 4.5 µl total volume.
2. Incubate 1 min at 80°C.
3. Incubate 5 min at 45°C.
4. Add 1.5 µl 20 u/µL MuMLV Reverse Transcriptase.
If necessary, dilute in 1:1 water: SALSA enzyme dilution buffer.
5. Mix.
6. Incubate:
   6.1. 15 min at 37°C,
   6.2. 98°C for 2 min (reverse transcriptase inactivation),
   6.3. Cool to 25°C.

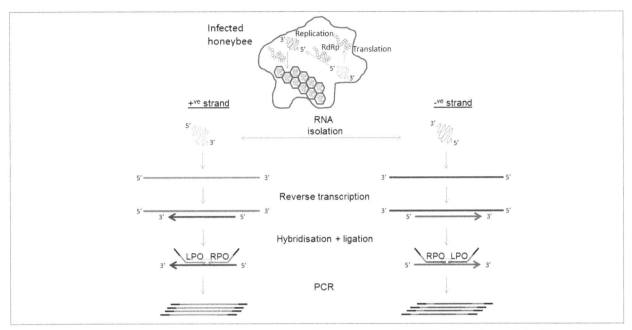

Fig. 6. Outline of the MLPA procedure for amplifying strand-specific ligated probes. LPO and RPO refer to the Left Probe Oligo and Right Probe Oligo respectively. RdRp refers to the viral RNA-dependent RNA polymerase.   Image © L De Smet.

Table 6. Primers and probes used for detection of positive and negative strand honey bee virus RNAs by MLPA. Adapted from De Smet *et al.* (2012).

VIRUS	STRAND	PRIMER	FUNCTION	SEQUENCE (5'-3')	SIZE (bp)
CBPV	+ (pos)	LDS22	(-)cDNA	GCCCCGATCATATAAGCAAA	88
		LDS23	(+)MLPA-LPO	gggttccctaagggttggaCCGTAGCTGTTTCTGCTGCGGT	
		LDS24	(+)MLPA-RPO	P-ACTCAGCTCAGCTCGACGCTCAGAtctagattggatcttgctggcac	
	- (neg)	LDS59	(+)cDNA	GAACATCCGGAACAGACGAT	88
		LDS60	(-)MLPA-LPO	gggttccctaagggttggaTCTGAGCGTCGAGCTGAGCTGAGT	
		LDS61	(-)MLPA-RPO	P-ACCGCAGCAGAAACAGCTACGGtctagattggatcttgctggcac	
DWV/KV VDV-1	+ (pos)	LDS8	(-)cDNA	TCACATTGATCCCAATAATCAGA	95
		LDS9	(+)MLPA-LPO	gggttccctaagggttggaTGACCGATTCTTTATGCAGCGAGCTCT	
		LDS10	(+)MLPA-RPO	P-TACGTGCGAGTCGTACTCCTGTGACAtctagattggatcttgctggcac	
	- (neg)	LDS31	(+)cDNA	GTGTGGTGCATCTGGAATTG	95
		LDS32	(-)MLPA-LPO	gggttccctaagggttggaGTTGTCACAGGAGTACGACTCGCA	
		LDS33	(-)MLPA-RPO	P-CGTAAGAGCTCGCTGCATAAAGAATCGGTtctagattggatcttgctggcac	
ABPV KBV IAPV	+ (pos)	LDS1	(-)cDNA (ABPV)	CAATGTGGTCAATGAGTACGG	104
		LDS2	(-)cDNA (KBV&IAPV)	TCAATGTTGTCAATGAGAACGG	
		LDS19	(+)MLPA-LPO	gggttccctaagggttggaCTCACTTCATCGGCTCGGAGCATGGATGAT	
		LDS4	(+)MLPA-RPO	P-ACGCACAGTATTATTCAGTTTTTACAACGCCCtctagattggatcttgctggcac	
	- (neg)	LDS62	(+)cDNA	TGAAACGGAACAAATCACCA	104
		LDS63	(-)MLPA-LPO	gggttccctaagggttggaCGAGCCGATGAAGTGTCTTGAGCCATGG	
		LDS64	(-)MLPA-RPO	P-GGGTATTGATCCTATTTGGAGTTTCCACATCATGtctagattggatcttgctggcac	
BQCV	+ (pos)	LDS16	(-)cDNA	CGGGCCTCGGATAATTAGA	122
		LDS21	(+)MLPA-LPO	gggttccctaagggttggaCTTCATGTTGGAGACCAGGTTTGTTTGCCGACTTACGGAA	
		LDS18	(+)MLPA-RPO	P-TGTCGTTAAACTCTAGGCTTTCCGGATGGCTTCTTCATGGtctagattggatcttgctggcac	
	- (neg)	LDS65	(+)cDNA	TTAAAAGCCCCGTATGCTTG	122
		LDS66	(-)MLPA-LPO	gggttccctaagggttggaTCAGCGCAACAGAAGCCATCCGGAAAGCCTAGAGTTTAACG	
		LDS67	(-)MLPA-RPO	P-ACATTCCGTAAGTCGGCAAACAAACCTGCCTTATCTGGTtctagattggatcttgctggcac	
SBPV	+ (pos)	LDS25	(-)cDNA	CGCAAACACGACGAATTTTA	131
		LDS26	(+)MLPA-LPO	gggttccctaagggttggaCGTTCAATGGTCGAGATAGAAGCCACAGTAGAAGTATTACGCGCT	
		LDS27	(+)MLPA-RPO	P-TCTTGTGTTTTGGCTTATGGGCGTGGGCCTGATCTTCATTCAGCtctagattggatcttgctggcac	
	- (neg)	LDS68	(+)cDNA	GGTGTCATAAACAGAATGACGAG	131
		LDS69	(-)MLPA-LPO	gggttccctaagggttggaTCAGCGCAACACTCAGGCCCACGCCCATAAGCCAAAACACAAGAA	
		LDS70	(-)MLPA-RPO	P-GCGCGTAATACTTCTACTGTGGCTTCTATCTCGCCTTATCTGGTtctagattggatcttgctggcac	
SBV	+ (pos)	LDS28	(-)cDNA	TGGACATTTCGGTGTAGTGG	140
		LDS29	(+)MLPA-LPO	gggttccctaagggttggaCGTTGATCCAATGGTCAGTGGACTCTTATACCGATTTGTTTAATGGTTGG	
		LDS30	(+)MLPA-RPO	P-GTTTCTGGTATGTTTGTTGACAAGAACGTCCACCTTCAGCCATTCAGCtctagattggatcttgctggcac	
	- (neg)	LDS71	(+)cDNA	CCTTACCTCTAGTAAGAAGACATTTGA	140
		LDS72	(-)MLPA-LPO	gggttccctaagggttggaTAAAAAACTACCGTGTAGTGGACGTTCTTGTCAACAAACATACCAGAAA	
		LDS73	(-)MLPA-RPO	P-CCCAACCATTAAACAAATCGGTATAAGAGTCCACTGAAAAGTCGGTGGAtctagattggatcttgctggcac	
β-Actin	+ (pos)	LDS58	(-)cDNA	TTTCATGGTGGATGGTGCTA	182
		LDS56	(+)MLPA-LPO	gggttccctaagggttggaGCAGGAAGTCGTTACCACCTGGCCCAC-GGAGCCAATTTCTCATGCTTGCCAACACTGTCCTTTCTGGAGGT	
		LDS57	(+)MLPA-RPO	P-ACCACCATGTATCCTGGAATCGCGAAAACGTGGTGTACCGGCTGTCTGGTATGTATGAG-TTTGTGGTGAtctagattggatcttgctggcac	
PCR		LDS11	PCR-Forward	gggttccctaagggttgga	n.a.
		LDS12	PCR-Reverse	gtgccagcaagatccaatctaga	

10.3.2.3. Hybridisation of MLPA half-probes

1. Add to the reverse transcription reaction and mix with care:
   1.1. 1.5 µl Probe-mix,
   1.2. 1.5 µl MLPA buffer.
2. Incubate:
   2.1. 1 min at 95°C,
   2.2. 16 h at 60°C in a PCR ThermoCycler.

10.3.2.4. Ligation of MLPA half-probes

1. Reduce the temperature of the thermal cycler to 54°C.
2. While at 54°C, add to each sample:
   2.1. 3 µl Ligase-65 buffer A,
   2.2. 3 µl Ligase-65 buffer B,
   2.3. 25 µl sterile water,
   2.4. 1 µl Ligase-65.
3. Mix well.
4. Incubate:
   4.1. 10-15 min at 54°C,
   4.2. 5 min at 98°C (inactivation of Ligase-65).
5. Cool on ice.

10.3.2.5. PCR amplification of MLPA probes

1. Mix in new tubes:
   1.1. 10 µl MLPA ligation reaction,
   1.2. 4 µl SALSA PCR buffer,
   1.3. 26 µl sterile water.
2. While the tubes are in the thermal cycler at 60°C, add to each tube:
   2.1. 2 µl SALSA PCR primers,
   2.2. 2 µl SALSA enzyme dilution buffer,
   2.3. 5.5 µl sterile water,
   2.4. 0.5 µl SALSA polymerase.
3. Incubate:
   3.1. 35 cycles [30 sec:95°C - 30 sec:60°C - 60 sec:72°C],
   3.2. 20 min:72°C.

10.3.3. Fragment analysis

The MLPA reaction products can be analysed on conventional slab electrophoresis, using a 4% agarose-TBE gel (De Smet *et al.,* 2012; Green and Sambrook, 2012), or using a high-resolution, semi-automatic electrophoresis system such as the BioAnalyzer (Aligent), Experion (Biorad), Qiaxcel (Qiagen) or MultiNA (Shimadzu), which are designed for separating short fragments. In all cases, interpretation of the results is simplified by loading a specific MLPA ladder, generated amplifying each of the MLPA targets individually from cloned controls and pooling these into a single ladder.

Agarose gel electrophoresis:

1. Prepare a 4% high resolution agarose gel in 1x TRIS-Borate-EDTA buffer (Green and Sambrook, 2012).
2. Mix:
   2.1. 10 µl aliquot of the MLPA reaction,
   2.2. 5 µl 4x Sample Buffer.
3. Load gel.
4. Run for 45-60 minutes at 75-90 volts.

Semi-automatic high-resolution gel electrophoresis:

1. Use gel system appropriate for 25-500 bp DNA fragments.
2. Follow manufacturers' instructions for sample preparation, loading, running and data analysis.

10.3.4. Controls

As for strand-specific RT-PCR, MLPA requires a large number of controls to rule out the possibility of artefactual results due to the methods used.

- A nucleic acid-free control (the two half-probes should not be able to ligate without a template).
- A reverse transcriptase-free control (the two half-probes should not be able to ligate on an RNA template).
- A cDNA primer-free control (identifies possible self-priming of RNA for cDNA synthesis).
- A half-probes-free control (only the two half-probes should generate an amplifiable template for PCR).
- A ligase-free control (the two half-probes should not be able to ligate without ligase0.
- A PCR primer-free control (the two half-probes should not be able to function as PCR primers either with each other or with the cDNA primer).

11. Virus variation

11.1. Introduction

Due to the importance of genetic variability to virus virulence and evolution, the detection and quantification of virus genetic variability as a trait in itself has been an interest throughout the history of virus research. Throughout the history of molecular diagnostics, new technologies have been adapted for the detection and quantification of polymorphisms or genetic variation, reviewed by Ahmadian and Lundeberg (2002). Below are a few of the more current methods.

11.2. Protocols

11.2.1. Nuclease protection assays (RPAs and SNPAs)

Nuclease protection assays are an efficient way to analyse the genetic complexities of natural populations of organisms (Kurath *et al.,* 1993; Arens, 1999; Wang and Chao, 2005). A labelled probe is hybridised to the nucleic acid sample of a population of organisms (usually viruses or other pathogens, sometimes related mRNA species) and then digested with RNAse (RNA probe) or S1-nuclease (DNA probe) which will cut the probe wherever there is a mismatch between probe and target. The resulting pattern of digested probe fragments, revealed by gel electrophoresis, is qualitatively and quantitatively indicative of the mismatch polymorphisms present in the nucleic sample. These procedures are called RNAse Protection Assay (RPA) and S1-Nuclease Protection Assay (SNPA).

> **Pros:** Entire populations can be screened for genetic complexity within the target sequence in a single reaction. The polymorphic sites can be mapped on the genome, through the sizes of the fragments produced.
>
> **Cons:** Assay is limited to about 300 bases, requiring many assays to cover a genome. Protocols are complex and subject to errors. The nature of the polymorphs requires further analysis.

11.2.2. Gel retardation assays (SSCP and DGGE)

Single Strand Conformation Polymorphism (SSCP) and Denaturing Gradient Gel Electrophoresis (DGGE) are two techniques that use electrophoresis to differentiate directly between variants in a population of sequences (Hauser *et al.,* 2000; Stach *et al.,* 2001). In SSCP the nucleic acids are made single-stranded, to fold into a preferred secondary structure. In DGGE, the nucleic acids are separated in an electrophoretic gel containing a salt gradient that will progressively denature the nucleic acids. In both methods, minor nucleotide differences between polymorphs in the population affect the migration of the DNA. Another technique with a similar philosophy is the heteroduplex mobility shift assay, where single nucleotide mismatches between a probe and target affect the mobility of the hybridised complex during electrophoresis, (Arens, 1999).

> **Pros:** Entire populations can be screened for genetic complexity within the target sequence in a single reaction.
>
> **Cons:** Assay is limited to about 300 bases, requiring many assays to cover a genome. Protocols are complex, sensitive to procedural accuracy and subject to errors. The nature of the polymorphs requires further analysis.

11.2.3. High Resolution Melting (HRM) analysis

Double stranded DNA can disassociate (or melt) into two single strands upon heating, and can re-associate (or hybridize) upon cooling, in a highly predictable fashion. This fundamental property of nucleic acids underpins all nucleic acid technologies. The principal parameters governing disassociation/hybridization are the length and composition of the DNA, the temperature and the salt concentration of the solution. Work in the 1950s demonstrated that the G-C pairing, with three hydrogen bonds, gave higher thermal stability than the A-T pairing, which has only two such bonds (Marmur and Doty, 1959). This made it possible to predict the temperature at which a DNA molecule would melt (T^m) from its length and base pair composition (Marmur and Doty, 1962). The discovery of DNA binding dyes such as SYBR-green, that fluoresce only when intercalated with double stranded but not single stranded molecules, provided a practical method to quantify the melting process based on a reduction in fluorescence during gradual heating, as the two DNA strands separated. This fluorescence-based detection was integrated with real-time PCR thermocyclers that can very precisely control the temperature of a DNA sample and collect fluorescence data between 10 and 200 times per °C, providing high-resolution melting curves that can distinguish single base pair differences between two PCR products (Wittwer *et al.,* 2003). This makes it possible to use High Resolution Melting (HRM) analysis to analyse the composition of mixed samples, *i.e.* samples containing two or more genetic variants of the same region, by comparing the melting curve of the mixed sample with those of the individual variants.

HRM analysis is a versatile method that can be applied to any sample that contains double stranded DNA, including cDNA or PCR products. The flexibility of HRM analysis has led to a diverse array of applications including pre-sequence screening, Single Nucleotide Polymorphism (SNP) typing, methylation analysis, microsatelite or Simple Sequence Repeat (SSR) marker screening (Arthofer *et al.,* 2011) and copy number quantification. Several of these techniques are covered in the *BEEBOOK* paper on molecular techniques (Evans *et al.,* 2013). Such applications of HRM are also relevant to virology, and the first record of a virological use of HRM analysis was to strain type West Nile virus (Papin *et al.,* 2004). Recently, HRM analysis has been used to monitor the relationship between varroa infestation and virus diversity (Martin *et al.,* 2012).

Many standard real-time PCR machines can be used for HRM analyses. Often an upgrade of the software package and the running of a calibration plate is all that is required to enable a real-time PCR machine to run HRM analyses. Since HRM is a highly technical and sensitive procedure that integrates reaction biochemistry with machinery and analysis software, the best advice is to follow the protocol, reagents and incubation profile recommended by manufacturer. The basic procedure is as follows:

1. Amplification: Amplify your chosen fragment from your experimental samples and cloned controls, using specific HRM reagents containing a saturating DNA DNA intercalating dye and the recommended incubation profile.

2. Replicates: Use a minimum of three technical replicates for each sample. The replicate melting profiles will be averaged and used to assess whether the sample is distinct from other samples/controls.

3. HRM: Immediately after amplification the PCR products are subjected to a high-resolution melting step, within the same tube, during which the decrease in fluorescence due to the transition of the DNA from double- to single-stranded shape is monitored.

4. Analysis: The melting curve of the experimental sample, containing a mixture of different variants, is compared to the melting curves of pure, cloned versions of each of the individual variants.

Pros: Simple; fast; flexible; cheap; sensitive; specific; low contamination risk.
Cons: Requires individual melting curves of (cloned) variants. Cannot identify nature of novel variants. Limited quantification of variants. Limited capacity to resolve complex mixture. Limited to very short genome fragments.

11.2.4. Sequencing

The most powerful means for detecting variation is sequencing, since every possible variant is identified and precisely mapped on the genome. There are several approaches that can be used. The purest and most expensive approach is to clone PCR products of the target(s) and sequence batches of individual clones. This also allows the relative frequencies of individual variants to be determined, even those variants occurring at very low frequencies. A second and cheaper approach is to sequence the PCR products directly and identify the polymorphisms at sites of ambiguity in the sequence (Forsgren *et al.,* 2009; Fig. 7). Since such double peaks can also be the result of sequencing artefacts, each polymorphic site has to be confirmed by a matching pattern when sequencing the complementary strand. Only major polymorphisms can be identified and quantitation is moderate, similar to HRM. The new, high volume automated sequencing methods (Next Generation Sequencing, or NGS) have the capacity to directly analyse complex DNA and RNA mixtures through sequencing followed by automated similarity searches. These methods are rapidly becoming cheaper and more accurate, mostly through massive multiplexing of reactions and samples. They are increasingly being used as a one-step diagnostic method capturing millions of different targets, thus benefiting also from economy of scale in the data generated. They have recently also been used in honey bee pathology studies (Cox-Foster *et al.,* 2007; Runckel *et al.,* 2011) and are covered in detail in the *BEEBOOK* paper on molecular techniques (Evans *et al.,* 2013).

Pros: Comprehensive; fast; flexible; accurate; sensitive; specific; low contamination risk; approximate quantification (NSG).
Cons: Expensive - precise - limited quantification (Sanger sequencing); Very expensive - approximate quantification (NGS).

12. Quality control

12.1 Introduction

Standardization of the diagnostic methods for detecting and quantifying bee viruses and of the interpretation of the results is the first requirement for improved harmonization of the data collected by different laboratories. The protocols and assays for different animal diseases registered with the World Organization for Animal Health (OIE; previously the Office International des Epizooties) ensure the global harmonization and standardization of detection methodologies. Moreover, the OIE provides criteria for the technical requirements and quality management in veterinary testing laboratories, in the form of a series of standards and recommendations that each laboratory should address in the design and maintenance of its quality management program. Valid laboratory results are essential for diagnosis, obtained by the use of good management practices, validated protocols and calibration methods as described in the ISO/IEC 17025 International Standard. By following these standards, a laboratory is able to obtain accreditation, linked to the international certification standard ISO 9001. OIE guidelines provide an interpretation of the ISO/IEC 17025 guidelines in the context of veterinary laboratories working with infectious diseases, including the validation of diagnostic assays, the production of international reference standards and laboratory competence testing. The European reference laboratory for honey bee diseases at ANSES in France is developing a set of standard diagnostic procedures for honey bee viruses, following these procedures and criteria.

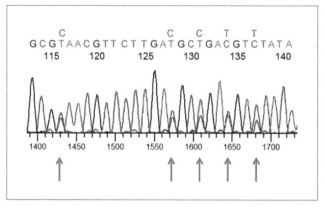

Fig. 7. Mixed virus sequences as revealed in sequence electropherogram. The mixed sequence can be resolved into component sequences using specifically designed software. Adapted from Forsgren *et al.,* 2009.

12.2. Assay selection and validation

Full validation of molecular techniques, as per guidelines issued by the World Organization for Animal Health (OIE) is a relatively new concept in the field of honey bee virus diagnosis. In general terms, a diagnostic protocol is designed in response to a particular diagnostic need and this is then developed into an optimized, documented and fixed procedure, using a series of intra-laboratory validation steps that demonstrate the reliability of results and the performance of the method. Recently, a new standard (XP U47-600) was developed by the French Standards Institute (AFNOR) concerning the minimum requirements for the development, validation and implementation of veterinary PCR-based diagnostic methods in animal health, based on the recommendations by the OIE and following the ISO/IEC 17025 criteria (NF, 2005; OIE, 2010). The validation procedure establishes the performance characteristics for each test method, such as sensitivity, specificity, detection and/or quantification limits.

The initial validation of a RT-qPCR assay involves two steps. The first concerns the validation of the qPCR assay itself, in terms of:

1. Analytical specificity.
2. The PCR detection limit (DL_{PCR}).
3. The PCR quantification limit (QL_{PCR}).
4. The linearity and efficiency of the qPCR assay.

The second step concerns the evaluation of the entire diagnostic protocol in terms of:

1. The method's detection limit (DL_{method}).
2. The diagnostic specificity and sensitivity on samples of known status.
3. The method's quantitation limit (QL_{method}) based on a validation range and accuracy profile.

In each of the two steps, various performance parameters are calculated, including measurement uncertainty (MU), deviations of repeatability and intermediate reliability.

The following is a step-by-step outline of how to develop an accredited RT-qPCR assay for the detection and quantitation of honey bee viruses, based on the successful development of such an assay for CBPV (Blanchard *et al.*, 2012).

12.2.1. Analytical specificity

12.2.1.1. Analysis in silico

Multiple nucleic acid sequences of the virus, obtained from public sequence databases and from a diverse range of biological and geographic sources, are compared *in silico* with each other and unrelated viruses using BLAST (Basic Local Alignment Search Tool; Altschul *et al.*, 1990), to identify regions of variability and conservation. See also the *BEEBOOK* paper on molecular techniques (Evans *et al.*, 2013).

A series of possible diagnostic assays are designed using bioinformatics tools, based on the particular diagnostic requirements for the method.

12.2.1.2. Experimental specificity

The specificity of the PCR assay is then tested experimentally, using **inclusivity** and **exclusivity** tests.

- **Inclusivity tests**

  Inclusivity tests assess the robustness of an assay, *i.e.* its ability to detect genetically diverse isolates. The PCR assay is evaluated against a panel of samples representing the full range of genetic diversity of the virus in question. This diversity is determined beforehand through phylogenetic analysis of bio-geographic isolates (see section 12.2.1.1.).

- **Exclusivity tests**

  Exclusivity tests assess the specificity of an assay, *i.e.* its ability to detect only the virus in question, and not any other viruses. The PCR assay is evaluated against a panel of viruses unrelated to the virus being tested, but which are found in the same environment or ecological area as the virus being tested.

12.2.2. PCR detection limit

The detection limit of a qPCR assay is the lowest number of nucleic acid targets in a given template volume that can be detected in at least 95% of replicate assays. The detection limit is established by performing at least three independent trials, each with trial consisting of three independent two-fold serial dilutions of a template of known concentration. At each dilution level in each serial dilution series, eight replicate qPCR assays are run, *i.e.* a total of 24 replicate assays at each dilution level. The detection limit is the highest dilution level (*i.e.* lowest amount of target nucleic acid template) giving at least 23 positive results from the 24 assays (95% of the replicates).

12.2.3. qPCR dynamic range and quantitation limit

PCR is an exponential (*i.e.* logarithmic) amplification process that is extremely consistent (*i.e.* predictable) over the entire reaction (35~40 cycles) and over a large range of initial target concentrations (at least 10^6-fold). This dynamic range and the quantitation limits of qPCR are determined using a 10-fold serial dilution series of known concentrations of (cloned) target DNA. A standard calibration curve is generated by linear regression of the quantification cycle (C_q) at which the PCR product is detected *vs.* the $\log_{10}$[target copy number]. The resulting algebraic equation:

$$C_q = a * \log_{10}[\text{target}] + b$$
(where '*a*' is the slope and '*b*' the intercept)

is then used to estimate the amount of target in a sample, given the C_q value (Bustin *et al.*, 2009). For accurate calibration of the curve and determining the error associated with data conversion, at least three independent trials of three independent 10-fold serial dilutions should be run. For each series and trial, the known amounts of target in each dilution are compared to the theoretical amounts estimated from the calibration curve, to obtain the individual bias, which is the averaged for all series and trials to obtain the mean bias (*mb*) at each dilution (an example is shown in Table 7). These values are then used to calculate the standard deviation of the obtained values (*SD*), and the uncertainty of the linearity is obtained by the formula $U_{LINi} = 2[\sqrt{SD^2 + mb^2}]$. The combined linearity uncertainly is defined for the entire calibration range by the formula $U_{LIN} = |\sqrt{\Sigma U_{LINi}^2 / k}|$ where k is the number of dilution levels. The quantitation limit of the assay is then set at the target concentration of the calibration range.

12.3. Method validation

12.3.1. Method detection limit

The method detection limit (DL_{method}) is the lowest amount of biological target in a sample that can be detected by the entire method (from processing through RT-qPCR). The DL_{method} is evaluated with biological reference samples obtained by spiking virus-free bee homogenates with known amounts of purified virus. At least two independent trials must be performed on two independent two-fold serial dilutions, with four replicate RNA extractions at each dilution level. The DL_{method} is the last dilution at which viral RNA can be detected in all replicates (100% frequency).

12.3.2. Method diagnostic sensitivity and diagnostic specificity

The diagnostic specificity and sensitivity is assessed on complete method analysis (processing through assay) of biological samples of known virus status (positive or negative). Diagnostic sensitivity is determined by the percentage of positive results among the known positive samples. Diagnostic specificity is determined by the percentage of negative results among the known negative samples.

12.3.3. Method quantitation limit and accuracy profile

The assessment of a method's quantitation limit is based on the construction and interpretation of an accuracy profile to estimate the precision and reliability of the values. Three independent trials must be performed on three independent 10-fold serial dilutions, including two replicate RNA extractions for each level of dilution. For each dilution series and each target amount, various parameters are determined from estimated target amounts, such as the inter-series variance and the repeatability variance, the sum of both giving the reliability variance. The standard deviation of the reliability (SDrl) is then obtained by the square root of the reliability variance. The mean bias is determined (difference between the theoretical value and the mean of the observed values). To construct the accuracy profile, the lower and upper tolerance interval limits of the quantitation method are determined using the following formula:

mean bias +/- 2 × SDrl

and compared to the acceptability limits defined by the laboratory, *e.g.* +/- 0.5 $\log_{10}$ (Blanchard *et al.,* 2012). The tolerance interval limits of the accuracy profile have to be within the acceptability limits, validating the method for the thus defined calibration range. The quantitation limit of the method is then determined by the first level load of the validated calibration range. An example of the confidence and acceptability limits of an RT-qPCR calibration curve is given Fig. 8, where the evaluated method is validated for a calibration range between 10^3 and 10^6 copies, with a quantitation limit of 10^3 copies.

12.4. Laboratory Validation

The final validation of a diagnostic method is through inter-laboratory proficiency tests, to evaluate the reproducibility and the overall uncertainty of the method, and to assess performance of other laboratories to conduct specifically this method (Birch *et al.,* 2004; Valentine-Thon *et al.,* 2001; Verkooyen *et al.,* 2003). To achieve this, candidate laboratories must submit to a training and accreditation programme.

12.4.1. Training and accreditation

The International Laboratory Accreditation Cooperation (ILAC) is an international cooperation of laboratory and inspection accreditation bodies created more than 30 years ago. It has published specific requirements and guides for laboratories and accreditation bodies. Under the ILAC system, ISO/IEC 17025 is to be used for accreditation. This procedure attests for the laboratory's technical competence and the reliability of its results. In each country, a sole national accreditation body is designated, as the French Accreditation Committee (COFRAC) in France, the Deutsche Akkreditierungsstelle GmbH (DAssK) in Germany or the Swedish Board for Accreditation and Conformity Assessment (SWEDAC) in Sweden. Performance of the method must thus be validated according to the OIE or AFNOR standards and approved by the national accreditation committee of each country. Furthermore, inter-laboratory proficiency tests should be carried out to evaluate the reproducibility of the method.

12.4.2. Inter-laboratory proficiency testing

The basic purpose of proficiency testing is to assess performance of laboratories in conducting specific method. Proficiency testing provides an opportunity to have an independent assessment of each laboratory's data compared to reference values or to the performance of other laboratories (*e.g.* Apfalter *et al.,* 2002). The participation of the laboratory to proficiency testing programs assesses if the laboratory's performances is satisfactory. In case of any potential problems within the laboratory, investigations to detect the difficulties are required. In order to successfully run proficiency test programs, the production and the distribution of reference materials (positive control, extraction control) are key points, as well as technical trainings of laboratories if necessary. In this framework, data harmonization could contribute to a better understanding of honey bee diseases and to a better diagnosis of pathological issues.

Table 7. Worked example of the estimation of primary and secondary statistics relating to the accuracy and confidence limits of a qPCR calibration curve. After Blanchard *et al.,* (2012).

Target amount (copies/reaction)	30	300	3000	30000
Theoretical value Log_{10} (Tv)	1.477	2.477	3.477	4.477
Measured value Log_{10} (Mv)	1.426	2.524	3.539	4.420
	1.490	2.443	3.507	4.469
	1.462	2.475	3.528	4.444
	1.492	2.439	3.509	4.468
	1.494	2.435	3.511	4.468
Bias (Mv - Tv)	-0.052	0.046	0.062	-0.057
	0.013	-0.034	0.030	-0.009
	-0.016	-0.002	0.051	-0.033
	0.014	-0.038	0.032	-0.009
	0.017	-0.043	0.034	-0.009
Sum of Mv	7.363	12.316	17.594	22.270
Mean Mv	1.473	2.463	3.519	4.454
Mean bias	-0.005	-0.014	0.042	-0.023
Standard deviation of Mv	0.029	0.037	0.014	0.022
U_{LINi}	0.060	0.080	0.088	0.063
U_{LINi^2}	0.004	0.006	0.008	0.004
U_{LIN}	0.074			

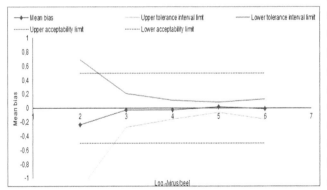

Fig. 8. Example of the mean bias, confidence interval, acceptability limits and quantitation limit for a RT-qPCR calibration curve. After Blanchard *et al.,* 2012.

13. Future perspectives

The future is bright for disease diagnosis and pathogen detection. The molecular biology revolution of the past quarter century has matured through the experimental, labour driven phase to high volume automated systems delivering reliable, high quality information. The revolution is likely to continue, with new methods being developed annually, increasing the options available to the diagnostic virologist. In the 1960's, 70's and 80's the development of semi-automated, sensitive serological assays precipitated a similar revolution in pathogen detection that made more insightful research into disease and epidemiology possible. The most pioneering honey bee virology revelations were made during this time, in particular the discovery and serological characterisation of most of the honey bee viruses that we know today. Several of these remain to be characterised at the nucleic acid level. The development of cheap, high throughput mass sequencing of genomes and transcriptomes has overtaken these efforts somewhat, leading to the identification of novel viral nucleic acid sequences in bee and mite samples that may very well represent the genomes of viruses that had already been discovered previously. Matching these historical virus discoveries and their serological data to these nucleic acid genomes is therefore an important and urgent task, to avoid confusion in bee virus classification and to make sure that the historical literature on these viruses remains relevant in the current molecular age.

The principal criteria for an ideal diagnostic system are sensitivity, accuracy, reliability, universality, simplicity, speed and cost. Most modern detection technologies are now sensitive enough to detect down to a single target molecule. This means that any future development will increasingly focus on quantitative detection (depending on the diagnostic requirements), with a concomitant change to a more integrated, quantitative disease management style. Accuracy of detection at the molecular level (and virus detection is largely molecular) depends essentially on the nature of the primary molecular recognition event, *i.e.* the interaction between target and probe. In this regard, nucleic acid-based detection has a considerable advantage over serological detection, since the kinetics of nucleic acid hybridisation is much more predictable and reliable than that of protein interactions. This also makes nucleic acid-based detection much more adaptable to changing requirements due to the discovery or emergence of new virus variants. The principal area of concern for molecular virus detection is reliability, *i.e.* avoiding misdiagnosis due to false-positive or false-negative results. The nucleic acid genomes of viruses are naturally highly variable and can evolve very quickly, while current molecular diagnostic methods are highly sensitive to minor variations in the nucleic acid target, making it prone to possible false-negative errors. This sensitivity is largely linked to the enzymes used for molecular detection and future developments in molecular virus diagnostics may therefore increasingly feature enzyme-free technologies (Liepold *et al.,* 2005).

The variability of virus genomes is an important component of a virus' adaptive response. It is in many ways a defining and unique characteristic for individual viruses. Other areas of virology now distinguish which viral forms offer increased pathogenicity, or which spread more easily. New methods that can directly describe and quantify this variability, such as HRMC, may become increasingly important in honey bee virology to clarify how the interactions between host factors, individual variants, combinations of variants or the variability as a whole, can induce a diseased state.

Disease is the result of a breakdown in a host's normal physiological state due to the presence or proliferation of a pathogenic agent. The simpler component of this interaction is the pathogen, and its detection. Future developments however, will increasingly focus on the host component of disease and the interplay between pathogen and host. This means that future technological direction in disease diagnosis will emphasise multiplexing, miniaturisation (Fiorini and Chui, 2005) and automation (Service, 2006; Belák *et al.,* 2009), to provide epigenetic data to better understand how the breakdown in the homeostasis between host and pathogen results in disease. Such information is important, since it can inform disease prevention, treatment and potential cures.

Finally, automation and increased demand for simpler, faster and cheaper technologies for routine diagnosis with wide applicability in low-tech settings (Higgins *et al.,* 2003; Schaad *et al.,* 2003) will ultimately drive the costs down to where disease surveillance and routine monitoring becomes cost-effective (Service, 2006), even in low priority areas like honey bee pathology.

14. Acknowledgements

The COLOSS (Prevention of honey bee COlony LOSSes) network aims to explain and prevent massive honey bee colony losses. It was funded through the COST Action FA0803. COST (European Cooperation in Science and Technology) is a unique means for European researchers to jointly develop their own ideas and new initiatives across all scientific disciplines through trans-European networking of nationally funded research activities. Based on a pan-European intergovernmental framework for cooperation in science and technology, COST has contributed since its creation more than 40 years ago to closing the gap between science, policy makers and society throughout Europe and beyond. COST is supported by the EU Seventh Framework Programme for research, technological development and demonstration activities (*Official Journal L 412, 30 December 2006*). The European Science Foundation as implementing agent of COST provides the COST Office through an EC Grant Agreement. The Council of the European Union provides the COST Secretariat. The COLOSS network is now supported by the Ricola Foundation - Nature & Culture.

15. References

AHMADIAN, A; LUNDEBERG, J (2002) A brief history of genetic variation analysis. *BioTechniques* 32: 1122-1124; 1126; 1128 passim.

ALLEN, M F; BALL, B V (1995) Characterisation and serological relationships of strains of Kashmir bee virus. *Annals of Applied Biology* 126(3): 471-484.
http://dx.doi.org/10.1111/j.1744-7348.1995.tb05382.x

ALLEN, M F; BALL, B V; WHITE, R; ANTONIW, J F (1986) The detection of acute paralysis virus in *Varroa jacobsoni* by the use of a simple indirect ELISA. *Journal of Apicultural Research* 25(2): 100-105.

ALTSCHUL, S F; GISH, W; MILLER, W; MEYERS, E W; LIPMAN, D J (1990) Basic local alignment search tool. *Journal of Molecular Biology* 215(3): 403-410.
http://dx.doi.org/10.1016/S0022-2836(05)80360-2

AMDIOUNI, H; MAUNULA, L; HAJJAMI, K; FAOUZI, A; SOUKRI, A; NOURLIL, J (2012) Recovery comparison of two virus concentration methods from wastewater using cell culture and real-time PCR. *Current Microbiology* 65(4): 432-437.
http://dx.doi.org/10.1007/s00284-012-0174-8

ANDERSON, D L; GIBBS, A J (1988) Inapparent virus infections and their interactions in pupae of the honey bee (*Apis mellifera* L.) in Australia. *Journal of General Virology* 69: 1617-1625.
http://dx.doi.org/10.1099/0022-1317-69-7-1617

ANDERSON, D L; GIBBS, A J (1989) Transpupal transmission of Kashmir bee virus and sacbrood virus in the honey bee (*Apis mellifera*). *Annals of Applied Biology* 114(1): 1-7.
http://dx.doi.org/10.1111/j.1744-7348.1989.tb06781.x

ANTÚNEZ, K; D'ALESSANDRO, B; CORBELLA, E; RAMALLO, G; ZUNINO, P (2006) Honey bee viruses in Uruguay. *Journal of Invertebrate Pathology* 91(1): 67-70.
http://dx.doi.org/10.1016/j.jip.2006.05.009

ANTÚNEZ, K; D'ALESSANDRO, B; CORBELLA, E; ZUNINO, P (2005) Detection of chronic bee paralysis virus and acute bee paralysis virus in Uruguayan honey bees. *Journal of Invertebrate Pathology* 90(1): 69-72.
http://dx.doi.org/10.1016/j.jip.2005.07.001

APFALTER, P; ASSADIAN, O; BLASI, F; BOMAN, J; GAYDOS, C A; KUNDI, M; MAKRISTATHIS, A; NEHR, M; ROTTER, M L; HIRSCHL, A M (2002) Reliability of nested PCR for detection of *Chlamydia pneumoniae* DNA in atheromas: results from a multicenter study applying standardized protocols. *Journal of Clinical Microbiology* 40(12): 4428-4434.
http://dx.doi.org/10.1128/JCM.40.12.4428-4434.2002

ARENS, M (1999) Methods for subtyping and molecular comparison of human viral genomes. *Clinical Microbioly Reviews* 12(4): 612–626.

ARTHOFER, W; STEINER, F M; SCHLICK-STEINER, B C (2011) Rapid and cost-effective screening of newly identified microsatellite loci by high-resolution melting analysis. *Molecular Genetics and Genomics* 286(3-4): 225-235.
http://dx.doi.org/10.1007/s00438-011-0641-0

BAILEY, L (1965) Paralysis of the honey bee, *Apis mellifera* Linnaeus. *Journal of Invertebrate Pathology* 7(2): 132-140.
http://dx.doi.org/10.1016/0022-2011(65)90024-8

BAILEY, L (1967) Acute bee-paralysis virus in adult honey bees injected with sacbrood virus. *Virology* 33(2): 368.
http://dx.doi.org/10.1016/0042-6822(67)90161-4

BAILEY, L (1968) The multiplication of sacbrood virus in the adult honey bee. *Virology* 36(2): 312-313. http://dx.doi.org/10.1016/0042-6822(68)90151-7

BAILEY, L (1976) Viruses attacking the honey bee. *Advances in Virus Research* 20: 271-304. http://dx.doi.org/10.1016/S0065-3527

BAILEY, L; BALL, B V (1978) Apis iridescent virus and "Clustering disease" of *Apis cerana*. *Journal of Invertebrate Pathology* 31(3): 368-371. http://dx.doi.org/10.1016/0022-2011(78)90231-8

BAILEY, L; BALL, B V (1991) *Honey bee pathology (2nd Ed.)*. Academic Press; London, UK. 193 pp.

BAILEY, L; BALL, B V; CARPENTER, J M (1980) Small virus like particles in honey bees associated with chronic paralysis virus and with a previously undescribed disease. *Journal of General Virology* 46: 149-155. http://dx.doi.org/10.1099/0022-1317-46-1-149

BAILEY, L; BALL, B V; PERRY, J N (1981) The prevalence of viruses of honey bees in Britain. *Annals of Applied Biology* 97: 109-118. http://dx.doi.org/10.1111/j.1744-7348.1981.tb02999.x

BAILEY, L; BALL, B V; PERRY, J N (1983a) Association of viruses with two protozoal pathogens of the honey bee. *Annals of Applied Biology* 103: 13-20. http://dx.doi.org/10.1111/j.1744-7348.1983.tb02735.x

BAILEY, L; BALL, B V; PERRY, J N (1983b) Honey bee paralysis: Its natural spread and its diminished incidence in England and Wales. *Journal of Apicultural Research* 22(3): 191-195.

BAILEY, L; BALL, B V; WOODS, R D (1976) An iridovirus from bees. *Journal of General Virology* 31(3): 459-461. http://dx.doi.org/10.1099/0022-1317-31-3-459

BAILEY, L; CARPENTER, J M; WOODS, R D (1979) Egypt bee virus and Australian isolates of Kashmir bee virus. *Journal of General Virology* 43(3): 641-647. http://dx.doi.org/10.1099/0022-1317-43-3-641

BAILEY, L; FERNANDO, E F W (1972) Effects of sacbrood virus on adult honey bees. *Annals of Applied Biology* 72: 27-35. http://dx.doi.org/10.1111/j.1744-7348.1972.tb01268.x

BAILEY, L; GIBBS, A J (1964) Acute infection of bees with paralysis virus. *Journal of Insect Pathology* 6(4): 395–407.

BAILEY, L; GIBBS, A J; WOODS, R D (1963) Two viruses from adult honey bees (*Apis mellifera* Linnaeus). *Virology* 21: 390-395. http://dx.doi.org/10.1016/0042-6822(63)90200-9

BAILEY, L; MILNE, R G (1969) The multiplication regions and interaction of acute and chronic bee-paralysis viruses in adult honey bees. *Journal of General Virology* 4(1): 9-14. http://dx.doi.org/10.1099/0022-1317-4-1-9

BAILEY, L; WOODS, R D (1974) Three previously undescribed viruses from the honey bee. *Journal of General Virology* 25: 175-186. http://dx.doi.org/10.1099/0022-1317-25-2-175

BAILEY, L; WOODS, R D (1977) Two more small RNA viruses from honey bees and further observations on sacbrood and acute bee-paralysis viruses. *Journal of General Virology* 37: 175-182. http://dx.doi.org/10.1099/0022-1317-37-1-175

BAKONYI, T; FARKAS, R; SZENDROI, A; DOBOS-KOVACS, M; RUSVAI, M (2002a) Detection of acute bee paralysis virus by RT-PCR in honey bee and *Varroa destructor* field samples: rapid screening of representative Hungarian apiaries. *Apidologie* 33(1): 63-74. http://dx.doi.org/10.1051/apido:2001004

BAKONYI, T; GRABENSTEINER, E; KOLODZIEJEK, J; RUSVAI, M; TOPOLSKA, G; RITTER, W; NOWOTNY, N (2002b) Phylogenetic analysis of acute bee paralysis virus strains. *Applied and Environmental Microbiology* 68(12): 6446-6450. http://dx.doi.org/10.1128/AEM.68.12.6446-6450.2002

BALL, B V (1985) Acute paralysis virus isolates from honey bee colonies infested with *Varroa jacobsoni*. *Journal of Apicultural Research* 24(2): 115-119.

BALL, B V (1989) *Varroa jacobsoni* as a virus vector. In *Cavalloro, R (Ed.). Present status of varroatosis in Europe and progress in the Varroa mite control*. EEC; Luxemburg. pp. 241-244.

BALL, B V; ALLEN, M F (1988) The prevalence of pathogens in honey bee colonies infested with the parasitic mite *Varroa jacobsoni*. *Annals of Applied Biology* 113: 237-244. http://dx.doi.org/10.1111/j.1744-7348.1988.tb03300.x

BALL, B V; OVERTON, H A; BUCK, K W (1985) Relationship between the multiplication of chronic bee paralysis virus and its associate particle. *Journal of General Virology* 66(7): 1423-1429. http://dx.doi.org/10.1099/0022-1317-66-7-1423

BECKER, S; FRANCO, J R; SIMARRO, P P; STICH, A; ABEL, P M; STEVERDING, D (2004) Real-time PCR for detection of *Trypanosoma brucei* in human blood samples. *Diagnostic Microbiology and Infectious Disease* 50(3): 193-199. http://dx.doi.org/10.1016/j.diagmicrobio.2004.07.001

BELÁK, S; THORÉN, P; LEBLANC, N; VILJOEN, G (2009) Advances in viral disease diagnostic and molecular epidemiological technologies. *Expert Reviews in Molecular Diagnostics* 9(4): 367-381. http://dx.doi.org/10.1586/ERM.09.19

BENJEDDOU, M; LEAT, N; ALLSOPP, M; DAVISON, S (2001) Detection of acute bee paralysis virus and black queen cell virus from honey bees by reverse transcriptase PCR. *Applied and Environmental Microbiology* 67(5): 2384-2387. http://dx.doi.org/10.1128/AEM.67.5.2384-2387.2001

BENJEDDOU, M; LEAT, N; ALLSOPP, M; DAVISON, S (2002) Development of infectious transcripts and genome manipulation of Black queen-cell virus of honey bees. *Journal of General Virology* 83(12): 3139-3146.

BERENYI, O; BAKONYI, T; DERAKHSHIFAR, I; KOGLBERGER, H; NOWOTNY, N (2006) Occurrence of six honey bee viruses in diseased Austrian apiaries. *Applied and Environmental Microbiology* 72: 2414-2420. http://dx.doi.org/10.1128/AEM.72.4.2414-2420.2006

BERGEM, M; NORBERG, K; AAMODT, R M (2006) Long-term maintenance of *in vitro* cultured honey bee (*Apis mellifera*) embryonic cells. *BMC Developmental Biology* 6(17). http://dx.doi.org/10.1186/1471-213X-6-17

BICKLE, P J; DOKSUM, K A (1981) An analysis of transformations revisited. *Journal of the American Statistical Association* 76(374): 296-311. http://dx.doi.org/10.1080/01621459.1981.10477649

BIRCH, L; ENGLISH, C A; BURNS, M; KEER, J T (2004) Generic scheme for independent performance assessment in the molecular biology laboratory. *Clinical Chemistry* 50(9): 1553-1559. http://dx.doi.org/10.1373/clinchem.2003.029454

BLANCHARD, P; SCHURR, F; OLIVIER, V; CELLE, O; ANTÚNEZ, K; BAKONYI, T; BERTHOUD, H; HAUBRUGE, E; HIGES, M; KASPRZAK, S; KOEGLBERGER, H; KRYGER, P; THIÉRY, R; RIBIÈRE, M (2009) Phylogenetic analysis of the RNA-dependent RNA polymerase (RdRp) and a predicted structural protein (pSP) of the Chronic bee paralysis virus (CBPV) isolated from various geographic regions. *Virus Research* 144(2): 334-338. http://dx.doi.org/10.1016/j.virusres.2009.04.025

BLANCHARD, P; OLIVIER, V; ISCACHE, A L; CELLE O; SCHURR, F; LALLEMAND, P; RIBIÈRE, M (2008) Improvement of RT-PCR detection of chronic bee paralysis virus (CBPV) required by the description of genomic variability in French CBPV isolates. *Journal of Invertebrate Pathology* 97(2): 182-185. http://dx.doi.org/10.1016/j.jip.2007.07.003

BLANCHARD, P; REGNAULT, J; SCHURR, F; DUBOIS, E; RIBIÈRE, M (2012) Intra-laboratory validation of chronic bee paralysis virus quantitation using an accredited standardised real-time quantitative RT-PCR method. *Journal of Virological Methods* 180(1-2): 26-31. http://dx.doi.org/10.1016/j.jviromet.2011.12.005

BLANCHARD, P; RIBIÈRE, M; CELLE, O; LALLEMAND, P; SCHURR, F; OLIVIER, V; ISCACHE, A L; FAUCON, J P (2007) Evaluation of a real-time two step RT-PCR assay for quantitation of Chronic bee paralysis virus (CBPV) genome in experimentally-infected bee tissues and in life stages of a symptomatic colony. *Journal of Virological Methods* 141: 7-13. http://dx.doi.org/10.1016/j.jviromet.2006.11.021

BONCRISTIANI, H F; DI PRISCO, G; PETTIS, J S; HAMILTON, M; CHEN, Y P (2009) Molecular approaches to the analysis of deformed wing virus replication and pathogenesis in the honey bee, *Apis mellifera*. *Virology Journal* 6: 1-9. http://dx.doi.org/10.1186/1743-422X-6-221

BOWEN-WALKER, P L; MARTIN, S J; GUNN, A (1999) The transmission of deformed wing virus between honey bees (*Apis mellifera* L.) by the ectoparasitic mite *Varroa jacobsoni* Oud. *Journal of Invertebrate Pathology* 73: 101-106. http://dx.doi.org/10.1006/jipa.1998.4807

BOYER, J C; HAENNI, A L (1994) Infectious transcripts and cDNA clones of RNA viruses. *Virology* 198(2): 415-426. http://dx.doi.org/10.1006/viro.1994.1053

BOX, G E P; COX, D R (1964) An analysis of transformations. *Journal of the Royal Statistical Society B* 26(2): 211-252. http://dx.doi.org/http://www.jstor.org/stable/2984418

BRUNETTO, M R; COLOMBATTO, P; BONINO, F (2009) Bio-mathematical models of viral dynamics to tailor antiviral therapy in chronic viral hepatitis. *World Journal of Gastroenterology* 15(5): 531-537. http://dx.doi.org/10.3748/wjg.15.531

BUSTIN, S A (2000) Absolute quantification of mRNA using real-time reverse transcription polymerase chain reaction assays. *Journal of Molecular Endocrinology* 25(2): 169-193. http://dx.doi.org/10.1677/jme.0.0250169

BUSTIN, S A; BEAULIEU, J F; HUGGETT, J; JAGGI, J; KIBENGE, R; FSBOLSVIK, P A; PENNING, L C; TOEGEL, S (2010) MIQE précis: Practical implementation of minimum standard guidelines for fluorescence-based quantitative real-time PCR experiments. *BMC Molecular Biology* 11:e74. http://dx.doi.org/10.1186/1471-2199-11-74

BUSTIN, S A; BENES, V; GARSON, J A; HELLEMANS, J; HUGGETT, J; KUBISTA, M; MUELLER, R; NOLAN, T; PFAFFL, M W; SHIPLEY, G L; VANDESOMPELE, J; WITTWER, C T (2009) The MIQE Guidelines: Minimum information for publication of quantitative Real-Time PCR experiments. *Clinical Chemistry* 55(4): 611-622. http://dx.doi.org/10.1373/clinchem.2008.112797

BUSTIN, S A; NOLAN, T (2004) Pitfalls of quantitative real-time reverse-transcription polymerase chain reaction. *Journal of Bimolecular Techniques* 15(3): 155-166.

CARRECK, N L; BALL, B V; MARTIN, S J (2010) The epidemiology of cloudy wing virus infections in honey bee colonies in the UK. *Journal of Apicultural Research* 49(1): 66-71. http://dx.doi.org/10.3896/IBRA.1.49.1.09

CARRIÈRE, M; PÈNE, V; BREIMAN, A; CONTI, F; CHOUZENOUX, S; MEURS, E; ANDRIEU, M; JAFFRAY, P; GRIRA, L; SOUBRANE, O; SOGNI, P; CALMUS, Y; CHAUSSADE, S; ROSENBERG, A R; PODEVIN, P (2007) A novel, sensitive, and specific RT-PCR technique for quantitation of hepatitis C virus replication. *Journal of Medical Virology* 79:155–160. http://dx.doi.org/10.1002/jmv.20773

CASADEVALL, A; PIROFSKI, L-A (1999) Host-pathogen interactions: Redefining the basic concepts of pathogenicity and virulence. *Infection and Immunity* 67(8): 3703-3713.

CASADEVALL, A; PIROFSKI, L-A (2001) Host-pathogen interactions: Attributes of virulence. *Journal of Infectious Diseases* 184(3): 337-344. http://dx.doi.org/10.1086/322044

CELLE, O; BLANCHARD, P; SCHURR, F; OLIVIER, V; COUGOULE, N; FAUCON, J P; RIBIÈRE, M (2008) Detection of chronic bee paralysis virus (CBPV) genome and RNA replication in various hosts: possible ways of spread. *Virus Research* 133: 280-284. http://dx.doi.org/10.1016/j.virusres.2007.12.011

CHANTAWANNAKUL, P; WARD, L; BOONHAM, N; BROWN, M (2006) A scientific note on the detection of honey bee viruses using real-time PCR (TaqMan) in varroa mites collected from a Thai honey bee (*Apis mellifera*) apiary. *Journal of Invertebrate Pathology* 91 (1): 69-73. http://dx.doi.org/10.1016/j.jip.2005.11.001

CHEN, Y P; EVANS, J D; FELDLAUFER, M F (2006a) Horizontal and vertical transmission of viruses in the honey bee (*Apis mellifera*). *Journal of Invertebrate Pathology* 92: 152-159.

CHEN, Y P; EVANS, J; HAMILTON, M; FELDLAUFER, M (2007) The influence of RNA integrity on the detection of honey bee viruses: molecular assessment of different sample storage methods. *Journal of Apicultural Research* 46(2): 81-87. http://dx.doi.org/10.3896/IBRA.1.46.2.03

CHEN, Y P; HIGGINS, J A; FELDLAUFER, M F (2005a) Quantitative real-time reverse transcription-PCR analysis of deformed wing virus infection in the honey bee (*Apis mellifera* L.). *Applied and Environmental Microbiology* 71(1): 436-441. http://dx.doi.org/10.1128/AEM.71.1.436-441.2005

CHEN, Y P; PETTIS, J S; COLLINS, A; FELDLAUFER, M F (2006b) Prevalence and transmission of honey bee viruses. *Applied and Environmental Microbiology* 72: 606-611.

CHEN, Y P; PETTIS, J S; EVANS, J D; KRAMER, M; FELDLAUFER, M F (2004a) Transmission of Kashmir bee virus by the ectoparasitic mite, *Varroa destructor*. *Apidologie* 35: 441-448. http://dx.doi.org/10.1051/apido:2004031

CHEN, Y P; PETTIS, J S; FELDLAUFER, M F (2005b) Detection of multiple viruses in queens of the honey bee *Apis mellifera* L. *Journal of Invertebrate Pathology* 90(2): 118-121. http://dx.doi.org/10.1016/j.jip.2005.08.005

CHEN, Y P; SMITH, I B; COLLINS, A M; PETTIS, J S; FELDLAUFER, M F (2004b) Detection of deformed wing virus infection in honey bees, *Apis mellifera* L., in the United States. *American Bee Journal* 144: 557-559.

CHEN, Y P; ZHAO, Y; HAMMOND, J; HSU, H T; EVANS, J; FELDLAUFER, M (2004c) Multiple virus infections in the honey bee and genome divergence of honey bee viruses. *Journal of Invertebrate Pathology* 87(2-3): 84-93. http://dx.doi.org/10.1016/j.jip.2004.07.005

CHERNESKY, M; JANG, D; CHONG, S; SELLORS, J; MAHONY, J (2003) Impact of urine collection order on the ability of assays to identify *Chlamydia trachomatis* infections in men. *Sexually Transmitted Diseases* 30(4): 345-347. http://dx.doi.org/10.1097/00007435-200304000-00014

CHUI, L; DREBOT, M; ANDONOV, A; PETRICH, A; GLUSHEK, M; MAHONY, J (2005) Comparison of 9 different PCR primers for the rapid detection of severe acute respiratory syndrome coronavirus using 2 RNA extraction methods. *Diagnostic Microbiology and Infectious Diseases* 53(1): 47-55. http://dx.doi.org/10.1016/j.diagmicrobio.2005.03.007

CLARK, T B (1978) A filamentous virus of the honey bee. *Journal of Invertebrate Pathology* 32: 332-340. http://dx.doi.org/10.1016/0022-2011(78)90197-0

COBEY, S W; TARPY, D R ; WOYKE, J (2013) Standard methods for instrumental insemination of *Apis mellifera* queens. In *V Dietemann; J D Ellis; P Neumann (Eds) The COLOSS* BEEBOOK, *Volume I: standard methods for* Apis mellifera *research. Journal of Apicultural Research* 52(4): http://dx.doi.org/10.3896/IBRA.1.52.4.

CORNMAN, S R; SCHATZ, M C; JOHNSTON, S J; CHEN, Y P; PETTIS, J; HUNT, G; BOURGEOIS, L; ELSIK, C; ANDERSON, D; GROZINGER, C M; EVANS, J D (2010) Genomic survey of the ectoparasitic mite *Varroa destructor*, a major pest of the honey bee *Apis mellifera*. *BMC Genomics* 11: 602. http://dx.doi.org/10.1186/1471-2164-11-602

CORONA, M; HUGHES, K A; WEAVER, D B; ROBINSON, G E (2005) Gene expression patterns associated with queen honey bee longevity. *Mechanisms of Ageing and Development* 126(11): 1230-1238. http://dx.doi.org/10.1016/j.mad.2005.07.004

COX, J L; HEYSE, J F; TUKEY, J W (2000) Efficacy estimates from parasite count data that include zero counts. *Experimental Parasitology* 96: 1-8. http://dx.doi.org/10.1006/expr.2000.4550

COX-FOSTER, D L; CONLAN, S; HOLMES, E; PALACIOS, G; EVANS, J D; MORAN, N A; QUAN, P L; BRIESE, T; HORNIG, M; GEISER, D M; MARTINSON, V; VANENGELSDORP, D; KALKSTEIN, A L; DRYSDALE, A; HUI, J; SHAI, J; CUI, L; HUTCHISON, S K; SIMONS, J F; EGHOLM, M; PETTIS, J S; LIPKIN, W I (2007) A metagenomic survey of microbes in honey bee colony collapse disorder. *Science* 318: 283-287. http://dx.doi.org/10.1126/science.1146498

CRAGGS, J K; BALL, J K; THOMSON, B J; IRVING, W L; GRABOWSKA, A M (2001) Development of a strand-specific RT-PCR based assay to detect the replicative form of hepatitis C virus RNA. *Journal of Virological Methods* 94(1-2): 111-120. http://dx.doi.org/10.1016/S0166-0934(01)00281-6

CRAILSHEIM, K; BRODSCHNEIDER, R; AUPINEL, P; BEHRENS, D; GENERSCH, E; VOLLMANN, J; RIESSBERGER-GALLÉ, U (2013) Standard methods for artificial rearing of *Apis mellifera* larvae. In *V Dietemann; J D Ellis; P Neumann (Eds) The COLOSS* BEEBOOK, *Volume I: standard methods for* Apis mellifera *research. Journal of Apicultural Research* 52(1): http://dx.doi.org/10.3896/IBRA.1.52.1.05

DAINAT, B; EVANS, J D; CHEN, Y P; NEUMANN, P (2011) Sampling and RNA quality for diagnosis of honey bee viruses using quantitative PCR. *Journal of Virological Methods* 174(1-2): 150-152. http://dx.doi.org/10.1016/j.jviromet.2011.03.029

DAINAT, B; KEN, T; BERTHOUD, H; NEUMANN, P (2009) The ectoparasitic mite *Tropilaelaps mercedesae* (*Acari, Laelapidae*) as a vector of honey bee viruses. *Insectes Sociaux.* 56: 40-43. http://dx.doi.org/10.1007/s00040-008-1030-5

DALL, D J (1987) Multiplication of Kashmir bee virus in pupae of the honey bee, Apis mellifera. Journal of Invertebrate Pathology 49 (3): 279-290. http://dx.doi.org/10.1016/0022-2011(87)90060-7

DE MIRANDA, J R (2008) Diagnostic techniques for virus detection in honey bees. In Aubert, M F A; Ball, B V; Fries, I; Moritz, R F A; Milani, N; Bernardinelli, I (Eds). Virology and the honey bee. EEC Publications; Brussels, Belgium. pp. 121-232.

DE MIRANDA, J R; CORDONI, G; BUDGE, G (2010a) The acute bee paralysis virus – Kashmir bee virus – Israeli acute paralysis virus complex. Journal of Invertebrate Pathology 103: S30-S47. http://dx.doi.org/10.1016/j.jip.2009.06.014

DE MIRANDA, J R; CHEN, Y; RIBIÈRE, M; GAUTHIER, L (2011) Varroa and viruses. In Carreck, N L (Ed.). Varroa – still a problem in the 21st century. International Bee Research Association; Cardiff, UK. pp 11-31.

DE MIRANDA, J R; DAINAT, B; LOCKE, B; CORDONI, G; BERTHOUD, H; GAUTHIER, L; NEUMANN, P; BUDGE, G E; BALL, B V; STOLTZ, D B (2010b) Genetic characterisation of slow paralysis virus of the honey bee (Apis mellifera L.). Journal of General Virology 91: 2524-2530. http://dx.doi.org/10.1099/vir.0.022434-0

DE MIRANDA, J R; DREBOT, M; TYLER, S; SHEN, M; CAMERON, C E; STOLTZ, D B; CAMAZINE, S M (2004) Complete nucleotide sequence of Kashmir bee virus and comparison with acute bee paralysis virus. Journal of General Virology 85: 2263-2270. http://dx.doi.org/10.1099/vir.0.79990-0

DE MIRANDA, J R; FRIES, I (2008) Venereal and vertical transmission of deformed wing virus in honey bees (Apis mellifera L.). Journal of Invertebrate Pathology 98: 184-189. http://dx.doi.org/10.1016/j.jip.2008.02.004

DE MIRANDA, J R; GENERSCH, E (2010) Deformed wing virus. Journal of Invertebrate Pathology 103: S48-S61. http://dx.doi.org/10.1016/j.jip.2009.06.012

DENHOLM, C H (1999) Inducible honey bee viruses associated with Varroa jacobsoni. PhD thesis, Keele University; UK. 225 pp.

DE SMET, L; RAVOET, J; DE MIRANDA J R; WENSELEERS, T; MORITZ R F A; DE GRAAF D C (2012) BeeDoctor, A versatile MLPA-based diagnostic tool for screening bee viruses. PLoS-ONE 7(10): e47953. http://dx.doi.org/10.1371/journal.pone.0047953

DIPRISCO, G; PENNACHCHIO, F; CAPRIO, E; BONCRISTIANI, H F; EVANS, J D; CHEN, Y P (2011) Varroa destructor is an effective vector of Israeli acute paralysis virus in the honey bee, Apis mellifera. Journal of General Virology 92: 151-155. http://dx.doi.org/10.1099/vir.0.023853-0

EBERT, D; BULL, J J (2003) Challenging the trade-off model for the evolution of virulence: is virulence management feasible? Trends in Microbiology 11: 15-20. http://dx.doi.org/10.1016/S0966-842X

EVANS, J D (2001) Genetic evidence for co-infection of honey bees by acute bee paralysis and Kashmir bee viruses. Journal of Invertebrate Pathology 78(4): 189-193. http://dx.doi.org/10.1006/jipa.2001.5066

EVANS, J D (2004) Transcriptional immune responses by honey bee larvae during invasion by the bacterial pathogen, Paenibacillus larvae. Journal of Invertebrate Pathology 85(2): 105-111. http://dx.doi.org/10.1016/j.jip.2004.02.004

EVANS, J D (2006) Beepath: An ordered quantitative-PCR array for exploring honey bee immunity and disease. Journal of Invertebrate Pathology 93(2): 135-139. http://dx.doi.org/10.1016/j.jip.2006.04.004

EVANS, J D; SCHWARZ, R S; CHEN, Y P; BUDGE, G; CORNMAN, R S; DE LA RUA, P; DE MIRANDA, J R; FORET, S; FOSTER, L; GAUTHIER, L; GENERSCH, E; GISDER, S; JAROSCH, A; KUCHARSKI, R; LOPEZ, D; LUN, C M; MORITZ, R F A; MALESZKA, R; MUÑOZ, I; PINTO, M A (2013) Standard methodologies for molecular research in Apis mellifera. In V Dietemann; J D Ellis; P Neumann (Eds) The COLOSS BEEBOOK, Volume I: standard methods for Apis mellifera research. Journal of Apicultural Research 52(4): http://dx.doi.org/10.3896/IBRA.1.52.4.11

EVANS, J D; HUNG, A C F (2000) Molecular phylogenetics and classification of honey bee viruses. Archives of Virology 145(10): 2015-2026. http://dx.doi.org/10.1007/s007050070037

EVISON, S E F; ROBERTS, K E; LAURENSON, L; PIETRAVALLE, S; HUI, J; BIESMEIJER, J C; SMITH, J E; BUDGE, G E; HUGHES, W O H (2012) Pervasiveness of parasites in pollinators. PLoS ONE 7(1): e30641. http://dx.doi.org/10.1371/journal.pone.0030641

EYER, M; CHEN, Y P; SCHAFER, M O; PETTIS, J S; NEUMANN, P (2009) Honey bee sacbrood virus infects adult small hive beetles, Aethina tumida (Coleoptera: Nitidulidae). Journal of Apicultural Research 48(4): 296-297. http://dx.doi.org/10.3896/IBRA.1.48.4.11

FIEVET, J; TENTCHEVA, D; GAUTHIER, L; DE MIRANDA, J R; COUSSERANS, F; COLIN, M E; BERGOIN, M (2006) Localization of deformed wing virus infection in queen and drone Apis mellifera L. Virology Journal 3: e16. http://dx.doi.org/10.1186/1743-422X-3-16

FIORINI, G S; CHIU, D T (2005) Disposable microfluidic devices: fabrication, function, and application. Biotechniques 38(3): 429-446. http://dx.doi.org/10.2144/05383RV02

FORSGREN, E; DE MIRANDA, J R; ISAKSSON, M; WIE, S; FRIES, I (2009) Deformed wing virus associated with Tropilaelaps mercedesae infesting European honey bees (Apis mellifera). Experimental and Applied Acarology 47: 87-97. http://dx.doi.org/10.1007/s10493-008-9204-4

FRIES, I; CAMAZINE, S (2001) Implications of horizontal and vertical pathogen transmission for honey bee epidemiology. Apidologie 32 (3): 199–214.

FUJIYUKI, T; MATSUZAKA, E; NAKAOKA, T; TAKEUCHI, H; WAKAMOTO, A; OHKA, S; SEKIMIZU, K; NOMOTO, A; KUBO, T (2009) Distribution of Kakugo virus and its effects on the gene expression profile in the brain of the worker honey bee Apis mellifera L. Journal of Virology 83: 11560-11568. http://dx.doi.org/10.1128/JVI.00519-09

FUJIYUKI, T; OHKA, S; TAKEUCHI, H; ONO, M; NOMOTO, A; KUBO, T (2006) Prevalence and phylogeny of Kakugo virus, a novel insect picorna-like virus that infects the honey bee (*Apis mellifera* L.), under various colony conditions. *Journal of Virology* 80: 11528-11538. http://dx.doi.org/10.1128/JVI.00754-06

FUJIYUKI, T; TAKEUCHI, H; ONO, M; OHKA, S; SASAKI, T; NOMOTO, A; KUBO T (2004) Novel insect picorna-like virus identified in the brains of aggressive worker honey bees. *Journal of Virology* 78: 1093-1100. http://dx.doi.org/10.1128/JVI.78.3.1093-1100.2004

GANTZER, C; MAUL, A; AUDIC, J M; SCHWARTZBROD, L (1998) Detection of infectious enteroviruses, enterovirus genomes, somatic coliphages, and *Bacteroides fragilis* phages in treated wastewater. *Applied and Environmental Microbiology* 64(11): 4307-4312.

GARDNER, S N; KUCZMARSKI, T A; VITALIS, E A; SLEZAK, T R (2003) Limitations of TaqMan PCR for detecting divergent viral pathogens illustrated by hepatitis A, B, C and E viruses and human immunodeficiency virus. *Journal of Clinical Microbiology* 41(6): 2417-2427.
http://dx.doi.org/10.1128/JCM.41.6.2417-2427.2003

GAUTHIER, L; RAVALLEC, M; TOURNAIRE, M; COUSSERANS, F; BERGOIN, M; DAINAT, B; DE MIRANDA, J R (2011) Viruses associated with ovarian degeneration in *Apis mellifera* L. queens. *PLoS Pathogens* 6: e16217.
http://dx.doi.org/10.1371/journal.pone.0016217

GAUTHIER, L; TENTCHEVA, D; TOURNAIRE, M; DAINAT, B; COUSSERANS, F; COLIN, M E; BERGOIN, M (2007) Viral load estimation in asymptomatic honey bee colonies using the quantitative RT-PCR technique. *Apidologie* 38: 426-436.
http://dx.doi.org/10.1051/apido:2007026

GENERSCH, E (2005) Development of a rapid and sensitive RT-PCR method for the detection of deformed wing virus, a pathogen of the honey bee (*Apis mellifera*). *Veterinary Journal* 169(1): 121-123. http://dx.doi.org/10.1016/j.tvjl.2004.01.004

GENERSCH, E; GISDER, S; HEDTKE, K; HUNTER, W B; MÖCKEL, N; MÜLLER, U (2013) Standard methods for cell cultures in *Apis mellifera* research. In *V Dietemann; J D Ellis; P Neumann (Eds) The COLOSS BEEBOOK, Volume I: standard methods for Apis mellifera research. Journal of Apicultural Research* 52(1): http://dx.doi.org/10.3896/IBRA.1.52.1.02

GILDOW, F E; D'ARCY, C J (1990) Cytopathology and experimental host range of *Rhopalosiphum padi* virus, a small isometric RNA virus infecting cereal grain aphids. *Journal of Invertebrate Pathology* 55(2): 245-257.
http://dx.doi.org/10.1016/0022-2011(90)90060-J

GISDER, S; AUMEIER, P; GENERSCH, E (2009) Deformed wing virus (DWV): viral load and replication in mites (*Varroa destructor*). *Journal of General Virology* 90: 463-467.
http://dx.doi.org/10.1099/vir.0.005579-0

GISDER, S; MOCKEL, N; LINDE, A; GENERSCH, E (2011) A cell culture model for *Nosema ceranae* and *Nosema apis* allows new insights into the life cycle of these important honey bee-pathogenic microsporidia. *Environmental Microbiology* 13(2): 404-413.
http://dx.doi.org/10.1111/j.1462-2920.2010.02346.x

GLOVER, R H; ADAMS, I P; BUDGE, G; WILKINS, S; BOONHAM, N (2011) Detection of honey bee (*Apis mellifera*) viruses with an oligonucleotide microarray. *Journal of Invertebrate Pathology* 107 (3): 216-219. http://dx.doi.org/10.1016/j.jip.2011.03.004

GRABENSTEINER, E; BAKONYI, T; RITTER, W; PECHHACKER, H; NOWOTNY, N (2007) Development of a multiplex RT-PCR for the simultaneous detection of three viruses of the honey bee (*Apis mellifera* L.): acute bee paralysis virus, black queen cell virus and sacbrood virus. *Journal of Invertebrate Pathology* 94(3): 222-225.
http://dx.doi.org/10.1016/j.jip.2006.11.006

GRABENSTEINER, E; RITTER, W; CARTER, M J; DAVISON, S; PECHHACKER, H; KOLODZIEJEK, J; BOECKING, O; DERAKHSHIFAR, I; MOOSBECKHOFER, R; LICEK, E; NOWOTNY, N (2001) Sacbrood virus of the honey bee (*Apis mellifera*): rapid identification and phylogenetic analysis using reverse transcription -PCR. *Clinical and Diagnostic Laboratory Immunology* 8(1): 93-104. http://dx.doi.org/10.1128/CDLI.8.1.93-104.2001

GREEN, M R; SAMBROOK, J (2012) *Molecular cloning: a laboratory manual (4th Ed.).* Cold Spring Harbor Laboratory Press; Cold Spring Harbor, USA. 2028 pp.

GUNJI, T; KATO, N; HIJIKATA, M; HAYASHI, K; SAITOH, S; SHIMOTOHNO, K (1994) Specific detection of positive and negative stranded hepatitis C viral RNA using chemical RNA modification. *Archives of Virology* 134(3-4): 293-302.
http://dx.doi.org/10.1385/0-89603-521-2:465

HARLOW, E; LANE D (1988) *Antibodies: A laboratory manual.* Cold Spring Harbor Laboratory Press; Cold Spring Harbor, New York, USA. 726 pp.

HAUSER, S; WEBER, C; VETTER, G; STEVENS, M; BEUVE, M; LEMAIRE, O (2000) Improved detection and differentiation of poleroviruses infecting beet or rape by multiplex RT-PCR. *Journal of Virological Methods* 89(1-2): 11-21.
http://dx.doi.org/10.1016/S0166-0934(00)00203-2

HIGGINS, J A; NASARABADI, S; KARNS, J S; SHELTON, D R; COOPER, M; GBAKIMA, A; KOOPMAN, R P (2003) A handheld real time thermal cycler for bacterial pathogen detection. *Biosensors and Bioelectronics* 18(9): 1115-1123. Erratum in: *Biosensors and Bioelectronics* 20, 663-664.
http://dx.doi.org/10.1016/S0956-5663(02)00252-X

HIGHFIELD, A C; EL NAGAR, A; MACKINDER, L C M; NOËL, L M-L J; HALL, M J; MARTIN, S J; SCHROEDER, D C (2009) Deformed wing virus implicated in overwintering honey bee colony losses. *Applied and Environmental Microbiology* 75(22): 7212-7220.
http://dx.doi.org/10.1128/AEM.02227-09

HORSINGTON, J; ZHANG, Z (2007) Analysis of foot-and-mouth disease virus replication using strand-specific quantitative RT-PCR. *Journal of Virological Methods* 144(1-2): 149-155. http://dx.doi.org/10.1016/j.jviromet.2007.05.002

HUMAN, H; BRODSCHNEIDER, R; DIETEMANN, V; DIVELY, G; ELLIS, J; FORSGREN, E; FRIES, I; HATJINA, F; HU, F-L; JAFFÉ, R; KÖHLER, A; PIRK, C W W; ROSE, R; STRAUSS, U; TANNER, G; TARPY, D R; VAN DER STEEN, J J M; VEJSNÆS, F; WILLIAMS, G R; ZHENG, H-Q (2013) Miscellaneous standard methods for *Apis mellifera* research. In *V Dietemann; J D Ellis; P Neumann (Eds) The COLOSS BEEBOOK, Volume I: standard methods for* Apis mellifera *research. Journal of Apicultural Research* 52(4): http://dx.doi.org/10.3896/IBRA.1.52.4.10

HUNG, A C F (2000) PCR detection of Kashmir bee virus in honey bee excreta. *Journal of Apicultural Research* 39(3-4): 103-106.

HUNG, A C F; SHIMANUKI, H (1999) A scientific note on the detection of Kashmir bee virus in individual honey bees and varroa mites. *Apidologie* 30(4): 353-354. http://dx.doi.org/10.1051/apido:19990414

HUNG, A C F; BALL, B V; ADAMS, J R; SHIMANUKI, H; KNOX, D A (1996a) A scientific note on the detection of American strains of acute paralysis virus and Kashmir bee virus in dead bees in one US honey bee (*Apis mellifera* L.) colony. *Apidologie* 27: 55–56.

HUNG, A C F; SHIMANUKI, H; KNOX, D A (1996b) Inapparent infection of acute bee paralysis virus and Kashmir bee virus in US honey bees. *American Bee Journal* 136: 874–876.

HUNTER, W B (2010) Medium for development of bee cell cultures (*Apis mellifera*: Hymenoptera: Apidae). *In Vitro Cellular & Developmental Biology-Animal* 46(2): 83-86. http://dx.doi.org/10.1007/s11626-009-9246-x

HUNTER, W; ELLIS, J; VANENGELSDORP, D; HAYES, J; WESTERVELT, D; GLICK, E; WILLIAMS, M; SELA, I; MAORI, E; PETTIS, J; COX-FOSTER, D; PALDI, N (2010) Large-scale field application of RNAi technology reducing Israeli acute paralysis virus disease in honey bees (*Apis mellifera*, Hymenoptera: Apidae). *PLoS Pathogens* 6: e1001160. http://dx.doi.org/10.1371journal.ppat.1001160

IQBAL, J; MÜLLER, U (2007) Virus infection causes specific learning deficits in honey bee foragers. *Proceedings of the Royal Society London* B274: 1517-1521. http://dx.doi.org/10.1098/rspb.2007.0022

JAKAB, G; DROZ, E; BRIGNETI, G; BAULCOMBE, D; MALNOË, P (1997) Infectious *in vivo* and *in vitro* transcripts from a full-length cDNA clone of PVY-N605, a Swiss necrotic isolate of potato virus Y. *Journal of General Virology* 78(12): 3141-3145.

KITAGISHI, Y; OKUMURA, N; YOSHIDA, H; NISHIMURA, Y; TAKAHASHI, J; MATSUDA, S (2011) Long-term cultivation of *in vitro Apis mellifera* cells by gene transfer of human c-myc proto-oncogene. *In Vitro Cellular and Developmental Biology-Animal* 47 (7): 451-453. http://dx.doi.org/10.1007/s11626-011-9431-6

KURATH, G; HEICK, J A; DODDS, J A (1993) RNase protection analyses show high genetic diversity among field isolates of satellite tobacco mosaic virus. *Virology* 194(1): 414-418. http://dx.doi.org/10.1006/viro.1993.1278

KUKIELKA, D; SÁNCHEZ-VIZCAÍNO, J M (2008) A sensitive one-step real-time RT-PCR method for detection of deformed wing virus and black queen cell virus in honey bee *Apis mellifera*. *Journal of Virological Methods* 147: 275–281. http://dx.doi.org/10.1016/j.jviromet.2007.09.008

KUKIELKA, D; SÁNCHEZ-VIZCAÍNO, J M (2009) One-step real-time quantitative PCR assays for the detection and field study of sacbrood honey bee and acute bee paralysis viruses. *Journal of Virological Methods* 161: 240–246. http://dx.doi.org/10.1016/j.jviromet.2009.06.014

LANFORD, R E; CHAVEZ, D; CHISARI, F V; SUREAU, C (1995) Lack of detection of negative-strand hepatitis C virus RNA in peripheral blood mononuclear cells and other extrahepatic tissues by the highly strand-specific rTth reverse transcriptase PCR. *Journal of Virology* 69(12): 8079-8083.

LANFORD, R E; SUREAU, C; JACOB, J R; WHITE, R; FUERST, T R (1994) Demonstration of *in vitro* infection of chimpanzee hepatocytes with hepatitis C virus using strand-specific RT/PCR. *Virology* 202(2): 606-614. http://dx.doi.org/10.1006/viro.1994.1381

LANZI, G; DE MIRANDA, J R; BONIOTTI, M B; CAMERON, CE; LAVAZZA, A; CAPUCCI, L; CAMAZINE, S M; ROSSI, C (2006) Molecular and biological characterization of deformed wing virus of honey bees (*Apis mellifera* L.). *Journal of Virology* 80: 4998-5009. http://dx.doi.org/10.1128/JVI.80.10.4998-5009.2006

LASKUS, T; RADKOWSKI, M; WANG, L F; VARGAS, H; RAKELA, J (1998) Detection of hepatitis G virus replication sites by using highly strand-specific Tth-based reverse transcriptase PCR. *Journal of Virology* 72(4): 3072-3075.

LAUBSCHER, J M; VON WECHMAR, M B (1992) Influence of aphid lethal paralysis virus and *Rhopalosiphum padi* virus on aphid biology at different temperatures. *Journal of Invertebrate Pathology* 60(2): 134-140. http://dx.doi.org/10.1016/0022-2011(92)90086-J

LAUBSCHER, J M; VON WECHMAR, M B (1993) Assessment of aphid lethal paralysis virus as an apparent population growth-limiting factor in grain aphids in the presence of other natural enemies. *Biocontrol Science and Technology* 3(4): 455-466. http://dx.doi.org/10.1080/09583159309355300

LEAT, N; BALL, B V; GOVAN, V; DAVISON, S (2000) Analysis of the complete genome sequence of black queen-cell virus, a picorna-like virus of honey bees. *Journal of General Virology* 81(8): 2111-2119.

LEE, P E; FURGALA, B (1967) Viruslike particles in adult honey bees (*Apis mellifera* L.) following injection with sacbrood virus. *Virology* 32: 11-17. http://dx.doi.org/10.1016/0042-6822(67)90247-4

LI, J L; PENG, W J; WU, J; STRANGE, J P; BONCRISTIANI, H F; CHEN, Y P (2011) Cross-species infection of deformed wing virus poses a new threat to pollinator conservation. *Journal of Economic Entomology* 104: 732–739. http://dx.doi.org/10.1603/EC10355

LIEPOLD, P; WIEDER, H; HILLEBRANDT, H; FRIEBEL, A; HARTWICH, G (2005) DNA-arrays with electrical detection: a label-free low cost technology for routine use in life sciences and diagnostics. *Bioelectrochemistry* 67(2): 143-150. http://dx.doi.org/10.1016/j.bioelechem.2004.08.004

LIN, L; LIBBRECHT, L; VERBEECK, J; VERSLYPE, C; ROSKAMS, T; VAN PELT, J; VAN RANST, M; FEVERY, J (2009) Quantitation of replication of the HCV genome in human livers with end-stage cirrhosis by strand-specific real-time RT-PCR assays: methods and clinical relevance. *Journal of Medical Virology* 81(9): 1569-1575. http://dx.doi.org/10.1002/jmv.21510

LOCKE, B; FORSGREN, E; FRIES, I; DE MIRANDA, J R (2012) Acaricide treatment affects viral dynamics in *Varroa destructor*-infested honey bee colonies via both host physiology and mite control. *Applied and Environmental Microbiology* 78: 227-235. http://dx.doi.org/10.1128/AEM.06094-11

LOMMEL, S A; MORRIS, T J; PINNOCK, D E (1985) Characterization of nucleic acids associated with Arkansas bee virus. *Intervirology* 23: 199-207. http://dx.doi.org/10.1159/000149606

MARMUR, J; DOTY, P (1959) Heterogeneity in deoxyribonucleic acids. 1. Dependence on composition of the configurational stability of deoxyribonucleic acids. *Nature* 183(4673): 1427-1429. http://dx.doi.org/10.1038/1831427a0

MARMUR, J; DOTY, P (1962) Determination of base composition of deoxyribonucleic acid from 1st thermal denaturation temperature. *Journal of Molecular Biology* 5(1): 109-118. http://dx.doi.org/10.1016/S0022-2836(62)80066-7

MAORI, E; LAVI, S; MOZES-KOCH, R; GANTMAN, Y; PERETZ, Y; EDELBAUM, O; TANNE, E; SELA, I (2007a) Isolation and characterization of Israeli acute paralysis virus, a dicistrovirus affecting honey bees in Israel: evidence for diversity due to intra- and inter-species recombination. *Journal of General Virology* 88: 3428-3438. http://dx.doi.org/10.1099/vir.0.83284-0

MAORI, E; PALDI, N; SHAFIR, S; KALEV, H; TSUR, E; GLICK, E; SELA, I (2009) IAPV, a bee-affecting virus associated with Colony Collapse Disorder can be silenced by dsRNA ingestion. *Insect Molecular Biology* 18(1): 55-60. http://dx.doi.org/10.1111/j.1365-2583.2009.00847.x

MAORI, E; TANNE, E; SELA, I (2007b) Reciprocal sequence exchange between non-retro viruses and hosts leading to the appearance of new host phenotypes. *Virology* 362(2): 342-349. http://dx.doi.org/10.1016/j.virol.2006.11.038

MARTIN, S J; HIGHFIELD, A C; BRETTELL, L; VILLALOBOS, E M; BUDGE, G E; POWELL, M; NIKAIDO, S; SCHROEDER, D C (2012) Global honey bee viral landscape altered by a parasitic mite. *Science* 336: 1304-1306. http://dx.doi.org/10.1126/science.1220941

McCLELLAND, E E; BERNHARDT, P; CASADEVALL, A (2006) Estimating the relative contributions of virulence factors for pathogenic microbes. *Infection and Immunity* 74(3): 1500-1504. http://dx.doi.org/10.1128/IAI.74.3.1500–1504.2006

McGUINESS, P H; BISHOP, G A; MCCAUGHAN, G W; TROWBRIDGE, R; GOWANS, E J (1994) False detection of negative-strand hepatitis C virus RNA. *Lancet* 343: 551-552. http://dx.doi.org/10.1016/S0140-6736(94)91509-1

MEEUS, I; SMAGGHE, G; SIEDE, R; JANS, K; DE GRAAF, D C (2010) Multiplex RT-PCR with broad-range primers and an exogenous internal amplification control for the detection of honey bee viruses in bumble bees. *Journal of Invertebrate Pathology* 105: 200-2003. http://dx.doi.org/10.1016/j.jip.2010.06.012

MINOR, P D (1985) Growth, assay and purification of picornaviruses. In *Mahy, B W (Ed.). Virology, a practical approach.* IRL Press; Oxford, UK. pp. 25–41.

MOON, J S; DOMIER, L L; MCCOPPIN, N K; D'ARCY, C J; JIN, H (1998) Nucleotide sequence analysis shows that *Rhopalosiphum padi* virus is a member of a novel group of insect-infecting RNA viruses. *Virology* 243(1): 54-65. http://dx.doi.org/10.1006/viro.1998.9043

MOORE, J; JIRONKIN, A; CHANDLER, D; BURROUGHS, N; EVANS, D J; RYABOV, E V (2011) Recombinants between deformed wing virus and Varroa destructor virus-1 may prevail in *Varroa destructor*-infested honey bee colonies. *Journal of General Virology* 92: 156-161 http://dx.doi.org/10.1099/vir.0.025965-0

NF EN ISO/IEC 17025 (2005) *General requirements for the competence of testing and calibration laboratories.* 40 pp.

OIE (2010) Manual of diagnostic tests and vaccines for terrestrial animals In *Chapter 1.1.1.4/ 5. Principles and methods of validation of diagnostic assay for infectious diseases.* Adopted in May 2009.

OLIVIER, V; BLANCHARD, P; CHAOUCH, S; LALLEMAND, P; SCHURR, F; CELLE, O; DUBOIS, E; TORDO, N; THIERY, R; HOULGATTE, R; RIBIÈRE, M (2008) Molecular characterization and phylogenetic analysis of chronic bee paralysis virus, a honey bee virus. *Virus Research* 132: 59–68. http://dx.doi.org/10.1016/j.virusres.2007.10.014

ONGUS, J R (2006) Varroa destructor virus 1: A new picorna-like virus in varroa mites as well as honey bees. PhD thesis, Wageningen University, Netherlands. 132 pp.

ONGUS, J R; PEETERS, D; BONMATIN, J M; BENGSCH, E; VLAK, J M; VAN OERS, M M (2004) Complete sequence of a picorna-like virus of the genus Iflavirus replicating in the mite *Varroa destructor*. *Journal of General Virology* 85: 3747-3755. http://dx.doi.org/10.1099/vir.0.80470-0

ONGUS, J R; ROODE, E C; PLEIJ, C W A; VLAK, J M; VAN OERS, M M (2006) The 5 ' non-translated region of Varroa destructor virus 1 (genus Iflavirus): structure prediction and RES activity in *Lymantria dispar* cells. *Journal of General Virology* 87: 3397-3407. http://dx.doi.org/10.1099/vir.0.82122-0

ONODERA, K (2007) Selection for 3'-end triplets for polymerase chain reaction primers. In *Yuryev, A (Ed.). PCR primer design. Methods in Molecular Biology™ series #402.* Humana Press; Totowa, New Jersey, USA. pp. 61-74.

PALACIOS, G; HUI, J; QUAN, P L; KALKSTEIN, A; HONKAVUORI, K S; BUSSETTI, A V; CONLAN, S; EVANS, J; CHEN, Y P; VAN ENGELSDORP, D; EFRAT, H; PETTIS, J; COX-FOSTER, D; HOLMES, E C; BRIESE, T; LIPKIN, W I (2008) Genetic analysis of Israel acute paralysis virus: distinct clusters are circulating in the United States. *Journal of Virology* 82(13): 6209-6217. http://dx.doi.org/10.1128/JVI.00251-08

PAPIN, J F; VAHRSON, W; DITTMER, D P (2004) SYBR green-based real-time quantitative PCR assay for detection of West Nile virus circumvents false-negative results due to strain variability. *Journal of Clinical Microbiology* 42(4): 1511-1518. http://dx.doi.org/10.1128/JCM.42.4.1511-1518.2004

PENG, W J; LI, J L; BONCRISTIANI, H F; STRANGE, J P; HAMILTON, M; CHEN, Y P (2011) Host range expansion of honey bee black queen cell virus in the bumble bee, *Bombus huntii. Apidologie* 42: 650–658. http://dx.doi.org/10.1007/s13592-011-0061-5

PEYREFITTE, C N; PASTORINO, B; BESSAUD, M; TOLOU, H J; COUISSINIER-PARIS, P (2003) Evidence for *in vitro* falsely-primed cDNAs that prevent specific detection of virus negative strand RNAs in dengue-infected cells: improvement by tagged RT-PCR. *Journal of Virological Methods* 113(1): 19-28. http://dx.doi.org/10.1016/S0166-0934(03)00218-0

PFAFFL, M W; HAGELEIT, M (2001) Validities of mRNA quantification using recombinant RNA and recombinant DNA external calibration curves in real-time RT-PCR. *Biotechnology Letters* 23(4): 275-282. http://dx.doi.org/10.1023/A:1005658330108

PIRK, C W W; DE MIRANDA, J R; FRIES, I; KRAMER, M; PAXTON, R; MURRAY, T; NAZZI, F; SHUTLER, D; VAN DER STEEN, J J M; VAN DOOREMALEN, C (2013) Statistical guidelines for *Apis mellifera* research. In *V Dietemann; J D Ellis; P Neumann (Eds) The COLOSS BEEBOOK, Volume I: standard methods for* Apis mellifera *research. Journal of Apicultural Research* 52(4): http://dx.doi.org/10.3896/IBRA.1.52.4.13

PIROFSKI, L A; CASADEVALL, A (2012) Q & A: What is a pathogen? A question that begs the point. *BMC Biology* 10: e6. http://www.biomedcentral.com/1741-7007/10/6

PLASKON, N E; ADELMAN, Z N; MYLES, K M (2009) Accurate strand-specific quantification of viral RNA. *PLoS ONE* 4(10): e7468. http://dx.doi.org/10.1371/journal.pone.0007468

PURCELL, M K; HART, S A; KURATH, G; WINTON, J R (2006) Strand-specific, real-time RT-PCR assays for quantification of genomic and positive-sense RNAs of the fish rhabdovirus, infectious hematopoietic necrosis virus. *Journal of Virological Methods* 132(1-2): 18-24. http://dx.doi.org/10.1016/j.jviromet.2005.08.017

RACANIELLO, V; BALTIMORE, D (1981) Cloned poliovirus complementary DNA is infectious in mammalian cells. *Science* 214 (4523): 916-919. http://dx.doi.org/10.1126/science.6272391

RADONIĆ, A; THULKE, S; MACKAY, I M; LANDT, O; SIEGERT, W; NITSCHE, A (2004) Guideline to reference gene selection for quantitative real-time PCR. *Biochemical and Biophysical Research Communications* 313(4): 856-862. http://dx.doi.org/10.1016/j.bbrc.2003.11.177

RENSEN, G; SMITH, W; RUZANTE, J; SAWYER, M; OSBURN, B; CULLOR, J (2005) Development and evaluation of a real-time fluorescent polymerase chain reaction assay for the detection of bovine contaminates in cattle feed. *Foodborne Pathogens and Disease* 2(2): 152-159. http://dx.doi.org/10.1089/fpd.2005.2.152

RHODES, E R; HAMILTON, D W; SEE, M J; WYMER, L (2011) Evaluation of hollow-fiber ultrafiltration primary concentration of pathogens and secondary concentration of viruses from water. *Journal of Virological Methods* 176(1-2): 38-45. http://dx.doi.org/10.1016/j.jviromet.2011.05.031

RIBIÈRE, M; BALL, B V; AUBERT, M F A (2008) Natural history and geographic distribution of honey bee viruses. In Aubert, M F A; Ball, B V; Fries, I; Morritz, R F A; Milani, N; Bernardinelli, I (Eds). *Virology and the honey bee*. EEC Publications; Brussels, Belgium. pp. 15–84.

RIBIÈRE, M; FAUCON, J P; PÉPIN, M (2000) Detection of chronic honey bee (*Apis mellifera* L.) paralysis virus infection: application to a field survey. *Apidologie* 31(5): 567-577. http://dx.doi.org/10.1051/apido:2000147

RIBIÈRE, M; OLIVIER, V; BLANCHARD, P (2010) Chronic bee paralysis virus. A disease and a virus like no other? *Journal of Invertebrate Pathology* 103: S120-S131. http://dx.doi.org/10.1016/j.jip.2009.06.013

RIBIÈRE, M; TRIBOULOT, C; MATHIEU, L; AURIÈRES, C; FAUCON, J P; PÉPIN, M (2002) Molecular diagnosis of chronic bee paralysis virus infection. *Apidologie* 33(3): 339-351. http://dx.doi.org/10.1051/apido:2002020

RODRIGUEZ, M A; CHEN, Y; CRAIGO, J K; CHATTERJEE, R; RATNER, D; TATSUMI, M; ROY, P; NEOGI, D; GUPTA, P (2006) Construction and characterization of an infectious molecular clone of HIV-1 subtype A of Indian origin. *Virology* 345(2): 328-336. http://dx.doi.org/10.1016/j.virol.2005.09.053

RORTAIS, A; TENTCHEVA, D; PAPACHRISTOFOROU, A; GAUTHIER, L; ARNOLD, G; COLIN, M E; BERGOIN, M (2006) Deformed wing virus is not related to honey bees' aggressiveness. *Virology Journal* 3: e61. http://dx.doi.org/10.1186/1743-422X-3-61

RUNCKEL, C; FLENNIKEN, M L; ENGEL, J C; RUBY, J G; GANEM, D; ANDINO, R; DERISI, J L (2011) Temporal analysis of the honey bee microbiome reveals four novel viruses and seasonal prevalence of known viruses, nosema, and crithidia. *PloS One* 6 (6): e20656. http://dx.doi.org/10.1371/journal.pone.0020656

SAGE, L (2004) Protein biochips go high tech. *Analytical Chemistry* 76 (7): 137A-142A. http://dx.doi.org/10.1021/ac0415408

SANTALUCIA, J (2007) Physical principles and visual-OMP software for optimal PCR design. In *Yuryev, A (Ed.). PCR primer design. Methods in Molecular Biology™ series #402.* Humana Press; Totowa, New Jersey, USA. pp. 3-33.

SCHAAD, N W; FREDERICK, R D; SHAW, J; SCHNEIDER, W L; HICKSON, R; PETRILLO, M D; LUSTER, D G (2003) Advances in molecular-based diagnostics in meeting crop biosecurity and phytosanitary issues. *Annual Review of Phytopathology* 41: 305-324. http://dx.doi.org/10.1146/annurev.phyto.41.052002.095435

SHAH, K S; EVANS, E C; PIZZORNO, M C (2009) Localization of deformed wing virus (DWV) in the brains of the honey bee, *Apis mellifera* Linnaeus. *Virology Journal* 6: e182. http://dx.doi.org/10.1.1186/1743-422X-6-182

SHEN, M Q; CUI, L W; OSTIGUY, N; COX-FOSTER, D (2005a) Intricate transmission routes and interactions between picorna-like viruses (Kashmir bee virus and sacbrood virus) with the honey bee host and the parasitic varroa mite. *Journal of General Virology* 86(8): 2281-2289. http://dx.doi.org/10.1099/vir.0.80824-0

SHEN, M Q; YANG, X L; COX-FOSTER, D; CUI, L W (2005b) The role of varroa mites in infections of Kashmir bee virus (KBV) and deformed wing virus (DWV) in honey bees. *Virology* 342:141-149. http://dx.doi.org/10.1016/j.virol.2005.07.012

SIEDE, R; BÜCHLER, R (2003) Symptomatische befall von drohnenbrut mit dem black queen cell virus auf hessischen bienenständen. *Berliner und Münchener Tierärztliche Wochenschrift* 116: 130-133.

SIEDE, R; BÜCHLER, R (2006) Spatial distribution patterns of acute bee paralysis virus, black queen cell virus and sacbrood virus in Hesse, Germany. *Wiener Tierärztliche Monatsschrift* 93(4): 90-93.

SIEDE, R; DERAKHSHIFAR, I; OTTEN, C; BERÉNYI, O; BAKONYI, T; KÖGLBERGER, H; BÜCHLER, R (2005) Prevalence of Kashmir bee virus in central Europe. *Journal of Apicultural Research* 44(3): 129. http://dx.doi.org/10.3896/IBRA.1.44.3.09

SINGH, R; LEVITT, A L; RAJOTTE, E G; HOLMES, E C; OSTIGUY, N; VANENGELSDORP, D; LIPKIN, W I; DEPAMPHILIS, C W; TOTH, A L; COX-FOSTER, D L (2010) RNA viruses in hymenopteran pollinators: evidence of inter-taxa virus transmission via pollen and potential impact on non-*Apis* hymenopteran species. *PLoS ONE* 5: e14357. http://dx.doi.org/10.1371/journal.pone.0014357

STACH, J E; BATHE, S; CLAPP, J P; BURNS, R G (2001) PCR-SSCP comparison of 16S rDNA sequence diversity in soil DNA obtained using different isolation and purification methods. *FEMS Microbiology Ecology* 36(2-3): 139-151. http://dx.doi.org/10.1111/j.1574-6941.2001.tb00834.x

STÅHLBERG, A; HÅKANSSON, J; XIAN, X; SEMB, H; KUBISTA, M (2004a) Properties of the reverse transcription reaction in mRNA quantification. *Clinical Chemistry* 50(3): 509-515. http://dx.doi.org/10.1373/clinchem.2003.026161

STÅHLBERG, A; KUBISTA, M; PFAFFL, M (2004b) Comparison of reverse transcriptases in gene expression analysis. *Clinical Chemistry* 50(9): 1678-1680. http://dx.doi.org/10.1373/clinchem.2004.035469

STOLTZ, D; SHEN, X R; BOGGIS, C; SISSON, G (1995) Molecular diagnosis of Kashmir bee virus infection. *Journal of Apicultural Research* 34(3): 153-160.

SUMPTER, D J T; MARTIN, S J (2004) The dynamics of virus epidemics in varroa-infested honey bee colonies. *Journal of Animal Ecology* 73: 51-63. http://dx.doi.org/10.1111/j.1365-2656.2004.00776.x

TANIGUCHI, T; PALMIERI, M; WEISSMANN, C (1978) A Qbeta DNA-containing hybrid plasmid giving rise to Qbeta phage formation in the bacterial host. *Annales de Microbiologie* 129B(4): 535-536.

TENTCHEVA, D; GAUTHIER, L; BAGNY, L; FIEVET, J; DAINAT, B; COUSSERANS, F; COLIN, M E; BERGOIN, M (2006) Comparative analysis of deformed wing virus (DWV) RNA in *Apis mellifera* and *Varroa destructor. Apidologie* 37: 41-50. http://dx.doi.org/10.1051/apido:2005057

TENTCHEVA, D; GAUTHIER, L; JOUVE, S; CANABADY-ROCHELLE, L; DAINAT, B; COUSSERANS, F; COLIN, M E; BALL, B V; BERGOIN, M (2004a) Polymerase chain reaction detection of deformed wing virus (DWV) in *Apis mellifera* and *Varroa destructor. Apidologie* 35 (4): 431-440. http://dx.doi.org/10.1051/apido:2004021

TENTCHEVA, D; GAUTHIER, L; ZAPPULLA, N; DAINAT, B; COUSSERANS, F; COLIN, M E; BERGOIN, M (2004b) Prevalence and seasonal variations of six bee viruses in *Apis mellifera* L. and *Varroa destructor* mite populations in France. *Applied and Environmental Microbiology* 70: 7185-7191. http://dx.doi.org/10.1128/AEM.70.12.7185-7191.2004

TERIO, V; MARTELLA, V; CAMERO, M; DECARO, N; TESTINI, G; BONERBA, E; TANTILLO, G; BUONAVOGLIA, C (2008) Detection of a honey bee iflavirus with intermediate characteristics between kakugo virus and deformed wing virus. *New Microbiologica* 31: 439-444.

TODD, J H; DE MIRANDA, J R; BALL, B V (2007) Incidence and molecular characterization of viruses found in dying New Zealand honey bee (*Apis mellifera*) colonies infested with *Varroa destructor. Apidologie* 38: 354-367. http://dx.doi.org/10.1051/apido:2007021

TOPLEY, E; DAVISON, S; LEAT, N; BENJEDDOU, M (2005) Detection of three honey bee viruses simultaneously by a single Multiplex Reverse Transcriptase PCR. *African Journal of Biotechnology* 4(8): 763-767.

VALENTINE-THON, E; VAN LOON, A M; SCHIRM, J; REID, J; KLAPPER, P E; CLEATOR, G M (2001) European proficiency testing program for molecular detection and quantitation of hepatitis B virus DNA. *Journal of Clinical Microbiology* 39(12): 4407-4412. http://dx.doi.org/10.1128/JCM.39.12.4407-4412.2001

VAN DER STEEN, J J M; CORNELISSEN, B; DONDERS, J;
BLACQUIÈRE, T; VAN DOOREMALEN, C (2012) How honey bees
of successive age classes are distributed over a one storey, ten
frames hive. *Journal of Apicultural Research* 51(2): 174-178.
http://dx.doi.org/10.3896/IBRA.1.51.2.05

VAN DER ZEE, R; GRAY, A; HOLZMANN, C; PISA, L;
BRODSCHNEIDER, R; CHLEBO, R; COFFEY, M F; KENCE, A;
KRISTIANSEN, P; MUTINELLI, F; NGUYEN, B K; ADJLANE, N;
PETERSON, M; SOROKER, V; TOPOLSKA, G; VEJSNÆS, F;
WILKINS, S (2012) Standard survey methods for estimating
estimating colony losses and explanatory risk factors in *Apis
mellifera*. In *V Dietemann; J D Ellis; P Neumann (Eds) The
COLOSS BEEBOOK, Volume I: Standard methods for* Apis mellifera
research. Journal of Apicultural Research 52(4):
http://dx.doi.org/10.3896/IBRA.1.52.4.18

VAN MUNSTER, M; DULLEMANS, A M; VERBEEK, M;
VAN DEN HEUVEL, J F J M; CLÉRIVET, A; VAN DER WILK, F
(2002) Sequence analysis and genomic organization of aphid
lethal paralysis virus: a new member of the family Dicistroviridae.
Journal of General Virology 83(12): 3131-3138.

VANENGELSDORP, D; LENGERICH, E; SPLEEN, A; DAINAT, B;
CRESSWELL, J; BAYLISS, K, NGUYEN, K B; SOROKER; V;
UNDERWOOD, R; HUMAN, H; LE CONTE, Y; SAEGERMAN, C
(2013) Standard epidemiological methods to understand and
improve *Apis mellifera* health. In *V Dietemann; J D Ellis, P
Neumann (Eds) The COLOSS* BEEBOOK: *Volume II: Standard
methods for* Apis mellifera *pest and pathogen research. Journal of
Apicultural Research* 52(4):
http://dx.doi.org/10.3896/IBRA.1.52.4.15

VERKOOYEN, R P; NOORDHOEK, G T; KLAPPER, P E; REID, J;
SCHIRM, J; CLEATOR, G M; IEVEN, M; HODDEVIK, G (2003)
Reliability of nucleic acid amplification methods for detection of
Chlamydia trachomatis in urine: results of the first international
collaborative quality control study among 96 laboratories. *Journal
of Clinical Microbiology* 41(7): 3013-3016.
http://dx.doi.org/10.1128/JCM.41.7.3013-3016.2003

WANG, T C; CHAO, M (2005) RNA recombination of hepatitis delta
virus in natural mixed-genotype infection and transfected cultured
cells. *Journal of Virology* 79(4): 2221-2229.
http://dx.doi.org/10.1128/JVI.79.4.221-2229.2005

WEBBY, R; KALMAKOFF, J (1998) Sequence comparison of the major
capsid protein from 18 diverse iridoviruses. *Archives of Virology*
143: 1949-1966.
http://dx.doi.org/10.1007/s007050050432

WEINSTEIN-TEXEIRA, E; CHEN, Y P; MESSAGE, D; PETTIS, J; EVANS,
J D (2008) Virus infections in Brazilian honey bees. *Journal of
Invertebrate Pathology* 99(1): 117-119.
http://dx.doi.org/10.1016/j.jip.2008.03.014

WHEELER, D E; BUCK, N; EVANS, J D (2006) Expression of insulin
pathway genes during the period of caste determination in the
honey bee, *Apis mellifera. Insect Molecular Biology* 15(5): 597-
602. http://dx.doi.org/10.1111/j.1365-2583.2006.00681.x

WILLIAMS, G R; ALAUX, C; COSTA, C; CSÁKI, T; DOUBLET, V;
EISENHARDT, D; FRIES, I; KUHN, R; MCMAHON, D P;
MEDRZYCKI, P; MURRAY, T E; NATSOPOULOU, M E; NEUMANN, P;
OLIVER, R; PAXTON, R J; PERNAL, S F; SHUTLER, D; TANNER, G;
VAN DER STEEN, J J M; BRODSCHNEIDER, R (2013) Standard
methods for maintaining adult *Apis mellifera* in cages under *in
vitro* laboratory conditions. In *V Dietemann; J D Ellis; P Neumann
(Eds) The COLOSS* BEEBOOK, *Volume I: standard methods for*
Apis mellifera *research. Journal of Apicultural Research* 52(1):
http://dx.doi.org/10.3896/IBRA.1.52.1.04

WILLIAMS, G R; ROGERS, R E L; KALKSTEIN, A L; TAYLOR, B A;
SHUTLER, D; OSTIGUY, N (2009) Deformed wing virus in western
honey bees (*Apis mellifera*) from Atlantic Canada and the first
description of an overtly-infected emerging queen. *Journal of
Invertebrate Pathology* 101: 77-79.
http://dx.doi.org/10.1016/j.jip.2009.01.004

WITTWER C T; REED, G H; GUNDRY, C N; VANDERSTEEN, J G;
PRYOR, R J (2003) High-resolution genotyping by amplicon
melting analysis using LCGreen. *Clinical Chemistry* 49(6): 853-
860. http://dx.doi.org/10.1373/49.6.853

WOLFE, R; CARLIN, J B (1999) Sample-size calculation for a
log-transformed outcome measure. *Controlled Clinical
Trials* 20: 547-554.
http://dx.doi.org/10.1016/S0197-2456(99)00032-X

YAÑEZ, O; JAFFÉ, R; JAROSCH, A; FRIES, I; MORITZ, R F A;
PAXTON, R J; DE MIRANDA J R (2012) Deformed wing virus and
drone mating in the honey bee (*Apis mellifera*): implications for
sexual transmission of a major honey bee virus. *Apidologie* 43: 17
-30. http://dx.doi.org/10.1007/s13592-011-0088-7

YANG, X; COX-FOSTER, D L (2005) Impact of an ectoparasite on the
immunity and pathology of an invertebrate: Evidence for host
immunosuppression and viral amplification. *Proceedings of the
National Academy of Science of the United States of America*
102: 7470-7475.
http://dx.doi.org/10.1073/pnas.0501860102

YUE, C; GENERSCH, E (2005) RT-PCR analysis of deformed wing virus
in honey bees (*Apis mellifera*) and mites (*Varroa destructor*).
Journal of General Virology 86: 3419-3424.
http://dx.doi.org/10.1099/vir.0.81401-0

YUE, C; SCHRÖDER, M; BIENEFELD, K; GENERSCH, E (2006)
Detection of viral sequences in semen of honey bees (*Apis
mellifera*): evidence for vertical transmission of viruses through
drones. *Journal of Invertebrate Pathology* 92: 105-108.
http://dx.doi.org/10.1016/j.jip.2006.03.001

YUNUS, M A; CHUNG, L M W; CHAUDHRY, Y; BAILEY, D;
    GOODFELLOW, I (2010) Development of an optimized RNA-based
    murine norovirus reverse genetics system. *Journal of Virological
    Methods* 169(1): 112-118.
    http://dx.doi.org/10.1016/j.jviromet.2010.07.006

YURYEV, A (2007) *PCR primer design. Methods in Molecular Biology™
    series #402*. Humana Press; Totowa, New Jersey, USA. 431 pp.

ZHANG, X; HE, S Y; EVANS, J D; PETTIS, J S; YIN, G F; CHEN, Y P
    (2012) New evidence that deformed wing virus and black queen
    cell virus are multi-host pathogens. *Journal of Invertebrate
    Pathology* 109: 156-159. http://dx.doi.org/10.1016/j.jip.2011.09.010

ZIONI, N; SOROKER, V; CHEJANOVSKY, N (2011) Replication of
    Varroa destructor virus 1 (VDV-1) and a Varroa destructor virus 1-
    deformed wing virus recombinant (VDV-1-DWV) in the head of
    the honey bee. *Virology* 417(1): 106-112.
    http://dx.doi.org/10.1016/j.virol.2011.05.009

Journal of Apicultural Research 52(4): (2013)

CPSIA information can be obtained
at www.ICGtesting.com
Printed in the USA
BVHW02s0553100518
515795BV00024B/114/P